Lecture Notes in Physics

Bisher erschienen/Already published

Vol. 1: J. C. Erdmann. Wärmeleitung in Kristallen, theoretische Grundlagen und fortgeschrittene experimentelle Methoden. II, 283 Seiten. 1969.

Vol. 2: K. Hepp, Théorie de la renormalisation. III, 215 pages. 1969.

Vol. 3: A. Martin, Scattering Theory: Unitarity, Analyticity and Crossing. IV, 125 pages. 1969.

Vol. 4: G. Ludwig, Deutung des Begriffs „physikalische Theorie" und axiomatische Grundlegung der Hilbertraumstruktur der Quantenmechanik durch Hauptsätze des Messens. 1970. Vergriffen.

Vol. 5: Schaaf, The Reduction of the Product of Two Irreducible Unitary Representations of the Proper Orthochronous Quantummechanical Poincare Group. IV, 120 pages. 1970.

Vol. 6: Group Representations in Mathematics and Physics. Edited by V. Bargmann. V, 340 pages. 1970.

Vol. 7: R. Balescu, J. L. Lebowitz, I. Prigogine, P. Résibois, Z. W. Salsburg, Lectures in Statistical Physics. V, 181 pages. 1971.

Vol. 8: Proceedings of the Second International Conference on Numerical Methods in Fluid Dynamics. Edited by M. Holt. 1971. Out of print.

Vol. 9: D. W. Robinson, The Thermodynamic Pressure in Quantum Statistical Mechanics. V, 115 pages. 1971.

Vol. 10: J. M. Stewart, Non-Equilibrium-Relativistic Kinetic Theory. III, 113 pages. 1971.

Vol. 11: O. Steinmann, Pertubation Expansions in Axiomatic Field Theory. III, 126 pages. 1976.

Vol. 12: Statistical Models and Turbulence. Edited by C. Van Atta and M. Rosenblatt. Reprint of the First Edition. VIII, 492 pages. 1975.

Vol. 13: M. Ryan, Hamiltonian Cosmology. VII, 169 pages. 1972.

Vol. 14: Methods of Local and Global Differential Geometry in General Relativity. Edited by D. Farnsworth, J. Fink, J. Porter, and A. Thompson. V, 188 pages.

Vol. 15: M. Fierz, Vorlesungen zur Entwicklungsgeschichte der Mechanik. V, 97 Seiten. 1972.

Vol. 16: H.-O. Georgii, Phasenübergang 1. Art bei Gittergasmodellen. IX, 167 Seiten. 1972.

Vol. 17: Strong Interaction Physics. Edited by W. Rühl and A. Vancura. V, 405 pages. 1973.

Vol. 18: Proceedings of the Third International Conference on Numerical Methods in Fluid Mechanics, Vol. I. Edited by H. Cabannes and R. Temam. VII, 186 pages. 1973.

Vol. 19: Proceedings of the Third International Conference on Numerical Methods in Fluid Mechanics, Vol. II. Edited by H. Cabannes and R. Temam. VII, 275 pages. 1973.

Vol. 20: Statistical Mechanics and Mathematical Problems. Edited by A. Lenard. VIII, 247 pages. 1973.

Vol. 21: Optimization and Stability Problems in Continuum Mechanics. Edited by P. K. C. Wang. V, 94 pages. 1973.

Vol. 22: Proceedings of the Europhysics Study Conference on Intermediate Processes in Nuclear Reactions. Edited by N. Cindro, P. Kulišic and Th. Mayer-Kuckuk. XIV, 329 pages. 1973.

Vol. 23: Nuclear Structure Physics. Proceedings 1973. Edited by U. Smilansky, I. Talmi, and H. A. Weidenmüller. XII, 296 pages. 1973.

Vol. 24: R. F. Snipes, Statistical Mechanical Theory of the Electrolytic Transport of Nonelectrolytes. V, 210 pages. 1973.

Vol. 25: Constructive Quantum Field Theory. The 1973 "Ettore Majorana" International School of Mathematical Physics. Edited by G. Velo and A. Wightman. III, 331 pages. 1973.

Vol. 26: A. Hubert, Theorie der Domänenwände in geordneten Medien. XII, 377 Seiten. 1974.

Vol. 27: R. K. Zeytounian, Notes sur les Ecoulements Rotationnels de Fluides Parfaits. XIII, 407 pages. 1974.

Vol. 28: Lectures in Statistical Physics. Edited by W. C. Schieve and J. S. Turner. V, 342 pages. 1974.

Vol. 29: Foundations of Quantum Mechanics and Ordered Linear Spaces. Advanced Study Institute, Marburg 1973. Edited by A. Hartkämper and H. Neumann. VI, 355 pages. 1974.

Vol. 30: Polarization Nuclear Physics. Proceedings 1973. Edited by D. Fick. IX, 292 pages. 1974.

Vol. 31: Transport Phenomena. Sitges International Schools of Statistical Mechanics, June 1974. Edited by G. Kirczenow and J. Marro. XIV, 517 pages. 1974.

Vol. 32: Particles, Quantum Fields and Statistical Mechanics. Proceedings 1973. Edited by M. Alexanian and A. Zepeda. V, 132 pages. 1975.

Vol. 33: Classical and Quantum Mechanical Aspects of Heavy Ion Collisions. Proceedings 1974. Edited by H. L. Harney, P. Braun-Munzinger, and C. K. Gelbke. VII, 311 pages. 1975.

Vol. 34: One-Dimensional Conductors GPS Summer School Proceedings, 1974. Edited by H. G. Schuster. VII, 371 pages. 1975.

Vol. 35: Proceedings of the Fourth International Conference on Numerical Methods in Fluid Dynamics, 1974. Edited by R. D. Richtmyer. V, 457 pages. 1975.

Vol. 36: R. Gatignol, Théorie Cinétique des Gaz à Répartition Discrète de Vitesses. II, 219 pages. 1975.

Vol. 37: Trends in Elementary Particle Theory. Proceedings 1974. Edited by H. Rollnik and K. Dietz. V, 472 pages. 1975.

Vol. 38: Dynamical Systems, Theory and Applications. Proceedings 1974. Edited by J. Moser. VI, 624 pages. 1975.

Vol. 39: International Symposium on Mathematical Problems in Theoretical Physics. Proceedings 1975. Edited by H. Araki. XII, 562 pages. 1975.

Vol. 40: Effective Interactions and Operators in Nuclei. Proceedings 1975. Edited by B. R. Barrett. XII, 339 pages. 1975.

Vol. 41: Progress in Numerical Fluid Dynamics. Proceedings 1974. Edited by H. J. Wirz. V, 471 pages. 1975.

Vol. 42: H II Regions and Related Topics. Proceedings 1975. Edited by D. Downes and T. L. Wilson. XII, 488 pages. 1975.

Vol. 43: Laser Spectroscopy. Proceedings 1975. Edited by S. Haroche, J. C. Pebay-Peyroula, T. W. Hänsch, and S. E. Harris. X, 466 pages. 1975.

Lecture Notes in Physics

Organic Conductors and Semiconductors

Lecture Notes in Physics

65

Organic Conductors and Semiconductors

Proceedings of the International Conference
Siófok, Hungary 1976

Edited by
L. Pál, G. Grüner, A. Jánossy, J. Sólyom

Springer-Verlag
Berlin Heidelberg GmbH 1977

Editors

L. Pál, Director of the Central Research Institute for Physics,
H-1525 Budapest, 114 P.O.B. 49 Hungary

G. Grüner, A. Jánossy, J. Sólyom, Central Research Institute for Physics,
H-1525 Budapest, 114 P.O.B. 49 Hungary

The Conference on Organic Conductors and Semiconductors was sponsored by

- the Hungarian Academy of Sciences
- the Central Research Institute for Physics, Budapest
- the Loránd Eötvös Physical Society, Budapest

ISBN 978-3-540-08255-2 ISBN 978-3-540-37342-1 (eBook)
DOI 10.1007/978-3-540-37342-1

Originally published by Springer-Verlag Berlin · Heidelberg · New York in 1977

Sole distribution rights for all socialist countries:
Akadémiai Kiadó, Budapest

CONTENTS

EDITORIAL NOTE

The Conference on Organic Conductors and Semiconductors was held at the resort of Siófok, Hungary from the 30th August to 3rd September 1976. It was sponsored by the Hungarian Academy of Sciences, the Central Research Institute for Physics and the Loránd Eötvös Physical Society. The main object of the conference was to bring together physicists and chemists from all parts of the world to discuss a field of rapidly growing importance: the physical and chemical properties of conducting organic materials. Although a number of papers dealt with the chemical aspects of the topic the main emphasis was on the physical properties. Nearly all papers dealt with quasi one-dimensional organic systems. The ordering of the papers follows that of the sessions of the Conference itself. Abstracts have been included of lectures where no full text was submitted.

We should like to point out that the camera-ready technique has been used in the production of this volume as this ensures rapid publication. Thus, editorial work has been limited to the correction of those misprints which might otherwise have interfered with the meaning. Illustrations have been reproduced in the form submitted.

We should like to express our thanks to the contributors for their cooperation in sending their manuscripts without delay.

This Editorial Note provides us with the opportunity of expressing thanks to the Organizing Committee and to all others concerned with arrangements which enabled the smooth running of the Conference.

Conference on Organic Conductors and Semiconductors, Siófok, Hungary 1976

SUMMARY REMARKS

JOHN BARDEEN

Department of Physics, University of Illinois, Urbana, Il, USA

This has been an extremely interesting conference. The organizers have done an excellent job in bringing together a large number of the leading investigators in the fascinating field of organic conductors. A unique feature has been the opportunity for scientists from different parts of the world to meet and discuss mutual problems on a personal basis.

The conference has been a very busy one with a large number of papers presented, but there has been time for discussion on both a formal and an informal basis. It would be impossible to give a summary of all of the very interesting experiments and ideas presented here; all that I will attempt to do will be to give my personal impressions of some of the different points of view and outstanding problems.

It will not be possible to acknowledge all those who have contributed importantly to the field and I will mention names only when necessary to identify a given piece of work or point of view. Detailed references may be found in the papers presented.

The initial impetus for the rapid expansion of research on organic conductors during the past three or four years was the hope of finding compounds with very high conductivity. Although the chances for conductivities in the range of good metals now appear remote, interest has continued to increase because of the fascinating problems of chemistry and physics involved in these materials. Chemists have been able to make many interesting compounds which conduct along stacks of planar molecules. They have studied the systematics of charge transfer between donor and acceptor stacks and the nature of charge transport along the stacks. The materials are of interest to theorists because fluctuations are much enhanced in 1D over 3D systems and because of the various types of collective modes that can occur, charge density waves (CDW), spin density waves (SDW) and superconductive pairing.

In spite of intensive work by experimentalists and theorists, many of the major questions in regard to these compounds are still unresolved and new ones have been opened up by experiment. It has turned out to be a very rich field indeed. Questions that attracted considerable discussion at this Conference are:

1) Nature of the electronic structure and transport in 1D systems with strong Coulomb repulsions ($U \gg 4t_{\parallel}$) and the connection with $4k_F$ lines in TTF-TCNQ shown by X-rays.

2) Possible enhancement of conductivity by Peierls-Fröhlich modes, including effects of pinning by impurities. The major question as to whether paraconductivity above T_c makes a significant contribution is still open. Another question is the role of solitons as elementary excitations and in conduction at low temperatures ($T \ll T_c$).

3) The role of interchain coupling. Renormalization group methods earlier applied to 1D systems have been extended to treat two or more interacting chains, with applications to the $4k_F$ problem. Extensions have been given of the Bak-Emory theory of the 38°, 49° and 54°K transitions in TTF-TCNQ in which Coulomb interactions between CDW's on different chains play an important role. Another question is how much 3D effects reduce fluctuations above T_c expected for 1D systems.

4) The effects of disorder introduced by impurities, inter- and intramolecular vibrations, random orientation of asymmetric molecules and alloying on conductivity and other properties of organic conductors. It has been shown that intramolecular vibrations contribute importantly to the electron-phonon interaction parameter λ and to $\sigma(\omega)$.

5) The problem of the Kohn anomaly and the central peak in $S(2k_F,\omega)$. Does the central peak arise from static deformations or are they dynamic so as to contribute to the conductivity and dielectric constant?

6) The possibility of superconductive pairing in organic conductors as suggested many years ago by W. A. Little. The discovery of superconductivity in $(SN)_x$ at very low temperatures created renewed interest in this problem, but recent work has shown that the band structure in this material is 3D rather than 1D. Theory indicates superconductivity in pseudo-1D systems with attractive backward scattering amplitude $(g_1-2g_2>0)$. Although one cannot expect true superconductivity in pseudo-1D systems, the observation of effects of pairing in enhancing the conductivity would be of great interest and attempts have been made to characterize suitable systems.

It will not be possible to discuss all of these interesting questions in these concluding remarks. Comments will be confined mainly to (1), (2), (4) and (6).

Although many new compounds have been discovered, the ones of most interest continue to be salts of TCNQ and compounds related to KCP, with TTF-TCNQ still attracting the greatest attention and continuing to exhibit the most unusual properties. Major developments during the past year have been the discovery in TTF-TCNQ by X-rays (INV 6) of lines showing CDW's at $4k_F$, and by neutron scattering (INV 16) the remarkable phase transitions at 54, 49 and 38 K.

In spite of the great number of measurements of the properties of this compound, including conductivity, dielectric constant, magnetic susceptibility, electron and nuclear spin resonance, specific heat, thermoelectric power, etc., many over wide temperature and frequency ranges, there is still no consensus as to how all the various pieces of the puzzle fit together. Even such a basic question as to whether most of the high temperature conduction is along the TTF or the TCNQ stacks or in hybridized orbitals of both remains open.

Because of the complexity of the actual problem, theorists are forced to idealize the problem to over-simplified models which emphasize one aspect of the problem or another. Earlier, definitive work had been done on collective modes in ideal 1D metals with scattering amplitudes for forward and backward directions using renormalization group methods. Work reported at this conference has extended such work to two or more interacting chains. Work was also reported on the more general problem of 3D phase transitions with Coulomb interactions and charge transfer matrix elements between chains. Fluctuations which occur in ideal 1D systems can be suppressed if coupling is mainly to 3D phonons, from 3D band structure or other interchain coupling effects. Charge transport either by individual electrons or by collective modes is much more strongly affected by disorder in 1D than in 3D systems.

Electronic Structure

Pseudo 1D crystals are often described by a tight binding Hubbard type model with matrix elements $t_{||}$ and $t_{\perp}$ ($t_{||} >> t_{\perp}$), an intramolecular Coulomb interaction, U, and an electron-phonon interaction parameter, λ. The quasi-particle states of the electrons can be described by the usual Bloch model with wave vector, k, provided that U is sufficiently small compared with the band width, $4t_{||}$. However, because of phase space considerations, particle-particle scattering is much more important in 1D than in 3D systems.

The states have an uncertainty in energy $\delta\varepsilon_k \sim \hbar/\tau$, where τ is the mean free time for scattering by impurities, phonons or particle-particle interaction: A Fröhlich-Peierls type transition, opening up a gap 2Δ with a distortion corresponding to $2k_F$, is possible only if the states are sufficiently well defined so that $\delta\varepsilon_k < \sim \Delta$. The same applies to the gap from hybridization of inverted bands on donor and acceptor chains. A 3D type band structure is appropriate if $\delta\varepsilon_k < \sim t_{\perp}$. To the extent that $\delta\varepsilon_k$ depends on thermal fluctuatio a 1D band structure may be appropriate at high temperatures and a 3D at low.

Problems of TTF-TCNQ

One reason that an understanding of TTF-TCNQ is so difficult is that $U/4t_{\parallel}$ is of the order of unity and $t_{\perp}$ is $\sim$0.05-0.10 $t_{\parallel}$ (S5), so that the 1D Bloch picture is only marginly valid. Further, with a m.f.p. of $\sim$ 100 Å, $\delta\varepsilon_k \sim 100\ k_B$ which, with $t_{\parallel} \sim 1000\ k_B$, is of the same order as Δ and of $t_{\perp}$. The parameters are in the borderline range where none of the simple pictures apply. The problem is a difficult one not only mathematically but also with regard to understanding the physics.

As discussed in the invited talk by Heeger (INV 9) the Pennsylvania group favors the picture of Lee, Rice and Anderson in which there are large 1D fluctuations in Δ extending up to room temperature and that the high conductivity above 60°K in TTF-TCNQ comes from the collective modes. The main arguments come from far infrared optical data from which one can derive the frequency dependent conductivity, $\sigma(\omega)$, and dielectric function, $\varepsilon(\omega)$. There is a drop in $\sigma(\omega)$ when ω drops below a frequency interpreted as the gap frequency, ω_g, followed by a rise at very low frequencies. There is corresponding rise in $\varepsilon(\omega)$ to very large values as $\omega \to 0$. The bulk of the contribution to the integral of $\sigma(\omega)$ comes from frequencies above 1000 cm^{-1}. If this picture is correct, the Peierls gap, 2Δ, would be large ($\sim$ 800 K). It is suggested that the width of the conductivity peak near $\omega=0$ is very small, perhaps only $\sim$ 1 cm^{-1}, although experimental data show only that it is less than $\sim$ 20 cm^{-1}. The conductivity and its temperature variation are essentially the same at microwave frequencies (B2) as at dc.

Other experiments suggest that 1D fluctuations may be greatly reduced by interchain effects and that the transition at 54°K is essentially that given by mean field theory. The conductivity at 60°K and above is not so large as to require a collective mechanism, but can be described by a band model with a m.f.p. varying as T^{-2} (or slightly more rapidly). This temperature variation may be due to particle-particle scattering. Some evidence in favor of this picture is the neutron and X-ray scattering data for the $2k_F$ anomaly which indicate that the $2k_F$ scattering decreases rapidly above the transition and disappears at $T \sim 150$ K. Specific heat and thermal conductivity data are consistent with this picture. Further evidence (H1) comes from temperature variations of the magnetic susceptibility in the neighborhood of the transitions at 38°, 49° and 54°K.

Since ordering of the TTF and the TCNQ chains occurs at different temperatures and evidence that most of the conductivity above the 54°K transition is along the TCNQ chain (C1), it appears unlikely that hybridized orbitals betweeen the chains are important (see also T1, S5). The question of how much of the conductivity is due to collective modes remains open. If the CDW fluctuations are more like m.f.t. than 1D, one must account for the large dip in $\sigma(\omega)$ below 1000 cm^{-1} on grounds other than the energy gap from pinned CDW's. A suggestion which perhaps deserves more careful study is whether or not a Fano antiresonance from coupling of $2k_F$ phonons with intramolecular vibrations could account for the data.

There was considerable discussion of the reasons for the $4k_F$ lines shown by X-ray data (INV 6) which do extend up to room temperature. One picture is that on at least one of the chains (most likely TTF) there is a large Coulomb repulsion ($U \gg 4t_{\parallel}$). One would then expect a spin wave at $2k_F$ and a CDW at $4k_F$ (T2). Another picture (INV 14, K2) is based on a model which involves coupling between the chains. Arguments based on Menyhárd-Sólyom renormalization group methods indicate that for a model with repulsive interactions, a $2k_F$ SDW will predominate at high temperatures and a $2k_F$ CDW at lower temperatures. That the $4k_F$ lines are indicative of the effects of repulsive interactions seems well accepted, but much remains to be done both theoretically and experimentally to understand transport and other properties of pseudo 1D systems of this sort.

Several other papers presented at the Conference also are concerned with effects of interchain coupling, some (A1, B3) to better understand the 38°, 49° and 54°K transitions in TTF-TCNQ and others (M2, M5) to extend the renormalization group methods to two or more chains.

The inverted band picture with hybridized bands appears to apply to TSeF-TCNQ, with both bands ordering at the same temperature, 29°K (T1). This may be due to a larger $t_{\perp}$ than in TTF-TCNQ or to longer correlation lengths and larger distortions on the donor chains (S3) or both. In HMTSF-TCNQ, the value of $t_{\perp}$ is even larger, so that it behaves at low temperatures much more like a 3D semimetal (INV 3, S5, J1). Pressures of the order of a few kbar are sufficient to increase $t_{\perp}$ and 3D effects significantly in a number of TCNQ compounds (INV 10).

Effects of Impurities and Disorder

The marked effects of disorder in pseudo-1D systems have been clarified by both experiment and theory. These include (a) transport in the absence of collective effects (INV 1, INV 13), (b) the role of impurities in pinning incommensurate CDW's and the effect on charge transport (F3), and (c) the relative effect of impurities on the Peierls and superconducting transition temperature.

Soviet scientists have been particularly interested in impurity effects in 1D metals. It was shown (INV 8) that impurities in a half-filled band give a singular enhancement in the density of states at the Fermi surface. This may be another manifestation of the well known impurity localization of states in 1D. This latter implies (INV 13) that at T=0, $\sigma(\omega)\to 0$ as $\omega\to 0$. With increasing temperature, phonons allow a hopping type transport from one localized site to another, with increasing conductivity. At still higher temperatures, phonons scatter the electrons with a corresponding decrease in $\sigma(\omega)$. The theory developed fits quantitatively with experiments on TCNQ salts with structural disorder.

There has long been an argument as to whether with impurity scattering present CDW's tend to enhance the conductivity over that which would be found in the absence of CDW's. In the Lee, Rice, Anderson picture the CDW's are fully developed at temperatures below room temperature and the relevant excitations are phase modes. The electron density is proportional to the gradient of phase. In an investigation of the effect of impurity pinning of such modes (F3), the frequency dependent low temperature conductivity, $\sigma(\omega)$, was found to have a peak at a characteristic pinning frequency. Thermal fluctuations depin the modes at a critical temperature, above which metallic behavior is expected. Below this temperature, the CDW's add to the resistivity, above to the conductivity. Interchain coupling and commensurability, which would also pin the modes, were not considered in this study. This work confirms earlier studies that CDW's can enhance the conductivity.

Possibility of Superconductive Pairing

Other studies, particularly by Soviet workers (INV 11) have looked into the possible effects of impurities and disorder on reducing the Peierls transition relative to the superconducting transition so that the latter can be observed. It is known that in 3D, nonmagnetic impurities have little effect on the superconducting T_c. However, it is found that in 1D, impurities depress both transition so that there is no advantage in introducing disorder.

Although possible from a theoretical point of view, superconductive pairing has not been observed in pseudo 1D systems. The 3D band structure plays an essential role in $(SN)_x$, so that it is not regarded as 1D. Arguments that have been given for the stability of supercurrents in 2D or 3D systems fail when applied to 1D. Individual electrons or phonons have sufficient momentum in 1D to change the collective motion, while this is not true in higher dimensions. This is perhaps most easily seen by considering a 1D chain in the form of a ring of length L. If there are n electrons, the Fermi wave vector $k_F = \pm\pi n/2L$.

Thus a change of wave vector by $2k_F$ can change the momentum by $\pi n/L$. This is just the momentum change which occurs if the quantum number describing superflow around the ring changes by one unit ($\frac{1}{2} n \times (2\pi/L)$). Thus all that one expects is high conductivity from collective effects. However, because of the relatively small number of conduction electrons per unit volume, it is unlikely that the conductivity would be large compared with that of good metals. These arguments are valid regardless of the cause of the pairing. To get true superconductivity, neighboring chains would have to be in phase, and this implies sufficient Josephson tunneling between chains, or a sufficiently large $t_\perp$.

Little's original suggestion was an exciton mechanism in which virtual excitons of polar molecules surrounding the conducting chain would replace or supplement the phonon mechanism. A careful reanalysis of this suggestion was presented at the Conference (D2). A structure was proposed in which a ligand system of highly polarizable molecules surrounds the spine. They find that pairing is possible at high temperatures, but only under very restricted conditions.

Concluding Remarks

These are just a few comments, obviously written from the viewpoint of a theorist, on the vast amount of work presented at the Conference. It is obvious that organic conductors is an active and fruitful field. To me, the main interest is to better understand how charge transfer and charge transport occur in organic systems. For this the study of the systematics of what determines the important variables, $t_{||}$, $t_\perp$, U, λ, etc. is of greatest importance. Charge transport by solitons or other excitations of CDW's is also of great interest. The problems have taken longer to unravel, and are also much richer, than I imagined a few years ago. We can expect continued progress and activity in the years to come.

REFERENCES

The lettering of the references of these Summary Remarks follows that of the Abstract book of the Conference. These references correspond to the papers presented here.*

INV 1 A.A.Abrikosov and I.A.Ryzhkin
INV 3 A.N.Bloch
INV 6 R.Comes and G.Shirane
INV 8 L.P.Gorkov
INV 9 A.J.Heeger
INV 10 D.Jerome
INV 11 A.I.Larkin and V.I.Mel'nikov
INV 13 E.I.Rashba, A.A.Gogolin and V.I.Mel'nikov
INV 14 T.M.Rice and P.A.Lee and R. A. Klemm
INV 16 R.Comes and G.Shirane

A 1 E. Abrahams, J.Sólyom and F.Woynarovich

B 2 A.J.Berlinsky, C.P.Barry, W.N.Hardy and L.Weiler
B 3 A.Bjelis and S.Barisic

C 1 P.M.Chaikin, R.L.Greene, R.Schumaker, S.Etemad and E.M.Engler

D 2 D.Davis, H.Gutfreund and W.A.Little

F 3 H.Fukuyama and P. A. Lee

H 1 R.M.Herman, M.B.Salamon, G. de Pasquali and G.Stucky

J 1 C.S.Jacobsen, K.Bechgaard and J.R.Andersen

K 2 R.A.Klemm, P.A.Lee and T.M.Rice

M 2 N.Menyhárd
M 5 L.Mihály and J.Sólyom

S 3 T.D.Schultz
S 5 G.Soda, D.Jerome, M.Weger, J.M.Fabre, L.Giral and K.Bechgaard

T 1 Y.Tomkiewicz
T 2 J.B.Torrance

*The papers Cl and B2 were announced but not presented at the Conference. For papers where only an abstract is presented see also the Proceedings of the NATO Advanced Study Institute on "Chemistry and Physics of One Dimensional Metals" Bolzano, Italy 1976, ed. H.J.Keller, Plenum Press (to be published).

I

ONE-DIMENSIONAL MODELS

Conference on Organic Conductors and Semiconductors, Siófok, Hungary 1976

QUANTUM SOLITONS IN ONE-DIMENSIONAL CONDUCTORS

A. LUTHER

NORDITA,
Blegdamsvej 17, DK-2100 Copenhagen Ø, Denmark

The eigenvalue spectrum of the continuum electron gas in one-dimension is calculated. New collective states are found and identified as solitons or bound states of these solitons. These new states carry current and can participate in transport. They are thermally activated at low temperatures, with a gap which is calculted for all values of the coupling strength. This exact calculation confirms a previous result based on integrating the renormalization group equations up to the coupling strength where a solution could be found.

The soliton gap, and the collective soliton excitations below the gap, provide a sensitive test of the one-dimensional character of a quasi-one-dimensional conductor. There can also be solitons associated with random potentials, and localization lengths to be identified with soliton gaps, which can be understood with the help of scaling laws.

Conference on Organic Conductors and Semiconductors, Siófok, Hungary 1976

INFRARED PROBLEMS AND PHASE TRANSITIONS OF CONTINUOUS ORDER IN LOW-DIMENSIONAL SYSTEMS

J. ZITTARTZ

Institut für Theoretische Physik, Universität Köln, Köln, FRG

Abstract: The infrared problem is the strong response, in particular logarithmic divergencies in perturbation theoretical treatments, of a system to perturbations which couple to low-lying energy excitations. Because of connections between phase space and momentum dispersion such problems arise normally in 1 and 2 space dimensions. In a unified description we show the mathematical equivalence of infrared behaviour for the following model situations: 1) 1-d Fermi systems with general interactions, 2) 1-d Bose systems with "Sine-Gordon" interaction, 3) the 2-d magnetic (harmonic) rotator model, 4) the 2-d classical Coulomb plasma.

Infrared singular behaviour is characterized by power law singularities in thermodynamic quantities with singular (or critical) exponents which usually vary continuously with system parameters. Depending on the physical interpretation of these parameters in the different situations, this implies: 1) and 2): singular ground-state properties, a special particle spectrum and instabilities for correlation functions; 3): the phase transition of continuous order for 2-d magnetic systems and other systems with continuous symmetry; 4) a smooth metal-insulator transition and special thermodynamic properties in the Coulomb plasma.

I. Introduction

Infrared problems in many particle systems have shown up sporadically within the last 10 or 20 years. However, only recently a unified picture and a unified mathematical description has evolved at least for a few of these infrared problems. In this context the phase transition of continuous order, recently proposed by Müller-Hartmann and Zittartz [1,2], seems to play an important role in systems which have a tendency towards bulk order. However, due to infrared fluctuations infinitly long range order, and thus the usual type of phase transition with non-vanishing order parameter, is not realized, but the system shows long range order and stays critical for a whole range of temperatures, for instance.

What is an infrared problem? If a system has low lying energy excitations in the form of a continuous spectrum, then it may react quite strongly to perturbations which couple to these excitations. In perturbation expansions one finds a power series in $\ln \lambda$ where λ is some characteristic small energy, temperature, magnetic field, small mass or small momentum. This logarithmic behaviour defines an infrared problem. Usually the logarithmic power series for the free energy or other thermodynamic quantities sum up to

> power law singularities of the form $\lambda^{\varkappa}$, where quite generally the singular exponents $\varkappa$ depend continuously on system parameters.

Where does one find infrared problems? In large systems we have translational invariance. Therefore low energy excitations will show a momentum dispersion $\varepsilon(q) \sim q^{\sigma}$ for $q \to 0$. Typical integrals in perturbation expansions have the form

$$\int_0 \frac{d^d q}{q^{\sigma}} |\text{matrix elem.}|^2 \sim \ln \ldots \ , \ d = \sigma \ .$$

This means that we have a chance to find a logarithmic (infrared) divergence and thus an infrared problem if dimension d and dispersion σ coincide. As the normal dispersions are $\sigma = 1$ for quantummechanical systems (massless particle-hole excitations in Fermi systems, massless Bose systems) and $\sigma = 2$ for classical systems with short range forces, we conclude:

> Infrared problems will show up in 1-d q.m. Fermi and Bose systems and in 2-d classical systems.

The connection of 1-d quantum field models and 2-d classical models is obvious because a quantum field theory in 1 space and 1 time dimension is equivalent to a Euclidean (or classical) field theory in 2 spatial dimensions.

In the next chapter we shall discuss the equivalence of a few model situations without going into the details of the derivations which have been and will be published elsewhere. In the last chapter the main results and physical properties which follow from the mathematical infrared problem will be discussed.

III. Equivalence of Models

Presumably many 1-d and 2-d models exhibit infrared behaviour and therefore will have quite interesting physical properties. We do not attempt to investigate all possible situations, but rather confine ourselves to a few models which are mathematically exactly equivalent. We start with

1) 1-d Fermi systems: To describe the fascinating experiments in quasi 1-d conductors and semiconductors one considers a 1-d electronic system. Besides a single particle energy band one has the interaction part [3,4,5]

$$H_{int.} = \frac{1}{2} \sum_{qss'} V_{ss'}(q) \; c^+ c^+ c \, c \quad , \tag{1}$$

where s denotes the spin, q the momentum transfer. At low temperatures only processes near the Fermi surface are relevant which leads to a parametrization in terms of a few parameters [3,4,5]. For small q-transfer we have the coupling $V(q \sim 0) = 2\pi\gamma$ relevant for the particle density degrees of freedom. These decouple from the spin density degrees of freedom where the two "backscattering" parameters u, v [5] enter:

$$V_{ss'}(q \sim \pm 2k_f) = 2\pi \left[u \, \delta_{s,s'} + v \, \delta_{s,-s'} \right] . \tag{2}$$

For the following it is convenient to introduce the quantities:

$$\eta = 4 \sqrt{\frac{1+u}{1-u}} \quad , \quad b = \frac{v}{2\pi \sqrt{1-u^2}} \quad . \tag{3}$$

Energy excitations in 1-d Fermi systems are effectively Bose excitations with zero mass. A suitable representation of Fermion field operators in terms of Bosons has been given by

Schotte[6] (see also [3]-[5]). With the restriction $|u| < 1$ this representation can be used to prove the exact equivalence of the 1-d Fermion problem to a 1-d Bose field theory. The procedure may be called "Bosonization". The particle density part of the Fermi system leads to a free Bose theory, the spin density part leads to a

2) 1-d Bose system with "Sine-Gordon" interaction[7].

The corresponding Euclidean field theory can then be viewed as a 2-d classical field theory and can be interpreted as the[8]

3) 2-d harmonic ferromagnetic rotator model.

This is a 2-d model of classical unit vectors (spins) on lattice sites r_i with Hamiltonian

$$H = \sum_{ij} \frac{1}{4} J_{ij} (\varphi_i - \varphi_j)^2 - \sum_i B \cdot \cos \varphi_i \quad , -\infty < \varphi_i < \infty \tag{4}$$

where φ_i is the angle against x-direction. The second part is the magnetic field energy, B the magnetic field. The first part describes a harmonic ferromagnetic interaction with short range coupling J_{ij}. This is a simple ferromagnetic situation in the sense that both interaction and field favour parallel spin alignment in positive or negative x-direction depending upon the sign of B. Qualitative changes in physical properties may occur only at B = 0. We then have the theorem:

The free energy is analytic both in B and T for $B \neq 0$ and the thermodynamic state is unique for $B \neq 0$.

The exact equivalence to situation 1) is expressed by denoting the two relevant parameters:

$$\eta = \frac{T}{2\pi c \cdot \sum_j J_{ij}} \quad , \quad b = \frac{B}{2T} \tag{5}$$

where c = 1/4 for n.n coupling. η and b are essentially the temperature and the magnetic field, respectively. The analyticity theorem tells us that (infrared) singularities in thermodynamic quantities may only occur at b = 0.

The partition function

$$Z(\eta, b) = \langle e^{2b\sum_i \cos\varphi_i} \rangle_{harm.} \tag{6}$$

is expanded in b, the b = 0 spin-spin correlation functions can be worked out (2,5,8) and the resulting expansion is interpreted as the ground-canonical partition function of the

4) 2-d classical Coulomb plasma:

$$Z_{gr.} = \sum_n \frac{z^{2n}}{(n!)^2} \int \frac{d^2r_1}{a^2} \cdots \frac{d^2r_{2n}}{a^2} \cdot \exp\{-\beta V_{2n}(r_1, \ldots, r_{2n})\} \tag{7}$$

$$V_{2n} = e^2 \sum_{\nu<\mu}^{2n} (-1)^{\nu-\mu} \cdot \ln \sqrt{\frac{a^2 + r_{\nu\mu}^2}{a^2}} \quad . \tag{8}$$

V_{2n} is the potential energy of a neutral 2-d Coulomb plasma with charge $\pm e$; for long range it exhibits the 2-d logarithmic Coulomb potential ($r \to \infty$) with a soft core cut-off at small distance a to avoid ultraviolet divergencies. The relevant parameters here are identified as

$$\eta = \beta e^2 \quad , \quad b = z \tag{9}$$

where z is the fugacity. The free energy $\beta F \Omega = -\ln \mathcal{Z}$ (Ω = volume) of the rotator model is now the pressure $p(z,\eta) = -F$, the particle density is $n = z \frac{\partial}{\partial z} \beta p$ and $p = p(n)$ will be the equation of state.

The plasma model is an infrared system par exellence because of the fact that the long range logarithmic potential between relevant degrees of freedom is built into the model at the outset; for the other situations 1) - 3) this becomes obvious only via the exact analogy to the plasma model, i.e. after performing the steps which lead to the representation (7), (8).

The infrared behaviour is most easily visualized by looking at the two particle correlation at z = 0:

$$\chi_2(r) = e^{-\beta V_2(r)} \simeq \left(\frac{a}{r}\right)^\eta , \quad r \to \infty .$$

This means that $\chi_2(r)$ (and all higher order correlation functions) show a power law decay with a continuously varying η-exponent ($0 \leq \eta < \infty$). This is reminiscent of critical point behaviour. Remark, however, that in the present case the varying η indicates a whole line of critical points.

The four model situations discussed so far are exactly equivalent by identifying the relevant parameters η and b in each case (Eq. (3),(5),(9)). There are other infrared situations which can be related approximately to the above situations. Without further discussion we mention: the Kondo problem(9), the Baxter model(10), critical point behaviour in general, the usual XY- or rotator model(2).

III. Results

From the analyticity theorem we know that infrared singularities, which imply an interesting physical structure, will appear only for $b \to 0$. The mathematical structure of the free energy in this limit can be worked out by exact scaling techniques and is described in detail in refs. (2,5,8). Here we only cite the main results without derivation.

The magnetic rotator model (situation 3)) shows the Phase transition of continuous order[1,2] with the following characteristic properties. The free energy is

$$F(b,T) = F(0,T) + f_{reg.} + f_{sing.} \tag{10}$$

where F(0,T) is the free energy of the pure harmonic part in (4) which is analytic in T and thus shows no specific heat singularity. For $b \to 0$ the regular part is a power series in b^2 while the singular part is

$$f_{sing.} = A(\eta) \cdot |b|^{\varkappa} \tag{11}$$

with the scaling law

$$\varkappa = \frac{4}{4-\eta} \quad (\eta < 4) \; , \; \text{"}\varkappa = \infty\text{"} \quad (\eta \geq 4) \; . \tag{12}$$

This means that for $\eta < 4$ (low temperatures) we have a power law singularity with a continuously varying exponent $\varkappa$ ($1 \leq \varkappa < \infty$). The "order" of the phase transition increases continuously to ∞ at T_∞, $\eta(T_\infty) = 4$. For $\eta > 4$ "$\varkappa = \infty$" indicates the absence of a power singularity, presumably the free energy is analytic as $b \to 0$, but it is at least infinitly often differentiable. Further properties:

a) $M_0 = - \partial F / \partial b |_0 = 0 \; (\varkappa > 1)$ at finite temperatures. This implies: <u>no</u> infinitly long range order.

b) The susceptibility $\chi_2 = - \partial^2 F / \partial b^2 |_0$ diverges below T_2 $(\eta(T_2) = 2)$ and is finite above T_2.

c) Higher derivatives $\chi_n = - \partial^n F / \partial b^n |_0$ diverge below T_n $(\eta(T_n) = n)$. These temperatures T_n increase with n and have an accumulation point at T_∞.

The physical picture of such a phase transition is that the system cannot sustain bulk order if the field b is reduced to zero. However, below T_∞ there remain infinitly long range fluctuations and thus a tendency towards ordering. All temperatures below T_∞ are therefore critical temperatures.

Accordingly the correlation length $\xi(b)$ diverges as $b \to 0$ below T_∞ $(\eta < 4)$ and we have[2,5]

$$\xi^{-1} = |b|^{\nu_b} \quad , \quad \nu_b = \frac{1}{2} \varkappa \; . \tag{13}$$

It should be mentioned that the analysis of ref. (2) concerns the usual classical XY- or rotator model and not just the harmonic version considered here. As shown in (2) the only change is the replacement

$$\eta_{harm.} \longrightarrow \eta = A(T)^{-1} \cdot \eta_{harm.} \tag{14}$$

where A is the characteristic dispersion coefficient for low lying energy excitations (for details see (2)).

We now turn to the

4) Plasma model: The analyticity theorem implies that the pressure p and particle density n as functions of η and z (9) are analytic for $z \neq 0$. However, infrared singularities show up for vanishing fugacity, i.e. in the dilute limit, implying that there exists no virial expansion. In fact, one obtains a quite interesting and unusual equation of state near $\eta = 2$ which will be discussed elsewhere (11). A further interesting property is a smooth metal-insulator transition (11,12). This is seen in the dielectric function $\varepsilon(q)$ which describes the response to an external electric field at finite particle density n. One obtains in the limit $q \to 0$:

$$\varepsilon(q) \sim \begin{cases} q^{-2} & 0 \leq \eta \leq 2 \\ q^{\eta-4} & 2 \leq \eta < 4 \end{cases}, \tag{15}$$

i.e. the usual metallic q^{-2}-response for high temperatures $(\eta = \beta e^2 \leq 2)$ and a quite unusual, but still metallic, response for $2 \leq \eta \leq 4$. For $\eta > 4$ we obtain

$$\lim_{q \to 0} \varepsilon(q) = 1 + \frac{g(\eta, n)}{\eta - 4}, \tag{16}$$

i.e. a dielectric (insulating) regime for low temperatures. In this regime the system may be visualized as being composed of neutral bound pairs (12,13).

Finally we describe the main properties of the

1) 1-d Fermi systems which follow from the above analysis of the infrared problem and which have been derived in ref. (5).

The free energy of situation 3) is now replaced by the groundstate energy E(u,v) for the spin-density part of the Fermion model (1-3); the thermodynamic state of the 2-d classical rotator model is replaced by the groundstate of the q.m. 1-d Fermion model. The analyticity theorem implies:

The groundstate is unique (non-degenerate) and E(v,u) is analytic both in v and u for $|u| < 1$ and $v \neq 0$.

Singularities in E show up for $v \to 0$ and with appropriate changes of the parameters are the same as discussed for the magnetic rotator model. The uniqueness of the groundstate implies the cluster decomposition property for correlation functions (vacuum expectation values) $\chi(x,t) \equiv \langle A^{+}(xt) \cdot A(00) \rangle$, namely

$$\lim_{(xt) \to \infty} \chi(xt) = |\langle A \rangle|^2 \tag{17}$$

where $\langle A \rangle$ is the groundstate expectation value. Furthermore the groundstate, being unique at $v \neq 0$, has the symmetries of the Hamiltonian. One concludes that operators A, which change the symmetries, have zero expectation values, $\langle A \rangle = 0$, implying that the corresponding correlation functions $\chi(xt)$ decay in the limit (17). This general result, for instance, implies that the correlation functions which describe the response towards charge density and spin density wave formation and towards formation of singlet or triplet superconducting pairs and which have been discussed in refs. (3,14), all decay to zero in the limit (17).

One final result concerns the particle spectrum of the 1-d Fermi system. The existence of a finite correlation length ξ for $b \neq 0$ (13) in the spin-spin correlation functions of the rotator model implies a gap in the energy spectrum, $\Delta = \xi^{-1}$, which for $v \to 0$ (or $b \to 0$) has the form (see (13))

$$\Delta(v) = c(\eta)\,|v|^{\frac{2}{4-\eta}} \tag{18}$$

with some coefficient $c(\eta)$. However, it should be clear that this gap only applies to energy eigenstates of the Fermion model which appear as intermediate states in Fermion correlation functions corresponding to the spin-spin correlation functions of the rotator model (or the Sine-Gordon model). In the Fermion model one considers also correlation functions, for instance, those describing pair formation discussed above, which have no analogue in the rotator model. Consequently the Fermion Hilbert space is larger. It remains to be seen whether there exists an energy gap in the whole Fermion Hilbert space; so far, its existence (18) has been established only for a sector.

References

(1) E. Müller-Hartmann and J. Zittartz: Phys. Rev. Lett. 33, 893 (1974); Z. Physik B22, 59 (1975)

(2) J. Zittartz: Z. Physik B23, 55, 63 (1976)

(3) A. Luther and V. Emery: Phys. Rev. Lett. 33, 589 (1974)

(4) S.T. Chui and P.A. Lee: Phys. Rev. Lett. 35, 315 (1975)
H.U. Everts and H. Schulz: Z. Physik B22, 285 (1975)

(5) J. Zittartz: Z. Physik B23, 277 (1976)

(6) K.D. Schotte: Z. Physik 230, 99 (1970)

(7) S. Coleman: Phys. Rev. D11, 2088 (1975)

(8) J. Zittartz: to be published

(9) P.W. Anderson and G. Yuval in: Magnetism Vol. V, ed. by H. Suhl, Academic Press N.Y., 1973

(10) R. Baxter: Ann. Phys. (N.Y.) 70, 193 (1972)

(11) J. Zittartz: to be published

(12) J. Zittartz and B.A. Huberman: Solid State Comm. 18, 1373 (1976)

(13) E.H.Hauge and P.C. Hemmer: Physica Norvegica 5, 209 (1971)

(14) P.A. Lee: Phys. Rev. Lett. 34, 1247 (1975);
H. Gutfreund and R.A. Klemm: Phys. Rev. B14, 1073 (1976)

Conference on Organic Conductors and Semiconductors, Siófok, Hungary 1976

SCALING THEORY OF A ONE-DIMENSIONAL FERMI GAS MODEL WITH TWO CHARACTERISTIC ENERGIES

J. SÓLYOM* and G. SZABÓ**

*Central Research Institute for Physics,
Budapest, Hungary
**Physics Department, Loránd Eötvös University,
Budapest, Hungary

The properties of a one-dimensional Fermi gas model with two characteristic energies, two bandwidth cut-offs, is studied. The direct electron-electron coupling and the phonon mediated effective coupling are cut off at energies E_o and ω_D, respectively, where E_o is the bandwidth of the electron energy band and ω_D is the Debye frequency. The model is treated in the framework of renormalization group approach. It is shown that this model can be mapped on the usual one cut-off model and the results obtained for that model can be applied.

The one-dimensional Fermi gas model has been intensively studied recently by several authors (1-6) using different methods, such as parquet diagram summation, the renormalization group and the bosonisation transformation. The models treated by these authors differ not only in the choice of the coupling constants, i.e. whether the backward scattering is included or not, but also in the choice of the physical cut-off parameter. The cut-off has to be introduced in any perturbational calculation to avoid the ultraviolet non-physical divergences. In the field theoretical treatment of the corresponding models (7) the cut-off plays no role since the Thirring model is renormalizable in field theoretical sense, in the statistical mechanical treatment, however, we will keep the cut-off.

There are two usual ways to introduce a cut-off in the one-dimensional Fermi gas model. One is to restrict the momenta or energies of all the electrons participating in a scattering process to be within a range of width $2k_c$ or $2\omega_D = 2vk_c$ around the Fermi points. Here v is the Fermi velocity. This type of cut-off is called bandwidth cut-off. Another possibility is to put a cut-off on the momentum or frequency transferred in the course of the scattering, the initial momentum or energy can however be anywhere. This is called transfer cut-off procedure. The importance of cut-off has been recently studied by Grest (8) who compared the perturbation theoretical results using different cut-off procedures.

In the Tomonaga model (9) there is no large momentum transfer interaction (no backward scattering) and usually a cut-off on the small momentum transfer interaction is introduced. In models with backward scattering a simple transfer cut-off does not lead to correct results. As was shown by Chui et al (10) in calculating the contribution of some of the diagrams a bandwidth cut-off has to be used. This calculation was later extended by Grest et al (11) to consider higher order corrections by making use of the renormalization group. The renormalization group treatment of the one-dimensional Fermi gas model (2) was first worked out for a bandwidth cut-off. Grest et al (11) have shown that the model with two cut-offs, one transfer and one bandwidth cut-offs, can be scaled to a one cut-off problem with a renormalized large momentum transfer interaction. Using this scaling argument the results of the earlier analysis can be applied to understand the low energy behaviour of the system with two cut-offs.

The question arises naturally whether a bandwidth or transfer cut-off model is nearer to the physical reality. If the effective electron-electron interaction is the result of a virtual phonon exchange, two different situations may appear. Depending on whether we study the polarization bubble type diagrams or Cooper pair type processes (see Fig. 1a and 1b, respectively), the elimination of the phonons from the problem leads to different types of effective interactions. In the first case the effective interaction has an energy transfer cut-off, given by the phonon energy. In the second case the effective interaction can be written in the form

$$V_{eff} \sim g^2 \left(\frac{1}{\varepsilon_k - \varepsilon_{k-q} - \omega_q} - \frac{1}{\varepsilon_{k'} - \varepsilon_{k'+q} + \omega_q} \right) \tag{1}$$

where ε_k and $\varepsilon_{k'}$ are the energies of the incoming electrons with momentum k and k', ε_{k-q} and $\varepsilon_{k'+q}$ are the energies of the scattered electrons and ω_q is the energy of the virtually exchanged phonon, g is the electron-phonon coupling. Considering the backward scattering with $q \sim 2k_F$, the coupling is attractive if $|\varepsilon_k - \varepsilon_{k-q}| \ll \omega_q$ and $|\varepsilon_{k'} - \varepsilon_{k'+q}| \ll \omega_q$. These conditions can only be satisfied if the energies of all the four participating electrons are within a range of width $2\omega_D$ around the Fermi energy, where ω_D is the Debye frequency. This consideration shows that depending on the type of interaction, transfer or bandwidth cut-off has to be used.

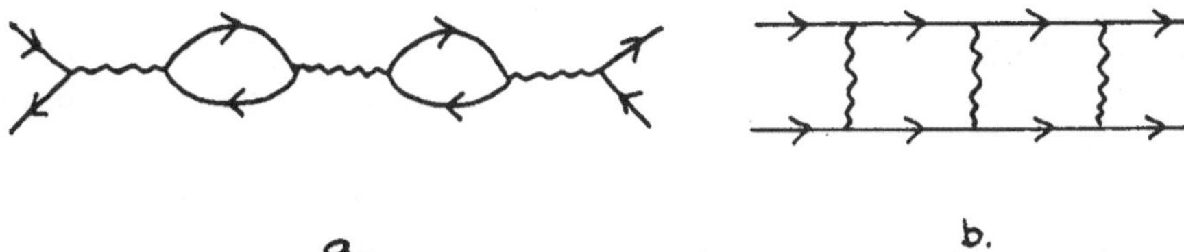

Fig. 1 Typical (a) zero-sound and (b) Cooper-pair type diagrams

In the model studied by Chui et al (10) and Grest et al (11) the cut-off in the interaction is always a momentum transfer cut-off, and a second cut-off comes from the finite electronic bandwidth. In this paper we study another model for the one-dimensional Fermi gas with two cut-offs, assuming two different kinds of interactions with different bandwidth cut-offs. The phonon mediated effective electron-electron coupling is cut-off at ω_D, the direct Coulomb type electron-electron coupling is cut-off at the bandwidth E_0. We will see that the results for the two models are rather similar and therefore we expect a reasonable description of the physical situation where the cut-off is partly transfer and partly bandwidth cut-off.

The Hamiltonian we want to study is as follows

$$H = H_0 + H_1 + H_2 , \tag{2}$$

$$H_0 = \sum_{k,\alpha} \varepsilon_k a^+_{k\alpha} a_{k\alpha} + \sum_{k,\alpha} \varepsilon_k b^+_{k\alpha} b_{k\alpha} , \tag{3}$$

$$\begin{aligned} H_1 = {} & \frac{g_{1ph}}{L} \sum{}' a^+_{k_1\alpha} b^+_{k_2\beta} a_{k_3\beta} b_{k_4\alpha} \\ & + \frac{g_{2ph}}{L} \sum{}' a^+_{k_1\alpha} b^+_{k_2\beta} b_{k_3\beta} a_{k_4\alpha} , \end{aligned} \tag{4}$$

$$H_2 = \frac{g_{1c}}{L} \sum a^+_{k_1\alpha} b^+_{k_2\beta} a_{k_3\beta} b_{k_4\alpha} + \frac{g_{2c}}{L} \sum a^+_{k_1\alpha} b^+_{k_2\beta} b_{k_3\beta} a_{k_4\alpha} , \tag{5}$$

where a^+_k (b^+_k) is the creation operator for an electron with momentum k around $+k_F$ ($-k_F$). The interaction term H_1 represents the phonon mediated effective electron-electron coupling and the momenta are restricted to the ranges ($k_F - k_c$, $k_F + k_c$) and ($-k_F - k_c$, $-k_F + k_c$), respectively. The term H_2 comes from the direct electron-electron coupling and acts between any electron within the band of width $2k_0$. For sake of simplicity the strength of the coupling is assumed to be independent of the momenta. The cut-offs k_c and k_0 in momentum space correspond to the energy cut-offs $\omega_D = v k_c$ and $E_0 = v k_0$, where v is the Fermi velocity.

The difference of our model and that of Chui et al (10) and Grest et al (11) can be best seen if we calculate the second order vertex corrections. Two such diagrams are shown in Fig. 2. In the momentum transfer model the analytic contribution corresponding to these diagrams is

$$\Gamma_a = - \frac{g_1^2}{2\pi v} \left(\ln \frac{\omega}{2\omega_D} - \frac{1}{2} i\pi \right), \tag{6}$$

and

$$\Gamma_b = \frac{g_1^2}{\pi v} \left(\ln \frac{\omega}{2E_0} - \frac{1}{2} i\pi \right), \tag{7}$$

respectively if $\omega \ll \omega_D$ and $\omega \ll E_0$. In our present model and in the same limit we get

$$\Gamma_a = - \frac{g_{1c}^2}{2\pi v} \left(\ln \frac{\omega}{2E_0} - \frac{1}{2} i\pi \right) - \frac{2 g_{1c} g_{1ph} + g_{1ph}^2}{2\pi v} \left(\ln \frac{\omega}{2\omega_D} - \frac{1}{2} i\pi \right), \tag{8}$$

and

$$\Gamma_b = \frac{g_{1c}^2}{\pi v}\left(\ln\frac{\omega}{2E_o} - \frac{1}{2}i\pi\right)$$
$$+ \frac{2g_{1c}\,g_{1ph} + g_{1ph}^2}{\pi v}\left(\ln\frac{\omega}{2\omega_D} - \frac{1}{2}i\pi\right). \qquad (9)$$

This difference will lead to a different mapping of this system on an equivalent one-cut-off model.

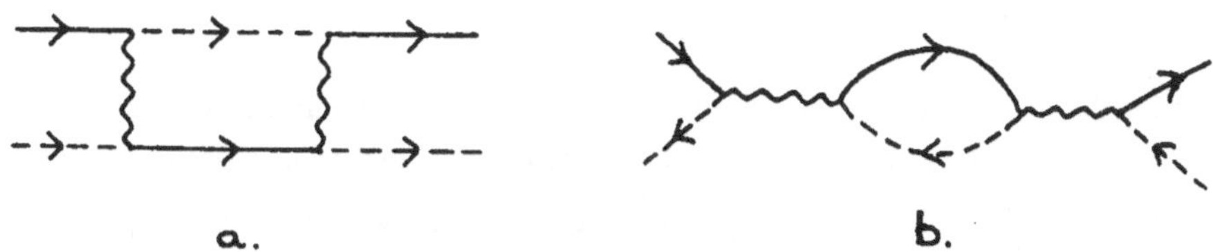

Fig. 2 Second order Cooper pair and zero-sound type diagrams. Solid (dashed) line stands for the propagation of electron with momentum around $+k_F$ $(-k_F)$.

We will proceed in a non-conventional way in finding this mapping and the corresponding scaling laws. The renormalization group transformation can be simply interpreted as a successive elimination of degrees of freedom near the cut-off at the expense of introducing a new "invariant" coupling. The Lie differential equations for the invariant coupling, the Green's function, the vertices and the response functions are known for the Fermi gas model with one cut-off (2,3). The physical intuition suggests that the same equations can be used for the two-cut-off case as well, the only difference being that the elimination of degrees of freedom happens in two distinct steps. Our strategy will be to show that with this assumption the perturbational results are reproduced correctly and then accepting that these relations are valid in higher orders, too, its consequences can be explored. The usual procedure is to derive the scaling equations from the perturbational expression. In the present problem, however, it seems more natural to start with the known scaling relations.

We are interested in the behaviour of the Green's function, vertex and response functions at energies much lower than the cut-off energies, $\omega \ll \omega_D \ll E_o$.

First we will eliminate the degrees of freedom between the two cut-offs by using the scaling equations of Ref. 2.

$$\frac{d g_1'(x)}{dx} = \frac{1}{x} \left\{ \frac{1}{\pi v} g_1'^2(x) + \frac{1}{2\pi^2 v^2} g_1'^3(x) + \ldots \right\}, \tag{10}$$

$$\frac{d g_2'(x)}{dx} = \frac{1}{x} \left\{ \frac{1}{2\pi v} g_1'^2(x) + \frac{1}{4\pi^2 v^2} g_1'^3(x) + \ldots \right\}. \tag{11}$$

Since only the Coulomb type coupling acts between electrons which have energies between ω_D and E_0, the elimination of degrees of freedom in this range will influence g_{1c} and g_{2c} only, g_{1ph} and g_{2ph} remain unchanged. The phase space between ω_D and E_0 is completely eliminated when the upper cut-off E_0 is scaled down to ω_D, which corresponds to $x = \omega_D / E_0$. The new couplings are denoted by $\overline{g}_{1c}$ and $\overline{g}_{2c}$

$$\overline{g}_{1c} = g_1'\left(\frac{\omega_D}{E_0}\right), \qquad \overline{g}_{2c} = g_2'\left(\frac{\omega_D}{E_0}\right), \tag{12}$$

and in solving eqs. (10) and (11) the boundary condition

$$g_1'(x=1) = g_{1c}, \qquad g_2'(x=1) = g_{2c} \tag{13}$$

has to be used. The outer phase space being eliminated, we are left with a one-cut-off problem, but since for energies lower than ω_D the phonon mediated coupling will also act, the effective couplings for the new one-cut-off problem are $\overline{g}_{1c} + g_{1ph}$ and $\overline{g}_{2c} + g_{2ph}$. The scaling relations can be further used by scaling the cut-off ω_D, but with a new boundary condition

$$g_1'(x=1) = \overline{g}_{1c} + g_{1ph}, \qquad g_2'(x=1) = \overline{g}_{2c} + g_{2ph}. \tag{14}$$

As a simple example we will calculate the density-density response function, $N(\omega)$. The Lie equation has been derived in Ref. 3 for the one-cut-off problem

$$\frac{\partial \ln \bar{N}(x)}{\partial x} = \frac{1}{x}\left\{ \frac{1}{\pi v}\left[2g_1'(x) - g_2'(x)\right] + \frac{1}{2\pi^2 v^2}\left[g_1'^2(x) - g_1'(x)g_2'(x) + g_2'^2(x)\right] + \dots \right\}, \tag{15}$$

where $x = \omega/\omega_D$ and

$$\bar{N}(\omega) = \pi v \omega \frac{\partial}{\partial \omega} N(\omega). \tag{16}$$

In the present problem N depends on ω/ω_D and ω/E_0 if $\omega \ll \omega_D \ll E_0$, but it depends on ω/E_0 only if $\omega_D \ll \omega \ll E_0$. Furthermore at $\omega = E_0$ $\bar{N}(\omega = E_0) = 1$ if the imaginary parts are neglected. Eq. (15) should be solved first in the range $\omega_D \ll \omega \ll E_0$ after having introduced on the right hand side the solution of eqs. (10) and (11) with the boundary condition of eq. (13).

$$\begin{aligned} \bar{N}\left(\frac{\omega}{E_0}\right) = 1 &+ \frac{1}{\pi v}(2g_{1c} - g_{2c}) \ln \frac{\omega}{2E_0} \\ &+ \frac{1}{2\pi^2 v^2}(2g_{1c} - g_{2c})^2 \ln^2 \frac{\omega}{2E_0} + \frac{3}{4\pi^2 v^2} g_{1c}^2 \ln^2 \frac{\omega}{2E_0} \\ &+ \frac{1}{2\pi^2 v^2}(g_{1c}^2 - g_{1c}g_{2c} + g_{2c}^2) \ln \frac{\omega}{2E_0} + \dots \end{aligned} \tag{17}$$

The value of $\bar{N}(\omega/E_0)$ when $\omega = \omega_D$ is denoted by

$$\bar{N}\left(\frac{\omega_D}{E_0}\right) = \bar{N}_0 . \tag{18}$$

In order to determine $N(\omega)$ for $\omega \ll \omega_D$, we have to solve eq. (15) again, but now with the boundary condition $\bar{N}(x=1) = \bar{N}_0$, and the invariant coupling should be calculated with the boundary condition in eq. (14).

The solution is similar to eq. (17), the unity being replaced by $\overline{N}_o$ and g_{1c} and g_{2c} being replaced by $\overline{g}_{1c} + g_{1ph}$ and $\overline{g}_{2c} + g_{2ph}$, respectively. Comparing this form with the straightforward perturbational expression which for $N(\omega)$ is

$$
\begin{aligned}
N(\omega) &= \frac{1}{\pi v} \ln \frac{\omega}{2E_o} + \frac{1}{2\pi^2 v^2}(2g_{1c} - g_{2c}) \ln^2 \frac{\omega}{2E_o} \\
&+ \frac{1}{2\pi^2 v^2}(2g_{1ph} - g_{2ph}) \ln^2 \frac{\omega}{2\omega_D} + \frac{1}{4\pi^3 v^3}(2g_{1c} - g_{2c})^2 \ln^3 \frac{\omega}{2E_o} \\
&+ \frac{1}{4\pi^3 v^3}(2g_{1ph} - g_{2ph})^2 \ln^3 \frac{\omega}{2\omega_D} \\
&+ \frac{1}{2\pi^3 v^3}(2g_{1c} - g_{2c})(2g_{1ph} - g_{2ph}) \ln^2 \frac{\omega}{2\omega_D} \ln \frac{\omega}{2E_o} \\
&- \frac{1}{12\pi^3 v^3}(g_{1c}^2 - 4g_{1c}g_{2c} + g_{2c}^2) \ln^3 \frac{\omega}{2E_o} \\
&- \frac{1}{12\pi^3 v^3}(g_{1ph}^2 - 4g_{1ph}g_{2ph} + g_{2ph}^2) \ln^3 \frac{\omega}{2\omega_D} \\
&- \frac{1}{6\pi^3 v^3}(g_{1c}g_{1ph} - 2g_{1c}g_{2ph} - 2g_{1ph}g_{2c} + g_{2c}g_{2ph}) \ln^3 \frac{\omega}{2\omega_D} \\
&+ \frac{1}{4\pi^3 v^3}(g_{1c}^2 - g_{1c}g_{2c} + g_{2c}^2) \ln^2 \frac{\omega}{2E_o} \\
&+ \frac{1}{4\pi^3 v^3}(g_{1ph}^2 - g_{1ph}g_{2ph} + g_{2ph}^2) \ln^2 \frac{\omega}{2\omega_D} \\
&+ \frac{1}{2\pi^3 v^3}(g_{1c}g_{1ph} - g_{1c}g_{2ph} - g_{1ph}g_{2c} + g_{2c}g_{2ph}) \ln^2 \frac{\omega}{2\omega_D} + \dots ,
\end{aligned}
$$

the renormalization group treatment gives the same results if $\overline{g}_{1c}$ and $\overline{g}_{2c}$ are expanded and the relation (16) is used. The other response functions are obtained similarly.

There is a slight modification in the calculation if the imaginary parts are not neglected. There is a discontinuity in the functions at $\omega = \omega_D$ and this should be taken into account in the boundary conditions. This does not modify the conclusions that the two-cut-off Fermi gas model can be mapped on a one-cut-off model with effective bare couplings $\bar{g}_{1c} + g_{1ph}$, $\bar{g}_{2c} + g_{2ph}$.

Considering now the low energy or low temperature behaviour of this system, it follows immediately that the phase diagram is the same as for the one-cut-off model (3, 12) with the only difference that the parameters are the new bare couplings $\bar{g}_{1c} + g_{1ph}$ and $\bar{g}_{2c} + g_{2ph}$. Since the Coulomb coupling is repulsive, the renormalization leads to its weakening and therefore $\bar{g}_{1c}$ and $\bar{g}_{2c}$ will have little effect. It can modify the value of the exponents characterizing the singular behaviour of the response functions, but the physics of the system is the same as that of the one-cut-off model.

The authors are indebted to Dr. G. Grest and Dr. A. Zawadowski for useful discussions.

REFERENCES

(1) Yu. A. Bychkov, L. P. Gorkov and I. E. Dzyaloshinsky, Zh. Eksp. Teor. Fiz. 50, 738 (1966) (Sov. Phys. - JETP 23, 489 (1966)).

(2) N. Menyhárd and J. Sólyom, J. Low. Temp. Phys. 12, 529 (1973).

(3) J. Sólyom, J. Low Temp. Phys. 12, 547 (1973).

(4) I. E. Dzyaloshinsky and A. I. Larkin, Zh. Eksp. Teor. Fiz. 65, 411 (1973) (Sov. Phys. - JETP 38, 202 (1974)).

(5) A. Luther and I. Peschel, Phys. Rev. B9, 2911 (1974).

(6) A. Luther and V. J. Emery, Phys. Rev. Lett. 33, 589 (1974).

(7) A. Luther, Phys. Rev. B. 14, 2153 (1976)

(8) G. S. Grest, Phys. Rev. B. 14, 5114 (1976)

(9) S. Tomonaga, Progr. Theor. Phys. 5, 544 (1950).

(10) S.-T. Chui, T. M. Rice and C. M. Varma, Solid State Comm. 15, 155 (1974).

(11) G. S. Grest, E. Abrahams, S.-T. Chui, P. A. Lee and A. Zawadowski, Phys. Rev. B14, 1225 (1976).

(12) P. A. Lee, Phys. Rev. Lett. 34, 1247 (1975).

Conference on Organic Conductors and Semiconductors, Siófok, Hungary 1976

ON THE EFETOV-LARKIN MODEL OF A STRONGLY INTERACTING ONE-DIMENSIONAL SYSTEM

M. FOWLER

Department of Physics, University of Virginia
Charlottesville, Virginia 22901

Efetov and Larkin have introduced a model Hamiltonian for a strongly interacting one-dimensional system, exactly soluble for a particular value of a coupling constant, and have demonstrated that the asymptotic behavior is the same as that of a long-wave boson Hamiltonian. Here we show that in the relevant energy range, their Hamiltonian can be transformed into such a long-wave Hamiltonian, over a range of values of the coupling.

Introduction

It is widely known that the asymptotic behavior of correlation functions for a physical system is determined by the long-wavelength excitations of the system. In the Renormalization Group technique, critical exponents are determined by transformations which lead to an effective Hamiltonian for long-wavelength excitations. Usually, this effective Hamiltonian is independent of the particular values of coupling constants in the basic (microscopic) Hamiltonian, and hence the critical exponents are similarly universal. For one-dimensional interacting systems, however, the coupling constants often change much more slowly under scaling, and even in the long-wavelength limit the system remembers the value of the basic coupling constant. Thus, the critical exponents are non-universal. Nevertheless, the asymptotic behavior of correlation functions is still presumably determined by long-wavelength excitations, although the relationship is not quite so transparent as before - dependence on the basic coupling constant is no longer a valid criterion for ignoring the shorter wavelength excitations of the system.

In an attempt to clarify the connection between long-wavelength excitations and correlation functions in a situation of this type, Efetov and Larkin[1] have studied a one-dimensional strongly interacting model which for a particular value of a coupling constant is exactly soluble. They find the asymptotic behavior for this particular value, and prove that it is identical to that predicted by a semiphenomenological long-wave boson Hamiltonian. We shall show here that their long-wave Hamiltonian can be deduced from their basic Hamiltonian by a sequence of transformations. Indeed, this remains true even away from the particular value of the coupling constant for which they solve their model. This eliminates the necessity of postulating the long-wave Hamiltonian, and, we believe, throws considerable light on the structure of the system, and the range of validity of the effective long-wave Hamiltonian.

The Efetov-Larkin Model

We first briefly review the work of Efetov and Larkin.[1] The basic Hamiltonian of their interacting one-dimensional system is

$$H = T \sum_{i,\alpha} a^{+}_{i,\alpha}(a_{i+1,\alpha} + a_{i-1,\alpha}) - V \sum_{i,\alpha} a^{+}_{i,\alpha} a^{+}_{i,-\alpha} a_{i,-\alpha} a_{i,\alpha} -$$

$$\sum_{\substack{i \neq j \\ \alpha,\beta}} V_{ij} a^{+}_{i\alpha} a_{i\alpha} a^{+}_{j\beta} a_{j\beta} \tag{1}$$

where a^{+}_{i} is an electron creation operator, T the band width, i the site index and α the spin index. The case examined is V >> T: that is, the same site electron-electron attraction is assumed to be so strong that every site contains two electrons or none. In this case, pair creation operators $b^{+}_{i} = a^{+}_{i+} a^{+}_{i-}$ can be introduced, and they obey

$$[b_i, b_j] = 0 \qquad i \neq j$$

$$b^{+}_{i} b_i + b_i b^{+}_{i} = 1 \qquad\qquad b^{2}_{i} = b^{+2}_{i} = 0 \tag{2}$$

In terms of these operators, the Hamiltonian is

$$H = \frac{1}{2m} \sum b^{+}_{i}(b_{i+1} + b_{i-1}) - \sum_{i \neq j} (V_{ij} - \frac{1}{8m} \delta_{|i-j|-1}) \rho_i \rho_j - 2V \sum b^{+}_{i} b_i \tag{3}$$

where $2m = VT^{-2}$ (correct to second order in T) and $\rho_i = b^{+}_{i} b_i$. The authors consider the particular case

$$V_{ij} = V^{0}_{ij} = \frac{1}{8m} \delta_{|i-j|-1} \tag{4}$$

for which value the model is exactly soluble, being equivalent to a noninteracting spinless fermion gas. The authors use this equivalence to find the asymptotic behavior of the pair and density correlation functions. (In fact, this case is just the spin one-half XY model with a magnetic field in the z direction. This is evident from (3), after noting from (2) that the operators b^{+}_{i}, b_i are isomorphic with raising and lowering operators for a spin one-half. The correlation functions of interest correspond to $\langle S^x S^x \rangle$ and $\langle S^z S^z \rangle$ and have been evaluated by McCoy and other authors[2] for a more general case, although the techniques of Efetov and Larkin for the density correlation function are more transparent.)

To elucidate the rôle of long-wavelength excitations in the asymptotic behavior of correlation functions, the authors write down a semiphenomenological Hamiltonian describing long-wave gapless excitations,

$$H = \int \left[\frac{(\rho^{(x)} - \bar{\rho})^2}{2K} + \frac{K v_s{}^2}{2} \left(\frac{\partial \varphi}{\partial x}\right)^2 \right] dx \tag{5}$$

where $\frac{\partial \bar{\rho}}{\partial \mu}$ is the compressibility, v_s the sound velocity. The density and phase operators satisfy

$$[\rho(x), \varphi(x')] = \delta(x-x') \tag{6}$$

By analogy with their previous work[3] on a one-dimensional superconductor with many bands, they write down the pair-pair correlation function as

$$G(R,\tau) = \bar{\rho} \left\langle T_\tau e^{2i[\varphi(R,\tau) - \varphi(0,0)]} \right\rangle \tag{7}$$

and evaluate the right hand side using Feynman's functional integral technique. They find that for Rv_s^{-1} and τ small compared with the inverse temperature, the asymptotic behavior has the form

$$\left| \frac{R}{v_s} + i\tau \right|^{-\alpha} \qquad \text{where } \alpha = 2(\pi K v_s)^{-1} \tag{8}$$

Similarly, exponents are evaluated for the density correlation function for k near zero and k near twice the fermi momentum (where low energy excitations again dominate).

For the particular value of the coupling constant mentioned above, it is easy to evaluate K, v_s and hence determine asymptotic behavior using the long-wave Hamiltonian. The results coincide with the exact solution, so the authors are able to conclude that in this case at least, and presumably in a more general situation, the long-wavelength excitations are indeed determining the asymptotic behavior. It is, of course, easy to see what these long-wavelength excitations correspond to for the particular case where the system is equivalent to a noninteracting spinless fermion gas. All excitations of such a system can be expressed as sums of particle-hole pairs. If the relevant energy scales are much smaller than the bandwidth, the fermion energy-momentum relationship can be taken to be linear. (It is only in this case, of course, that the system is equivalent to a boson system having a well defined energy-momentum relationship.) Thus, in this case, the long-wave Hamiltonian written down by Efetov and Larkin simply corresponds to the Tomonaga bosons of the noninteracting fermi gas. Moreover, the various correlation functions for the fermi gas are most easily found by using the boson representation for the fermion operators, and formulated in this way the equivalence of the two approaches is evident.

Generalizing from the value $V_{ij} = V_{ij}^0$ in (4), but assuming V_{ij} operates only between nearest neighbors, the XY Hamiltonian becomes a Heisenberg-Ising (HI) Hamiltonian. Some correlation properties of this model have been found exactly,[4] using techniques developed by Baxter[5] and others. These could be used to check Efetov and Larkin's hypothesis over a wider class of models. However, it is simpler and more useful for our purposes to examine the methods developed by Luther and Peschel[6] for the HI model. They demonstrate that to first order in J_z (that is, $V_{ij} - V_{ij}^0$) it is equivalent to a Luttinger model,[7] at least for the leading asymptotic behavior of correlation functions. The exponents are determined by using the boson Hamiltonian for the Luttinger model, and the boson representation of fermion operators.[8] Thus, for the relevant energy range, the Efetov-Larkin Hamiltonian (3) has again been transformed into a long-wave boson Hamiltonian, corresponding to (5), and the exponents arrived at in this way agree with theirs.

The remainder of this section is devoted to a more detailed derivation of (5) from (3). We first transform (3) into a Luttinger Hamiltonian, following Luther and Peschel,[6] to first order in the coefficient J_z of $\rho_i\rho_j$. (This is, in fact, the limit of accuracy of the Luther-Peschel replacement of the HI Hamiltonian by a Luttinger model. The equivalence breaks down completely as J_z approaches unity where the HI model is singular.) We drop the constant magnetic field term, because when the Hamiltonian is cast in the Luttinger form it is evident that this term merely alters the fermi level and cannot affect the exponents.

Putting $2mV_{ij} - \frac{1}{4} = J_z$, and dropping the overall factor $-(2m)^{-1}$ gives

$$H = -\sum_i \left(S_i^x S_{i+1}^x + S_i^y S_{i+1}^y + J_z S_i^z S_{i+1}^z \right) \tag{9}$$

For $J_z = 0$, as stated above, H is equivalent to a noninteracting gas of spinless fermions, which for low energy phenomena is equivalent to a noninteracting Luttinger model.

For $J_z \neq 0$, Luther and Peschel show that H corresponds to an interacting Luttinger model (in the standard notation)

$$H = \frac{2\pi v_F}{L} \sum [\rho_1(k)\rho_1(-k) + \rho_2(-k)\rho_2(k)] - \frac{4J_z}{L} \sum_k \rho_1(k)\rho_2(-k) \tag{10}$$

They use the boson representation of fermion operators to evaluate the leading asymptotic terms in spin correlation functions. For example,

$$\left\langle S^{+}(x)S^{-}(0)\right\rangle \propto \left\langle \exp\left(-\frac{\phi_1(x)}{2} - \frac{\phi_2(x)}{2}\right)\exp\left(\frac{\phi_1(0)}{2} + \frac{\phi_2(0)}{2}\right)\right\rangle$$

$$\text{where } \phi_{1,2}(x) = \pm\frac{2\pi}{L}\sum_k \rho_{1,2}(k)e^{-ikx}k^{-1} \tag{11}$$

with a cutoff factor corresponding to the bandwidth in the sum over k. Since the Hamiltonian is bilinear in boson operators and the ϕ's are linear, the expectation value is straightforward to evaluate, and

$$\langle S^{+}(x)S^{-}(0)\rangle \propto x^{-\theta}$$

$$\theta = \frac{1}{2}\left(\frac{\pi-2J_z}{\pi+2J_z}\right)^{\frac{1}{2}} \approx \frac{1}{2} - \frac{J_z}{\pi} \tag{12}$$

For $J_z = 0$, the result $\theta = \frac{1}{2}$ corresponds to - and agrees with - the pair-pair correlation exponent of Efetov and Larkin (and, of course, the exact results of McCoy et al[2]). For $J_z \neq 0$, (12) gives (to first order) the appropriate generalization of their result.

Our next step is to transform the Luttinger Hamiltonian (10) into the Efetov-Larkin long-wavelength Hamiltonian (5). This particular equivalence is actually mentioned by Efetov and Larkin in a later section of their paper. The transformation is accomplished by putting

$$\rho(x) = \rho_1(x) + \rho_2(x), \quad \nabla\varphi(x) = \pi(\rho_1(x) - \rho_2(x)) \tag{13}$$

Substituting into (5), we arrive at (10) with

$$\frac{1}{K} + Kv_s{}^2\pi^2 = 2V_F\pi$$

$$\frac{1}{K} - Kv_s{}^2\pi^2 = -4J_z \tag{14}$$

Thus we have established the equivalence (for small J_z) between (3) and (5). Note from (13) and (11) that

$$\varphi(x) = \frac{\phi_1(x) + \phi_2(x)}{2} \tag{15}$$

As a check, we find the exponent corresponding to (12) from (7) using the functional integral method. Our φ from (13) is actually the phase of the pair operator rather than of the single fermion operator, so (7) loses the factor of two in the exponent. Hence, the result differs from (8) by a factor of four, that is,

$$\left\langle e^{i\varphi(x)} e^{-i\varphi(0)} \right\rangle \propto x^{-\alpha}, \quad \alpha = (2\pi K v_s)^{-1} \tag{16}$$

We see from (14) that to first order in J_z,

$$\alpha = \frac{1}{2} - \frac{J_z}{\pi} \tag{17}$$

in agreement with (12).

The other critical exponents found by Efetov and Larkin, for the density with $k \gtrsim 0$ and $k \gtrsim 2k_F$, generalized to nonzero J_z can be easily shown by similar techniques to agree with the appropriate small J_z results of Luther and Peschel.

References

1. K. B. Efetov and A. I. Larkin, Correlation Functions in One Dimensional System with Strong Interaction, ZhETF 69, 764 (1975).

2. See e.g. B. M. McCoy, Phys. Rev. 173, 531 (1968); B. M. McCoy, E. Barouch and D. B. Abraham, Phys. Rev. A4, 2331 (1971).

3. K. B. Efetov and A. I. Larkin, ZhETF 66, 2290 (1974), Sov. Phys. JETP 39, 1129 (1974).

4. J. D. Johnson, S. Krinsky and B. M. McCoy, Phys. Rev. A8, 2526 (1973).

5. R. J. Baxter, Ann. Phys. (N. Y.) 70, 193 (1972).

6. A. Luther and I. Peschel, Phys. Rev. B12, 3908 (1975).

7. J. M. Luttinger, J. Math. Phys. 4, 1154 (1963); D. C. Mattis and E. H. Lieb, J. Math. Phys. 6, 304 (1965).

8. A. Luther and I. Peschel, Phys. Rev. B9, 2911 (1974); D. Mattis, J. Math. Phys. 15, 609 (1974).

Conference on Organic Conductors and Semiconductors, Siófok, Hungary 1976

CLUSTER EXPANSION METHODS APPLIED TO A I-d FERMION SYSTEM

H. SCHULZ

Institut für Theoretische Physik Technische Universität Hannover, 3000 Hannover, FRG

The thermodynamics of a 1-d Fermi system can be perfectly mapped onto the thermodynamics of a two-component classical real gas on the surface of a cylinder. The relationship between these two infrared problems (cf. Zittartz's contribution) is exploited as follows. We treat the classical plasma by a modified Mayer cluster expansion method (the lowest order term corresponding to the Debye Hückel theory), and obtain an exponentially activated behavior of the specific heat (cf. Luther's contribution) of the original quantum gas by simply reinterpreting the meaning of thermodynamic variables.

For linearized bare spectrum $\varepsilon_k = v_f(|k| - k_f)$ the Hamiltonian H of the 1-d Fermi system can be represented by Bose-operators [1],

$$H = H_\rho + H_\sigma \quad , [H_\rho , H_\sigma] = 0 \ . \qquad (1)$$

We concentrate on the subsystem H_σ , which involves spin-density operators only and exclusively contains the interaction potential $2\pi v_f \gamma(q)$ for backward scattering $q \sim \pm 2k_f$. γ is denoted by $\gamma_\parallel$ for spin-non-flip processes and by

$\gamma_\perp$ for spin-flip. The model is that of Luther and Emery [1], before they introduce new quasi-fermion operators to obtain a gap at $\gamma_\parallel = -3/5$. The free energy F_σ corresponding to H_σ can be calculated directly [2]. Canonical transformation, expansion in powers of $\gamma_\perp$ and evaluation of the quantum mechanical averages lead to the exact formal solution

$$F_\sigma = E_0 - \frac{L}{v}\frac{\pi}{6}T^2 - T\ln Z_\perp \quad , \tag{2}$$

$$Z_\perp = \sum_{N=0}^{\infty}\left(\frac{1}{N!}\right)^2\left(\frac{w}{a^4}\right)^N \int_0^L d^{2N}x \int_0^{\beta v} d^{2N}y \; e^{-A\sum_{\nu>\mu=1}^{2N}(-1)^{\nu+\mu} g(\vec{r}_\nu - \vec{r}_\mu)} \quad , \tag{3}$$

where L is the length of the system, $v = v_f \cdot \sqrt{1-\gamma_\parallel^2}$, $w = (\gamma_\perp v_f / 2\pi v)^2$, $A = 4\sqrt{(1+\gamma_\parallel)/(1-\gamma_\parallel)}$ and a is the finite momentum cutoff, by which the linearized model is to be supplemented [1],[3],[6].

Obviously the sum (3) is the grand partition function of a classical plasma (charges $\pm e$) in an area $L\beta v$. One only has to interpret A as e^2/Θ (Θ = temperature of the plasma) and w as $(a/\lambda)^4 e^{\bar{\mu}/\Theta}$ (λ = thermal De Broglie wavelength, $\bar{\mu}$ = chemical potential for pairs). The function $g(\vec{r})$, to be interpreted as the interaction potential, is not simply a logarithm as in Chui and Lee's zero-temperature treatment [3]. In place of that it solves (up to soft core modification) the Poisson equation on a cylinder with circumference βv ,

$$\Delta g(\vec{r}) = -2\pi \cdot \left(\frac{1}{\pi}\frac{a}{x^2+a^2}\right)\sum_n \delta(y + n\beta v) \tag{4}$$

and thus explicitly depends on the Fermion temperature.

Cluster theoretical treatment of the plasma leads to results in the canonical ensemble, especially to the (dimensionless) excess free energy

$$s = -\frac{a^2}{L\beta v\,\Theta}\left(\mathcal{F} - \mathcal{F}_0\right) \tag{5}$$

Once this function has been calculated, one obtains $Z_\perp$ by Legendre transformation $-\Theta \ln Z_\perp = \mathcal{F} - \bar{\mu} N$; and the Fermion free energy (2) as a function of spin-flip coupling is given by the parameter representation

$$\left.\begin{aligned} F_\sigma &= E_0 - \frac{Lv}{a^2}\left\{\frac{\pi}{6}\left(\frac{a}{\beta v}\right)^2 + \frac{\sigma}{2\pi A} + (1-\sigma\partial_\sigma) s\right\} \\ w &= \left(\frac{\sigma}{4\pi A}\right)^2 e^{-4\pi A \partial_\sigma s} \end{aligned}\right\} \quad (6)$$

where the (dimensionless) plasma density $\sigma = (2N/L\beta v)a^2 2\pi A$ acts as the parameter, $0 < \sigma < \infty$. Details of the application of cluster theory [4] to the above cylinder plasma will be published elsewhere [5]. Here we only list the main results. The excess free energy is obtained as a series

$$s = s_c + s_2 + s_3 + \dots \quad . \qquad (7)$$

the first term of which (s_c) corresponds to the Debye-Hückel theory. We obtain

$$s_c = -\frac{a^2}{2\pi}\int_0^\infty dq \left[\omega + \frac{2}{\beta v}\ln(1-e^{-\beta v \omega}) - q - \frac{2}{\beta v}\ln(1-e^{-\beta v q})\right] \quad (8)$$

with

$$\omega = \sqrt{q^2 + \sigma e^{-|aq|} a^{-2}} \quad , \qquad (9)$$

which is obviously the free energy of a non-interacting Bose system with gap minus that one for linear spectrum. The fourth term in (8) is readily evaluated to give $-(\pi/6)(a/\beta v)^2$. It compensates all the regular temperature dependence in F_σ , (6). Consider for example the case of very low temperature ($a/\beta v \to 0$) and small coupling $\gamma_\perp$. Then (8), (9), (6) lead to

$$F_\sigma^{DH} = E_0 - \frac{L}{v}\frac{4-A}{8\pi A}\Delta^2 - \frac{L}{v}\sqrt{\frac{2}{\pi}}\, T^{3/2}\Delta^{1/2} e^{-\Delta/T} \qquad (10)$$

$$\Delta = \frac{v}{a}\cdot(b\gamma_\perp)^{\frac{2}{4-A}} , \quad b = \left(\tfrac{1}{2}e^{\gamma}\right)^{A/2}\left(4+\tfrac{1}{4}A^2\right) . \qquad (11)$$

The Δ^2 -contribution to the ground state energy agrees with the leading term of Zittartz's exact calculation [6] . The exponent in the formula (11) for the gap energy agrees with Luther's recent investigation [7] . It can be shown that the Debye Hückel description becomes exact as one approaches the plane $A=0$ ($\gamma_\parallel = -1$) in the T-$\gamma_\perp$-A -diagram and the "plane" $\gamma_\perp \to \infty$ (in this limit there is a gap for arbitrary $A > 0$). Once a gap has developed in Debye Hückel approximation, it persists also when the higher order corrections (7) are included - at least in the range $0 < A < 2$ $(-1 < \gamma_\parallel < -3/5)$. Within pure thermodynamics we cannot detect the nature of the states, see however Everts' contribution. The gap Δ probably corresponds to the lowest bound soliton [7] , see Luther's contribution.

[1] A. Luther and V.J. Emery, Phys. Rev. Lett. 33, 589 (1974).

[2] H.U. Everts and H. Schulz, Z. Physik B 22, 285 (1975).

[3] S.-T. Chui and P.A. Lee, Phys. Rev. Lett. 35, 315 (1975).

[4] H.L. Friedman, Ionic Solution Theory, New York Interscience Publishers 1962.

[5] H. Schulz, Z. Physik (to be published)

[6] J. Zittartz, Z. Physik B 23, 277 (1976).

[7] A. Luther, Phys. Rev. B (to be published).

Conference on Organic Conductors and Semiconductors, Siófok, Hungary 1976

MAGNETIC SUSCEPTIBILITY OF A ONE-DIMENSIONAL FERMION-SYSTEM

H.U. EVERTS and W. KOCH

Institut für Theoretische Physik Technische Universität Hannover,
3000 Hannover, FRG

Abstract: The equivalence of the backward scattering model of a 1-D fermion system to a 2-D classical Coulomb gas is employed to calculate the dynamical magnetic susceptibility of the 1-D fermion system. The result points towards the existence of bound magnon states in a 1-D fermion system.

Several authors [1], [2], [3] have shown that the spin part of the so-called backward scattering model of a 1-D fermion system [4] is thermodynamically equivalent to a 2-D Coulomb gas. In this note we shall describe how this equivalence can be employed to calculate the dynamical wavevector dependent magnetic susceptibility of a 1-D fermion system within the backward scattering model.

In boson respresentation the spin part of the Hamiltonian of this model is given by [4]

$$H_\sigma = 2\pi v_F \Big\{ \frac{1}{L} \sum_{0<q} \left[\sigma_1 \sigma_1^+ + \sigma_2^+ \sigma_2 - \gamma_{\parallel} (\sigma_1 \sigma_2^+ + h.c.) \right]$$

$$+ \gamma_\perp \int_0^L dx \left[(\psi_{1\uparrow}^+ \psi_{2\downarrow}^+ \psi_{1\downarrow} \psi_{2\uparrow})(x) + h.c. \right] \Big\} , \qquad (1)$$

where

$$\psi^{+}_{j\uparrow}(x)\,\psi_{j\downarrow}(x) = (2\pi a)^{-1}\, e^{i\phi_j(x)} \quad , \quad j = 1,2$$

$$\phi_j(x) = \frac{2\pi i}{L}\sum_{0<q} \frac{e^{-aq/2}}{q}\left(e^{iqx}\,\sigma^{+}_j(q) - h.c.\right). \qquad (2)$$

$\sigma_1(q)$, $\sigma_2(q)$ are spin density operators,

$$\sigma_j(q) = 2^{-1/2}\left(\rho_{j\uparrow}(q) - \rho_{j\downarrow}(q)\right),$$

which obey boson-like cummutation relations [4]. The subscript $j = 1,2$ refers to the two linear branches of the unperturbed ($\gamma_{\parallel} = \gamma_{\perp} = 0$) excitation spectrum of the model. a is a momentum cut-off.

For the particular value $\gamma_{\parallel} = -3/5$ of the spin non-flip coupling constant one can easily calculate the magnetic susceptibility by using the diagonalization procedure of Ref. [4] . For low frequencies and small wave vectors, $\omega < vq \ll \Delta$, the result is as follows:

$$\chi^{\gamma_{\parallel}=-3/5}(q,\omega) = \frac{2}{\pi v}\left\{\frac{(vq)^2}{(vq)^2-\omega^2}\sqrt{\frac{2\pi\Delta}{T}}\,e^{-\frac{\Delta}{T}} + \frac{4}{3}\,\frac{(vq)^2}{(2\Delta)^2+\omega^2}\right\}, \qquad (3)$$

where $v = v_F\,(1-\gamma_{\parallel}^2)^{1/2}$, $\Delta = \frac{v_F}{a}|\gamma_{\perp}|$.

This clearly reflects the gap in the spin excitation spectrum, which, according to Ref. [4], is a free fermion spectrum for this particular value of $\gamma_{\parallel}$. The first term describes the response of the thermally excited fermions to the magnetic field, the second is due to excitations across the gap. By calculating $\chi(q,\omega)$ for other values of $\gamma_{\parallel}$ one may thus hope to gain insight into the variation of the spectrum as a function of $\gamma_{\parallel}$. The equivalence of the backward scattering model to the 2-D Coulomb gas offers a possibility of performing such a calculation.

As in Refs. [1], [2], [3] the connection of $\chi(q,\omega)$ with quantities defined for the 2-D Coulomb gas is established by expanding the statistical operator and the time development operator in the formal expression

$$\chi(q,\omega_n) = 2\langle T\{(\sigma_1+\sigma_2)(q,\tau)\,(\sigma_1^+ + \sigma_2^+)(q,0)\}\rangle \tag{4}$$

with respect to the coupling constant $\gamma_\perp$:

$$\chi = \chi_0\{1 + \sum_{N=1}^{\infty} \gamma_\perp^{2N} \chi_{2N}\} .$$

The quantum mechanical averages which define the expansion coefficients χ_{2N} can be evaluated by well known methods [5] and χ can be cast into the form

$$\chi(q,\omega_m) = \chi_0(\vec{q})\{1 - \frac{4\pi A}{\vec{q}^2}\langle\langle \rho(\vec{q})\,\rho(-\vec{q})\rangle\rangle\} . \tag{6}$$

Here, $\langle\langle \rho(\vec{q})\,\rho(-\vec{q})\rangle\rangle$ is the static density correlation function of a 2-D neutral Coulomb gas constrained to the surface of a cylinder of circumference T^{-1} and of length L/v [3]. The symbol $\langle\langle \cdots \rangle\rangle$ denotes the grand canonical average. The fugacity z of the Coulomb gas and its interaction strength A are determined by the coupling constants of H_c :

$$z = \frac{v_F}{2\pi v}|\gamma_\perp| \quad , \quad A = 4\left[\frac{1+\gamma_\parallel}{1-\gamma_\parallel}\right]^{1/2} . \tag{7}$$

Furthermore, $\vec{q} = a\,(q, \frac{\omega_n}{v})$

Obviously, $1 - (4\pi A/\vec{q}^2)\langle\langle \rho(\vec{q})\,\rho(-\vec{q})\rangle\rangle$ is just the inverse of the dielectric function of the Coulomb gas. χ_0 is the susceptibility of the noninteracting ($\gamma_\perp = 0$) system:

$$\chi_0(\vec{q}) = \frac{A}{\pi v}\,\frac{(vq)^2}{(vq)^2 + \omega_n^2} . \tag{8}$$

Now, the task consists of calculating $\langle\langle \rho(\vec{q})\, \rho(-\vec{q}) \rangle\rangle$. To accomplish this we have used the modified Mayer cluster expansion [3], [6] to determine the two-particle correlation functions g_{++}, g_{+-} between equal and opposite charges. $\langle\langle \rho(\vec{q})\, \rho(-\vec{q}) \rangle\rangle$ is then obtained from

$$\langle\langle \rho(\vec{q})\, \rho(-\vec{q}) \rangle\rangle = 1 + \frac{1}{4\pi A} \{ g_{++}(\vec{q}) - g_{+-}(\vec{q}) \} \quad . \tag{9}$$

It turns out that the coupling strength A is the only physical parameter that appears explicitly in the cluster series and that for $A < 2$ all cluster integrals converge. It is therefore possible to arrange the cluster series as a power series in A. To lowest order in A one finds

$$g^{\text{D.H.}}_{\alpha\beta}(\vec{q}) = 1 + e_\alpha e_\beta \frac{2\pi A}{1 + \underset{\sim}{\vec{q}}^{\,2}} \quad , \quad \underset{\sim}{\vec{q}} = \vec{q}/\Delta_1 \quad , \quad e_\lambda = \pm 1 \,, \tag{10}$$

where $$\Delta_1 = \frac{v}{a} f(A)\, |\gamma_\perp|^{\frac{2}{4-A}} \quad , \quad f(A \to 0) = \text{const.} \tag{11}$$

is the gap that occurs in the thermodynamic functions of Ref. [3]. (9) is just the Debye-Hückel expression for the two-particle correlation function. Combining (6), (9), (10) and (11) one obtains finally

$$\chi^{\text{D.H.}}(q,\omega) = \frac{A}{\pi v} \frac{(vq)^2}{\Delta_1^2 + (vq)^2 - \omega^2} \quad . \tag{12}$$

This result, which should be valid for $A \ll 1$, i.e. $\gamma_\parallel \simeq -1$, has several puzzling features.

Firstly, in the static limit $\chi^{\text{D.H.}}$ vanishes as q goes to zero. This would imply that the static uniform susceptibility of the fermion system were zero at any temperature, which is clearly an unphysical result. The inclusion of higher order terms of the cluster series does not improve the result in this respect.

Secondly, one recognizes that $\chi^{D.H.}(q,\omega)$ is the response function of a system of noninteracting bosons with gap. Thus it differs qualitatively from the result (3) even at $T=0$.

The latter puzzle may find its solution in a recent investigation of the spectrum of H_σ by Luther [7]. There it is shown that for $A<2$ an increasing number of excitation branches appear below the fermion branch which is present for all values of $A<4$. The states which correspond to these branches are to be viewed as bound spin-wave states in the model under consideration. From the equivalence of H_σ and the Hamiltonian of the Sine-Gordon theory it can be inferred [7], [8] that the lowest bound magnon branch attains boson character as A approaches zero. In all probability $\chi^{D.H.}(q,\omega)$ represents the contribution of this lowest bound magnon branch to the magnetic susceptibility. Correspondingly, the gap Δ_1, appearing in $\chi^{D.H.}(q,\omega)$, is to be interpreted as the minimum excitation energy of this lowest branch.

It remains to be seen why the approach which we have pursued in the work outlined here does not yield the contributions of the other branches, in particular that of the highest fermion-type branch, to the susceptibility.

References:

1. S.T. Chui and P.A. Lee, Phys. Rev. Lett. 35, 315 (1975).
2. J. Zittartz, Z. Physik B 23, 55 (1976).
3. H. Schulz, preceding paper in this volume and to be published in Z. Physik.

4. A. Luther and V.J. Emery, Phys. Rev. Lett. 33, 589 (1984).

5. K.D. Schotte, Z. Physik 230, 99 (1970).

6. H.L. Friedman, "Ionic Solution Theory", Interscience (New York 1972).

7. A. Luther, preprint, Nordita 76/18.

8. R. Dashen, B. Hasslacher and A. Neveu, Phys. Rev. D 11, 3424 (1975).

Conference on Organic Conductors and Semiconductors, Siófok, Hungary 1976

EXCITATIONS IN THE ANTIFERROMAGNETIC HEISENBERG CHAIN

A. SÜTÖ and P. FAZEKAS

Central Research Institute for Physics, Budapest, Hungary

Excitations of the antiferromagnetic Heisenberg chain of N atoms with spin 1/2 are examined. It is found that the energy of the lowest lying excitations for a total spin S and wave number q converges to $\pi/2|\sin q|$ as $N\to\infty$, if $S \lesssim \ln N$.

The chain of 1/2 spins coupled by nearest neighbour antiferromagnetic Heisenberg interaction is one of the most widely investigated quantum-mechanical model systems. Since the basic work of Bethe[1] in which he gave the proper classification of the eigenstates and solved the problem for finite closed chains many attempts have been made to obtain results for the infinite system. Hulthén[2] found a suitable transformation of Bethe's non-linear equations to a linear integral equation and calculated the ground state energy in the $N\to\infty$ limit. His method proved to be very useful to obtain the energy of some other eigenstates. It was applied by des Cloizeaux and Pearson[3] /dCP/ to obtain the lowermost excited states for any wave number. These states are triplets and form a branch which is described by the function,

$$^{1}\epsilon_{\infty}(q) = \frac{\pi}{2}\,|\sin q| \qquad (-\pi \leq q < \pi)$$

of q, the wave number of the state. On this scale the ground state has q=0.

Ovchinnikov[4] used Hulthén's method to the more complicated case of the singlet excitations and he found the lowest lying ones to generate a curve

$$^{0}\epsilon_{\infty}(q) = \pi\,|\sin q/2| \qquad (-\frac{\pi}{2} \leq q < \frac{\pi}{2})$$

It is surprising that this curve terminates in the middle of the Brillouin zone.

All these works follow the original treatment of Bethe and the calculation of the energy of the chosen eigenstate is exact in the $N \rightarrow \infty$ limit. The question remains however whether the states were properly chosen: is the state, treated by Hulthén, really the ground state; are the triplet and singlet excitations the lowermost ones for a given wave number. For the ground state, Yang and Yang[5] proved that Hulthén' s choice was correct. No similar mathematical proof is available for the triplet excitations, however we have sufficient numerical evidence[3] that the given branch is really the lowest lying one. No comparable numerical check had been published for the singlet excitations. Later we learned, however, that an inconsistency between numerical data and the predicted analytic form[4] had been noticed by Majumdar and coworkers /unpublished, private communication/ and Bonner and Sutherland reach similar conclusions in a recent preprint.

Using numerically calculated excitation energies for small chains up to ten atoms we tried to find a numerical support to Ovchinnikov's result /Fazekas and Sütő[6]/. We plotted the energies of the singlet states versus the wave number, q for N=6, 8 and 10 and drew the lower edge of the spectrum in each case. These curves look very like to a half period of a sine and a linear extrapolation, to 1/N=0, of the energies at q=0, π and the maximum energy of the curves suggests that the limit is probably ${}^{1}\epsilon_{\infty}(q)$, the dCP limit, instead of the function proposed by Ovchinnikov. Thus the problem of the singlet excitations needs further consideration.

After this investigation the question naturally arose whether the lowest-lying states of higher multiplicity form distinct branches or they also accumulate to the dCP limit. This question has been found to have a definite answer:

Let us consider the antiferromagnetic Heisenberg hamiltonian

$$\mathcal{H} = \sum_{i=1}^{N-1} \left(\vec{S}_i \vec{S}_{i+1} - \frac{1}{4} \right) + \left(\vec{S}_N \vec{S}_1 - \frac{1}{4} \right)$$

Let $\hat{E}$ be the ground state energy of the system and E the energy of the lowest lying state with a given total spin-quantum--number, S>1, and wave number, q. Then

$$^{S}\epsilon_N(q) \equiv E - \hat{E} = \frac{\pi}{2}|\sin q| + O\left(\frac{S}{\ln N/S}\right)$$

$$\left(\frac{2\pi S}{N} \leq |q| \leq \pi\right)$$

As a consequence, one would obtain that

$$^{S}\epsilon_N(q) \rightarrow \frac{\pi}{2}|\sin q|$$

if $S/\ln(N/S) \rightarrow 0$ as $N \rightarrow \infty$ and, on the other hand, $^{S}\epsilon_N(q) \rightarrow \infty$ if, for example, $S=N^{\alpha}$, for any $0<\alpha<1$. For somewhat more details the reader is referred to another publication /Sütő[7]/. In the present context the second term on the right hand side should not be regarded as the error of the calculation but an estimate of how fast the energy levels of large but finite systems converge to the result valid in the $N \rightarrow \infty$ limit. There are reasons to believe, however, that we have obtained an overestimate and the convergence is quicker than logarithmic. First of all, it has been known for the triplet excitations[3] and we found numerically for the singlet excitations[6] that the gap to the lowermost of these excitations goes to zero as 1/N. Griffiths[8] obtained the lowest energy eigenstate for a general $S \geq 0$, which belongs to $q=0$ for S even and $q=\pi$ for S odd, and came up with the result for the excitation energy $\sim \frac{S^2}{N}\left(1-\frac{1}{2\ln N/S}\right)$. Inside the Brillouin zone the convergence may be slower than this but probably not so much as our result would indicate. Note that according to this, states with $S<\sqrt{N}$ will accumulate to the dCP limit as $N \rightarrow \infty$. A detailed investigation still remains to be done. Its result should provide further insight into the question of the lack of an antiferromagnetic symmetry-breaking in the linear chain. An investigation, along similar lines, of the appearance of symmetry-breaking in S=1/2 isotropic systems will be published elsewhere[9].

Acknowledgements. The authors are indebted to Prof. C. K. Majumdar for correspondence and to Prof. Bill Sutherland for sending his preprint prior to publication. A.S. thanks Prof. A.A. Ovchinnikov for several discussions.

References

1. Bethe, H., Z. Physik 71, 205 /1931/
2. Hulthén, L., Ark.Mat.Astr.Fys. 26A, 1 /1938/
3. Des Cloizeaux, J. and Pearson, J.J., Phys. Rev. 128, 2131 /1962/
4. Ovchinnikov, A.A., Zh. Eksp. Teor. Fiz. 56, 1354 /1969/ Soviet Physics JETP 29, 727 /1969/
5. Yang, C.N. and Yang, C.P., Phys. Rev. 150, 321 /1966/
6. Fazekas, P. and Sütő, A., Solid State Commun. 19, 1045 /1976/
7. Sütő, A., Solid State Commun. 20, to be published
8. Griffiths, R.B., Phys. Rev. 133, A768 /1964/
9. Sütő, A. and Fazekas, P., Philosophical Magazine, to be published

DISPLACEMENT CORRELATION OF CLASSICAL IONS IN THE INCOMMENSURATE FRÖHLICH MODEL OF A ONE-DIMENSIONAL METAL

E. EISENRIEGLER

Institut für Festkörperforschung der Kernforschungsanlage Jülich, Jülich, FRG

The one-dimensional Fröhlich model is investigated for $k_B T \ll \delta_0 \ll E_F$ with δ_0 the mean field Peierls gap at T = 0. Treating the ions classically and neglecting band structure and Umklapp effects the electronic energy for a given ion distortion is determined by Bogolyubov-type equations with a complex order parameter.

For $|p| \gg \delta_0 / v_F$ (p is the wavevector deviation from $2k_F$) the (electron-renormalized) phonons are shown to be harmonic with a universal dispersion curve, i.e. independent of T and the possible presence of long wavelength ($|p| < \delta_0 / v_F$) phonon excitations. For an infinitely long-ranged Peierls distortion the harmonic phonon excitations are calculated for arbitrary wavevector and temperature. If $|p| < \delta_0 / v_F$, one obtains a mixing and energy splitting of the two phonons $2k_F + p$ and $-2k_F + p$ leading to the so-called harmonic phase and amplitude modes. This phonon coupling decreases rapidly for $|p| \gtrsim \delta_0 / v_F$ and both dispersions merge into the universal dispersion.

The long wavelength ($|p| < \delta_0 / v_F$) phonon excitations in general are described approximately by a Ginzburg-Landau (GL) energy functional where the important fluctuations occur in the phase of the order parameter. For wavelengths small compared to the GL phase-coherence length ($|p| \gg k_B T / v_F$), the harmonic phase (and amplitude) mode become well defined modes and determine the correlation function. Actually, the GL and harmonic results for the correlation function coincide for $k_B T / v_F \ll |p| \ll \delta_0 / v_F$ in leading order in the small quantity $k_B T / \delta_0$.

The theory is compared with X-ray and neutron scattering experiments on KCP.

1. MODEL HAMILTONIAN AND ITS RELATION TO KCP

Consider the 1D Fröhlich model (1)

$$H = \frac{f_E}{2} \sum_n s_n^2 + \sum_i \left\{ \frac{p_i^2}{2m} + \sum_n v(x_i - na - s_n) \right\} \quad (1)$$

to represent a single chain in a quasi one-dimensional conductor like KCP (2). Here s_n is the displacement of the n'th ion from its equilibrium position na, f_E is an Einstein restoring force, p_i and x_i are momentum and position operators of the i'th electron with mass m, and v is an effective electron-ion interaction potential. The ion kinetic energy is suppressed in (1). If band-structure and Umklapp effects (3) are neglected ($\tilde{v}_k = 0$ for $|k| > \frac{\pi}{a}$) one obtains

$$H = \frac{1}{L} \sum_p \frac{f_E a}{w_p^2} |\hat{\Delta}_p|^2 + \sum_i \left\{ \frac{p_i^2}{2m} + (e^{ik_F x_i} \Delta(x_i) + hc) \right\} \quad (2)$$

where

$$w_p = (2k_F + p)\tilde{v}_{2k_F+p} \; , \; \hat{\Delta}_p = iw_p \hat{s}_{2k_F+p} \; , \; 0 < 2k_F + p < \frac{\pi}{a} \quad (3)$$

are the electron phonon interaction constant and a modulation factor of the $2k_F$-potential respectively and $\sim$, $\wedge$ denote Fourier transforms of continuous and discrete functions in 1D space with periodicity length L. We assume $2k_F$ to be in the first Brillouin zone (3) such that $w_o \neq 0$.

In (2) the originally discrete ions act effectively as a continuum and there is no driving mechanism for a possible Peierls distortion Δ towards phase locking or commensurability with respect to the lattice ("Incommensurate Fröhlich Model") contrary to Ref. (5). The relevance of the model to KCP can be estimated from the approximations it implies:

a/ Neglect of band structure effects: Optical experiments (7) indicate nearly free electron behaviour along the chain; X-ray experiments (8) show no commensurability of Peierls-wavelength with lattice constant.

b/ Neglect of interchain interaction: confine to $T > T_{3D}$.

c/ Assumption that (Peierls gap δ_o at T = 0) $\ll$ E_F.

d/ Assumption that $T \ll \delta_o$.

Conditions b/, c/, d/ - in turn - imply $T_{3D} \ll T \ll \delta_o \ll E_F$. This can be reasonably fulfilled in KCP: $120 \ll T/K \ll 10^3 \ll 3.10^4$.

e/ Classical treatment of ions.

f/ Neglect of electron-electron interaction.

g/ Neglect of impurities: probably the most serious approximation.

The present treatment concentrates on the electron phonon interaction as the most fundamental aspect of a Peierls lattice instability. For this purpose the Fröhlich Hamiltonian is the simplest relevant model. We calculate the ion displacement correlation for the Incommensurate Model (Eq. (2), i.e. complex Δ) including thermodynamic fluctuations of the ion positions.

2. EQUAL TIME CORRELATION FUNCTION

For our classical ions, one obtains

$$<|\delta_{2k_F+p_o}|^2> \sim <|\chi_{p_o}|^2> = Z^{-1} \prod_p \int d^2\chi_p \, |\chi_{p_o}|^2 \exp\left\{ - \mathcal{H}_{eff}[\Delta]/T \right\} \tag{4}$$

with the effective ion Hamiltonian

$$\mathcal{H}_{eff}[\Delta] = \frac{1}{L} \sum_p \frac{f_E a}{w_p^2} |\chi_p|^2 + \int dE \; N(E,\Delta)(-T) \; \ell n \left\{ 1+e^{-(E-E_F)/T} \right\} \tag{5}$$

consisting of bare phonons and their renormalization due to the electrons. Here N is the electronic density of states in the presence of a potential Δ .

We now have 2 subproblems:

(i) Electronic problem: the need to calculate N for thermally probable fluctuations Δ

(ii) Phononic problem: the need to perform functional integration $\prod \int d^2\Delta$ over fluctuations.

3. H_{eff} FOR PROBABLE Δ's (PROBLEM (i))

Relevant potentials for electrons ($|p| \ll k_F$, $|\Delta| \ll E_F$) imply nearly free electron wavefunctions for energies $|E_\alpha - E_F| \ll E_F$

$$\psi_\alpha(x) = e^{ik_F x} u_\alpha^+(x) + e^{-ik_F x} u_\alpha^-(x) \tag{6}$$

with the u's varying slowly on the length k_F^{-1}. The Schrödinger equation for ψ turns to Bogolyubov-type equations for the u's:

$$\begin{pmatrix} \partial_x/i\ , & \Delta \\ \Delta^*, & -\partial_x/i \end{pmatrix}\begin{pmatrix} u_\alpha^+ \\ u_\alpha^- \end{pmatrix} = (E_\alpha - E_F)\begin{pmatrix} u_\alpha^+ \\ u_\alpha^- \end{pmatrix},\quad v_F = 1\ . \tag{7}$$

a/ Distortion Δ of lowest H_{eff} is Peierls distortion (1, 9)

$$\Delta = \delta_T\, e^{i\phi} = \text{const},\ \frac{f_E a}{w_o^2} = \frac{1}{2\pi}\,\ell n\,\frac{8E_F}{\delta_o}\ ,\ T_p = \frac{\delta_o}{1.76} \tag{8}$$

Note the degeneracy in ϕ, the phase of Δ .

b/ For Δ varying slowly on the lenght δ_o^{-1} one can (in formal similarity to the case of a superconductor) perform a Ginzburg-Landau (GL) expansion (4) to the Bogolyubov equations. For T $\ll \delta_o$ this yields

$$\mathcal{H}_{eff}[\Delta] = \int\frac{dx}{2\pi}\left\{(|\Delta|-\delta_o)^2 + \frac{(|\Delta|')^2}{12\delta_o^2} + \frac{(\phi')^2}{4}\right\} \tag{9}$$

c/ To include Fourier amplitudes χ_p with $p > \delta_o$ (see Fig. 1) we separate $\Delta = \Delta_o + d\Delta$ in a slowly and rapidly varying part, containing p's smaller and larger than $p_c \sim$ some δ_o's respectively. One can show (4) that

$$\mathcal{H}_{eff}[\Delta] - \mathcal{H}_{eff}[\Delta_o] = \frac{1}{L}\sum_{|p|>p_c} f_p|\chi_p|^2 \tag{10}$$

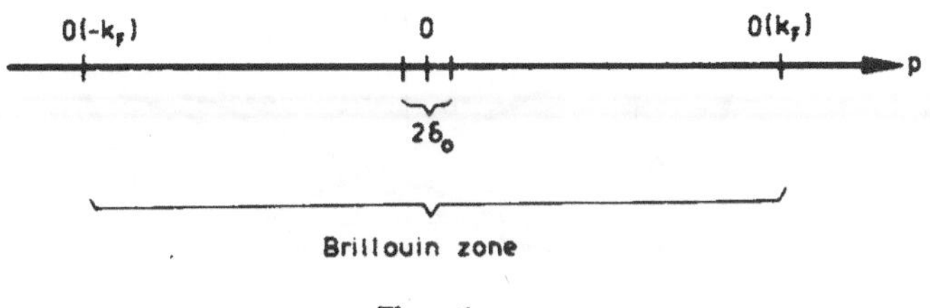

Fig. 1

with f_p independent of Δ_o. This considerably simplifies the fctl. integral over fluctuations (problem II) as we shall see later.

The physics behind statement (10) is explained most easily by considering the special case Δ_o = constant. The electronic states $|\alpha\rangle$ belonging to Δ_o are then Bloch electrons and we have to consider the change in free energy to second order

$$\sum_{\alpha\alpha'} \langle\alpha|\underline{dV}|\alpha'\rangle \; (\langle\alpha|\underline{dV}|\alpha'\rangle)^* \; \frac{f_\alpha - f_{\alpha'}}{E_\alpha - E_{\alpha'}} \tag{11}$$

in the perturbation

$$dV = \frac{1}{L} \sum_p e^{i(2k_F+p)x} \; i w_p \; s_{2k_F+p} + hc\,. \tag{12}$$

This is illustrated in Fig. 2 showing the plane wave contributions to the Bloch states $|\alpha\rangle$ and $|\alpha'\rangle$. For given α and α' there are obviously three different types of processes leading to couplings

$$s_{2k_F+p} \; s^*_{2k_F+p'} \, , \quad s_{-2k_F+p} \; s^*_{2k_F+p'} \, , \quad s_p \; s^*_{2k_F+p} \tag{13}$$

the latter two of which are important only if α or α' have an energy $|E_{\alpha,\alpha'} - E_F| < O(\delta_o)$ (for the empty circle and square in Fig. 2 to exist), and therefore occur with appreciable weight in (11) only for $|p| < O(\delta_o)$. The last process is negligible even then since the electron phonon coupling $p\tilde{v}_p$ is small.

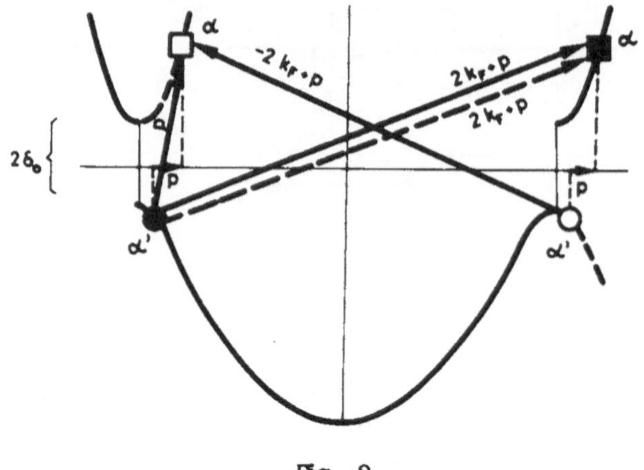

Fig. 2

For $|p| \gg \delta_o$ only the first process (leading to (10)) remains and the overwhelming pairs α, α' contributing to it (for given p) have energies $|E_{\alpha,\alpha'} - E_F| \gg O(\delta_o)$, i.e. are plane waves such that the prefactor f_p in (10) is independent of δ_o.

Therefore the harmonic expansion of $\mathcal{H}_{eff}[\Delta]$ around Δ_o = const

$$\mathcal{H}_{eff}[\Delta] - \mathcal{H}_{eff}[\Delta_o] = \frac{1}{\lambda} \sum_{p>0} \{ |D_p^{\parallel}|^2 f_p^{\parallel} + |D_p^{\perp}|^2 f_p^{\perp} \} \tag{14}$$

for arbitrary p defines modes (4)

$$\begin{pmatrix} D_p^{\parallel} \\ D_p^{\perp} \end{pmatrix} = \frac{1}{\sqrt{2}|\Delta_o|} \begin{pmatrix} \Delta_o^*, & \Delta_o \\ -i\Delta_o^*, & i\Delta_o \end{pmatrix} \begin{pmatrix} \chi_p \\ \chi_{-p}^* \end{pmatrix} \tag{15}$$

which are mixtures of plane wave modes with dispersion (4)

$$\begin{aligned} \pi f_p^{\parallel} &= \ln \frac{|\Delta_o|}{\delta_o} + \frac{\sqrt{1+x^2}}{x} \ln(\sqrt{1+x^2}+x) \\ \pi f_p^{\perp} &= \ln \frac{|\Delta_o|}{\delta_o} + \frac{x}{\sqrt{1+x^2}} \ln(\sqrt{1+x^2}+x) \end{aligned} \quad , \; x = \frac{p}{2|\Delta_o|} \tag{16}$$

for $T \ll \delta_o$. These dispersions are shown in Fig. 3 for $|\Delta_o| = \delta_o$.

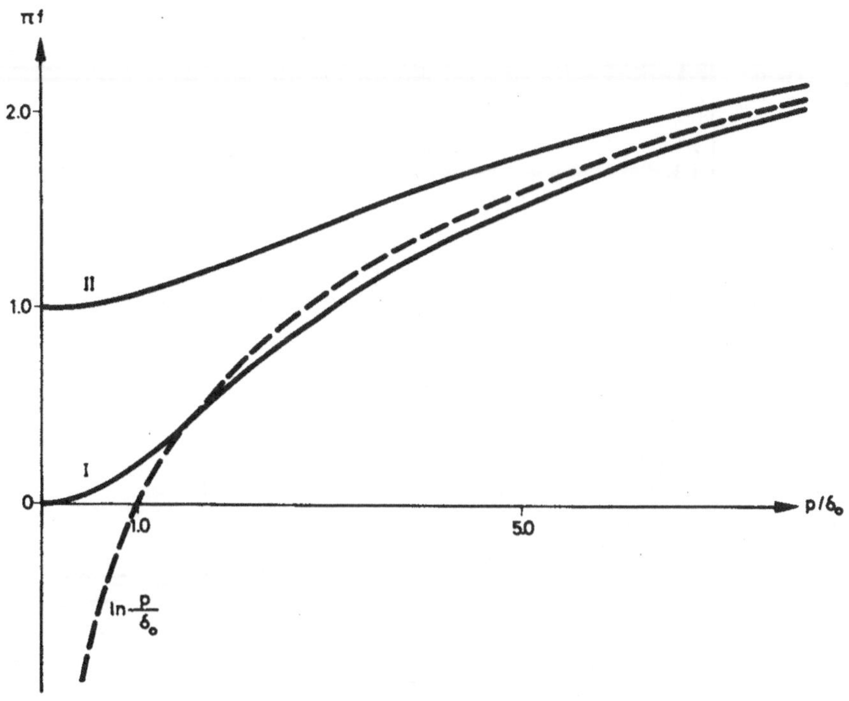

Fig. 3

For $|p| \ll \delta_o$ and $|\Delta_o| = \delta_o$ the modes coincide with the amplitude and phase mode discussed in (10) and the result (16) in this limit (up to $\mathcal{O}(p^2)$.) is in agreement with the GL expression (9) (11). For $|p| \gg \delta_o$ the modes degenerate, i.e. the plane wave modes are good modes and the dispersion f is independent of $|\Delta_o|$ in agreement with the claim in Eq. (10).

4. AVERAGE OVER DISPLACEMENT FLUCTUATIONS (PROBLEM (ii))

According to Eq. (10) the plane wave modes $\tilde{\Delta}_p$ for $|p| > p_c$ are <u>independent</u> harmonic modes (decoupled from one another and from those with $|p| < p_c$). Therefore the functional integral (4) gives

$$\langle |s_{2k_F+p_o}|^2 \rangle \sim \frac{T}{f_{p_o}} \qquad \text{for } p_o > p_c \tag{17}$$

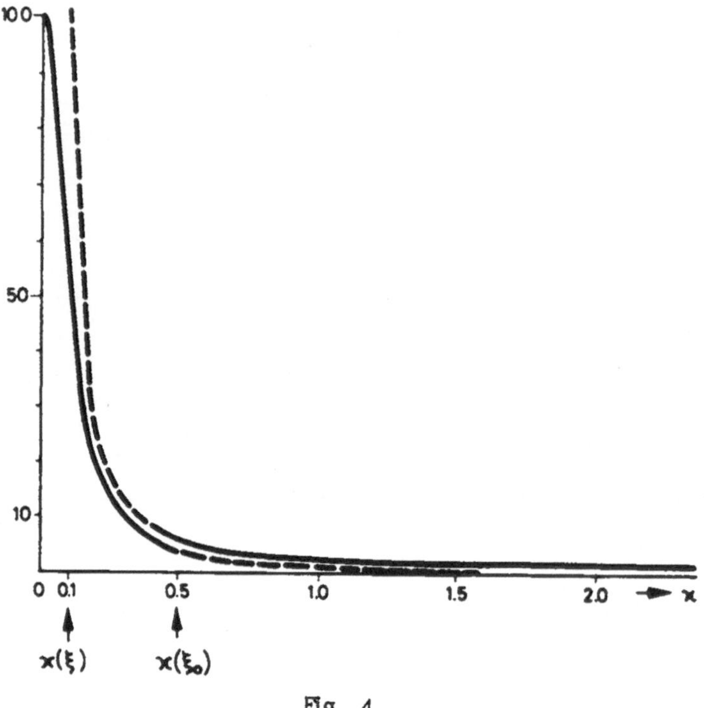

Fig. 4

For $|p_o|<p_c$ all modes $|p|>p_c$ cancel from the numerator and denominator in (10) and one is left with a functional integral over long wavelength fluctuations ($|p|<p_c$) for which one approximately can use the GL expression (9). For $T \ll \delta_o$ only phase fluctuations are important (GL → XY model) and yield (4)

$$<|s_{2k_F+p_o}|^2> \sim \int dx\ e^{ip_o x}\ \delta_o^2\ <e^{-i\phi(x)}e^{i\phi(o)}> =$$

$$= \delta_o^2\ \frac{2\pi T}{(\pi T)^2+p_o^2} \tag{18}$$

For $\delta_o \gg p_o \gg \xi^{-1} \equiv \pi T$ this agrees with the harmonic expression $T/2f^{\perp}_{p_o}$ (Eq. (16)) and therefore

$$<|s_{2k_F+p_o}|^2> \sim \frac{1}{2}\left(\frac{T}{f^{\parallel}_{p_o}} + \frac{T}{f^{\perp}_{p_o}}\right) \text{ for } \xi^{-1} \ll p_o \tag{19}$$

interpolates between (17) and (18) for $T \ll \delta_0$ (see Fig. 4 for $\tau T = \delta_0/5$). The correlation length ξ (300 K) can be compared with experiment (8) with the result: ξ (EXP) > 60, ξ (Ref. 12) = 7, ξ (Present) = 30 Pt - Pt spacings.

ACKNOWLEDGEMENT

It is a pleasure to thank Prof. G. Eilenberger for numerous illuminating discussions.

REFERENCES

(1) H. Fröhlich, Proc. R. Soc., A223, 296 (1954)
(2) K. Krogmann, Angew. Chem. Int. Ed. Engl. 8, 35 (1969)
(3) In KCP, $2k_F$ is outside the first Brillouin zone and the $2k_F$ phonon is an Umklapp phonon. The present treatment can be easily generalized (see Ref. 4) to include this effect.
(4) An extended version of the present paper is being published as a JÜL-report and is available from the author upon request.
(5) M. J. Rice, S. Strässler, Solid State Comm., 13, 125 and 1931 (1973)
(6) Proc. German Physical Society Conference on One-Dimensional Conductors, Saarbrücken, Germany, 1974, ed. by H. G. Schuster (Springer, Berlin, 1975)
(7) P. Bruesch cf. Ref. 6.
(8) R. Comes cf. Ref. 6.
(9) C. G. Kuper, Proc. R. Soc., A227, 214 (1955)
(10) P. A. Lee, T. M. Rice, P. W. Anderson, Solid State Comm., 14, 703 (1974)
(11) The corresponding result in Ref. 10 for $\omega_t^2 \sim f^{\parallel}$ contains a mistake, the prefactor 4/3 should correctly read 1/3. This is important for understanding that $f_p^{\parallel,\perp}$ merge for $|p| \gg \delta_0$.
(12) P. A. Lee, T. M. Rice, P. W. Anderson, Phys. Rev. Lett., 31, 462 (1973)

II

QUASI-ONE-DIMENSIONAL MODELS

Conference on Organic Conductors and Semiconductors, Siófok, Hungary 1976

HIGH TEMPERATURE STRUCTURAL INSTABILITY IN QUASI-ONE-DIMENSIONAL CONDUCTORS

S. BARIŠIĆ

Institute of Physics of the University, Zagreb, Yugoslavia

Abstract: The most important steps on the way from the simple tight-binding band approximation to the simple fluctuation theory for high-temperature quasi-one-dimensional conductors are discussed. The tight-binding formulation is reviewed with particular emphasis on the strong anisotropic effects. This formulation is then set into the context of the many-body theory. It appears that at sufficiently high temperatures the parquet degeneracy of the many-body theory might break. The ensuing simplification leads to a single-order parameter phase transition theory which is reviewed and completed in respect of the problem of quasi-one-dimensional conductors.

1. Introduction

This paper deals with the recently discovered quasi-one-dimensional conductors KCP ($K_2Pt(CN)_4\ Br_{0.3} \cdot 3H_2O$) and TTF — TCNQ, but touches also some more general problems. The common feature of the mentioned conductors is that they show[1, 2] a structural instability at rather high temperatures $k_B T_c \sim \sim \hbar\,\omega_0(\vec{\varkappa})$, where $\omega_0(\vec{\varkappa})$ is the bare frequency of the phonon $\vec{\varkappa}$, the softening of which is related to the instability.

The first purpose of the present paper is to set the approximate theory which was developed in a series of papers[3-5], into a more general context of the many body theory. The thinking behind the rather drastic simplifications of Refs.[3-5] is exposed in Sections 2 and 3. Our further aim here is to present our previous results, scattered in literature, in an unified text, in order to emphasize their logical interdependence. This is done in Sections 3 and 4. The Section 4 contains some previously unpublished calculations added here for the sake of completeness.

2. *Tight-binding background*

Both KCP and TTF-TCNQ are built of parallel linear chains. The chains are quite well separated, and so the electrons propagate over a long distance along a conducting chain before hopping on to a neighbouring chain. In a perfect conductor this hopping from chain to chain will be coherent and can be described by the well-known tight-binding band approximation. The propagation along the chains also occurs by hopping. The corresponding hopping frequency is smaller than that of the circular movement of electrons around the nucleus on a given site, i. e. the tight-binding very probably applies also to the intrachain propagation.

The general theory of the electron-phonon interaction in the tight-binding approximation has been described in our previous works[3]. It was shown that if the electron spectrum is described by tight-binding, then the bare phonon spectrum should be regarded roughly as including all the renormalizations unrelated to the electron propagation. The extra renormalization due to the band formation can be added through an appropriate bare (rigid-ion) electron-phonon coupling by taking also into account the Coulomb interaction of the displaced charge density waves (CDW). Simultaneously with the bare phonon frequencies $\omega_0(\vec{q})$, which vanish in the long-wave-length limit, the bare electron-phonon matrix element also vanishes in this limit. This illustrates the fact that in the tight-binding limit the *bare* electron-phonon coupling is *short-ranged.*

The fact that the bare coupling is short-ranged, or more precisely, that its anisotropy is roughly the same as that of the electronic band, is important in order to understand the very existence of quasi-one-dimensional materials. If the electron band is quasi-one-dimensional, such will also be the bare electron-phonon matrix element. E. g. band

$$\varepsilon(\vec{k}) = 2J\{\cos k_{\|} d_{\|} + \eta[\cos(k_{\perp} d_{\perp})_1 + \cos(k_{\perp} d_{\perp})_2]\} \tag{1}$$

corresponds to the bare matrix element for the polarization $\vec{\varepsilon}^{\,\lambda}_{\vec{q}}$

$$I^{\lambda}_{\vec{k},\vec{k}+\vec{q}} = 2iq_0 J\left\{\frac{\vec{d}_{\|}}{d_{\|}}\vec{\varepsilon}^{\,\lambda}_{\vec{q}}\,[\sin k_{\|} d_{\|} - \sin(k_{\|} + q_{\|}) d_{\|}] + \right.$$

$$\left. + \eta'\left(\frac{\vec{d}_{\perp}}{d_{\perp}}\right)_{1,2} \vec{\varepsilon}^{\,\lambda}_{\vec{q}}\,[\sin k_{\perp} d_{\perp} - \sin(k_{\perp} + q_{\perp}) d_{\perp}]_{1,2}\right\}, \tag{2}$$

where q_0 and η' are the appropriate coefficients. Equs. (1,2) with $\eta, \eta' \ll 1$ correspond to KCP, while the geometry of TTF-TCNQ is more complicated; but even so the tight-binding formulation remains tractable. Here we shall ignore the geometrical problems currently under consideration by A. Bjeliš and the author and keep in mind only Equs. (1) and (2). Since the small overlap ηJ is more sensitive to the change in intersite distance than the large, one, J, η' is probably larger than η, but this difference can safely be ignored in qualitative considerations.

To Equs. (1,2) we have to add the Fourier transform of the bare Coulomb matrix element

$$U(\vec{q}) = V_0 + \frac{e^2}{d_{\|}} \{a(q_{\|},0) + [a(q_{\|}, d_{\perp}, 0) \cos q_{\perp} d_{\perp}]_{1,2}\}. \tag{3}$$

This matrix element is associated with the CDW occurring on the lattice sites of the crystal, i. e. we neglect here the presumably small displacements of ions accompanying the CDW. In Equ. (3), V_0 is the on-site Coulomb interaction, while

$$a(q_{\|},0) = -\log[2(1 - \cos q_{\|} d_{\|})] \tag{4}$$

is the intrachain Madelung constant, and

$$a(q_{\|}, d_{\perp}, 0) \sim \frac{e^{-q_{\|} d_{\perp}}}{\sqrt{q_{\|} d_{\perp}}} \tag{5}$$

is the interchain Coulomb coupling. Expressions (3—5) are valid, provided that $q_{\|} d_{\perp} \gtrsim 1$ (with a simple modification[4]) in Equ. (5) for $q_{\|} \approx \pi/d_{\|}$).

It may be of use to make here a short physical digression. The quantity $a(q_{\|}, d_{\perp}, 0)$ has a simple physical interpretation. It is proportional[4]) to a potential created at the distance $d_{\perp}$ from a chain carrying a CDW with the wavelength $2\pi/q_{\|}$. At distances $d_{\perp}$ larger than the CDW wavelength, many periods contribute to the potential. Since the overall charge of the CDW is zero, the resulting potential is (exponentially) small. However, the range of a, $q_{\|}^{-1}$, is always larger than the range of the overlap integrals J, given by q_0^{-1} of Equ. (2), which is an atomic distance.

Returning to our formulation, we realize that the problem is completely defined if the frequencies $\omega_0(\vec{q})$, the anharmonic interactions of bare phonons, and the higher-order electron-phonon couplings are known. The frequencies $\omega_0(\vec{q})$ are also expected to be very anisotropic because, as mentioned above, they roughly correspond to the forces in a neutral insulator. The inclusion of electron-phonon couplings beyond the linear term (2) presents a considerable problem[3]), and so we should only mention that for most purposes all these terms may be combined into a phenomenological anharmonic coupling.

3. *Many-body formulation*

General. The many-body theory for the problem defined above has been worked out diagrammatically[6,7]) by neglecting both the anharmonic couplings of bare phonons and the electron-phonon couplings beyond the linear term (e.

g. Equ. (2)). The omission of these terms leads to an important formal simplification, i. e. the whole theory can be formulated[6] in terms of an effective electron-electron coupling

$$\gamma(\vec{k}, \vec{k}', \vec{q}, \omega) = U(\vec{q}) - g_{\vec{k},\vec{q}}\, g_{\vec{k}',-\vec{q}}\, D_0(\vec{q}, \omega), \tag{6}$$

where

$$g_{\vec{k}\vec{q}} = \left(\frac{\hbar}{2NM\,\omega_0(\vec{q})}\right)^{\frac{1}{2}} I_{\vec{k},\vec{k}+\vec{q}}. \tag{7}$$

Here D_0 is the bare phonon Green function while $I_{\vec{k},\vec{k}+\vec{q}}$ is given by Equ. (2). As usual, the phonon mediated electron-electron interaction is retarded in time.

Following Ref.[6] we express the renormalized phonon Green function D

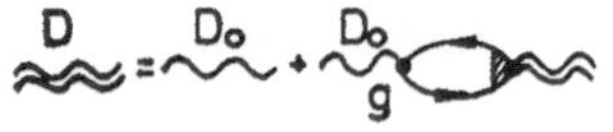

D = D₀ + D₀ g (8)

in terms of effective electron-electron interaction

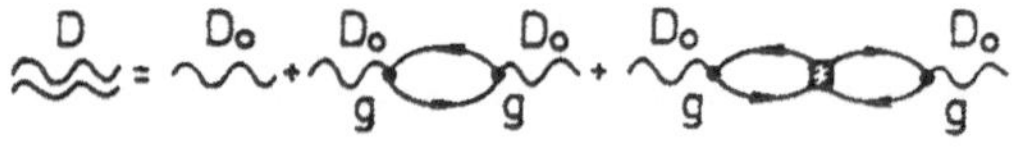

D = D₀ + D₀ g D₀ g + D₀ g D₀ g (9)

The shaded square represents the renormalized vertex involved in the usual Bethe-Salpeter equation

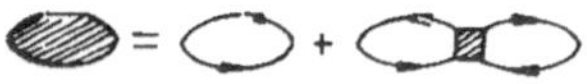

(10)

shown here rather schematically for a two-particle two-points correlation function. The renormalized vertex in Equs. (9,10) represents the sum of the usual »electron-electron« diagrams, with the elementary interaction (6).

At a finite temperature the elementary bubbles[3] in Equs. (9, 10)

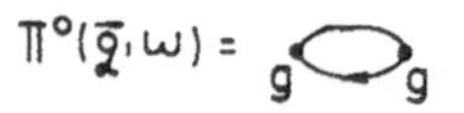

$$\Pi^0(\vec{q}, \omega) = \quad g \; g \tag{11}$$

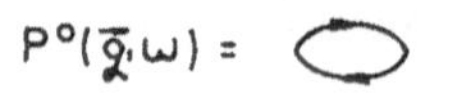

$$P^0(\vec{q}, \omega) = \tag{12}$$

$$Q^0(\vec{q}, \omega) = \quad g \tag{13}$$

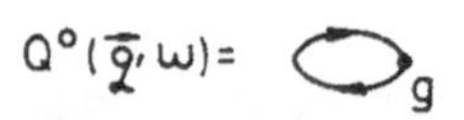

are regular for any q, ω. Therefore, exempting the trivial singularities in D_0, any divergence in the correlation functions (9, 10) must come from the singularity of the renormalized vertex itself.

Space effects. The singularity of the vertex will not always be reflected in the same way in a susceptibility to the staggered stress (9) or the electric field (10). This is due to the k-structure (band structure) of quantity $g_{\vec{k}\vec{q}}$ as given by Equs. (7) and (2). This $\vec{k}$ dependence causes the quantity Q° ($\vec{q}$, ω) to vanish for certain highly symmetric (simply commensurate) deformations and/or for a half filled band. (A detailed discussion of Q° is given in Ref[3]). In order to illustrate the possible consequence of the disappearance of Q°, with P° and Π° finite, we split the vertex into the part which starts and/or finishes with a phonon line (e. g. diagrams a, c of Fig. 1) and the part containing all other diagrams (e. g. diagrams 1b, d). The first part will be multiplied by a small quantity Q° in Equ. (10) but not in Equ. (9). Thus, if the part starting with a phonon line contains a strong singularity, this singularity will be strongly reflected only in the phonon correlation function. This is the case in which a small charge density wave (CDW) accompanies a large lattice deformation. One can also easily imagine the opposite situation where the leading divergence occurs in that part of the vertex which starts with the Coulomb line. Then, again because of the possibly small Q°, or else, because of the small g, the deformation will be small with respect to the appropriately normalized CDW.

As a rule, the complication with Q° may be ignored[3] when dealing with incommensurate deformations and partially filled bands. Then the $\vec{k}$ structure of $g_{\vec{k}\vec{q}}$ may be roughly neglected everywhere, especially in the dangerous quanitity Q°. Π°, P° and Q° become roughly proportional[4]. Now, there is no difference in discussing Equs. (9) and (10). Deformations and CDWs are simply proportional. The temperature at which the vertex diverges is the common transition temperature for structural (Peierls) and dielectric (CDW) instability. However, it is important to distinguish between the phonon and the Coulomb mechanism of instability, according to which one of the two possibly attractive terms, (7) or (4,5), dominates the vertex (6). Not only will the ratio of the deformation and CDW be different in these two limits, but even the whole aspect of the many-body theory may change. The latter point is further discussed in the next sub-section.

Retardation effects. If the phonon contribution dominates the bare vertex (6), the retardation effects associated with heavy ions can play an important role in the many-body theory. In order to develop this point in greater detail let us, for reasons of clarity, ignore the Coulomb contribution to the bare vertex (6). Some simple vertex corrections are shown in Fig. 1. These particular diagrams are chosen because in the one-dimensional case ($\eta = \eta' = 0$) they all yield the same*) $g^6 \log^2 T$ contribution to the vertex, provided that the retardation effects are neglected. Such a degenerate situation is usually named parquet.

---

*) Note however that the two logarithmic contributions are incompatible with the momentum $2k_F$ on all wavy lines of Fig. 1c. This can be remedied by exchanging say, the lower ends of the electron lines in the Cooper bubble, but then the spins in the bubble can not be different.

The retardation effects in the parquet become important[6] for temperatures $k_B T > \hbar\omega_0(\vec{q})$. All Matsubara frequencies ($\hbar\omega_n = 2n\, k_B T$) except one ($n = 0$) in D_0's will then exceed $\omega_0(\vec{q})$. All vertex diagrams can then be conveniently divided into two classes. Let the first class incorporate the diagrams which contain integrations over the phonon frequency (e. g. diagrams 1b, c, d). Diagrams such as 1a will then belong to the second class. Due to the energy conservation in the electron-phonon vertex, each phonon line in Fig. 1a carries only and the same fre-

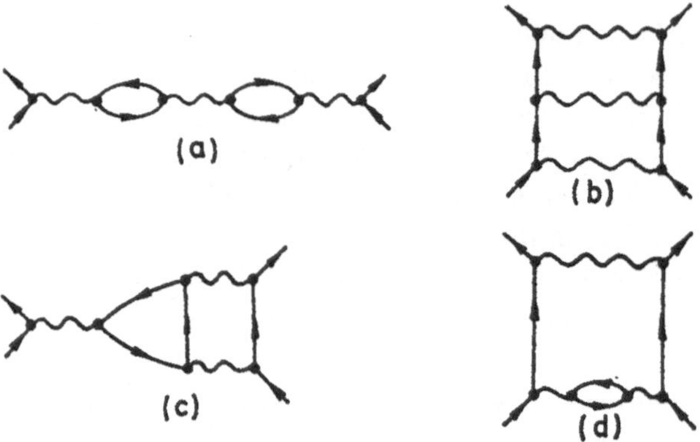

Fig. 1. Four (low order) parquet diagrams, degenerate at low temperatures.

quency, i. e. the external frequency of the entering electron-hole pair. In a phase transition we are interested in the static response $\omega_n = 0$. For this external frequency the retardation effects are entirely absent from (1a). In contradistinction, the retardation effects are expected to reduce the contribution of the diagrams belonging to the first class, through summation over the presumably predominantly small D_0 functions. The retardation effects on diagrams of the type (1b, d) involved in the Eliashberg equations have been studied in considerable detail by a number of authors[3,8,9]. As pointed out in Ref.[3], phonons with frequencies $\hbar\,\omega^{RPA}(\vec{q}) < k_B T$ are not very active in superconductivity, even if the external frequency in Eliashberg equations vanishes. The stronger condition $k_B T > \hbar\,\omega_0(\vec{q}) > \hbar\,\omega^{RPA}(\vec{q})$ is probably sufficient to ensure that all diagrams of the first class (1c etc.) are negligible with respect to those of the second class, provided that η is not too large. In fact, η primarily reduces[9] the diagrams of the Migdal (RPA) channel (Fig. 1a).

To sumarize:

- within the phonon mechanism the parquet degeneracy of Fig. 1 breaks for $k_B T > \hbar\,\omega_0$ if η is small, and
- only the RPA vertex corrections remain.

RPA theory. The usual quantity used in phonon theories is the phonon self--energy defined here by Equ. (8). This exact phonon selfenergy can easily be expressed in terms of the electron-electron vertex by eliminating the Green function D between Equs. (8) and (9). Introducing further the RPA and $g_{\vec{k}\vec{q}} \simeq g_{\vec{q}}$ simplifications, we find immediately that

$$\Pi_{\mathrm{RPA}} = \frac{-g^2 P^\circ (1 + \Gamma_0 P^\circ)}{1 - g^2 D_0 (1 - \Gamma_0 P^\circ)}, \tag{14}$$

where Γ_0 represents the RPA vertex, obtained by the summation of all diagrams of type 1a,

$$\Gamma_0 = \frac{-g^2 D_0}{1 - g^2 D_0 P^\circ}. \tag{15}$$

Inserting Equ. (15) into Equ. (14), we obtain

$$\Pi_{\mathrm{RPA}} = -g^2 P^0 = \Pi^0. \tag{16}$$

Equ. (14) has the same algebraic structure as has the exact equation for Π, except that in RPA the matrices are replaced by numbers. From this it is clear that Equ. (16) for Π could have been obtained alternatively by saying that both the bare vertex in the denominator and the renormalized vertex in the nominator of Equ. (14) are small due to the smallness of D_0 for $k_B T > \hbar \omega_0 (\vec{q})$. (The Coulomb contribution is already omitted here). But to remain consistent, we have to regard the external lines D_0 in Equs. (8) or (9) as also being small, i. e. to arrive to $D \simeq D_0$, far from the untrivial singularity in D.

It was therefore important to show above that Equ. (16) holds not only when Γ is small, but also when it is large and given by Equ. (15). Γ_0 is large when

$$1 \approx g^2 D_0 P^\circ. \tag{17}$$

Of course, condition (17) coincides with the equation which gives the singularity in D and is obtained by combining Equs. (16) and (8). This coincidence agrees with our general statement (the end of sub-section *General*) that the nontrivial singularity in D can arise only from the singularity in Γ.

Equ. (17) may have one or more (complex) solutions $\omega (\vec{q}, T)$ for a given $\vec{q}$ and T (working now with a retarded $D (\vec{q}, \omega)$. The requirement $\omega (\vec{q}, T) = 0$ gives $T = T(\vec{q})$, while the largest of these temperatures $T(\vec{\varkappa}) = T_{\mathrm{MF}}$ plays the role of the (mean-field) transition temperature. In more physical terms, $D^{-1} (\vec{q}, 0)$ calculated within RPA is proportional to the harmonic deformation energy[3]. This energy is minimal at $\vec{q} = \vec{\varkappa}$. The value of the minimum is proportional to

$T - T_{MF}$, while the expansion of this energy in terms of $\vec{q} - \vec{\varkappa}$ determines[4] the longitudinal and transverse correlation length, $\xi_{0\parallel}$ and $\xi_{0\perp}$, respectively.

The limit $\eta = \eta' = 0$ in Equ. (17) has been thoroughly studied in Refs.[10,4]. Let us denote the characteristic frequency of Π_{RPA} by ω_c. For $T \ll T_F$ (Fermi temperature) $\hbar\,\omega_c \sim \lambda_{\vec{q}}^{-1/2}\, k_B T$, where $\lambda_{\vec{q}} = -\, n_F g^2/\omega_0\,(\vec{q})$. $\omega_c \gg \omega_0$ means practically $k_B T \gg \hbar\,\omega_0$ since usually $\lambda_{\vec{q}} \sim 1$. In the latter inequality we recognize a condition for the applicability of the RPA theory. On other hand, $\omega_c > \omega_0$ ensures that Equ. (17) has only one (adiabatic) solution. In the opposite limit $\omega_c < \omega_0$, Equ. (17) has two solutions. Close enough to T_{MF}, which is given by

$$T_{MF} \approx 2.28\; T_F\, e^{-1/\lambda_{\vec{\varkappa}}} \tag{18}$$

irrespective of the value of ω_c/ω_0, the two solutions occur rather close to the real ω-axis. The first solution corresponds to $\omega \approx \omega_0$. In the vicinity of this frequency, Π_{RPA} is small, i. e. $D \approx D_0$ (despite the fact that Γ_0 is large). Thus, this solution corresponds essentially to the trivial D_0 pole in Equ. (9). But since $\omega_c < \omega_0$, $\Pi_{RPA}\,(\omega)$ varies rapidly on the scale fixed by ω_0 becoming large for $\omega = 0$. This time the large Γ_0 corresponds to a large renormalization Π_{RPA} and to the instability at $T = T_{MF}$. The two solutions give rise to a three peak structure of the neutron scattering cross--section. But $\omega_c \ll \omega_0$ also means $k_B T_{MF} < \hbar\,\omega_0$, the limit in which the RPA results are incomplete. This might be one of the reasons why the fit[1] of the neutron cross-section obtained on KCP by the RPA results is not too successful.

The requirement $T_{MF} \ll T_F$ was omitted in Refs.[3,11] with Nb_3 Sn-like compounds in mind, but keeping $\eta = \eta' = 0$.

In Refs.[9,12] $k_B T_{MF}$ is again taken as being much smaller than T_F but $\eta \neq 0$, while $\eta' = 0$. The most unstable mode occurs at $\vec{\varkappa} = [2k_F, \pi/d_\perp, \pi/d_\perp]$, where k_F is the Fermi wave vector of the corresponding $\eta = 0$ band. Such $\vec{\varkappa}$ corresponds to the best nesting of the Fermi surfaces. T_{MF} is still given roughly by Equ. (18). The $\eta \neq 0$ model[9,12] leads to interchain correlations. The longitudinal and transverse correlation lengths are

$$\xi_{0\parallel} \simeq \frac{\hbar v_F}{k_B T} \tag{19}$$

$$\xi_{0\perp} \approx \eta\, k_F\, d_\perp\, \xi_{0\parallel}, \tag{20}$$

respectively. Not surprisingly, the length and time scale are related here, $\xi_{0\parallel} \sim \sim v_F/(\lambda^{1/2}\,\omega_c)$.

The interchain distance (10Å) and the character of the d-functions forming the band in KCP lead to a very small η. Thus, although the most unstable mode is actually observed for the $\vec{\varkappa}$ found above, the measured values[1] of $\xi_{0\perp}$ are too large to be reconciled with Equ. (20). For KCP one must go back to the Coulomb mechanism of interchain coupling, Equ. (5).

The situation in TTF-TCNQ and similar materials is less clear, because of the considerable geometrical complications. It is not clear at present whether our model[4] is sufficient or not to describe the empirical findings[2] in these con ductors.

The Coulomb coupling (3) is always present and is probably important[4] in both types of materials mentioned above. Extending simply the foregoing discussion to the combined vertex (6), we shall find that the Coulomb coupling has to be combined with a phonon term in the RPA channel; in other parquet diagrams the phonon contribution may presumably be neglected for $k_B T \gg \hbar \omega_0 (\vec{q})$. Then, if the phonon coupling in the RPA channel is larger than the Coulomb term, the lowest order Coulomb correction of the purely phonon RPA result is again given by the RPA diagrams. This suggests that we can work with the RPA channel not only up to the zeroth order in Coulomb coupling but also up to the first order. Obviously, the quantitative elaboration of this idea requires a more detailed investigation of the retardation effects in the parquet.

If in the presence of a small Coulomb term the RPA procedure is accepted, the corresponding results are obtained by simply replacing $g^2 D_0$ with $g^2 D_0 - U$ in Equs. (15) and (17) (the RPA theory in which g keeps its k dependence is described in Ref.[3]). Taking $\eta = \eta' = 0$ and considering again the limit $T_{MF} \ll T_F$, with

$$T_{MF} \approx 2.28\, T_F \exp \left[1/n_F \left(\frac{g^2}{\omega_0} + U(\vec{\varkappa}) \right) \right] \tag{21}$$

we recover the most unstable mode $\vec{\varkappa}$ at $\vec{\varkappa} = [2k_F, \pi/d_\perp, \pi/d_\perp]$. $q_\perp = \pi/d_\perp$ makes the best use of the attractive coupling (5). Expanding again in terms of $(\vec{q} - \vec{\varkappa})$, we also also recover Equ. (18), but Equ. (19) is replaced by

$$\xi_{0\perp}^2 / d_\perp^2 \approx \frac{2n_F e^2}{d_\parallel} \, a\,(q_\parallel, d_\perp, 0) \log^2 \frac{2T_F}{T} \tag{22}$$

with a from Equ. (5).

Equ. (22) seems to fit reasonably well the experimental findings[1] in KCP. Also, the Coulomb model is currently being extended to include the geometrical complications encountered in TTF-TCNQ. Although the preliminary results[4] seem to agree with the phase sliding ($\varkappa_\perp = \varkappa_\perp (T)$) observed[2] in TTF-TCNQ, the respective work has not yet produced any firm conclusions.

Obviously, the mechanism which gives the larger interchain correlation, i. e. the one which within the simple model defined by Equs. (1–5) produces the larger $\xi_{0\perp}$ between Equs. (20) and (22), is the one which should be retained physically as the cause of the interchain correlations. However, we wish to warn once more against using the too large (negative) α's of Equs. (4) and/or (5) in the RPA theory. It is true that Equs. (20, 21) remain well defined[12] in the limit $|U(\vec{q})| \gg |g^2/\omega_0|$, and have a simple physical meaning of the CDW instability in Equ. (10) with nothing much happening to the Peierls Equ. (9). However, the strong Coulomb mechanism requires full many-body treatment, beyond RPA, as discussed in Refs,[6,7,14,15]. It should also be kept in mind that some of our basic approximations, especially the omission [3] of the electron-phonon coupling through Coulomb forces in Equ. (3), may become questionable*) in the strong Coulomb--weak hopping limit.

Transitory remarks. The RPA or even the parquet approximation are mean-field theories in the sense that even in the one-dimensional system they lead to phase transition at finite temperature T_{MF}. Of course, the physical systems under consideration are not one-dimensional and the phase transition is allowed to occur at some finite T_c. However, T_c may be considerably lower than T_{MF}. If $k_B T_c$ is still larger than $\hbar\omega_0$, the arguments which led us to retain the RPA diagrams remain valid close to $T_c < T_{FM}$. The shift from T_{MF} to T_c may be attributed to the terms mentioned in Section 2 as giving rise to an effective anharmonic coupling. These terms are omitted in the discussion of this Section. Also, the diagrams beyond RPA may contribute[16] to the effective anharmonic coupling, but this point too would require a careful consideration regarding the retardation effects.

This line of thinking brings us to a rather simple Ginzburg-Landau-like model in which an effective anharmonic coupling of deformations is added to the harmonic deformation energy D_{RPA}^{-1}. In a one-dimensional conductor such a single--order parameter model would lead to $T_c = 0$. But in the range $k_B T < \hbar\omega_0$, the results of the model are not applicable to the one-dimensional conductor. In this range of temperatures the situation is parquet like[6,14,15], i. e. it involves two- (or more) order parameters at the same time. For the quasi-one-dimensional conductors, however, the model may apply to the whole range of temperatures if it leads to $k_B T_c > \hbar\omega_0$. Below the three-dimensional transition at T_c the situation may continue to be non-parquet like, down to low temperatures.

In the following section we shall investigate the Ginzburg-Landau (GL) single-order parameter model. Although this model applies to low-dimensional conductors with the reservations made above, it is relevant for other, simpler low--dimensional systems. One important approximation (mean-field treatment of interchain coupling), which is usually used without justification in this[17,18] or more complicated models, is shown to lead to correct results, at least within the Ginzburg-Landau model.

* This point arose during a discussion with G. Grüner.

4. Ginzburg-Landau theory

General considerations. The starting point of this Section is the spatially anisotropic G-L functional

$$f(r) = a' \left[\frac{T - T_{MF}}{T_{MF}} |\psi|^2 + \xi_{0\|}^2 \sum_{i=1}^{d'} \left| \frac{\partial \psi}{\partial x_i} \right|^2 + \xi_{0\perp}^2 \sum_{i=d'+1}^{d} \left| \frac{\partial \psi}{\partial x_i} \right|^2 \right] + b |\psi|^4. \quad (23)$$

Here ψ is the n-component order parameter, and we assume

$$\frac{\xi_{0\perp}^2}{\xi_{0\|}^2} = a_{GL} \ll 1. \quad (24)$$

We keep here the one-dimensional notations (18—22) but do not specify d and d' for the moment.

The G-L functional (23) has to be combined with appropriate cut-off wave-lengths. These wave-lengths need not be the same in d' longitudinal and $\Delta = d-d'$ transverse directions. Let us denote them with $Q_{\|}^{-1}$ and $Q_{\perp}^{-1}$, respectively. The cut-offs describe that portion of the Brillouin zone to which harmonic part of the expansion (23) applies. As will be shown below and as argued previously[5,12,19] the question of cut-offs is quite important and thus requires some discussion.

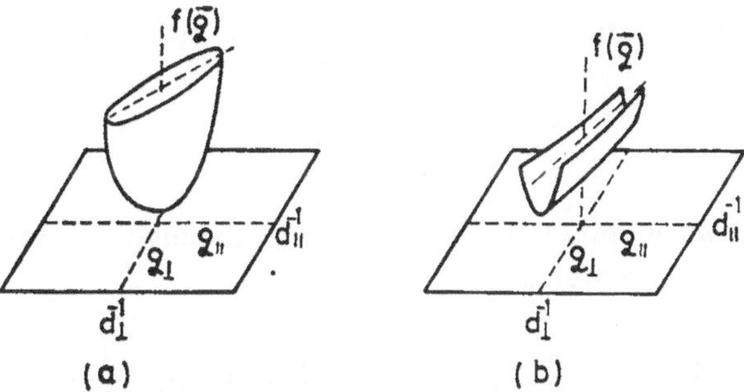

Fig. 2. Two opposite limits for the behaviour of the Fourier transform $f(\vec{q})$ of the harmonic free energy given by Equ. (23). The Fourier componet $\psi(\vec{q})$ of $\psi(\vec{r})$ is kept independent of $\vec{q}$ in the figure.

If both $\xi_{0\|}^{-1}$ and $\xi_{0\perp}^{-1}$ are smaller than the corresponding Brillouin zone dimensions, it is natural to truncate the harmonic part of Equ. (23) at the same energy in all directions (Fig. 2a).

This leads to

$$\frac{a_{GL} Q_{\perp}^2}{Q_{\|}^2} = 1. \quad (25)$$

We shall note that the limit of small a_{GL} can be achieved in the regime (25) only by taking $Q_{||}$ small ($\xi_{0||}$ large) since $Q_{\perp}$ cannot exceed the Brillouin zone dimension. When a_{GL} of Equ. (24) is decreased by decreasing $\xi_{0\perp}$, with $\xi_{0||}$ remaining fixed, $Q_{\perp}$ reaches its maximal value determined by the Brillouin zone and then stays constant (Fig. 2b); Equ. (25) breaks. In conclusion, the limit of small $\xi_{0\perp}$ is physically associated with a_{GL}-independent cuf-offs $Q_{||}$ and $Q_{\perp}$ and the limit of large $\xi_{0||}$, $\xi_{0\perp}$ with Equ. (25). Let us start by considering the former.

Crossover limit. This is the situation in which the system (23) exhibits a crossover behavior. Above a certain crossover temperature $T^*(a_{GL})$, it behaves according to the d'-dimensional critical indices; below it, the critical behavior is essentially d-dimensional (for d and d' see Equ. (23)). When a_{GL} tends towards zero, both $T^*(a_{GL})$ and the critical temperature $T_c(a_{GL})$ tend continuously towards the critical temperature of the d'dimensional system. For brevity, we shall deal here only with situations in which the dimensionality d' is so low that there is no phase transition at finite temperature associated with this dimension. That is to say that we are considering here the crossover from the unshaded into the shaded region of the n, d' diagram of Fig. 3. But instead of extending our homogeneity argument[5] to this whole region, we will pay particular attention to the $n \to \infty$ limit, where the otherwise lacking constants of proportionality can be calculated in detail.

It seems therefore advisable to try and clarify the analogy between the familiar free Bose gas[5] and the $n \to \infty$ G-L system. We shall do it by calculating $T_c(a_{GL})$ for the latter case. In the limit $n \to \infty$ the anharmonic term in Equ. (23) can be treated in the Hartree approximation. The transition temperature is then determined by

$$a' \frac{T_{MF} - T_c}{T_{MF}} = n\, \Sigma_c. \tag{26}$$

Neglecting the contribution to the self-energy Σ_c coming from wave vectors beyond the cut-offs $Q_{||}$, $Q_{\perp}$, we obtain

$$\Sigma_c \approx \frac{b\, k_B\, T_c}{(2\pi)^d\, a'\, \xi_{0||}^2} \int^{Q\perp} d^{\Delta} k_{\Delta} \int^{Q_{||}} d^{d'} k_{d'} \frac{1}{k_{d'}^2 + a_{GL} k_{\Delta}^2}. \tag{27}$$

Here $\Delta = d - d'$, and $k_{d'}$ and k_{Δ} are the wave vectors in d' and Δ dimensions, respectively. We notice the analogy between Equ. (27) and the equation (5) determining the transition temperature of the free anisotropic Bose gas: with $a_{GL} \to 0$, Σ_c tends towards the constant a'/n which here plays a role which is analogous to the total number of free bosons in the boson equation. For $T_c \ll T_{MF}$, Equ. (27) can conveniently be rewritten as

$$\frac{a'}{n} = \frac{b\, S_{d'}\, S_{\Delta}}{(2\pi)^d\, a'\, \xi_{0||}^2}\, k_B\, T_C\, a_{GL}^{\frac{d'-2}{2}} \int_0^{Q\perp} k_{\Delta}^{d-3}\, dk_{\Delta} \int_0^{Q_{||}/k_{\Delta} a_{GL}^{1/2}} \frac{x^{d'-1}\, dx}{1 + x^2} \tag{28}$$

with

$$S_p = 2\pi^{p/2}\,\Gamma(p/2).$$

In the last integration the upper limit is large for any finite $k_\Delta \leq Q_\perp$, provided that

$$\frac{Q_{\parallel}^2}{Q_\perp^2\, a_{\mathrm{GL}}} \gg 1, \tag{29}$$

i. e. that Equ. (25) is not fulfilled. In other words, the cut-off $Q_{\parallel}$ disappears from Equ. (28) if the increase of $Q_\perp^2$ with decreasing $\xi_{0\perp}^2$ is slower than a_{GL}^{-1}. This is particularly the case for small $\xi_{0\perp}$ when $Q_\perp$, as mentioned before, is roughly independent of a_{GL}. The integrations (28) are easily carried out when the upper limit in the second integral is set equal to infinity (a_{GL} small). The result is convergent, provided that $d > 2$ and $2 > d' > 0$. In analogy with the result[5)] for the Bose gas we find that

$$k_B T_C = \frac{4a'\,\xi_{0\parallel}^2}{n\,b\,Q_\perp^{d-2}}\;\frac{(2\pi)^{d-1}(d-2)\sin\dfrac{\pi d'}{2}}{Sd'\,S_\Delta}\;a_{\mathrm{GL}}^{\frac{2-d'}{2}}. \tag{30}$$

Equ. (30) coincides in fact with a more general expression

$$T^* \sim T_C \sim \xi_{0\parallel}^2\, a_{\mathrm{GL}}^{\frac{2-d}{2}}, \tag{31}$$

which K. Uzelac and the author propose, extending the $d' = 1$ homogeneity argument[15)] to the shaded region of Fig. 3. The latter argument is more general because it is independent of n, assumed large in Equ. (30). Since for T^* and T_c the exact cross-over exponents, i. e.

$$\psi = \psi = \frac{2}{2-d'} \tag{32}$$

are independent of n, Equs. (30) and (31) coincide in this respect in the case of an arbitrary n. The same cross-over exponent was obtained previously[17, 18)] for $d' = 1$, $n = 2$ applying the mean-field approximation to the interchain coupling $\xi_{0\perp}^2$ in Equ. (23). We can see here that in the Ginzburg-Landau model this approximation leads to the exact result (32).

Equ. (30) gives the prefactor lacking in Equ. (31), provided that only the lowest-order anharmonic coupling b is retained. Of course, the dependence on n in Equ. (30) should not be taken too seriously if n is of the order of one. For $d' = 1$, $n > 1$, Equ. (30) reduces to

$$\frac{T_c}{T_b^\perp} = \frac{3\,(2\pi)^{d-1}(d-2)}{n\,S_{d-1}}\,\xi_{0\perp}\,Q_\perp \tag{33}$$

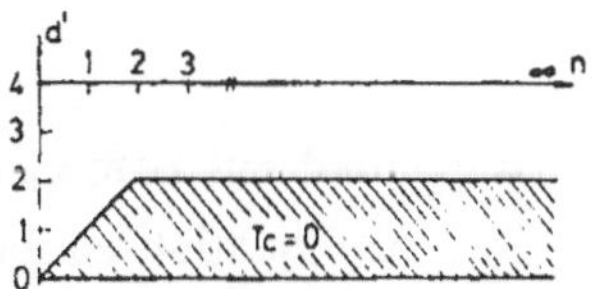

Fig. 3. Parts of the d', n plane for which the $\xi_{0\perp} = 0$ model of Equ. (23) has respectively finite or zero transition temperature. The $d' = n$ line has not been firmly established.

which defines the characteristic temperature scale

$$k_B T_b^{\perp} = \frac{4}{3} \frac{a'^2 \xi_{0\parallel}}{b Q_{\perp}^{d-1}}. \tag{34}$$

The same temperature $T_b^{\perp}$ also occurs for $d' = 1$, $n = 1$ when[5,19]

$$\frac{T_c}{T_b^{\perp}} \sim |\log \xi_{0\perp} Q_{\perp}|^{-1}. \tag{35}$$

We shall note that $T_b^{\perp}$ has nothing in common with T_{MF}, which becomes an unimportant parameter for $T_c \ll T_b^{\perp}$.

Symmetric limit. This is the limit in which Equ. (25) applies. Let us start again by considering the simple $n \to \infty$ equation. With Σ_c defined by Equ. (26), and introducing a new variable

$$q_\Delta = \sqrt{a_{GL}}\, k_\Delta \tag{36}$$

into Equ. (27) and taking into account Equ. (25), we shall find that

$$\Sigma_c \approx \frac{b k_B T_c}{a' \xi_{0\parallel}^2 Q_{\parallel}^{2-d}} \frac{S_d}{d-2} a_{GL}^{-\frac{\Delta}{2}}, \tag{37}$$

provided that $d > 2$. It should be noted that with $Q_\parallel \sim 1/\xi_{0\parallel}$ (small), Σ_c is roughly independent of $\xi_{0\parallel}$. In fact we should always ensure that the retained contribution to Σ_c is really dominant. Equs. (26) and (37) can easily be solved for T_c

$$T_c \approx \frac{T_{MF}}{1 + \frac{T_{MF}}{T_b^{\parallel}} a_{GL}^{-\frac{\Delta}{2}}} \tag{38}$$

where

$$k_B T_b^{\parallel} = \frac{a' \xi_{0\parallel}^2}{n b Q_{\parallel}^{d-2}} \frac{(2\pi)^d (d-2)}{S_d}. \tag{39}$$

As pointed out before, Equ. (38) is roughly independent of $\xi_{0\parallel}$ (large).

An equation similar to Equ. (38) is found in the linearized renormalization group theory[21]. In order to show that, we shall follow the prescription of Ref.[19] and symmetrize the expression (23) with cut-offs (25), and change the variables (36) in the direct space. Thus introducing

$$x'_\Delta = \sqrt{a_{GL}}\, x_\Delta \tag{40}$$

we have

$$F[\psi] = \int_{\Omega} \mathrm{d}^d \vec{r} f(\vec{r}) = \int_{\Omega'} \mathrm{d}^d \vec{r}' f'(\vec{r}'). \tag{41}$$

Here[19)]

$$\vec{r}'' = \{x_1 \ldots x_{d'}, x'_{d'+1} \ldots x'_{\Delta} \ldots x'_d\}, \quad f'(\vec{r}') = a_{\mathrm{GL}}^{\frac{\Delta}{2}} f(\vec{r}), \quad \Omega' = a_{\mathrm{GL}}^{\frac{\Delta}{2}} \Omega. \tag{42}$$

The new function $f'(\vec{r}')$ has a symmetric gradient term $\left|\dfrac{\partial \psi(\vec{r}')}{\partial \vec{r}'}\right|^2$ and is associated, by Equ. (25), with the same cut-off $Q_{||}$ in all directions.

Applying now the usual $\varepsilon = 4\text{-}d$ expansion procedure to the determination of the critical surface (line) in the parameter space defined by Equ. (42), we find

$$T_c = \frac{T_c^*}{1 + w \dfrac{b\, k_B\, T_{MF}}{\xi_{0||}^2\, a'^2\, Q_{||}^{2-d}}\, a_{\mathrm{GL}}^{-\frac{\Delta}{2}}}. \tag{43}$$

Here

$$w \sim \frac{(n+2)\left(1 + \dfrac{n+2}{2n+16}\,\varepsilon\right)}{2 + \left(1 - \dfrac{n+2}{n+8}\right)\varepsilon}, \tag{44}$$

and T_c^* depend on the recurrence step used in the renormalization procedure[21)]. Equ. (43) is valid for n arbitrary, $4 > d \gg 2$, and with $T_{MF} \approx T_c^*$. Equ. (38) is valid for large n, $4 > d > 2$ and T_{MF} arbitrary. For large n and $T_c^* \approx T_{MF}$, Equs. (28) and (48) coincide (and determine the lacking, n-independent factor in w of Equ. (44)). It is interesting that the constant $T_b^{||}$ of Equ. (39) approximates the corresponding quantity of Equ. (43) rather faithfully over the whole range of n.

As argued previously[19)], the symmetrizable system (23) and (25) obeys the d-dimensional critical laws. Only the scales are changed according to relations (40) and (42). The critical range is increased and the fluctuation effects are enhanced. In particular, the fluctuation specific heat is increased. Both temperature dependent correlation lengths $\xi_{||}(T)$ and $\xi_{\perp}(T)$ follow the same temperature law because they are deduced from the correlation length $\xi_{||}(T)$ by the inverse transformation (40). Their ratio is therefore independent of temperature,

$$\frac{\xi_{\perp}^2(T)}{\xi_{||}^2(T)} = \frac{\xi_{0\perp}^2}{\xi_{0||}^2} = a_{\mathrm{GL}}, \tag{45}$$

in contrast to the cross-over case where this ratio is strongly temperature dependent.

Conclusion. In this Section we have seen that according to the choice of cut-offs the large anisotropy ($a_{\mathrm{GL}} \ll 1$) may correspond to two qualitatively different situations, the cross-over or the symmetric limit. In both limits the shift of the transition temperature from its mean-field value may be considerable, depending on the parameters. The clear-cut difference between the two limits occurs in the critical behaviour itself.

At sufficiently high temperatures KCP falls into the cross-over limit. $\xi_{0\perp}$, measured[11], or calculated from Equ. (22), is considerably smaller than $d_\perp$. However, $\xi_{0\perp}$ itself is (non-critically) temperature dependent and at T_c is estimated to be of of the order of $d_\perp$. However, the temperature dependence given by Equ. (22) (and also by Equs. (19, 20)) should not be taken too seriously for $T < T_{MF}$. This equation neglects (at least) the electron-self-energy effects[13,22,23] (the Peierls pseudogap). Still, it appears probable that close to T_c, KCP falls somewhere between the two extreme limits considered here.

The symmetric limit corresponds very probably to the BCS superconductivity in the conducting polymer[24] $(SN)_x$. It seems that there the ratio of the longitudinal and transverse correlation lengths is temperature independent, as in Equ. (45).

The situation in TTF-TCNQ is potentially much richer than that in KCP. Transverse correlations in this material are probably considerably stronger in one than in the other of the two transverse directions. This can lead to a complicated 3—2—1 dimensional cross-over for $n = 2$. Empirically[2], however, the fluctuation tails do not seem to be as large in this material as they are in KCP. It might well be that the structural transition in TTF-TCNQ falls into the symmetric limit, with only the fluctuations enhanced. It would be interesting therefore to have more data about the temperature behaviour of all three, or at least of two correlation lengths.

5. *Final remarks*

In this paper we have tried to describe the most important steps on the way from the simple tight-binding approximation to the simple fluctuation model for quasi-one-dimensional conductors. A major crossing on this way is certainly traversed in earlier sub-sections, where are given the arguments in favour of the single-order parameter phase transition theory for high-temperature quasi-one-dimensional conductors.

In order to emphasize the main line of reasoning, many details were deliberately left out and some problems neglected. Such are, for example, the electron-self-energy effects which merit at least to be mentioned here. The lowest-order self-energy correction gives rise to the Peierls pseudogap above T_c[13,22,23]. The Peierls pseudogap is present in the electron spectrum even when the phonon retardation effects are included[23]. The role of the Peierls pseudogap in parquet corrections or even in the Eliashberg-BCS equation has not yet been carefully investigated. Related with it is the problem of the role played by the long-range

($q_{||} \approx 0$) part of the interaction[6]. This question has been discussed in detail, for example in Ref[7], using the unretarded long-range forces. In this connection it should be noted that the quasi-one dimensional equation (3) obtained here for $q_{||}\, d_{\perp} \gg 1$ becomes strongly three-dimensional in the »long range« limit $q_{||}\, d_{\perp} \ll 1$. In the ensuing many-body theory[7], the decisive ratio seems to be $e^2/\hbar v_F$ or equivalently $\frac{e^2/d_{\perp}}{k_B T_{MF}}$ versus $\frac{\xi_{0||}}{d_{\perp}}$. $e^2 \ll \hbar v_F$ was shown[7] to correspond to the one--dimensional-like »infrared catastrophe« situation; the other limit has not been examined in detail. Although $\xi_{0||}/d_{\perp}$ is large in KCP and probably also in TTF--TCNQ, it is still very likely that $\frac{e^2}{d_{\perp}}/k_B T_{MF}$ is even larger in these materials. It might be therefore reasonably expected that the $q_{||} \approx 0$ part of the forces does not play a very important role in the theory of these materials.

Obviously, many of the ideas presented or mentioned here are qualitative and require a more careful examination. Therefore the present paper should not be misunderstood as an attempt to supply a final answer to the question of basic physics in KCP and TTF-TCNQ; it only indicates a possibly fruitful line of research and gives some results along this line.

Acknowledgment

Much of the work discussed in this paper was carried out in collaboration with J. Friedel, J. Labbé, A. Bjeliš, K. Šaub and K. Uzelac. The initative for the work discussed in Section 3 came from discussions with P. Nozières and L. P. Gor'kov. The author also feels indebted to many of his colleagues for their helpful remarks.

References

1) R. Comès, M. Lambert, H. Launois and H. R. Zeller, Phys. Rev. **B8** (1973) 571, B. Renker, L. Pintschovius, W. Glaser, H. Rietschel, R. Comès, L. Liebert and W. Drexel, Phys. Rev. Lett. **32** (1974) 876, J. W. Lynn, M. Yizumi, G. Shirane, S. A. Werner, and R. B. Saillant, Phys. Rev. **B.12** (1975) 1154;

2) F. Denoyer, R. Comès, A. F. Garito and A. Y. Heeger, Phys. Rev. Lett. **35** (1975) 445, G. Shirane, Review paper at International Conference on Low Lying Lattice Vibrational Modes, San Juan, Puerto Rico, (1975);

3) S. Barišić, Phys. Rev. **B5,** (1972) 932 and 941, S. Barišić, Ann. de Physique **7** (1972) 23;

4) S. Barišić and K. Šaub, J. Phys. **C.6** (1973) L367, A. Bjeliš, K. Šaub and S. Barišić, N. Cimento **23B** (1974) 102, S. Barišić, Proceedings of the German Physical Society Conference on »One Dimensional Conductors«, Saarbrücken (1974) K. Šaub, S. Barišić and J. Friedel, Phys. Lett. **56A** (1976) 302;

5) S. Barišić and K. Uzelac, J. de Physique, **36** (1975) 325 and 1267;

6) Y. A. Byckov, L. P. Gor'kov and J. E. Dzjaloshinskii, Sov. Phys. JETP **23** (1966) 489, ŽETF **50,** (1966) 738, L. P. Gor'kov and I. E. Dzjaloshinskii, ŽETF **67** (1974) 397, J. E. Dzjaloshinskii and A. Y. Larkin, Sov. Phys. JETP **34** (1972) 422 (ŽETF 61, 791);

7) J. E. Dzjaloshinskii and A. Y. Larkin, ŽETF **65** (1973) 411;

8) P. B. Allen, Sol. St. Commun. **12** (1973) 379, G. Bergmann and D. Rainer, Z. Physik **263** (1073) 59;

9) B. Horovitz and A. Birnboim, Sol. St. Commun. **19** (1976) 91;

10) S. Barišić, A. Bjeliš and K. Šaub, Sol. St. Commun. **13** (1973) 1119; B. Horovitz, M. Weger and H. Gutfreund, Phys. Rev. **B9** (1974) 1246;

11) S. Barišić, Sol. St. Commun. **9** (1971) 1507;

12) B. Horovitz, H. Gutfreund and M. Weger, Phys. Rev. **B8** (1975) 3174;

13) B. R. Patton and L. Y. Sham, Phys. Rev. Lett. **31** (1973) 631;

14) N. Menyhárd and J. Sólyom, J. Low Temp. Phys. **12** (1973) 529;

15) J. Sólyom J. Low Temp. Physics **12** (1973) 547, L. Mihály and Y. Sólyom, Int. Conf. on Statistical Physics, Budapest (1975) LT; 14, Otaniemi Finland;

16) D. Alleender, Y. W. Bray and J. Bardeen, Phys. Rev. **B9** (1974) 119;

17) D. Y. Scalapino, Y. Imry and P. Pincus, Phys. Rev. **B11** (1975) 2042;

18) P. Manneville, J. Physique **36** (1975) 301;

19) S. Barišić and S. Marčelja, Sol. St. Commun. **7** (1969) 1395;

20) W. Dieterich, Z. Phys. **270** (1974) 239;

21) F. Y. Wegner, Sükajarvi Summer School, Finland 1973, G. Toulouse and P. Pfeuty, Introduction au Groupe de Renormalisation, Presses Universitaires de Grenoble, 1975;

22) P. A. Lee, T. M. Rice and P. W. Anderson, Phys. Rev. Lett **31** (1973) 462, M. J. Rice, and S. Strässler, Sol. St. Commun. **13** (1973) 1389;

23) A. Bjeliš and S. Barišić, J. Physique Lettres **36** (1975) 169;

24) L. J. Azevodo, W. G. Clark, D. G. Deutscher, R. L. Greene, G. B. Street and L. J. Luter, Sol. St. Commun. (1975);

Conference on Organic Conductors and Semiconductors, Siófok, Hungary 1976

ON THE CHARGE DISTRIBUTION AND THE LATTICE DISTORTION OF QUASI-ONE-DIMENSIONAL SYSTEMS

A.A. OVCHINNIKOV, V. Ya. KRIVNOV, V.E. KLYMENKO, I.I. UKRAINSKY

Karpov Institute of Physical Chemistry, Moscow, USSR

Abstract

Electrostatic energy and electronic structure of donor-acceptor molecular crystals based on tetracyanquinodimethane is considered. The ground state energy of quasi--one-dimensional systems consisting of electron donors (TTF, NMP, Rb) and acceptors (TCNQ) is calculated. It is shown that for TTF-TCNQ, NMP-TCNQ, and Rb-TCNQ crystals the dominant contribution is made by the electrostatic energy. The electrostatic energy (or Madelung energy) E_M for these systems has been calculated numerically by the Evald method using the X-ray data on the structure of corresponding crystals and molecular systems. The charge distributions in the molecules and ions under consideration (i.e. $TCNQ^{(o,-)}$, $NMP^{(o,+)}$, and $TTF^{(o,+)}$) have been obtained from the PPP, CNDO/2 and INDO calculations.

It is shown that the total electron transfer from donors (D) to acceptors (A) is energetically disadvantageous for the TTF-TCNQ and NMP-TCNQ crystals. For these systems the Madelung energy per a pair D^+A^- is smaller than the difference of the donor ionization potential I and the electron affinity of acceptor A : $E = E_M + I - A < 0$.

The dependence of the electrostatic energy of NMP-TCNQ and TTF-TCNQ crystals on the charge transfer (ρ) from donors to acceptors is also studied. The results of these calculations allow us to determine the degree of the charge transfer ρ , namely: $1/2 < \rho < 3/4$ for TTF-TCNQ, $1/5 < \rho < 1/2$ for NMP-TCNQ, and $\rho = 1$ for Rb-TCNQ. The analysis of the magnetic properties of NMP-TCNQ and TTF-TCNQ complexes makes it possible to determine the value of ρ more precisely. In the case of non-complite charge transfer ($\rho < 1$) the ground state of such systems corresponds to the Wigner crystal. The total energy of the ground state of these kind is essentially lower (by a few electron-volt per a pair DA) than that for the metallic state (uniform charge distribution along the stacks ... DDD ... and ... AAA ...). The dynamics of the charged defects in such a Wigner crystal is considered. It is shown that at very small ρ the electron-phonon interaction must be taken into account and in this case the system of the charged defects consist of a number of nearly independent polarons (or solitons) which seems to be in the superconducting state.

Now as a result of new experimental studies we know that the lattice structure of quasi-one-dimensional high-conducting systems, such as TTF-TCNQ or KCP is more complicated than it seemed initially. Moreover this structure depends on the temperature, and the parameters of superlattice change with temperature altaration /1/. The simplest way to explain this distortion is to connect this phenomenon with Peierls instability of the lattice of separated chains, and, consequently, new size of lattice cell will be $\frac{2\pi\hbar}{2p_F}$, p_F being Fermi momentum of one-dimensional Fermi gas. But the presence of many lattices in the same crystal which have different sizes of cell makes this explanation doubtful. On the other hand Kohn anomaly in a spectrum of inelastic neutron scattering can appear independently of reasons causing of the lattice distortion and even without actual distortion if there is certain back-ground for making corresponding vibration mode more soft. For instance the electron correlations do not cause real lattice distortion although it can be a reason for appearance of Kohn anomaly. Thus, we must find another ground for these distortions. The main purpose of our report is to call your attention to the existence of some unusual mechanism of similar distortions. Further we shall show that if ρ, the number of electrons per one acceptor or holes per one donor, is less than 1, such a system in a dielectric phase, will certainly have the superlattice. But its size will be rather $\frac{2\pi\hbar}{p_F} \sim \frac{1}{\rho}$ than $\frac{2\pi\hbar}{2p_F} \sim \frac{2}{\rho}$, the usual Peierls value.

At first I would like to consider value ρ more attentively /2,3/. Now it is clear that ρ cannot be 1 for high-conducting systems because, as it will be shown below, the value of interaction (or to be more exact repulsion) of two electrons situated in the same molecule is too large to allow this system to be metal at any temperature. The system undergoes a metal-insulator Mott transition.

According to our quantum chemistry calculations and calculations which were fulfilled by other authors the interaction energy of two electrons located on one TCNQ molecule is equal to about 4 ev. The accuracy of such a calculation in quantum chemistry is usually of order of 10%. The mistakes can arise during the calculation of polarisation energy of two electrons in a crystal. Below we shall discuss the method of estimation of this value.

Let me briefly describe the whole situation. The full Hamiltonian of such a system, say, TTF-TCNQ crystal, can be written as following

$$H = H_A + H_D + T_{AD} + H_M \quad , \tag{1}$$

where H_A is Hubburd-type Hamiltonian of separated acceptor chains, H_D is the same for donor chains, T_{AD} is an energy of quantum electron tunneling from A to D, and, finally, H_M is electrostatic energy of this crystal. We must determine the energy of full Hamiltonian with conditions

$$\frac{1}{N_0} \hat{N}_A = \frac{1}{N_0} \hat{N}_D = \rho \tag{2}$$

where $\hat{N}_A$ and $\hat{N}_D$ are, respectively, the operators of full number of electrons in A chains and holes in D-chains, N_0 - is a full number of A or D molecules in this crystal.

This Hamiltonian is too complecated to be fully investigated in general case. However, there is a hierarchy of the terms of this Hamiltonian. We can write down the system of inequalities of such a view

$$E_{ind} \gg E_{est} \gg E_{tun} \tag{3}$$

where E_{ind} is the energy of individual molecules TCNQ or TTF (or others). The first inequality means that molecules keep their electronic structure in crystal, i.e. the influence of crystal field is very small in comparison with inner molecular forces. The calculation of electronic structure (and geometrical one) usually can be done by one of the quantum chemistry methods, for instance, CNDO and all methods have given the same results (in any case with respect to the inner charge distribution/. E_{est} is electrostatic energy which will be further in question and E_{tun} is a tunneling energy which is of order of 0,1 ev. This latter influences magnetic properties of the system. So the degree of charge transfer from donors to acceptor, ρ depends mainly on the electrostatic term.

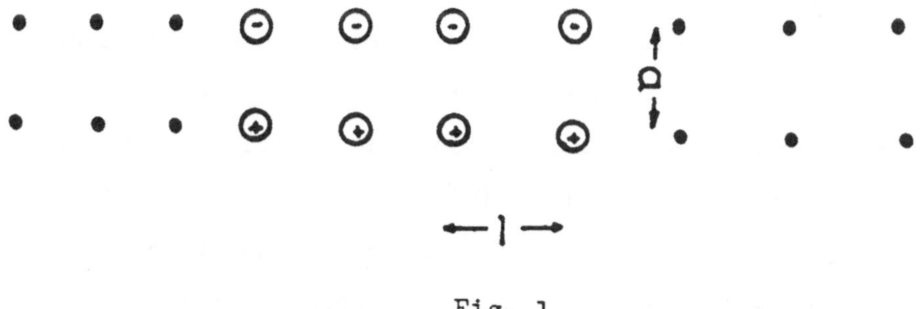

Fig. 1

The necessity of the determination of the optimal value of ρ is clearly seen from the following simple consideration. Let us consider two infinite and parallel chains of the point charges q and -q (Fig.1).

The electrostatic energy of this system per one neutral pair of molecules can be written as follows:

$$E_M = \frac{q^2}{a}\left[-1 + 2g(a,l)\right] \tag{4}$$

where

$$g(a,l) = \sum_{n=1}^{\infty}\left[\frac{a}{nl} - \frac{a}{\sqrt{n^2l^2+a^2}}\right] \tag{5}$$

From these expressions it follows that the sign of E_M can be arbitrary dependent on the value a/l, $E_M < 0$, when $l/a \gg 1$. Now we calculate the change of the total electrostatic energy of this system when the charge pair q and -q is destructed. Using this notation we obtain

$$\Delta E = q^2/a\left[1 - 4g(a,l)\right] \tag{6}$$

The value of $\Delta E < 0$ when $1 < 4g(a,l)$. Thus, the total energy of the infinite system of point charges can become lower when the number of charge pairs q and -q decreases even though $E_M < 0$. Concequently, for given values of lattice constants it is possible that the minimum of the total energy corresponds to $\rho < 1$.

Electrostatic energy of this crystal has been evaluated by means of Evald method. In these calculations each TTF and TCNQ molecule (or NMP and TCNQ, or Rb-TCNQ, etc) has been treated as the system of point charges. Now we turn to the results of these calculations.

In the case $\rho = 1$ the crystal consists only of the ions $TCNQ^{(-)}$ and $TTF^{(+)}$. In order to evaluate how the charge distribution in these ions affects the value of the Madelung energy E_M calculations with slightly different distributions were carried out and we saw that the Madelung energy does not essentially depend on this distribution, namely E_M (for TTF-TCNQ system) is of order of $2 \pm 0,2$ ev. The polarisation of the ions results in lowering E_M by about 10%. The small effects of the polarisation can be explained by the presence of the center of inversion in the space group of TCNQ-TTF crystal at $\rho = 1$. But we must remember that lowering of the total energy can be larger if $\rho \neq 1$.

When the charge transfer is not complete, i.e. $\rho < 1$, the crystals contain both ions and neutral molecules. In this case, generally speaking, for a given value of ρ we can place the ions and neutral molecules in the crystal lattice by a number of different ways. It is clear that the real situation corresponds to the configuration with the lowest total energy. Some distributions of ions and molecules used in present calculations are given in the next figures (Figs. 2 and 3).

In this picture (Fig.2) the arrangement of the ions and neutral molecules in the lattice of the TTF-TCNQ crystal is shown for $\rho = 1/2$ (a) and $\rho = 3/4$ (b) Δ, O - the centers of ions $TCNQ^{(-)}$ and $TTF^{(+)}$ and ▲, ● are the centers of the corresponding molecules. The next figure (Fig.3) shows us the arrangement of ions and molecules for $\rho = 1/2$ (a) and $\rho = 2/3$ (b).

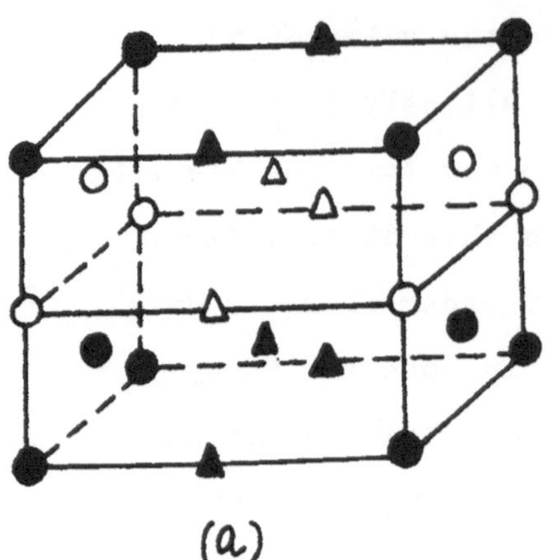

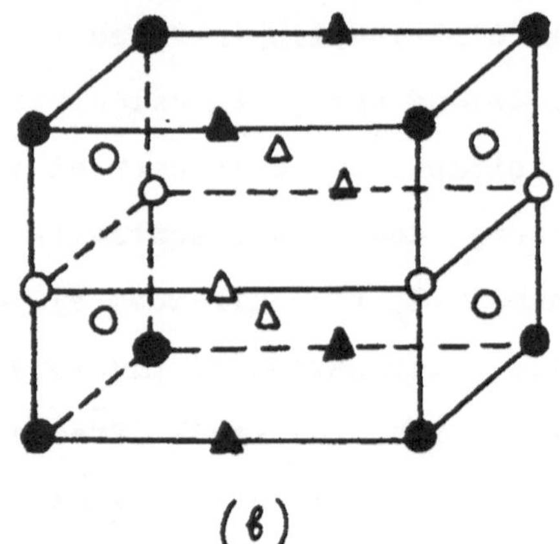

Fig. 2

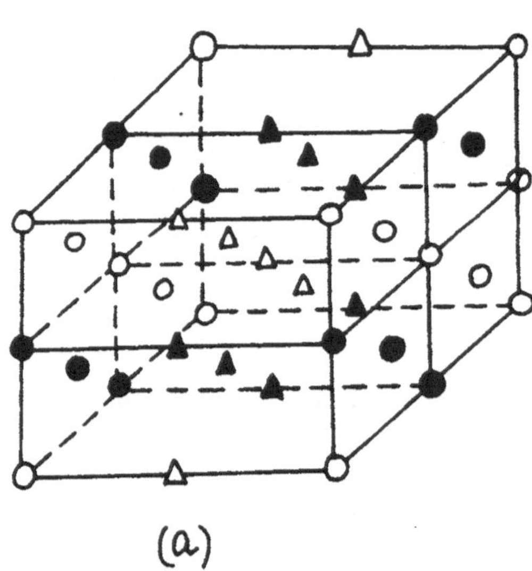

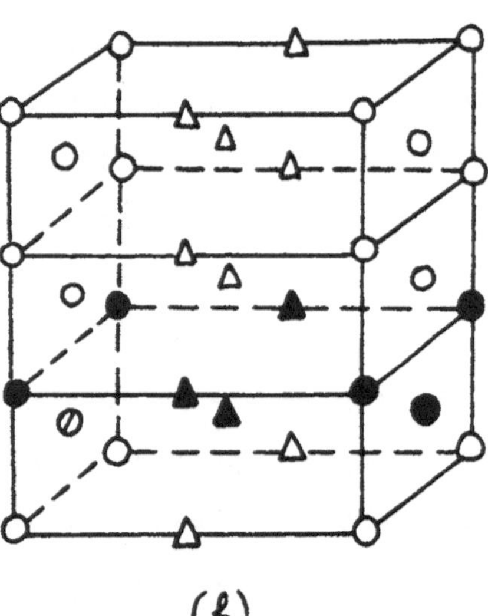

Fig. 3

These distribution have been chosen in such a way that the distances between the ions of the same signs are maximal for the given value of ρ . We note also that for ρ =1/3 (or 2/3) and 3/4 (or 1/4) the symmetry of crystalline field is disturbed, namely, the center of symmetry disappears . As a result, an additional polarisation energy appears.

On the next figure (Fig.4) the electrostatic energy of the TTF-TCNQ crystal per a pair of TCNQ-TTF molecules

$$E(\rho) = \rho \left[E_M(\rho) + I - A \right] \qquad (7)$$

is plotted versus ρ . Here I is an ionisation potential of TTF and A is an affinity of TCNQ.

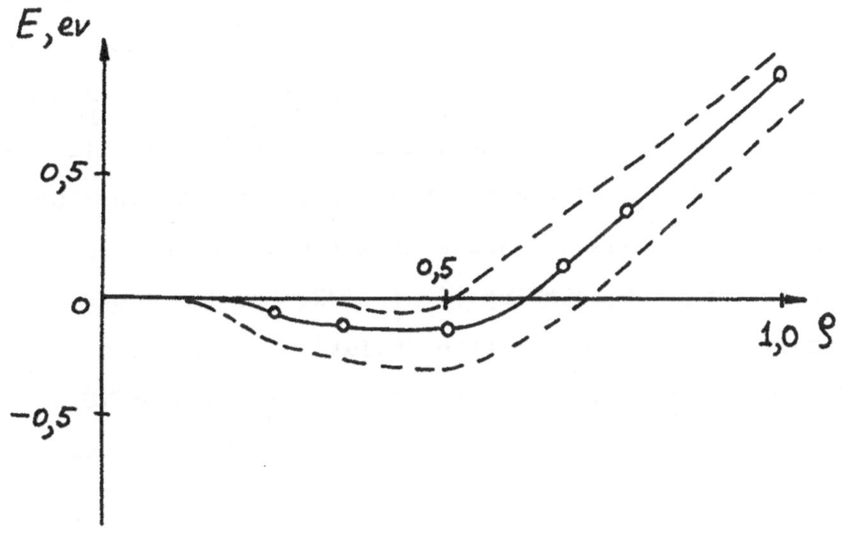

Fig. 4

Using the early experimental data on the ionisation potential of TTF and the electron affinity of TCNQ we assume that I-A ~ 3 ev. The solid curve is drawn through the points obtained by computer calculations of $E_M(\rho)$ for ρ = 1, 3/4, 2/3, 1/2, 1/3, 1/4, 0. The dotted lines indicate the upper limite of possible errors of the calculations. As is seen from this graphs , the energy minimum corresponds to the interval $1/4 < \rho < 3/4$.

The similar results were obtained for NMP-TCNQ and Rb-TCNQ crystals. But in case of NMP-TCNQ the energy minimum is shifted to the region of small ρ and in Rb-TCNQ case the minimum is achieved at $\rho = 1$. And the latter shows us that Rb-TCNQ is a semiconductor with a big electronic gap, that is in a full agreement with experimental data.

It's necessary to do some remarks about this curve. Here it should be noted that the function $E(\rho)$ in the electrostatic approximation (1) is not a function in the usual sense and its continuity is broken up at any rational point $\rho = q/p$ (where p and q are mutually simple numbers). This discontinuity appears by the following reason. At zero temperature the system of electrons and holes is arranged in the Wigner crystal with a cell consisting of p TCNQ (or TTF) molecules. Let us study $E_M(\rho)$ with a little deviation of ρ from q/p i.e. $\rho = q/p + \varepsilon$, where $\varepsilon \ll 1$. Assume that $\varepsilon > 0$. It means that we should introduce into the initial Wigner lattice $N\varepsilon$ charged defects - additional pairs of D^+A^- (N is the number of molecules of the D or A types in the crystal). The energy of production of one charged defect (at small ε defects do not interact) is δ_+. Hence, at small $\varepsilon > 0$ $E(\rho) = E(q/p) + \varepsilon\delta_+$. At $\varepsilon < 0$ we must introduce into the initial lattice the $N\varepsilon$ neutral defects, i.e. we must discharge $N|\varepsilon|$ pairs of D^+A^-. If the energy of production of one neutral defect is δ_- , then in the neighbourhood $\rho = q/p$ $E(\rho) = E(q/p) + \varepsilon\,\delta_-$. Generally speaking, $\delta_+ \neq \delta_-$ since these are energies for production of different defects. Thus, the continuity of $E(\rho)$ is broken up at this point. Further,

$$\lim_{\rho \to q/p+0} \frac{\partial E}{\partial \rho} = \delta_+ \qquad \text{and} \qquad \lim_{\rho \to q/p-0} \frac{\partial E}{\partial \rho} = \delta_- ,$$

i.e. the corresponding derivatives of the expression (1) by ρ are connected with the energies of both charged and neutral defects in the initial Wigner lattice and, consequently, with the gap in the spectrum of the Fermi excitations. At temperatures different from zero and in the presence of small resonance integral between molecules β as well these discontinuities are smoothed, but the gap in the spectrum of the Fermi excitations may remain up to certain critical value of β .

Another remark is connected with the method of the estimation of polarisation energy. Our curves were drawn in the assumption that $I-A \sim 3$ ev, but last experimental data shows that I-A is more close to 4 ev. In this case the curve would not have a minimum and ρ_{min} would be zero. But it turns out that if we take into account the polarisation energy the minimum arises again. The calculation made by Ukrainsky and Mironow has been carried out according to the following scheme. The first step is the calculation of the ground state energy and charge distribution of isolated neutral molecules and ions by CNDO method. The second one is the calculation of Madelung sum of this crystal at experimental location of molecules. The next one consists in evaluating the electric field potential on the molecules and ions. After determining of that the energy and charge distribution of isolated molecules and ions has been calculated with taking into account above mentioned electric field. An so on some times untill the self-consistant energy and charges were obtained. It has turn out the curve has been nearly the same if I-A ~ 4 ev.

In connection with this calculation very important quwstion arises. Does this method give full polarisation energy or not? The correct answer is yes, if the temperature of the system is zero and we neglect quantum resonance tunneling transfer of electrons from A to D, i.e. the terms of the type

$$\beta\left(\sum_{n,6} a^{+}_{n6}\, a_{n+16} + c.c\right) \qquad (8)$$

in full Hamiltonian, and "no" otherwise, because by using such a method we cannot take into account possible displacement of charges from molecules (or ions) to another ones though we took completely into account the displacmmnt of charges within separated molecules (ions). The last effect would be very essential if the energy gap for a creation of charge defects in this Wigner crystal is small at $T > 0$ and $\beta \neq 0$. The influence of this effect on the polarisation energy cannot be estimated now for the real crystal but there is an exact model which shows us that, when $T=0$ and β is close to the creation energy of charged defect, a phase trasition "dielectric-metal" or "Wigner crystal-metal" is possible. I mean the model of the lattice spinless Fermi gas having the Hamiltonian

$$H = \beta \sum_{n=1}^{N} \left(a_n^+ a_{n+1} + c.c. \right) + \gamma \sum_{n=1}^{N} a_n^+ a_n a_{n+1}^+ a_{n+1} \tag{9}$$

At $\gamma > 2|\beta|$ Wigner crystal is formed in the system and if $\gamma \leq 2|\beta|$ the system is a metal /4/. Now I believe that the energy parameter of real system , such as TCNQ-TTF and other high conducting donor-acceptor crystal are close to the critical ones. So the energy gap for charged excitation of this system is very small and, consequently, the transition temperature to the metallic state is small too. Terefore we can speak about this transition point as a melting point of Wigner crystal.

Another consequence of Wigner crystal creation in dielectric phase is the appearance of lattice distortion because when ρ equals, say, 2/3, the electric field in three neighbour cells of lattice is different from each other. So in this case the period of lattice m will be 3. Generally speaking, at $\rho = q/p$ we have m=p. In accordance with the experimental data for TTF-TCNQ system the constant of superlattice in direction b is very large, may be 17 ($\rho = 2/17$). In this case there are different ways of charge location which have approximately the same energy and they can appear at different temperature. So it is one of the possible explanations of a number of phase transition points in TTF-TCNQ system.

The consideration of magnetic properties of these crystals /5/ allows us to obtain more accurate ρ .

The method of such a consideration is the following. For given ρ and a corresponding Wigner crystal we have determined the spin Hamiltonian of the system by using the second or the fourth order of perturbation theory with respect to β, quantum tunneling energy. The value of exchange integral in this Hamiltonian is of order of β^2/Δ, where Δ is an energy of defect which is connected with the displacement of a charge from its equilibrium position to neighbour molecule or ion. This Hamiltonian has been investigated for systems $TCNQ_2$-TTF and NMP-TCNQ by using spin wave method. The result is pictured on the next figure (Fig.5) at $\rho = 2/3$.

We can see that in the region of small temperature both theoretical and experimental paramagnetic susceptibility tends to zero that it gives us the way to determine the value of exchange integral. But in region of large temperature (large in comparison with β^2/Δ) there is essential difference between theoretical and experimental curves. It can be understood if we assume that β is of order of Δ and the energy gap for creation of defect is small, so that perturbation theory is not correct. Similar investigation of NMP-TCNQ crystal gives us most probable value for ρ, $\rho = 1/8$.

In connection with this calculation the question has arisen: Is this calculation correct when β is of order of Δ ? We answer "yes" because in any case the scale of Coulomb energy and energy connected with a quantum transportation is very different. And if we neglect the latter we catch rough

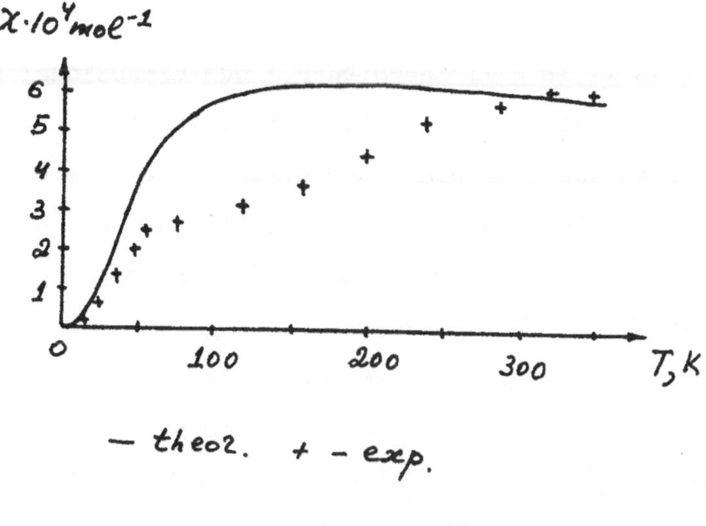

Fig. 5

features of this phenomenon. Thus, one of important features of such a system is the distortion of the lattice at low temperature.

It is also interesting to consider the situation when $\beta \sim \Delta$. In this case the previous calculations gives us only the region of most probable ρ but to make it more accurate we must conclude in consideration the term with β . When $\beta \gg \Delta$ we have metallic state of the system. As a result the reasons for the lattice distortion disappear and the electron-phonon interaction becomes to be very important because the gas of charged defects looks like weakly interacting gas of spinless Fermi particles.

In this connection let us consider the Fröhlich-type Hamiltonian. As it is well known suah a one-dimensional gas has no stability with respect to the transition to the Peierls dielectric or to the superconductor state. If we use the self-consistent method the gap in the electronic spectrum of Fermi excitations will be of order of $\varepsilon_F e^{-1/g^2}$ for Peierls distorted phase and $\hbar\omega e^{-1/g^2}$ for superconducting one. where g^2 being the constant of electron-phonon interaction. Since the correction in the ground state energy is proportional to the square of this gap one may conclude that the Peierls phase is energetically more profitable than the superconducting one if $\varepsilon_F > \hbar\omega$ and vica versa. Though, there exists the opinion that Peierls distorted system is also the superconducting one [2]. We shall show that this conclusion is not true for the lattice Fermi gas in the adiabatic limit. In particular early Fröhlich's /6/ and Bardeen's et ae /7/ works were fulfilled just in that approximation.

The lattice electron-phonon Hamiltonian has such a view

$$H = \sum_{n,6} \left(\beta_n a^+_{n6} a_{n+1 6} + c.c.\right) + \frac{\varkappa}{2}\sum_n \xi_n^2 + \frac{M}{2}\sum_n \dot{u}_n^2 , \qquad (10)$$

$$\xi_n = u_{n+1} - u_n ,$$

u_n being a displacement of nth center, M is its mass and $\beta_n = \beta + \beta' \xi_n$.

In adiabatic approximation, i.e. $\omega_q = 0$ or mass M of the molecule is infinite we can find the eigenvalue and eigenfunction of the system exactly. Indeed to find the ground state a energy we must minimize the full energy with respect to all displacements u_n. It is easy to see that

$$\xi_n = -\frac{\beta'}{\varkappa}\sum_{\sigma}\langle a^{+}_{n\sigma}a_{n+1\sigma} + c.c.\rangle, \tag{11}$$

where the average must be calculated by using the eigenfunction of the ground state. Substituting this expression for ξ_n to the full Hamiltonian we obtain the Hartree-like Hamiltonian, which can be exactly soluted if we accept that

$$\xi_n = \{\ldots\ldots \xi_1, \xi_2, \ldots, \xi_p, \xi_1, \xi_2, \ldots, \xi_p, \ldots\} \tag{12}$$

where $\rho = 1/p$, a rational number. There are just p self-consisting solutions for values ξ_n which can be obtained by shifting the all ξ_n; $\xi_n \to \xi_{n+1} \to \xi_{n+2} \ldots$, etc. The full wave function Ψ is the linear combination such a view

$$\Psi = \sum_{K} e^{\frac{i2\pi K}{p}} \Psi_K \tag{13}$$

The Ψ has the true properties with respect to the translation group. But the average of the current operator $\hat{J} = i\sum_{n,\sigma}(a^{+}_{n\sigma}a_{n+1\sigma} - c.c.)$ is equal to zero for an infinite system. Thus in adiabatic limit this system cannot be a super-conductor. It is interesting to clear the properties of this system at very large p, or small ρ. When $1/\rho \gg 1/g^2$ ($g^2 = \frac{\beta'^2}{\beta\varkappa}$, the constant of electron-phonon interaction), the Fermi gas consists of the lattice of polarons, each of them can be determined by the solution of the problem of one electron interacting with the lattice. In case of the continuous chain (or the string) the corresponding to equation Schrödinger will reduce to the non-linear one

$$i\frac{\partial\psi}{\partial t}=-\frac{\hbar^2}{2m}\frac{\partial^2\psi}{\partial x^2}-g^2|\psi|^2\psi \tag{14}$$

for one electron and to the system of the non-linear Schrödinger equations for many electron

$$i\frac{\partial\psi_n}{\partial t}=-\frac{\hbar^2}{2m}\frac{\partial^2\psi_n}{\partial x^2}-g^2\Big(\sum_m|\psi_m(x)|^2\Big)\psi_n(x)\,. \tag{15}$$

The solution of this system can be obtained exactly but for us it is enough to investigate it in the limit of large and small g^2. In the first case the electrons form the lattice with a period $1/\rho\, a_0$ (a_0 is a cell size) and each electron is surrounded by a static phonon cloud. Such a particle may be called a soliton or a polaron.

The wave function has a view

$$\psi_n(x)=\frac{mg^2}{2\,\mathrm{ch}\, mg^2\left(x-\frac{na_0}{\rho}\right)} \tag{16}$$

If the whole system moves with the velocity V the value g^2 must be renormalized as follows

$$g^2\to g^2\frac{1}{1-V^2/c^2} \tag{17}$$

where c being the sound velocity. The energy of repulsion of two solitons can be estimated by evaluating of the overlap of two solitons situated at the distance R from each other

$$V(R) \approx \int_{-\infty}^{\infty} \psi_R(x)\,\psi_0(x)\,dx \cong R\,e^{-mg^2 R} \quad (18)$$

On the other hand if this energy for $R \sim \frac{1}{p} a_0$ is too small in comparison with non-adiabatic energy of solitons their lattice will be destroyed. To obtain the corresponding criterion we evaluate the width, $4\beta_H$, of the polaron zone. Accordingly to Holstein / 9 / we have

$$\beta_H = \beta e^{-S}, \quad S = \frac{\pi \beta'^2 \sqrt{M}}{2 \varkappa^{3/2}} \quad (19)$$

Thus, in the first order in the non-adiabatic energy the system of polarons (solitons) may be described by the Hamiltonian

$$H_{pol} = \beta_H \sum_{n=1}^{N} \left(b_n^+ b_{n+1} + b_{n+1}^+ b_n \right) + \sum_{n', n=1}^{N} V(R_{n,n'})\, b_n^+ b_n b_{n'}^+ b_{n'} \quad (20)$$

where b_n^+ and b_n are the Pauli operators of polarons. In this case the critarion for the transition of polaron solid system to the liquid one is

$$V\left(\frac{a_0}{p}\right) < 2\beta_H$$

In the liquid phase this system seems to be a superconductors because its correlation function

$$\langle b_R^+ b_0 \rangle \cong G(R) \quad (21)$$

falls off with distant R in according to a power law and the integral $\int G(R)\,dR$ is divergent /8/. On the other hand to judge about it with the assurance we must consider the influence of impurities.

References

1. R.Comes, G.Shirane, S.M.Shapiro, A.F.Garito, A.J.Heeger, Preprint, BNL-21110; 1976.
2. V.E.Klimenko, V.Ya.Krivnov, A.A.Ovchinnikov, I.I.Ukrainskii, A.F.Shvets. Zh. Exp. Theor. Phys. 69, 240 (1975).
3. I.I.Ukrainskii, V.E.Klymenko, A.A.Ovchinnikov, Preprint, ITP-75-89E, 1975, Phys. Stat. Sol. (in press), 1976.
4. A.A.Ovchinnikov, Zh. Exp. Theor. Phys., 64, 342 (1973).
5. V.Ya.Krivnov, A.A.Ovchinnikov. Zh. Exp. Theor. Phys. (Letters), 21, 696 (1975).
6. H.Fröhlich, Phys. Rev. 79, 845 (1950) Proc. Roy. Soc., A223, 296 (1954).
7. J.Bardeen, D.Allender, J.W.Bray, Phys. Rev. B9, 119/1974/.
8. K.B.Yefetov, A.I.Larkin, Zh. Exp. Theor. Phys. 69, 764 (1975).
9. T.Holstein, Ann. Phys. 8, 343, 1959.

Conference on Organic Conductors and Semiconductors, Siófok, Hungary 1976

QUASI-ONE-DIMENSIONAL SYSTEMS*

D.J. SCALAPINO

Department of Physics, University of California,
Santa Barbara, California 93106, USA

The free energy of a Ginzburg-Landau field describing a system of weakly coupled chains in a plane is identified with the ground-state energy of a linear array of quantum anharmonic oscillators. The equivalent Hamiltonian is simplified for both the real and complex fields using a truncated basis of states. Results for both the real and complex fields will be discussed. In addition, the behavior of the specific heat and inverse correlation length for finite numbers of weakly coupled chains will be discussed.

*Work supported by the National Science Foundation.

Conference on Organic Conductors and Semiconductors, Siófok, Hungary 1976

THE ROLE OF INTERCHAIN COUPLING IN LINEAR CONDUCTORS

T.M. RICE and P.A. LEE

Bell Laboratories Murray Hill, New Jersey 07974

and

R.A. KLEMM*

Bell Laboratories, Murray Hill, New Jersey 07974 and Stanford University, Stanford, California 94305, USA

ABSTRACT

Renormalization group methods are applied to a set of coupled linear chains in a model in which direct hopping between chains is excluded. Attention is focused on models in which intrachain interactions are repulsive. It is shown first for two chains and then for N chains that a repulsive δ-function intrachain interaction scales with decreasing temperature to the attractive fixed point. In the N-chain model this attractive fixed point is characterized by a charge density wave whose phase is coherent between chains. A variety of response functions, spin and charge density waves at wave vector $2k_F$ (k_F is the Fermi wave vector), the uniform magnetic susceptibility, excitonic response functions, and $4k_F$ response functions are calculated numerically. A comparison is made to the behavior of TTF-TCNQ (tetrathiafulvalene-tetracyanoquinodimethane) at temperatures above 60 K.

* Work at Stanford University supported by US Army Research Council

I. INTRODUCTION

There is now general agreement on the theory of a single metallic chain.[1-6] In practice any material is composed of an array of chains with some interchain coupling between them. If the backward scattering matrix element of the intrachain interaction is attractive, a single chain scales to the strong coupling limit and the charge density wave and singlet superconductor response functions have power law divergences at low temperatures. If the interchain interaction is weak then it is very reasonable to treat the interchain interaction within the mean-field approximation.[7] However, for repulsive interactions a single chain scales to a weak coupling limit. In this case a method which treats interchain and intrachain interactions on the same footing is desirable. Such methods have been introduced for this problem by Gor'kov and Dzyaloshinskii[8] and by Mihály and Sólyom.[9]

The case of repulsive interactions is especially interesting since there is now increasing evidence that the suggestion that strong intrachain Coulomb interaction predominates in TTF-TCNQ, made by Torrance and coworkers[10] and by Ovchinnikov and coworkers[11] several years ago, is correct. For this reason we have explored the problem of coupled chains with repulsive interactions and details will be published elsewhere.[12]

II. RENORMALIZATION GROUP METHOD

We will employ the renormalization group method. In this approach one assumes that a given Hamiltonian with values of the energy cutoff E_F and coupling constants g can be transformed into a Hamiltonian with a new cutoff E_F' and coupling constants g'. The differential forms of these scaling relationshipe for the coupling constants can be calculated by a perturbation expansion. The lowest order form of these equations was obtained by Gor'kov and Dzyaloshinskii[8]

$$\frac{\partial g_{1ij}}{\partial \xi} = \frac{2}{2\pi v} \sum_k g_{1ik} g_{1kj} + \frac{1}{2\pi v} g_{1ij}(g_{2ij} - g_{2ii}); \frac{\partial g_{2ij}}{\partial \xi} = \frac{1}{2\pi v} g_{2ij}^2 \quad (1)$$

where g_{1ij}, g_{2ij} are the backward and forward scattering matrix elements between electrons on chain i and j. Note no actual transfer of particles between chains is included. The Fermi velocity is denoted by v and $\xi = \ln(E_F'/E_F)$.

The fixed point behavior of Eq. (1) for a single chain has been examined by Menyhárd and Sólyom.[3] If $g_{1ii} > 0$ then as $\xi \to -\infty$, or equivalently the temperature $T \to 0$, then g_{1ii} scales to zero. As this is a weak coupling fixed point the perturbation expansion on the rhs of Eq. (1) remains valid. By contrast if $g_{1ii} < 0$ then $g_{1ii} \to -\infty$ at a finite value of ξ. Clearly the perturbation expansion breaks down near this strong coupling fixed point. The boundary between the domains of these two fixed points is determined simply by the sign of g_{1ii}.

III. TWO CHAINS

Consider first the two chain problem and for simplicity let us make both chains equivalent. Then four coupled equations (1) can be reduced to three by substituting the variables $U = g_{1ii}/2\pi v$, $W = g_{1ij}/2\pi v$, $V = (g_{1ii}-2g_{2ii}+2g_{2ij})/2\pi v$ leading to

$$\frac{\partial U}{\partial \xi} = 2U^2+2W^2$$

$$\frac{\partial V}{\partial \xi} = 4W^2$$

$$\frac{\partial W}{\partial \xi} = W(3U+V) \qquad (2)$$

The fourth combination $\Lambda = g_{1ii}-2g_{2ii}-2g_{2ij}$ remains invariant. Equation (2) can be further simplified by dividing out the ξ variable

$$\frac{dU}{dV} = \frac{U^2+W^2}{2W^2}$$

$$\frac{dW}{dV} = \frac{3U+V}{4W} \qquad (3)$$

This pair of homogeneous equations can be further simplified by introducing the variables $x(= U/V)$ and $y(= W/V)$ to yield

$$\frac{dx}{dy} = \frac{2x^2+2y^2-4xy^2}{3xy+y-4y^3} \quad . \qquad (4)$$

The two fixed points of this equation are $x=y=0$ and $x = |y| =1$. The first of these corresponds to $g_1 \to 0$ and is the generalization of weak coupling fixed point for repulsive interactions in a single chain. The second is a strong coupling fixed point similar to that which occurs in the case of attractive interactions in a single chain. Although the properties of fixed points which occur are simply generalizations of the single chain, the criterion which determines the boundary between the domain of each fixed point -- this boundary is also known as the separatrix -- is different. This can be seen by plotting at each point the value of the rhs of Eq. (4). The curve $x=y^2$ is a special solution of Eq. (4). Alternatively one can recombine Eq. (2) to obtain

$$\frac{d}{d\xi}(W^2-UV) = 2U(W^2-UV) \tag{5}$$

This quantity, by Eq. (5), cannot change sign in the scaling process and thus forms the separatrix. The necessary conditions to scale to the weak coupling fixed point are $UV > W^2$ and $U > 0$ and $V > 0$. When expressed in terms of the original coupling constants it is trivial to show that for δ-function interactions on the chains and between the chains one always scales to the strong coupling fixed point irrespective of the sign of either interaction. Thus one can start in the two chain problem with a purely repulsive behavior and scale to the attractive regime. At first sight this is a surprising result. However, Sólyom[3] showed that in a single chain if $2g_{2ii} > g_{1ii} > 0$ the charge density wave (CDW) response is divergent as $\xi \to -\infty$. In the two chain problem the interchain coupling through the Hartree term can then drive the system to the strong coupling fixed point.

IV. N-CHAIN

In an array of N-chains a similar behavior is found to occur. In fact a simple approximation which describes well the region of repulsive interactions is to restrict the range of the interchain coupling to nearest neighbor interactions only. A similar analysis to that outlined above for the two-chain case can be carried out. However, more generally Eq. (1) can be integrated numerically. It is convenient to Fourier transform Eq. (1) in the chain index and then numerically integrate the resulting pair of integro-differential equations. In this way the spatial and temperature dependence of the effective coupling constants can be obtained. An examination of the numerical solutions shows that if we start with repulsive nearest neighbor interactions then as ξ decreases g_1 decreases and the interaction remains mostly nearest neighbor until quite near the fixed point. There is one significant difference between the two and N-chain problems. In the former case the interaction became truly attractive but in the latter case $g_1(\vec{q})$ diverges only for a specific value of $\vec{q}$ at the fixed point. The spatial representation of the coupling constant remains finite and within a single chain one finds g_{2ii} may remain positive whereas g_{1ii} usually becomes small and negative.

Before applying this model to TTF-TCNQ one needs first to generalize it to different types of chain and then to estimate appropriate starting values for the coupling constants. In TTF-TCNQ alternate chains are composed of electrons (TCNQ) and holes (TTF). This affords no essential complication since one can transform to a particle representation which treats the holes as particles. Under this transformation the coupling constant between chains changes sign. As an example the Coulomb interaction will be attractive between electrons and holes. There is a large amount of arbitrariness in the choice of the starting values of the coupling constants. The backward scattering matrix element between particles on neighboring chains is, however, expected to be small, $\exp(-2k_F d)$ in dimensionless units, where d is the interchain separation. However, the other coupling constants are expected to be ~ 1 in dimensionless units. We shall discuss below numerical results using parameters $g_{1ii} = g_{2ii} = 0.6\ (2\pi v)$; $g_{1i,i+1}\ (= w_1) = -10^{-2}\ (2\pi v)$ and $g_{2i,i+1}\ (= w_2) = -0.5\ (2\pi v)$. Using these values the set of Eq. (1) has been integrated numerically and the temperature dependence of the effective coupling constants obtained. With these starting values the system scales to the strong coupling fixed point at $\xi \approx -4$.

From a knowledge of the temperature dependent effective coupling constants, the temperature dependence of a variety of response functions can be obtained. As an example consider the $2k_F$ charge and spin density waves (CDW, $N(k,\vec{q},\omega)$ and SDW, $\chi(k,\vec{q},\omega)$), where k is measured from $2k_F$ and $\vec{q}$ is the wave vector perpendicular to the chains. As suggested first by Zawadowski[3] it is the derivatives

$$\overline{N} = \pi v\ \partial N/\partial\xi;\ \ X = 2\pi v\ \partial X/\partial\xi \tag{6}$$

which obey scaling relationships. The lowest order perturbation expansion for these response functions leads to the equation

$$\frac{\partial \ln \overline{N}_{ij}}{\partial\xi} = \frac{4}{2\pi v} g_{1ij} - \frac{2}{2\pi v} g_{2ii}\delta_{ij}$$

$$\frac{\partial \ln \overline{X}_{ij}}{\partial\xi} = \frac{-2}{2\pi v} g_{2ii}\delta_{ij} \tag{7}$$

At the fixed point only the CDW response function diverges corresponding to a transition to a phase with CDW on each chain which are ordered with respect to each other. Other response functions remain basically one dimensional. In Fig. 1 the numerical results are plotted for N and X. Also shows there are excitonic response functions corresponding to the coupling of electrons on the TCNQ chain and holes on the TTF chain. Note these excitonic response functions are difficult to observe since the coupling constant to an external probe involves the direct overlap of TCNQ and TTF wave function -- a very small quantity.

Recently it has been observed experimentally that the CDW response at $4k_F$ persists to much higher temperature than that at $2k_F$.[13] The $4k_F$ CDW response function can be treated also by these techniques. For a single chain we find the result

$$\pi(4k_F+k,\omega,T) \sim \left(\frac{\max(vk,\omega,T)}{E_F}\right)^{2+4(g_1-2g_2)/2\pi v} \qquad (8)$$

This result agrees with that obtained previously by Emery[14] using a different technique to lowest order in g_1 and g_2. Note the $4k_F$ response does not diverge in the weak coupling limit, rather a substantial bare Coulomb coupling is required to cause a divergence. Interchain coupling has little or no effect on $\pi(4k_F)$ except when $N(2k_F)$ becomes strongly divergent and a $4k_F$ response can appear as a harmonic of a $2k_F$ CDW. At higher temperature the temperature dependence of Eq. (8) is quite different to that of a simple harmonic of $N(2k_F)$. In fact under the same conditions that equation is enhanced one finds that $\chi(2k_F)$ is also enhanced. This suggests that one should regard the $4k_F$ CDW response as a harmonic of the $2k_F$ SDW and as such it can be enhanced to temperatures higher than the $2k_F$ CDW is.

Repulsive Coulomb interactions on a chain will lead to an enhancement of the uniform Pauli susceptibility χ_p. As we have seen g_{1ii} weakens as temperature is lowered. This in turn leads to reduction in χ_p as temperature is lowered. The impurity scattering time, τ, that enters the resistivity will also have a temperature dependence, in contrast, to usual metals. One can show that

$$\tau^{-1} = 4cV_{imp}^2 \bar{N}(\vec{q}=0,T)/\pi^2 \tag{9}$$

where c is concentration and V_{imp} the potential of the impurities. With the starting parameters quoted above we find that τ^{-1}, and therefore the resistivity ρ, will have a minimum albeit a shallow one. The inclusion of electron-phonon scattering could sharpen this minimum but even so would not lead to a rapid enough temperature variation in $\rho(T)$ to explain the experiments on TTF-TCNQ.

In conclusion, a model with moderate to strong intrachain Coulomb interaction can qualitatively explain the experimental results on TTF-TCNQ including a phase transition to a CDW phase. However, it is not possible within the simple model described here to achieve quantitative agreement between theory and experiment.

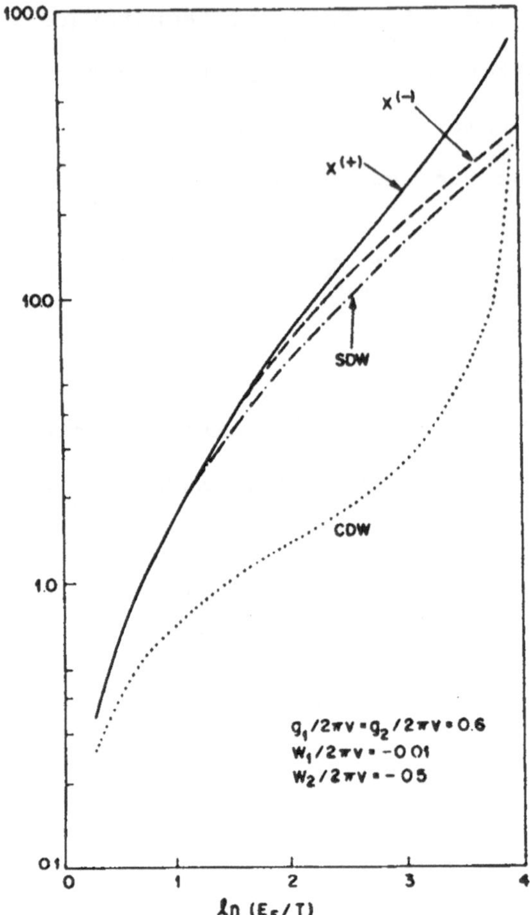

Fig. 1 The temperature dependence of response functions corresponding to charge and spin density waves of period $2k_F$ (CDW and SDW). The density of states factors of $(2\pi v)^{-1}$ for SDW and $(\pi v)^{-1}$ for CDW have been removed. The excitonic response functions ($X^{\pm}$) correspond to coupling of electrons and holes in different chains in even (+) and odd (-) combinations. [For more details see ref. 12.]

REFERENCES

1. Yu. A. Bychkov, L. P. Gor'kov and I. E. Dzyaloshinskii, Zh. Eksp. Teor. Fiz. 50, 738 (1966) [Sov. Phys.-JETP 23, 489 (1966)].
2. I. E. Dzyaloshinskii and A. I. Larkin, Zh. Eksp. Teor. Fiz. 61, 791 (1971) [Sov. Phys.-JETP 34, 422 (1972)].
3. N. Menyhárd and J. Sólyom, J. Low Temp. Phys. 12, 529 (1973); J. Sólyom, ibid. 12, 547 (1973).
4. H. Fukuyama, T. M. Rice, C. M. Varma and B. I. Halperin, Phys. Rev. B10, 3775 (1974).
5. A. Luther and V. J. Emery, Phys. Rev. Lett. 33, 589 (1974).
6. P. A. Lee, Phys. Rev. Lett. 34, 1247 (1975); S.-T. Chui and P. A. Lee, Phys. Rev. Lett. 35, 315 (1975).
7. D. J. Scalapino, Y. Imry and P. Pincus, Phys. Rev. B11, 2042 (1975).
8. L. P. Gor'kov and I. E. Dzyaloshinskii, Zh. Eksp. Teor. Fiz. 67, 397 (1974) [Sov. Phys.-JETP 40, 198 (1975)].
9. L. Mihály and J. Sólyom (preprint).
10. J. B. Torrance, B. A. Scott and F. B. Kaufman, Sol. State Comm. 17, 1369 (1975); J. B. Torrance and B. P. Silverman, Bull. Am. Phys. Soc. 20, 498 (1975) and preprint.
11. V. E. Klymenko, V. Ya. Krivnov, A. A. Ovchinnikov, L. Z. Ukrainsky and A. F. Shvets, to be published.
12. P. A. Lee, T. M. Rice and R. A. Klemm, Phys. Rev. (in press).
13. R. Comes (this volume).
14. V. J. Emery, Phys. Rev. Lett. 37, 107 (1976).

Conference on Organic Conductors and Semiconductors, Siófok, Hungary 1976

THREE – DIMENSIONAL ORDERING IN THE SYSTEM OF WEAKLY COUPLED CHAINS

L. MIHÁLY and J. SÓLYOM

Central Research Institute for Physics, Budapest, Hungary

The one-dimensional Fermi gas model has been studied by many authors /1-10/ with the aim to have a better understanding of the low temperature behaviour of quasi-one-dimensional systems, like charge transfer salts. Most of the quasi-one-dimensional systems show not only signs of large one-dimensional fluctuations, which this model can account for, but undergo a real phase transition which this model cannot describe. A finite transition temperature is obtained if a weak coupling between the linear chains is taken into account. In this paper we consider such a system consisting of a set of linear chains with strong intrachain and weak interchain electron-electron couplings. Similar models have been studied by Gorkov and Dzyaloshinsky /11/ and Klemm and Gutfreund /12/ using a different approximation scheme. A comparison with their results will be given at the end of this paper.

The model will be specified by starting from the strictly one-dimensional case. The most general Hamiltonian which has been used contains five different interaction terms with five coupling constants. These interactions are shown diagrammatically in Fig. 1
The solid (dashed) lines represent electrons with momentum near $+k_F$ ($-k_F$) and α is the spin index. The energy spectrum of the electrons is usually approximated by a linear dispersion relation with Fermi velocity v_F.

There are many ways to generalize this model to a set of weakly interacting chains. First of all we will restrict ourselves to the case when the propagation of the electrons is along the chains only and no interchain hopping is allowed. Next, we will neglect the exchange processes and assume that the electrons travelling along the chains can exchange momentum but not the spin. With these restrictions we should still consider five types of couplings. The interaction will

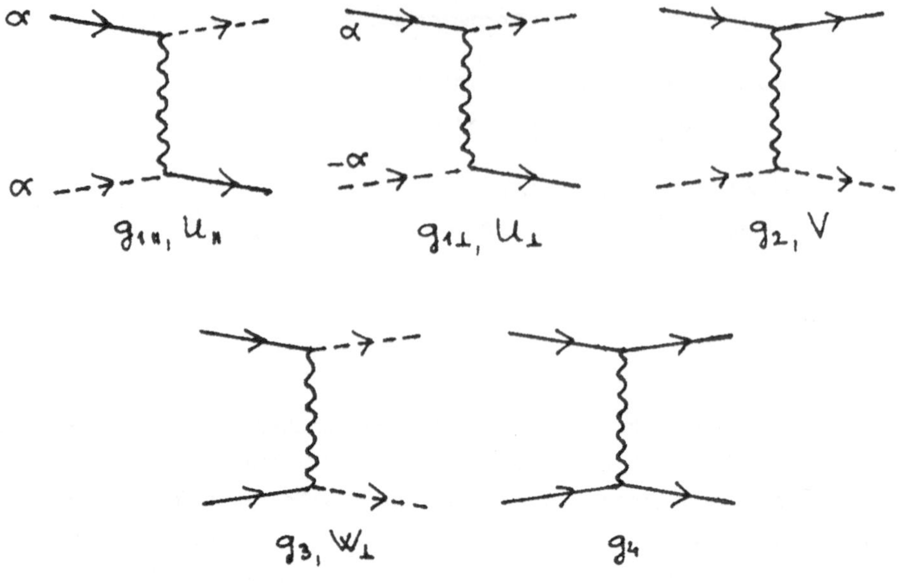

Fig. 1.

depend on the distance between the chains and we will not restrict ourselves to first neighbour interaction.

Further simplification can be achieved if the band is not half-filled, in which case the Umklapp processes (g_3) can be neglected. The term with g_4 can be incorporated into a Fermi velocity renormalization and it will not be considered any longer. Finally the backward scattering term (g_1) will be assumed to be spin independent, $g_{1\perp} = g_{1\parallel} = g_1$.

The model /13/ is thus characterized by the dispersion relation

$$\varepsilon_k = v_F \left(|k_\parallel| - k_F\right), \tag{1}$$

where $k_\parallel$ is the momentum component parallel to the chain direction, by the coupling constants g_{1ij}, g_{2ij} where i and j label the chains, and by a bandwidth cut-off ω_D.

The possibility of long range order can be best studied by examining the eventual divergences in the response functions. We have investigated three of them, the charge density, spin density and Cooper pair fluctuations. A divergence of them indicates the appearence of charge density waves, spin density waves or superconducting state. These temperature dependent response functions can be obtained by the analytic continuation to the upper ω halfplane of the correlation functions

$$N(k,q,\omega_\nu) = -\int_0^{1/T} d\tau\, e^{i\omega_\nu\tau} \sum_{i,j} e^{iq(R_i-R_j)}$$

$$\times \langle T_\tau \{ \sum_\alpha \int \frac{dp}{2\pi} c^+_{ip\alpha}(t)\, c_{ip+k\alpha}(t) \tag{2}$$

$$\times \sum_\beta \int \frac{dp'}{2\pi} c^+_{jp'\beta}(0)\, c_{jp'-k\beta}(0) \} \rangle ,$$

$$\chi(k,q,\omega_\nu) = -\int_0^{1/T} d\tau\, e^{i\omega_\nu\tau} \sum_{i,j} e^{iq(R_i-R_j)}$$

$$\times \langle T_\tau \{ \int \frac{dp}{2\pi} c^+_{ip\uparrow}(t)\, c_{ip+k\downarrow}(t) \tag{3}$$

$$\times \int \frac{dp'}{2\pi} c^+_{jp'\uparrow}(0)\, c_{jp'-k\downarrow}(0) \} \rangle ,$$

$$\Delta(\omega_\nu) = -\int_0^{1/T} d\tau\, e^{i\omega_\nu\tau} \sum_{i,j} \langle T_\tau \{ \int \frac{dp}{2\pi} c_{ip\uparrow}(t)\, c_{i-p\downarrow}(t) \tag{4}$$

$$\times \int \frac{dp'}{2\pi} c^+_{j-p'\uparrow}(0)\, c^+_{jp'\downarrow}(0) \} \rangle .$$

Here $\omega_\nu = 2\pi\nu T$, k denotes the momentum component parallel to the chains and q is the perpendicular component. The vector R_i gives the position of the ith chain. The singularity appears first for $k = 2k_F$ and $\omega = 0$ and only these particular values will be considered.

The renormalization group treatment is very suitable to study this system. It allows to take into account in a systematic way the leading, next to leading, etc. logarithmic corrections. The physical picture behind the renormalization group transformation is that the change in the energy scale and the simultaneous change in the coupling constants maps the original system onto a similar system in which the Green's functions, vertices and response functions have the same analytic structure, they only differ in a multiplicative factor.

These relations are usually formulated for the dimensionless Green's function $d(\omega)$ and the dimensionless vertices $\tilde{\Gamma}_{1ij}$, $\tilde{\Gamma}_{2ij}$, defined by

$$G(\omega) = G^\circ(\omega) \cdot d(\omega) , \tag{5}$$

$$\Gamma^{(ijlm)}_{\alpha\beta\gamma\delta} = g_{1ij}\,\tilde{\Gamma}_{1ij}\,\delta_{\alpha\gamma}\,\delta_{\beta\delta}\,\delta_{il}\,\delta_{jm} - g_{2ij}\,\tilde{\Gamma}_{2ij}\,\delta_{\alpha\delta}\,\delta_{\beta\gamma}\,\delta_{im}\,\delta_{jl}\,. \tag{6}$$

Scaling the cut-off ω_D to ω_D', the new scaled couplings are to be determined from the relations

$$d\left(\frac{x}{\omega_D'},\, g_{1ij}',\, g_{2ij}'\right) = Z\, d\left(\frac{x}{\omega_D},\, g_{1ij},\, g_{2ij}\right), \tag{7}$$

$$\tilde{\Gamma}_{1ij}\left(\frac{x}{\omega_D'},\, g_{1ij}',\, g_{2ij}'\right) = Z_{1ij}^{-1}\, \tilde{\Gamma}_{1ij}\left(\frac{x}{\omega_D},\, g_{1ij},\, g_{2ij}\right), \tag{8}$$

$$\tilde{\Gamma}_{2ij}\left(\frac{x}{\omega_D'},\, g_{1ij}',\, g_{2ij}'\right) = Z_{2ij}^{-1}\, \tilde{\Gamma}_{2ij}\left(\frac{x}{\omega_D},\, g_{1ij},\, g_{2ij}\right), \tag{9}$$

$$g_{1ij}' = g_{1ij}\, Z^{-2}\, Z_{1ij}\,, \tag{10}$$

$$g_{2ij}' = g_{2ij}\, Z^{-2}\, Z_{2ij}\,, \tag{11}$$

where x is any energy-like variable, such as ω, vk or T.

These scaling relations can be written in a differential form and we get for the dimensionless couplings

$$\gamma_{1ij} = \frac{g_{1ij}'}{2\pi v_F}\,, \qquad \gamma_{2ij} = \frac{g_{2ij}'}{2\pi v_F}\,, \tag{12}$$

the following equations.

$$\frac{\partial \gamma_{1ij}}{\partial \xi} = 2\Big\{ \sum_k \gamma_{1ik}\gamma_{1kj} + \gamma_{1ij}(\gamma_{2ij} - \gamma_{2ii}) + \sum_k \gamma_{1ij}(\gamma_{1ik}^2 + \gamma_{2ik}^2 - \gamma_{2ik}\gamma_{2kj}) + \gamma_{1ij}(\gamma_{2ij} - \gamma_{2ii})\gamma_{1ii} + \ldots \Big\}, \tag{13}$$

$$\frac{\partial \gamma_{2ij}}{\partial \xi} = \gamma_{1ij}^2 (1 + \gamma_{1ii}) + 2\sum_k \gamma_{1ik}^2 (\gamma_{2ij} - \gamma_{2kj}) + \ldots , \tag{14}$$

where $\xi = \ln(\omega_D'/\omega_D)$. When these effective couplings are used in calculating the temperature dependence of the response functions, $\xi = \ln(\omega_D'/\omega_D)$ is replaced by $\xi = \ln(T/\omega_D)$ and therefore in this sense we can speak about the temperature dependence of the effective couplings.

The scaling equations (13) and (14) can be solved near the fixed point in a self-consistent way with the assumption that γ_{2ij} becomes independent of $R_i - R_j$. Then, eq. (13) contains γ_{1ij} only and it can be solved. The solution can be best found in Fourier representation with

$$\gamma_1(q) = \sum_i \gamma_{1ij}\, e^{-iq(R_i - R_j)} , \tag{15}$$

and the scaling equation has the form

$$\frac{\partial \gamma_1(q,\xi)}{\partial \xi} = 2\Big\{ \gamma_1^2(q,\xi) + \gamma_1(q,\xi)\frac{1}{N}\sum_{q'} \gamma_1^2(q',\xi) + \ldots \Big\} \tag{16}$$

The solution of this equation has been analyzed for three different situation. The bare, unrenormalized coupling is exponentially decreasing with the distance between the chains, or it is an attractive or repulsive weak nearest neighbour coupling with a strong intrachain interaction.

In all the cases the renormalized coupling becomes less and less distance dependent and it has an infinite range at the critical temperature. For the case of exponentially decreasing attractive interaction between the chains, the renormalized coupling is also exponentially decreasing, the characteristic length is however, increasing with decreasing temperature and diverges at the critical temperature.

In the case when the bare coupling is a nearest neighbour one, the renormalized couplings are non-vanishing for next nearest neighbours, etc. Near the transition temperature all the chains are equally strongly coupled. The difference between the attractive and repulsive interactions is that in the first case all the chains are coupled attractively, while in the second case the coupling is alternatively repulsive or attractive. The situation is shown schematically in the next figure:

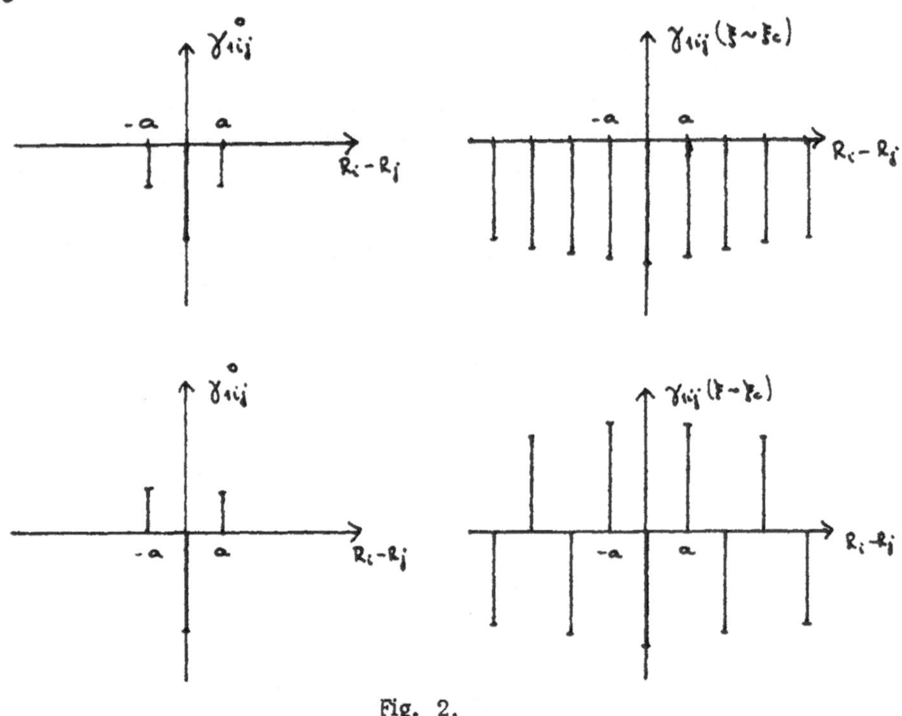

Fig. 2.

Knowing the behaviour of the renormalized couplings as a function of temperature, the response functions can be easily calculated. They obey a simple differential equation

$$\frac{\partial \ln \bar{N}(q,\xi)}{\partial \xi} = 4\,\gamma_1(q,\xi) + 2\,\frac{1}{N}\sum_{q'} \gamma_1^2(q',\xi) + \ldots , \tag{17}$$

$$\frac{\partial \ln \bar{\chi}(q,\xi)}{\partial \xi} = 2\frac{1}{N}\sum_{q'} \gamma_1^2(q',\xi) + \cdots , \tag{18}$$

$$\frac{\partial \ln \bar{\Delta}(\xi)}{\partial \xi} = 2\frac{1}{N}\sum_{q'} \gamma_1(q',\xi) + 2\frac{1}{N}\sum_{q'} \gamma_1^2(q',\xi) + \cdots , \tag{19}$$

where

$$\bar{N}(q,\xi) = \pi v_F \frac{\partial N(q,\xi)}{\partial \xi} , \tag{20}$$

$$\bar{\chi}(q,\xi) = 2\pi v_F \frac{\partial \chi(q,\xi)}{\partial \xi} , \tag{21}$$

$$\bar{\Delta}(\xi) = -\pi v_F \frac{\partial \Delta(\xi)}{\partial \xi} . \tag{22}$$

Inserting the solution for the effective coupling $\gamma_1(q, \xi)$ it turns out, that χ and Δ are non-singular and therefore the system cannot have a magnetic or superconducting state. The density-density response function is, however, divergent indicating the occurrence of a charge density wave state. The wave vector where $N(q, \xi)$ diverges is $q = 0$, $k = 2k_F$ where q is the momentum component perpendicular to the chains and k is the parallel component, if the bare coupling is an attractive exponentially decreasing or nearest neighbour coupling. In the case of repulsive nearest neighbour coupling the divergence appears at $q = Q$, $k = 2k_F$ where Q is at the Brillouin zone boundary. The $q = 0$ case corresponds to a situation in which the charge density waves are in the same phase for all the chains. If $q = Q$, the charge density waves on the neighbouring chains are in opposite phase.

It should be pointed out that a strong intrachain repulsion will modify the results. Above a certain strength no phase transition is obtained in the present approximation. The situation is similar for an exponentially decreasing repulsive interaction, too. The present model has no phase transition in the approximation considered here. This feature can be understood by recalling that the single chain problem has a superconducting type instability for strong repulsive intrachain backward scattering. The Coulomb-type coupling between the chains does not favor a three-dimensional superconducting ordering and the system has no phase transitions. A model with interchain hopping /11, 12/ can have a superconducting state.

Finally, we have to comment on the approximations used in this calculation. The first approximation was to neglect the forward scattering term g_2, the other was to consider the leading and next to leading logarithmic corrections only, i.e. a second-order renormalization was performed and the higher order corrections were neglected.

In studying the effect of g_2 we can compare our results with the calculations of Lee, Rice and Klemm /14, 15/, who have studied the two chain and N-chain problems, including the intrachain and interchain forward scattering. They conclude that the renormalized backward scattering can be changed from repulsive interaction to attractive one by the forward scattering terms and therefore the domain of attraction of the attractive fixed point is sensitive to g_2. This does not modify, however, the structure of the solution and the conclusion that only a charge density wave state can occur in this model. It is interesting to notice that the change of sign of the couplings, which is not allowed in the single chain problem, can take place without g_2 as well. Similarly to a large and positive $\gamma_2(q)$, a large attractive nearest neighbour coupling can renormalize a repulsive intrachain backward scattering to the attractive fixed point.

Gorkov and Dzyaloshinsky /11/ have used the parquet approximation in their study of quasi-one-dimensional systems. Our calculation goes one step forward and we have taken into account next to leading logarithmic corrections as well. From the comparison the importance of these corrections can therefore be inferred. The corrections are not negligible and this indicates that the next corrections, which have been neglected in this calculation, are not small either. Therefore the numerical values obtained for the renormalized coupling or for the response functions are not to be taken literally. Since, however, both approximations lead to the same physical results, namely they indicate the appearance of a charge density wave state, this statement is probably valid in a more rigorous treatment, too. The results of Klemm and Gutfreund /12/, who treated the interchain coupling in mean field approximation, also support this conclusion.

REFERENCES

/1/ Ju. A. Bychkov, L. P. Gorkov and I. E. Dzyaloshinsky, Zh. Eksperim. i Teor. Fiz., 50, 738 (1966) (English transl. Soviet Phys. - JETP 23, 489 (1966))

/2/ I. E. Dzyaloshinsky and A. I. Larkin, Zh. Eksperim. i Teor. Fiz. 61, 791 (1971) (English transl. Soviet Phys. - JETP 34, 422 (1972))

/3/ N. Menyhárd and J. Sólyom, J. Low Temp. Phys., 12, 529 (1973)

/4/ J. Sólyom, J. Low Temp. Phys., 12, 547 (1973)

/5/ A. Luther and I. Peschel, Phys. Rev. B9, 2911 (1974)

/6/ A. Luther and V. J. Emery, Phys. Rev. Letters 33, 589 (1974)

/7/H. Fukuyama, T. M Rice, C. M. Varma and B. I. Halperin, Phys. Rev. B10, 3775 (1974)

/8/ P. A Lee, Phys. Rev. Letters, 34, 1247 (1975)

/9/ H. U. Everts and H. Schulz, Z. Physik B22, 285 (1975)

/10/M. Kimura, Progr. Theoret. Phys. 53, 955 (1975)

/11/L. P. Gorkov and I. E. Dzyaloshinsky, Zh. Eksperim. i. Teor. Fiz. 67, 397 (1974)

/12/R. A. Klemm and H. Gutfreund, Phys. Rev. B14, 1086 (1976)

/13/L. Mihály and J. Sólyom, J. Low Temp. Phys., 24, 579 (1976)

/14/ A. Lee, T. M. Rice and R. A. Klemm, paper published in this volume

/15/R. A. Klemm, P. A. Lee and T. M. Rice, paper published in this volume

Conference on Organic Conductors and Semiconductors, Siófok, Hungary 1976

ORDER IN COUPLED METALLIC CHAINS

R.A. KLEMM[†]

Stanford University, Stanford, California 94305, USA

and

H. GUTFREUND

The Hebrew University of Jerusalem, Jerusalem, Israel

Abstract

We consider a model of the quasi-one-dimensional conductors in which the electrons propagate on a lattice of identical metallic chains. The intrachain interactions are g_1, g_2, and g_4 and the interchain interactions are v_1, v_2, and v_4. The calculations are carried out for $g_1=0$ and $g_1/\pi v_F = -6/5$, which corresponds to the Luttinger and Luther-Emery lines. When only low-momentum transfer interchain interactions are considered, there is no phase transition. With interchain backscattering, the system may undergo a charge density wave phase transition, but no other type of phase transition. In particular, there cannot be a superconducting phase transition at a finite temperature without some interchain hopping. When interchain tunnelling processes are included, all types of possible long range orderings are possible, and the system undergoes a phase transition of the type which for the particular values of the intrachain interactions, the purely one-dimensional response function is the most divergent as $T \rightarrow 0$.

†Present address: Department of Physics, Iowa State University, Ames, Iowa 50010 U.S.A.

I. Introduction

As it is well known that short range forces cannot give rise to a phase transition in one dimension, the existence of such phase transitions is due at least partly to interchain coupling. There are several types of possible interchain coupling, and we have chosen to investigate those types most relevan to the quasi-one-dimensional conductors. In particular, we wish to investigate which types of phase transitions are possible with only interchain Coulomb interactions or single-particle interchain hopping.

II. The Single Chain

We shall not discuss the model in great detail, as such a discussion is given elsewhere.[1] We consider the system to be composed of a lattice of one-dimensional electron gases, each gas of the Luther-Emery form.[2] That is, the electrons have a linear spectrum of the Luttinger form, and the interactions g_1 (backscattering) and g_2 (forward scattering). We shall also allow them to have a g_4, or forward scattering between two electrons on the same side of the Fermi surface. These interactions are shown in Fig. (1). The two-particle response behavior of this system is summarized in Fig. (2). Each of the four divergent response functions has a region in g_1/g_2 space where it is predominate over the others. The addition of g_4 does not alter this picture qualitatively, but only changes the details of the picture. In Fig. (3), we have shown the regions of different response behavior for $g_2=g_4$.

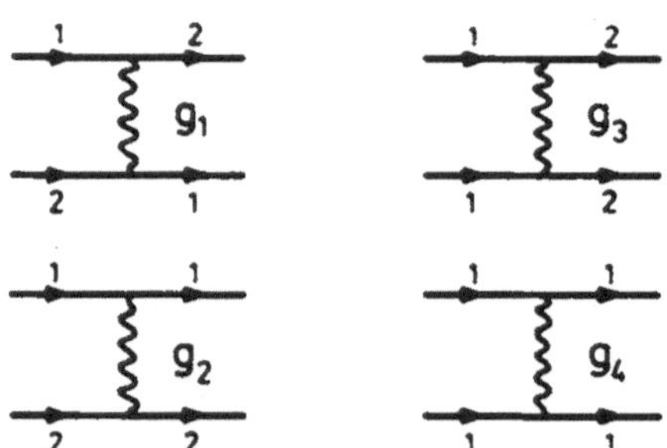

Fig. 1 Shown are diagrammatic representations of the four possible intrachain interactions. The indices 1 and 2 refer to different sides of the Fermi surface. The umklapp process, g_3, is only important for a half-filled band and is therefore not considered.

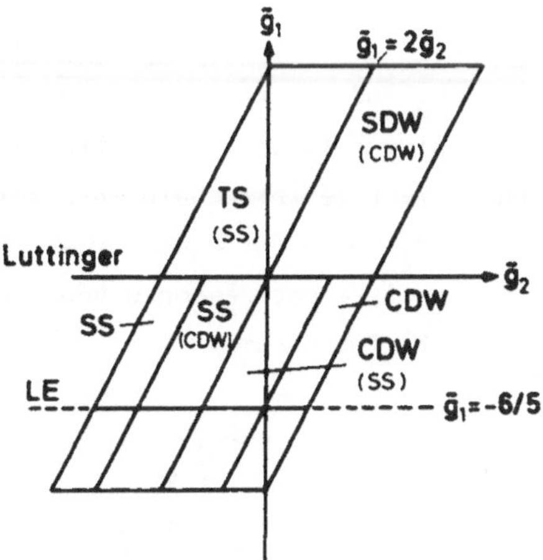

Fig. 2 Shown is a plot in g_1/g_2 space for $g_4=0$ of the regions for which the Bogoliubov transformations are valid of the various types of two-particle one-dimensional ordering. The predominate response function in each region is indicated by bold, large lettering.

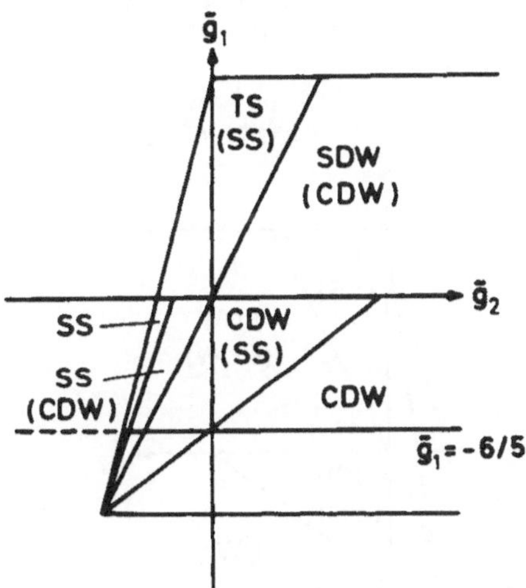

Fig. 3 Shown is a plot in g_1/g_2 space for $g_4=g_2$ of the regions of different types of one dimensional two-particle ordering. Note that the boundary for positive g_2 for which the Bogoliubov transformation becomes invalid is extended to infinity.

III. Interchain interactions

The nearest-neighbor interchain interactions we shall consider are shown in Fig. (4). The interactions v_2 and v_4 are interchain forward scattering interactions, and these do not give rise to a phase transition of any type, but serve only to renormalize the effective intrachain interactions. In Fig. (5), we have shown the regions in g_1/g_2 space for $g_2=g_4$, $v_2=v_4$, and $|v_2| = \frac{1}{2}|g_2|$. The boundaries of the regions of different response behavior are distorted, but the qualitative picture is still the same.

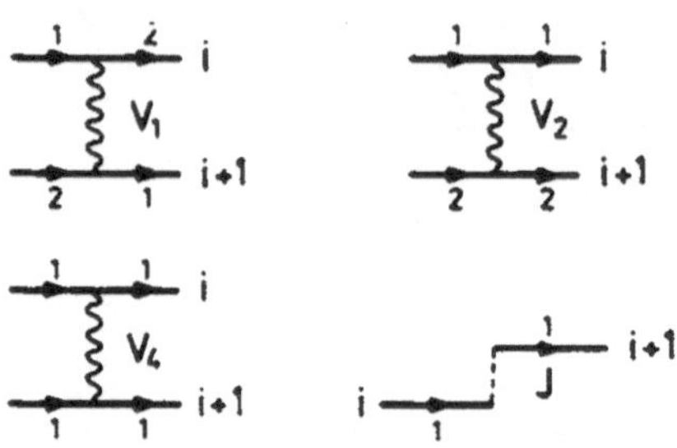

Fig. 4 Shown are diagrammatic representations of the nearest-neighbor interchain interactions. The interchain Coulomb interactions are v_1, v_2, and v_4 and J is the strength of the single particle interchain tunnelling process.

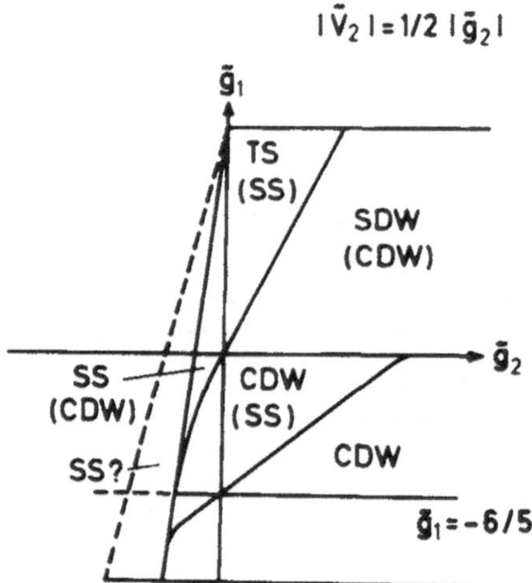

Fig. 5 Shown is a plot in g_1/g_2 space for $g_4=g_2$, for $v_2=v_4$, for $|v_2|=\frac{1}{2}|g_2|$, and for $v_1=0$. The resulting regions of one-dimensional two-particle ordering are shown. Note that the boundary for $g_2<0$ for which the Bogoliubov transformation is valid is displaced to the right, and there is no region where the SS response is the only divergent one. We may speculate, however, that in the region for $g_1<0$ to the left of this boundary and to the right of the (dotted) boundary in the absence of v_2 and v_4, that the SS response is the only divergent response.

If we now consider v_1, the interchain backscattering interaction, we find that it couples the chains with regard to charge density wave order only. Thus, v_1 can drive a charge density wave phase transition, but no other type of phase transition. The transition temperature can be found by treating the v_1 interaction in mean-field theory, and it depends upon v_1 as a power law, the power depending upon the effective intrachain interactions. Investigation of the fluctuations reveals that this v_1 problem is similar to walks on a lattice of particle-hole pairs, and it turns out that the self-avoiding random walks from chain to chain of particle-hole pairs are the most important. The transition temperature obtained by this self-avoiding random walk procedure has the same power law dependence upon v_1, but is reduced in magnitude by a constant depending only upon the structure of the lattice of chains.

When interchain tunneling processes are included, the system may undergo a phase transition of any type; it will undergo a transition to that state which for the particular g_1/g_2 values has the most divergent one-dimensional response. Thus, we can say, for example, that for $g_1 < 0$, that in the region $\tilde{g}_2 - \frac{1}{2}\tilde{g}_1 > 0$ (in Fig. (2)), the interchain backscattering will drive a charge density wave (CDW) phase transition, and the interchain tunneling will also favor a CDW phase transition over a singlet superconducting (SS) phase transition. In the region where the SS response is the only divergent one, the interchain tunnelling drives the phase transition. In the intermediate region where the purely one-dimensional SS response is more divergent than the CDW response, there is a competition between the interchain backscattering giving rise to a CDW phase transition, and the tunnelling favoring a phase transition of the SS type. We note that recent neutron data[3] on $(SN)_x$ showing the presence of a weak Kohn anomaly is indicative that $(SN)_x$ must lie in this region of competition between the SS and CDW responses. There is enough $2k_F$ excitations to see the Kohn anomaly, but the tunnelling favors the SS phase transition.

References

1. H. Gutfreund and R. A. Klemm, Phys. Rev. B 14, 1073 (1976); R. A. Klemm and H. Gutfreund, Phys. Rev. B 14, 1086 (1976).
2. A. Luther and V. J. Emery, Phys. Rev. Lett. 33, 589 (1974).
3. L. Pintschovius, H. Wendel, and H. Kahlert, these proceedings.

Conference on Organic Conductors and Semiconductors, Siófok, Hungary 1976

THE TWO-CHAIN PROBLEM: A MODEL OF TTF-TCNQ

R.A. KLEMM

Bell Laboratories, Murray Hill, New Jersey 07974 and
Stanford University, Stanford, California 94305, USA

and

P.A. LEE and T.M. RICE

Bell Laboratories, Murray Hill, New Jersey 07974, USA

Abstract

The two-chain, inverted band model for TTF-TCNQ is solved by the exact methods developed by Luther and Emery and by Chui and Lee. On each chain, there are intrachain interactions g_1 and g_2, and there are interchain interactions w_1 and w_2. We investigate all of the possible two-particle correlation functions, of which the new divergent ones not present in the single chain problem are of the excitonic insulator type. In addition, we investigate essentially all of the many possible four-particle correlation functions. For the single chain, there are two divergent ones; the $4k_F$ response found by Emery, and a four-particle condensation of two Cooper pairs. For the two-chain problem, there are many others, the most unusual of which are of the excitonic molecule type. In addition, the $4k_F$ response is enhanced by the interchain coupling, and is divergent for weak intrachain coupling as well as for strong coupling. Using renormalization group techniques, we find that the essential features of the temperature dependence of the relative size of the $4k_F$ and $2k_F$ excitations recently observed can be explained in terms of a crossover from a region in which the $4k_F$ excitation is dominant to one in which the $2k_F$ charge density wave response is dominant.

I. Introduction

Recent observation of the $4k_F$ as well as the $2k_F$ scattering in TTF-TCNQ[1] has generated a great deal of excitement. As regards the theorist, this is particularly important, as it enables him at last to be able to say something about the size and sign of the effective electron-electron interaction strengths, which until now had only been the subject of conjecture. In order to help clarify this situation, and to investigate the possibility of new additional types of excitations, we have studied a two-chain model for TTF-TCNQ in which the electrons on the TTF chains have a band inverted relative to the normal TCNQ band.

II. The Model

We consider the free Hamiltonian to be of the form

$$\mathcal{H}_{oa} = \sum_{k,s} v_F k (a^\dagger_{1,ks} a_{1,ks} - a^\dagger_{2,ks} a_{2,ks}) \tag{1}$$

$$\mathcal{H}_{ob} = \sum_{k,s} (-v_F k)(b^\dagger_{1,ks} b_{1,ks} - b^\dagger_{2,ks} b_{2,ks}) \tag{2}$$

where $a_{i,ks}(b_{i,ks})$ annihilates an electron of the a, or TCNQ, (b, or TTF) chain with spin $s = \pm 1$, and momentum k relative to the Fermi momentum k_F on the $i = 1,2$ branch. For simplicity, we assume the Fermi velocities to be the same in magnitude, although opposite in sign. Although this is not exactly the case in TTF-TCNQ, we expect that this model will contain most of the essentia features of that and similar charge-transfer systems. Using the transformation $b_{1,ks} \to c^+_{2,-ks}$, etc., Eq. (2) may be written as

$$\mathcal{H}_{ob} = \sum_{k,s} v_F k (c^\dagger_{1,ks} c_{1,ks} - c^\dagger_{2,ks} c_{2,ks}) \tag{3}$$

Thus, an electron on the inverted band TTF chain corresponds to a hole with a normal band structure, and vice-versa. The interactions are shown in Fig. (1). The intrachain interactions g_1 and g_2 are backward and forward scattering interactions, as in the single chain problem, and are assumed to be the same on both chains. The interchain interactions are w_1 and w_2. These interactions are between electrons, and we remark that the transformation to the c-notation changes the sign of w_1 and w_2 (although the sign of w_1 is unimportant).

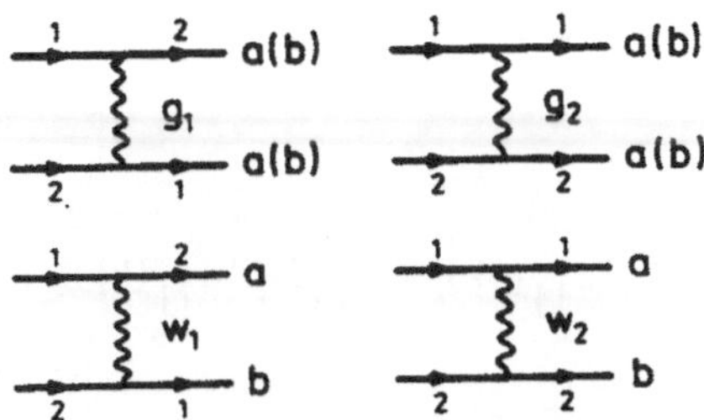

Fig. 1 Diagrammatic representations of the interactions. The quantities a and b are chain indices, and 1 and 2 indicate sides of the Fermi surface.

We now write the operators a and c in terms of the charge density and spin density operators ρ and σ, such as[2,3]

$$\rho_{ia}(k) = \frac{1}{\sqrt{2}} \sum_{q,s} a^{\dagger}_{i,k+q\,s}\, a_{i,q\,s}$$

and (4)

$$\sigma_{ia}(k) = \frac{1}{\sqrt{2}} \sum_{q,s} s\, a^{\dagger}_{i,k+q\,s}\, a_{i,q\,s}$$

and similarly for $\rho_{ic}(k)$ and $\sigma_{ic}(k)$, and we define

$$\rho^{\pm}_{i}(k) = \frac{1}{\sqrt{2}} [\rho_{ia}(k) \pm \rho_{ic}(k)]. \tag{5}$$

Using the Bose representation of the fermion field used by Luther and Emery,[3] we write the Hamiltonian in terms of the σ_{ia}, σ_{ic}, ρ^{+}_{i}, and ρ^{-}_{i} operators. We then perform four Bogoliubov transformations,[2] mixing σ_{1a} and σ_{2a} with the (imaginary) angle ψ,

$$\sigma_{1a}(k) \to \sigma_{1a}(k) \cosh \psi + \sigma_{2a}(k) \sinh \psi, \tag{6}$$

and we similarly mix σ_{1c} and σ_{2c} with ψ, ρ^{+}_{1} and ρ^{+}_{2} with φ_{+}, ρ^{-}_{1} and ρ^{-}_{2} with φ_{-}. We set

$$\tanh 2\varphi_{\pm} = - \tilde{g}_2 + \tilde{U}_{\parallel}/2 \mp \tilde{w}_2$$

and (7)

$$\tanh 2\psi = \tilde{U}_{\parallel}/2,$$

where $\tilde{g} \equiv g/\pi v_F$. The Hamiltonian may be written in its most useful form,

$$\mathcal{H} = \mathcal{H}_{\sigma_a} + \mathcal{H}_{\sigma_c} + \mathcal{H}_{\rho+} + \mathcal{H}_{\rho-} + \mathcal{H}', \tag{8}$$

where

$$\mathcal{H}_{\sigma_j} = \frac{2\pi}{L} v'' \sum_{k>0} (\sigma_{1j}(k)\sigma_{1j}(-k) + \sigma_{2j}(-k)\sigma_{2j}(k)) + \frac{U_\perp}{(2\pi\alpha)^2} \int dx(\Theta_j^2(x) + \Theta_j^{+2}(x)), \tag{9}$$

$$\mathcal{H}_{\rho\pm} = \frac{2\pi}{L} v'_\pm \sum_{k>0} (\rho_1^\pm(k)\rho_1^\pm(-k) + \rho_2^\pm(-k)\rho_2^\pm(k)), \tag{10}$$

and

$$\mathcal{H}' = \frac{w_1}{(2\pi\alpha)^2} \int dx \{P^2(x) + P^{+2}(x)\}\{\Theta_a(x) + \Theta_a^+(x)\}\{\Theta_c(x) + \Theta_c^+(x)\}, \tag{11}$$

and

$$\Theta_j(x) = \exp [\sum_k A_k(x)\frac{1}{\sqrt{2}} e^{\psi}(\sigma_{1j}(k) + \sigma_{2j}(k))], \tag{12}$$

$$P(x) = \exp [\sum_k A_k(x) \tfrac{1}{2}e^{\varphi_-}(\rho_1^-(k) + \rho_2^-(k))], \tag{13}$$

$A_k(x) = \frac{2\pi}{kL} e^{-ikx - a|k|/2}$, $j = a,c$ is the chain index, α^{-1} is the momentum transfer cutoff, $v'' = v_F \operatorname{sech} 2\psi$, and $v'_\pm = v_F \operatorname{sech} 2\varphi_\pm$. We have followed the notation of Luther and Emery, denoting the parallel and antiparallel spin intrachain interactions by $U_\parallel$ and $U_\perp$, respectively, but we will eventually set $U_\parallel = U_\perp = g_1$. Clearly, the Hamiltonian is diagonal in the ρ^+ variables. By our choice of the Fermi velocities opposite in magnitude and the interactions g_1 and g_2 on the two chains equal, we have removed two degrees of freedom from the problem, leaving 6.

We remark that Eq. (9) contains the Hamiltonians for two classical Coulomb gases of charge $q_\sigma = 2e^{\psi}$ and fugacity $U_\perp/(2\pi\alpha)^2$. Eq. (11) couples these two Coulomb gases to a third classical Coulomb gas of charge $q_{\rho-} = \sqrt{2}e^{\varphi_-}$ and fugacity $w_1/(2\pi\alpha)^2$. In a classical 2D Coulomb gas, the potential between two charges of magnitude q is $q^2 \ln(r/\alpha)$, where $r^2 = (\alpha + \tau)^2 + x^2$

is the square of the distance between the two charges, and α is the "soft" core. This coupling of the three Coulomb gases can be seen by writing the partition function as an expansion in powers of w_1,[4]

$$Z = \sum_{n=0}^{\infty} \left(\frac{w_1}{(2\pi\alpha)^2}\right)^{2n} \frac{1}{(n!)^2} \prod_{i=1}^{2n} \int d^2x_i \exp[q_{\rho^-}^2 \sum_{i<j}^{2n} s_i s_j \ln r_{ij}/\alpha]$$

$$\cdot \langle T_\tau \prod_{i=1}^{2n} [\Theta_a(x_i\tau_i) + \Theta_a^\dagger(x_i\tau_i)] S_a \rangle_{\mathcal{H}^\circ_{\sigma_a}}$$

$$\cdot \langle T_\tau \prod_{i=1}^{2n} [\Theta_c(x_i\tau_i) + \Theta_c^\dagger(x_i\tau_i)] S_c \rangle_{\mathcal{H}^\circ_{\sigma_c}} \tag{14}$$

where the spin-density operators are all in the Heisenberg representation with respect to the free parts of $\mathcal{H}\sigma_a$ and $\mathcal{H}\sigma_c$, and

$$S_{a(c)} = T_\tau \exp \int dx d\tau \, (\Theta^2_{a(c)}(x\tau) + \Theta^{\dagger 2}_{a(c)}(x\tau)) \tag{15}$$

Thus, the expectation values in Eq. (14) involve all possible correlation functions of the σ_a and σ_c Coulomb gases. In the Coulomb gas picture, this means adding charges of $\pm q_\sigma/2$ to the vacuum and "screening" with all possible combinations of $\pm q_\sigma$ charges such that the total charge of the system is zero.

Examination of the partition function reveals that for $g_1 < 0$, both the σ_a and σ_c spin degrees of freedom have a gap in their spectra, and for $\tilde{g}_2 - \tilde{g}_1/2 + \tilde{w}_2 > -3/5 - 8|\tilde{w}_1|/25$ (to lowest order in w_1), the ρ^- modes also have a gap. For $g_1 > 0$, the spin degrees of freedom do not have a gap, but for $g_1 > 0$ and $\tilde{g}_2 - \tilde{g}_1/2 + \tilde{w}_2 > -|\tilde{w}_1|$ (to lowest order in w_1), the ρ^- modes have a gap in their spectra. We define the four regions of various gaps to be regions I-IV, in a counterclockwise fashion starting from the region in which both the ρ^- and σ modes have gaps [see Fig. (2)]. We note that in Fig. (2) we have assumed w_1 finite so as to couple the chains and to induce gaps, but negligibly small so as not to require a third dimension to the figure.

III. Two-Particle Correlation Functions

For the single chain ($w_1 = w_2 = 0$), there are seven two-particle correlation functions, four of which are divergent. The divergent ones are the CDW (charge-density wave), SDW (spin-density wave), SS (singlet superconductor), and TS (triplet superconductor) response functions[5-7], and are shown in Fig. (3). The

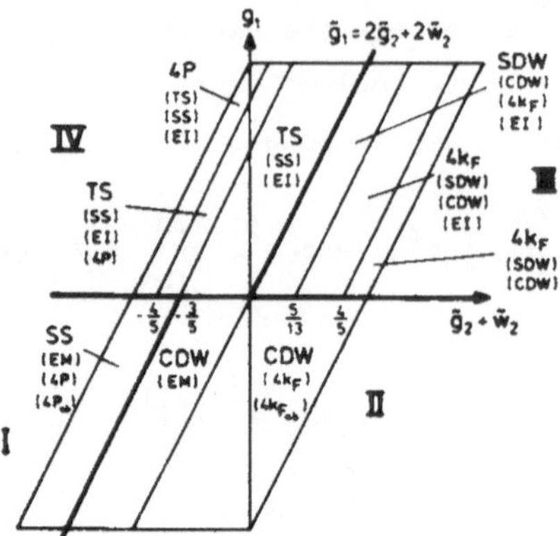

Fig. 2 Plot of the regions for which the various response functions diverge. The dark lines separate the four regions of various gaps. In each region, the predominate response is indicated by the largest lettering, and the other divergent ones are ranked according to the strength of their divergences.

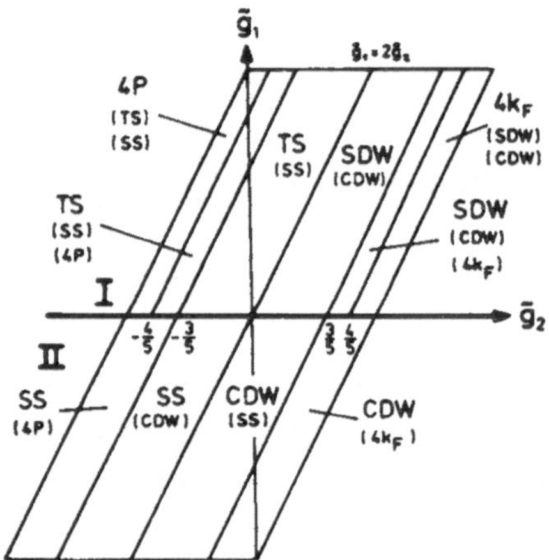

Fig. 3 Plot of the regions in g_1/g_2 space for the single chain. In addition to the four divergent two-particle responses, there are also two divergent four-particle responses. In each region, the predominate response is indicated by the larger lettering, and the other divergent ones are ranked according to size.

q=0 density-density response function, and the q=0 spin-density response function, and the singlet $2k_F$ pairing function are all non-divergent.[7] For the two-chain problem, there are the same seven response functions on each chain, but in addition there are eight two-particle response functions involving both chains. We define a response function of the type $\mathcal{O}$ to be of the form

$$\chi_{\mathcal{O}\mathcal{O}}(k,w) = -i \int_0^\infty dt \int_{-\infty}^\infty dx\, e^{-ikx + iwt} \langle [\mathcal{O}(xt), \mathcal{O}^\dagger(00)] \rangle \; . \tag{16}$$

The eight additional functions are due to the following operators:

$\psi_{1a+}\, \psi^\dagger_{1b\pm}$: q = 0 excitonic insulator (EI)

$\psi_{1a+}\, \psi_{1b\pm}$: q = $2k_F$ interchain Cooper pairing

$\psi_{1a+}\, \psi^\dagger_{2b\pm}$: q = $2k_F$ excitonic insulator

$\psi_{1a+}\, \psi_{2b\pm}$: q = 0 interchain Cooper pairing.

Of these eight interchain two-particle correlation functions, only the q=0 excitonic insulator (EI) responses are divergent. There are both symmetric and antisymmetric forms of EI (from the sum and difference of the two operators), and the magnitude of the symmetric form is larger. The regions in which the various two-particle correlation functions are divergent are shown in Fig. (2). We note that EI diverges for $g_1 \geq 0$ and $\tilde{g}_2 - \tilde{g}_1/2 + \tilde{w}_2 < 4/5$ (for finite but small w_1). In region I, the SS is the only divergent two-particle response. In region II, the CDW is the only divergent response, as all of the others have gaps in their frequency dependence. In region III, the CDW, SDW, and EI responses are all divergent, but the SDW is the most divergent. In region IV, the TS, SS, and EI responses are all divergent, but the TS is the most divergent two-particle response.

Thus, with regard to two-particle response functions, the main difference between the single chain and the two-chain systems is the appearance of the EI response function for $g_1 \geq 0$. Also, the SS response function for $g_1 < 0$ is suppressed in the region $\tilde{g}_2 - \tilde{g}_1/2 + \tilde{w}_2 > -3/5$, due to the presence of the gap in the ρ^- spectrum.

IV. Four-Particle Correlation Functions

For the single chain, there are two divergent four-particle response functions other than those composed of a divergent two-particle response operator and a density operator. The first was found by Emery,[8] and is a $4k_F$ response of the operator $\psi_{1+}\psi_{1-}\psi^{\dagger}_{2+}\psi^{\dagger}_{2-}$, which diverges for $\tilde{g}_2 - \tilde{g}_1/2 > 3/5$, and is more divergent than the SDW and CDW response functions for $g_1 \geq 0$ and $\tilde{g}_2 - \tilde{g}_1/2 > 4/5$ (see Fig. (3)). There is also the four-particle q=0 condensation of two Cooper pairs, described by the operator $\psi_{1+}\psi_{1-}\psi_{2+}\psi_{2-}$, which diverges in the region $\tilde{g}_2 - \tilde{g}_1/2 < -3/5$, and is more divergent than the TS and SS response functions for $g_1 > 0$ and $\tilde{g}_2 - \tilde{g}_1/2 < -4/5$ (see Fig. (3)). One may interpret the $4k_F$ response function as a combination CDW-CDW or SDW-SDW, and the q=0 function as a combinatio TS-TS or SS-SS function. However, in the regions in which these functions are more divergent than the two-particle functions are, this means that the tendency is for all particle-hole (or particle-particle) pairs to themselves form pairs.

With the two-chain inverted band model, there are many possible four-particle response functions, most of which are not divergent. For brevity, we shall only describe the divergent ones. These are shown in Fig. (2). If we compare Figs. (2 and (3), we see that in region III, the range of the divergence of the $4k_F$ response function is much greater than for the single chain problem. This is due to the ρ^- modes being "frozen out" without giving rise to a gap. Thus, the inverted bands model greatly enhances the $4k_F$ response. We note that the $4k_F$ response function is the most divergent for $\tilde{g}_2 - \tilde{g}_1/2 + \tilde{w}_2 > 5/13$, roughly a factor of 2 less than in the single chain case. We also note that the extension of the divergence of the $4k_F$ response into region II is also enhanced, and in that region there is also the interchain $4k_F$ response function described by the operator $\psi_{1a+}\psi^{\dagger}_{2a}\psi^{\dagger}_{2b\pm}\psi_{1b\pm}$, which diverges for $g_1 < 0, \tilde{g}_2 - \tilde{g}_1/2 + \tilde{w}_2 > 0$. In region II, there are also two other response functions, both of which are excitoni molecules (EM). These responses, which involve exciting two particles from one chain to the other, are described by the operators $\psi_{2a-}\psi_{1a+}\psi^{\dagger}_{2b-}\psi^{\dagger}_{1b+}$, which diverge for $g_1 < 0$ and $\tilde{g}_2 - \tilde{g}_1/2 + \tilde{w}_2 < 0$, and $\psi_{1a-}\psi_{1a+}\psi^{\dagger}_{1b-}\psi^{\dagger}_{1b+}$, which diverge for $\tilde{g}_2 - \tilde{g}_1/2 + \tilde{w}_2 < 0$ (and thus extends into region IV). In region I, the divergent four-particle response functions are the single chain four-particle condensation of Cooper pairs, both EM response functions, and an interchain condensation of two Cooper pairs, described by the operator $\psi_{1a-}\psi_{2a+}\psi_{1b-}\psi_{2b+}$.

However, in region I, none of these four-particle response functions are as divergent as the two-particle SS response function. In region IV, we have not concluded an exhaustive study of the hundreds of possible response functions. However, it appears that the only divergent ones are the single chain four-particle condensation of Cooper pairs and the second type of excitonic molecule, both of which diverge for $\tilde{g}_2 - \tilde{g}_1/2 + \tilde{w}_2 < -3/5$, and are stronger than the TS and SS response functions for $\tilde{g}_2 - \tilde{g}_1/2 + \tilde{w}_2 < -4/5$ (for finite but negligible w_1 and w_2).

V. Discussion

A comparison of Figs. (2) and (3) shows that the predominate new results of this two-chain model not present in the single chain model are the presence of $4k_F$ excitations for weak coupling strengths, and the appearance of the excitonic insulator and excitonic molecule response functions. In order to understand the temperature dependence of the interaction strengths, we have performed the first-order renormalization group calculations for the invariant couplings, and find

$$\frac{\partial \tilde{g}_1}{\partial \xi} = \tilde{g}_1^{\,2} + \tilde{w}_1^{\,2}$$

$$\frac{\partial \tilde{w}_1}{\partial \xi} = \tfrac{1}{2}\tilde{w}_1 (3\tilde{g}_1 + \tilde{v}) \qquad (17)$$

$$\frac{\partial \tilde{v}}{\partial \xi} = 2\tilde{w}_1^{\,2}$$

and

$$\frac{\partial}{\partial \xi} (\tilde{g}_1 - 2\tilde{g}_2 + 2\tilde{w}_2) = 0$$

where $\tilde{v} = 3\tilde{g}_1 - 2\tilde{g}_2 - 2\tilde{w}_2$, and $\xi = \ln(T/E_F)$. Similar equations have also been found by Mihály and Sólyom.[9] We observe that the last equation in Eq. (17) corresponds to Eq. (10), in that the ρ^+ part of the Hamiltonian can be diagonalized. From Eq. (17), for $\tilde{v} < 0$, the scaling is to strong coupling. Thus, if we start in region III, we scale to region II as we go down in temperature. This implies crossing from positive g_1 to negative g_1, which was not allowed in the single chain case. Similarly, if we start in region I, we also scale to region II and strong coupling. However, if we start in region IV, we stay in region IV, and scale to weak coupling.[10] Thus, from any of the regions I-III, we scale to region II and strong coupling.

The implications for TTF-TCNQ are that since only the $4k_F$ excitations are seen at high temperatures, we expect that this corresponds to the part of region III where $\tilde{g}_2 - \tilde{g}_1/2 + \tilde{w}_2 > 5/13$, where the $4k_F$ response function is domina This is roughly similar to the conclusions of the single chain model, except that the interactions may be weaker in our case. However, as we scale down in temperature, there is a crossover to region II where the SDW response (which had been strong and divergent in region III) develops a gap, and the CDW respons function becomes more divergent than the $4k_F$ response function. Whether this crossover to negative g_1 could explain the observed results that the $4k_F$ to $2k_F$ crossover occurs well above the ID to 3D crossover[1] is hard to determine, as the temperatures involved are all within a factor of two of each other.

In addition, there may conceivably be the possibility of observing q=0 resp functions in TTF-TCNQ or similar charge-transfer compounds. At high temperature if the interactions are weak enough (but not so weak that the $4k_F$ response funct is not the most divergent), the EI response function is divergent. Also, at low temperatures (or just above the 54^oK transition), there may be a possibility of observing the EM q=0 response function, as the line determining the boundary of the EM response function scales to the right in Fig. (2). Whether or not it sca fast enough to the right in order to intersect the effectively scaled interactic strength before the phase transition takes place, is hard to determine. A simil scaling of the boundary of more divergent $4k_F$ response relative to $2k_F$ response in the upper half $\tilde{g}_1/(\tilde{g}_2 + \tilde{w}_2)$ plane (in which the boundary scales to the right faster than the interaction scales to the right) could possible explain the crossover to $2k_F$ dominant response above the temperature where the interchain interaction become strong, for negative g_1.

References

†Present address: Dept. of Physics, Iowa State University, Ames, Iowa 50010

1. J. P. Pouget, S. K. Khanna, F. Denoyer, R. Comès, A. F. Garito, and A. J. He Phys. Rev. Lett. 37, 437 (1976).
2. D. C. Mattis and E. H. Lieb, J. Math. Phys. 6, 304 (1965).
3. A. Luther and V. J. Emery, Phys. Rev. Lett. 33, 589 (1974).
4. S.-T. Chui and P. A. Lee, Phys. Rev. Lett. 35, 315 (1975).
5. H. Gutfreund and R. A. Klemm, Phys. Rev. B 14, 1073 (1976).
6. N. Menyhárd and J. Sólyom, J. Low Temp. Phys. 12, 529 (1973); J. Sólyom, ibid. 12, 547 (1973).
7. H. Fukuyama, T. M. Rice, C. M. Varma, and B. I. Halperin, Phys. Rev. B 10, 3775 (1974).
8. V. J. Emery, Phys. Rev. Lett. 37, 107 (1976).
9. L. Mihály and J. Sólyom, J. Low Temp. Phys. 24 579 (1976).
10. P. A. Lee, T. M. Rice, and R. A. Klemm, unpublished.

Conference on Organic Conductors and Semiconductors, Siófok, Hungary 1976

ON THE PROBLEM OF PHASE TRANSITIONS IN QUASI-ID METALLIC SYSTEMS

N. MENYHÁRD

Central Research Institute for Physics, Budapest, Hungary

ABSTRACT

The problem of phase transitions is investigated in the Gorkov-Dzyaloshinskii model of weakly coupled metallic chains using the multiplicative renormalization group method. It is found that while the type of the order that develops in the system is determined by the interchain interaction, the critical temperature is strongly influenced by the intrachain couplings.

In this paper a weakly coupled system of metallic chains will be investigated. The model we use was introduced by Gorkov and Dzyaloshinskii (1) and also investigated by Mihály and Sólyom (2) with the result that for attractive interactions a charge density wave (CDW) type phase transition occurs in the system at a finite temperature. Our aim is to study the effect of the 1D fluctuations on the phase transition. It is found that the one dimensional fluctuations can strongly increase the critical temperature of the 3D phase transition provided the character of the 1D fluctuations matches that of the 3D phase transition (i. e. the same type of response functions are enhanced and singular, respectively).

The problem is characterized by two scales of energy: the longitudinal intrachain coupling energy (g_1, g_2) and the interchain coupling energy (g_1^{ij}, g_2^{ij}, where i, j denote the sites of the chains in the plane perpendicular to the chain direction). The subscripts 1 and 2 specify the backward and forward scattering coupling constants, respectively.

The bare vertex of the coupled system is as follows (see also Fig. 1)

$$\Gamma^{(0)\alpha\beta\gamma\delta}_{ijkl} \underset{(g_1^{ii}=g_2^{ii}=0)}{=} (g_1^{ij}+g_1\delta_{ij})\delta_{ik}\delta_{il}\delta_{\alpha\gamma}\delta_{\beta\delta} - (g_2^{ij}+g_2\delta_{ij})\delta_{il}\delta_{jk}\delta_{\alpha\delta}\delta_{\beta\gamma}$$

Fig. 1 Diagrammatic representation of the elementary interactions. i, j, k, l denote the positions of chains in the plane perpendicular to the chain direction, α, β, γ, δ are spin indices. Electrons propagating with longitudinal momenta near $+k_F$ ($-k_F$) are represented by full (dotted) lines.

The Gel-Mann-Low type renormalization group approach (3) is used to calculate the invariant couplings $g_1'^{ij}$, $g_2'^{ij}$, g_1' and g_2' (corresponding to the bare couplings g_1^{ij}, g_2^{ij}, g_1 and g_2, respectively), which in turn determine the characteristic susceptibilities (charge density wave (CDW), spin density wave (SDW) and superconducting (SC) type ones) of the system (4).

We want to describe the system under the conditions when a) the bare interaction along the chains is much stronger than between the chains, i.e. $|\sum_i g_1^{ij}|$, $|\sum_i g_2^{ij}| \ll |g_1|$, $|g_2|$, b) the intrachain coupling constants are such that the CDW susceptibility of the 1D system is divergent as $T \to 0$ (i.e. $g_1 < 0$ or $g_1, g_2 > 0$ and $(2g_2 - g_1) > 0$), c) the temperature is lying in a range where the 1D and 3D correlations are of comparable strength as sketched on Fig. 2.

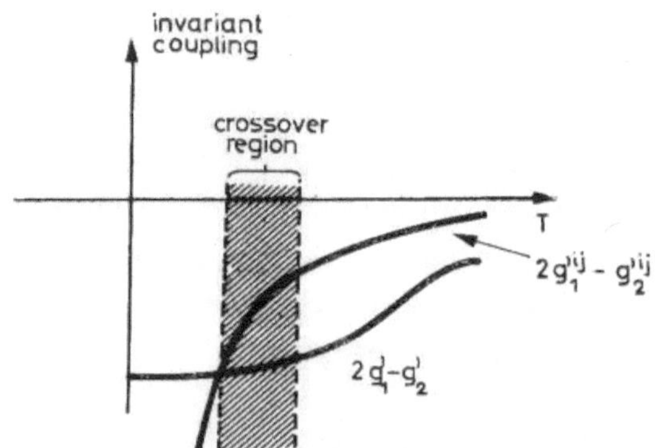

Fig. 2 Sketch of the invariant couplings $2g_1'^{ij} - g_2'^{ij}$ and $2g_1' - g_2'$. The crossover region, where they are of about the same strength is hatched.

To this end, in deriving the Lie equations for the invariant couplings the following approximations are made

i/ the intrachain invariant couplings are taken into account in the second order renormalization approximation (3),

ii/ the interchain invariant couplings are accounted for to lowest order,

iii/ $g_2'^{ij}$ is neglected. This is a self-consistent assumption near T_c (1, 2) and for $T \lesssim \omega_D$. Due to the inclusion of g_2', a qualitatively correct result is also expected in between.

iv/ $g_1'^{ii}$ is neglected compared to g_1', g_2'. This is justifiable provided $(T - T_c)/T_c < \exp(-|g_1'|)$. Thus we get (5)

$$\frac{\partial g_1'^{ij}}{\partial \xi} = 2\sum_m g_1'^{im} g_1'^{mj} + 2 g_1'^{ij}\left[2g_1' - g_2' + g_1'^2 + g_2'^2 - g_1' g_2'\right] \qquad i \neq j \tag{1}$$

$$\frac{\partial g_2'^{ij}}{\partial \xi} = 0 \tag{2}$$

$$\frac{\partial g_1'}{\partial \xi} = 2(g_1')^2 + 2(g_1')^3 \tag{3}$$

$$g_2' = \frac{1}{2} g_1' + g_2 - \frac{1}{2} g_1 \tag{4}$$

(the coupling constants are in units of $2\pi v$, v being the Fermi velocity) with $\xi = \ln T/\omega_D$. ω_D characterizes the cut-off of the coupling constants g_1, g_2 (3). The last two equations are the same as those derived by Sólyom and the author (3) when investigating the onset of long range order in a single metallic chain as $T \to 0$. g_1' and g_2' being uninfluenced by the interchain couplings to the order we calculate, the expression in the square brackets on the right hand side of Eq. (1) is determined by the solution of the 1D problem. According to Sólyom (4)

$$2(2g_1' - g_2' + g_1'^2 + g_2'^2 - g_1' g_2') = \partial \ln \bar{\chi}_{1D}^{CDW} / \partial \xi$$

where $\bar{\chi}_{1D}^{CDW} = \pi v \partial \chi_{1D}^{CDW} / \partial \xi$ and χ_{1D}^{CDW} is the CDW susceptibility of the 1D electron gas. Thus Eq. (1) can be rewritten in a form which after

Fourier transforming with respect to $(R_i - R_j)$ reads

$$\frac{\partial g_1'^{\perp}(q_\perp,\xi)}{\partial \xi} = 2\left(g_1'^{\perp}(q_\perp,\xi)\right)^2 + g_1'^{\perp} \frac{\partial \ln \bar{\chi}_{1D}^{CDW}}{\partial \xi} \tag{1a}$$

Eq. (1a) is solved to yield

$$g_1'^{\perp}(q_\perp,T) = \frac{g_1^{\perp}(q_\perp)\,\bar{\chi}_{1D}^{CDW}(T)}{1 - 2\pi v\, g_1^{\perp}(q_\perp)\, \chi_{1D}^{CDW}(T)} \tag{5}$$

where

$$g_1^{\perp}(q_\perp) = \sum_k g_1^{ik} e^{iq_\perp(R_i - R_k)}$$

Now we assume that $T/T_{1D} \ll 1$, where T_{1D} is the characteristic temperature below which strong short range correlations exist in the 1D system. In this regime and for $g_1 < 0$, Eq. (3) gives

$$g_1' = -1 + (T/T_{1D})^2 + \ldots \tag{6}$$

where $T_{1D} = \omega_D |g_1|^{1/2} \exp\left(-1/2\,|g_1|\right)$. It is to be noted here that Eq. (6) should not be taken too seriously as it has been obtained in a temperature region which is beyond the range of validity of Eqs (3) and (4). All the higher order terms would be needed to find the correct value of g_1' at $T = 0$. Nevertheless, comparison with certain exact results (6) shows that the expressions obtainable using Eq. (6) for the response functions are qualitatively correct. Especially we can write

$$\bar{\chi}_{1D}^{CDW} = A(T/T_{1D})^{-\gamma}; \qquad \left(\chi^{CDW} = -\frac{1}{\pi v \gamma}\bar{\chi}^{CDW}\right) \tag{7}$$

where A and γ depend on the coupling constants and γ is near unity. Using Eq. (6) $\gamma = 3/2$ (4) and $A = |g_1|^{-3/2}$ follows.

Eq. (5) becomes infinite when

$$2\pi v g_1(q_\perp) \chi_{1D}^{CDW}(T) = 1$$

which defines a transition temperature$(|g_1^\perp| \equiv |g_1^\perp(q_0)| = \max |g_1^\perp(q_\perp)|)$

$$T_c = T_{1D} \left[\frac{2A}{\gamma} |g_1^\perp| \right]^{1/\gamma} \tag{8}$$

where Eq. (7) for χ_{1D}^{CDW} has been used. The immediate vicinity of T_c is clearly beyond the range of validity of Eqs (1) - (4).

The dependence of T_c on the interchain coupling constant is characterized by the 1D susceptibility exponent, a result obtained also by Klemm and Gutfreund (7) who studied the phase transitions in a quasi 1D metal on the basis of the two exactly solvable models of the 1D electron gas (6), (8) following Scalapino et al.'s (9) treatment of other pseudo-one-dimensional systems.

It can be easily seen that the singular behaviour of $g_1^{\prime\perp}$ appears as a pole in the CDW susceptibility of the coupled system $\chi_{3D}^{CDW} \sim [T - T_c]^{-1}$, while the other three response functions do not get enhanced at finite temperatures. This indicates the tendency of the system to undergo a CDW-type phase transition.

The close relationship between the invariant coupling and the CDW susceptibility is due to the structure of the interchain interaction in the variables, i, j, k, l namely, $g_1^{ij} \delta_{ik} \delta_{jl}$ (see also Fig. 1). Other interactions are also conceivable (10). The exchange interaction $g_2^{ij} \delta_{ik} \delta_{jl}$ will not favour CDW formation but instead enhances SDW type bubbles. The corresponding effective coupling is $g_{2Ex}^{\prime\perp} \sim g_{2Ex}^{\perp} \chi_{3D}^{SDW}$. In quasi 1D systems the interchain exchange interaction is generally small, moreover the SDW susceptibility of the individual 1D chains is not enhanced (4, 6) (except for the small enhancement in the repulsive case). Consequently, no antiferromagnetic ordering is expected, at least in the framework of the present model of quasi-1D metallic systems. Superconducting order could result from a pair-hopping-type interaction (10) (e.g. a combination of direct interaction with single particle hopping).

Finally let us briefly discuss the case g_1, $g_2 > 0$, $(2g_2 - g_1) > 0$. The solution of Eq. (3) is (3) $g_1' = g_1/(1 + 2g_1\xi)$ i.e. the invariant couplings remain weak in the whole temperature range. As a consequence the 1D fluctuations are much smaller, the CDW susceptibility of the 1D system (4)

$$\bar{\chi}_{1D}^{CDW}(T) \sim (T/\omega_D)^{-2g_2+g_1}$$

shows a much weaker singularity as $T \to 0$ than in the $g_1 < 0$ case ($|g_1|, |g_2| \ll 1$ is always supposed). Eq. (8) gives

$$T_c = \omega_D \left(2|g_1^{\perp}|/(2g_2-g_1)\right)^{1/(2g_2-g_1)}$$

and thus the critical temperature is also much smaller than in the previous case. It is worth noting in this context that the second order renormalization group treatment of the same model by Mihály and Sólyom (2), according to which $T_c = \omega_D \exp\left[1/2((g_1 + g_1^{\perp})\right]$, leads to the absence of a phase transition for g_1, $g_2 > 0$ provided $g_1 > |g_1^{\perp}|$. The reason for this discrepancy lies in their neglect of g_2', in the calculation (11).

ACKNOWLEDGEMENTS

I am indebted to P. Szépfalusy and A. Zawadowski for fruitful discussions and valuable remarks.

REFERENCES

(1) L. P. Gorkov, and L E. Dzyaloshinskii, Zh. Eksp. Teor. Fiz. 67, 397 (1974)

(2) L. Mihály and J. Sólyom, J. Low Temp. Phys., 24, 579 (1976)

(3) N. Menyhárd and J. Sólyom, J. Low Temp. Phys. 12, 529 (1973)

(4) J. Sólyom, Low Temp. Phys. 12, 547 (1973)

(5) N. Menyhárd, to be published in Solid State Communications

(6) A. Luther and V. J. Emery, Phys. Rev. Letters 33, 589 (1974)

(7) R. A. Klemm and H. Gutfreund, Phys. Rev. B, to be published

(8) A. Luther and I. Peschel, Phys. Rev. B9, 2911, (1974); D. C. Mattis and E. H. Lieb, J. Math. Phys. 6, 304 (1965)

(9) D. J. Scalapino, Y. Imry and P. Pincus, Phys. Rev. B11, 2042 (1975)

(10) N. Menyhárd, J. Phys. A8, 1982 (1975)

(11) For a detailed discussion of the effect of $g_2^{\prime\, ij}$ (i = j included) in the repulsive case see P. A. Lee, T. M. Rice and R. A. Klemm, to be published

Conference on Organic Conductors and Semiconductors, Siófok, Hungary 1976

THE EXCITON MODEL OF SUPERCONDUCTIVITY IN LINEAR CHAINS - REVISITED

D. DAVIS

Corning Glass Works, Corning, New York;

H. GUTFREUND

The Racah Institute of Physics, Hebrew University, Jerusalem, Israel;

W.A. LITTLE

Physics Department, Stanford University, California, USA

We present a detailed calculation of the transition temperature of a model, filamentary excitonic superconductor. The proposed structure consists of a linear chain of transition-metal atoms to which is complexed a ligand system of highly polarizable dye molecules. The model is discussed in the light of recent developments in our understanding of one-dimensional metals. We show that for the structure proposed, the momentum dependence of the exciton interaction results in the superconducting state being favoured over the Peierls state, and in vertex corrections to the electron-exciton interaction which are small. The calculation of the transition temperature is based on what we believe to be reasonable estimates of the strength of the excitonic interaction, Coulomb repulsion and band structure. For the particular model proposed, transition temperature of several hundred degrees are calculated. However, we find superconductivity only in systems in which the excitonic medium is within a covalent bond length of, and completely surrounds the conductive spine. This imposes severe constraints on the structure of any excitonic superconductor.

Conference on Organic Conductors and Semiconductors, Siófok, Hungary 1976

EXCITONIC "SUPERFLUIDITY" IN LOW-DIMENSIONAL CRYSTALS

I.O. KULIK and S.I. SHEVCHENKO

Physico-Technical Institute of Low Temperatures, UkSSR Academy of Sciences, Kharkov 310164, USSR

Consider a system comprising low-dimensional elements (sheets, threads, etc.) possessing metallic conductivity of different types (electron- and hole-like) n - p - n - p - ... Owing to Coulomb interaction, V_{12}, between the elements, one obtains a picture of excitonic pairing of electrons and holes which are separated in space. If the tunneling matrix element (transfer integral) between adjacent sheets (layers), T_{12}, is much smaller than the energy gap, Δ, the properties of such electron-hole condensate may be "superfluid" rather than insulating [1]. The longitudinal response of the system is characterized by alternating currents flowing along the layers [2]

$$\vec{j}_n = -\vec{j}_{n+1} = -\frac{ne^2}{Mc}(\vec{A}_n - \vec{A}_{n+1})$$

$\vec{A}_n$ being the value of the vector potential at the n-th layer. This results in a (small) diamagnetic susceptibility, χ, which is however strongly H-dependent in a manner similar to the superconducting case. The response of an exciton-paired p - n junction to the transverse field (V) is insulating, provided that V satisfies condition $eV < \hbar\omega_c$ where $\omega_c = 2\left(\frac{\Delta_c T_{12}}{N(0)V_{12}}\right)^{1/2}$ is the frequency of collective oscillations in the system. The T_{12}-term in the Hamiltonian fixes the phase of in a manner similar to [1], but in case of $eV > \hbar\omega_0$ quasi-Josephson behaviour results, namely, the transverse current $j_{n,n+1}$ is an oscillating function of time with the frequency

The above considerations may be relevant to some peculiarities of structural transformations in charge-transfer low-dimensional conducting systems TTF-TCNQ [3].

1. R.R. Guseinov, L.V. Keldish. ZhETF, 63, 2255 (1972).
2. S.I. Shevchenko, I.O. Kulik. Letters to ZhETF, 23, 171 (1976).
3. F. Denoyer, F. Comès, A.F. Garito, A.J. Heeger. Phys. Rev. Lett., 35, 445 (1975).

III

IMPURITY EFFECTS AND DISORDER IN ONE DIMENSION

Conference on Organic Conductors and Semiconductors, Siófok, Hungary 1976

THE IMPURITY EFFECT ON THE PHASE TRANSITIONS IN THE QUASI-ONE-DIMENSIONAL CONDUCTORS

A.I. LARKIN and V.I. MEL'NIKOV

L.D. Landau Institute for Theoretical Physics of the Academy of Sciences of the USSR, Moscow, USSR

In the majority of quasi-one-dimensional compounds the superconducting transition is not observed, since at higher temperatures there takes place a phase transition into a dielectric state. It has been suggested that suppression of the dielectric transition by impurities may be beneficial for producing superconductivity. This concept is based on the absence of the non-magnetic impurity influence on the superconducting transition in three-dimensional conductors.

Below it is shown that in quasi-one-dimensional conductors the impurities suppress both the dielectric and superconducting transitions. This is related to the fact that the BCS formula for the superconducting transition temperature cannot be applied to quasi-one-dimensional conductors. For the three-dimensional case the transition temperature is determined by the density of electron states, its dependence on the impurity concentration being weak. In contrast, in the quasi-one-dimensional case this temperature depends on the amplitude of the electron pair jump from one thread to another and from the type of the one-dimensional correlation function.

We shall consider two types of impurities for the one-dimensional case. If the impurity potential is smooth, the process of scattering on them proceeds quasi-classically. In this case no real scattering takes place and the impurity effect may be reduced to the appearance of a random phase of the electron wave function. As has been shown by Zawadowski (1), such impurities do not affect the thermodynamics of the one-dimensional system, in which, however, no phase transitions exist. The finite temperature of the transition arises due to three-dimensional effects which establish the coherent state in the whole volume. The impurities cause the phase shift on each thread, and, as a result, the coherence drops and the transition temperature diminishes.

In case the impurity potential is rather short-ranged, the scattering with the alternative electron momentum should not be neglected. Such impurities produce a significant effect upon the superconducting transition temperature, which turns to zero at a certain concentration of impurities.

In a mean field approximation based on the electron interaction on different lines we obtain the following equation for the dielectric transition temperature (2):

$$1 = V \int \Pi(x) e^{2 i p_F x} dx , \tag{1}$$

where V is the electron interaction on the neighbouring threads, $\Pi(x)$ is the density correlator for the electrons of one thread.

Expression (1) involves the Fourier components of the density operators with the momentum $2\, p_F$. To take account of the quasi-classical impurities one should multiply $\Pi(x)$ by $\exp\left[-\frac{2i}{v_F}\int_0^x U(x')dx'\right]$, where $U(x)$ is a random potential caused by impurities. After averaging over impurities this factor acquires the form $\exp(-x/\ell)$, ℓ being the electron free path.

When the electron-electron interaction $g \ll 1$, is weak, one may observe a region of small $x \ll a \exp(1/g)$ (a is the lattice constant), in which interaction is inessential. In this region

$$\Pi(x) = \frac{2T \cos 2p_F x}{v_F^2\, \mathrm{sh}(2\pi T |x| / v_F)} \tag{2}$$

At low temperatures the right-hand side of Eq. (1) is equal to $V(\pi v_F)^{-1} \ln(\ell/a)$,therefore, provided $\ell < a \exp(\pi v_F / V)$ Eq. (1) has no solution at any temperatures, and there is no phase transition. This result concerns the case when the electron interaction on different threads is of the same order of magnitude as that on one thread. Under these conditions the result coincides with that obtained by Bulaevskii (3).

In another limiting case when the electron interaction on different threads V is weaker than the interaction $g v_F$ on one thread, the following formula for correlator $\Pi(x)$ calculated with this interaction taken into account should be inserted into formula (1) (2):

$$\Pi(x) \approx \frac{1}{v_F x} \left(\frac{x}{a \exp(1/g)}\right)^{2-1/\kappa} \tag{3}$$

where the parameter α depends on the interaction g and $1-\alpha \simeq g/2$ at $g \ll 1$.

Then for the free path ℓ at which the dielectric transition temperature turns to zero we get the equation

$$1 \approx \frac{V}{v_F (1-\alpha)} \left(\frac{\ell}{a \exp (1/q)} \right)^{2-1/\alpha} \tag{4}$$

Thus, at a sufficiently small free path the dielectric transition vanishes.

The superconducting transition in a quasi-one-dimensional case may take place only when electron jumps from one thread into another are allowed. In this case the transition temperature can be found from formula (38) in paper (2):

$$1 = W \int G(x, \tau) dx d\tau \tag{5}$$

where W is the amplitude of the electron pair jump from one thread into another, $G(x, \tau)$ is the correlator of superconducting pairs on one thread.

In the case of quasi-classical scattering the correlator $G(x, \tau)$ does not change. Hence, as was in the pure case (2), we obtain the following relation between the transition temperature and parameter W :

$$(T_c/\Delta)^{2-\alpha} = W v_F / \varepsilon_F^2 \tag{6}$$

where Δ is the gap determined by the BCS formula.

The amplitude W decreases with an increase in the impurity concentration. This follows from the expression

$$W_{ij} = J^2 \int d\omega dx \langle F_{i\omega}(x) F^+_{j\omega}(x') \rangle , \tag{7}$$

where J is the amplitude of the electron jump from one thread into another, i and j are the numbers of the neighbouring threads, $F_{i\omega}(x)$ is the Gor'kov function, the brackets denote the averaging over the impurity positions.

If the quasi-classical impurities are taken into account function $F_{j\omega}(x)$ is multiplied by $\exp(\frac{i}{v_F} \int_0^x U_j(x) dx)$, where $U_j(x)$ is the impurity potential on the j-th thread. On averaging over the impurities for each thread we obtain:

$$W \approx \frac{J^2}{v_F^2}\int \frac{d\omega}{2\pi}\,\frac{\Delta^2}{\omega^2+\Delta^2}\int \exp\left[-2|x|\sqrt{\omega^2+\Delta^2}/v_F - |x|/\ell\right]dx \approx$$
$$\approx J^2\Delta\tau/v_F\,, \qquad (\tau = \ell/v_F) \tag{8}$$

It follows from formulas (6) and (8) that with an increase in the concentration of quasi-classical impurities the superconducting transition temperature decreases, but remains finite at any impurity concentration.

The impurities which scatter the electrons backwards produce a profound effect upon the superconducting transition temperature, since they change not only W in accordance with formula (8), but also the law of decreasing function $G(x,\tau)$ with distance.

As in the pure case, we suggest the behaviour of $G(x,\tau)$ at large distances to be determined by slow fluctuations of the phase:

$$G(x,\tau) = \left(\int e^{-F[\varphi]}D\varphi\right)^{-1}\int e^{i\varphi(x,\tau)-i\varphi(0,0)}e^{-F[\varphi]}D\varphi\,, \tag{9}$$

where the functional $F[\varphi]$ is calculated in the assumption that the local superconducting state is established and equals

$$F[\varphi] = \frac{K}{2}\int\left[\left(\frac{\partial\varphi}{\partial\tau}\right)^2 + v_s^2\left(\frac{\partial\varphi}{\partial x}\right)^2\right]dx\,d\tau\,, \tag{10}$$

where K is the density of states on the Fermi level, Kv_s^2 being the superfluid density.

From formulas (9) and (10) it follows that at large distances we have:

$$G(x,\tau) \approx (v_s/\Delta)^{\alpha}(\Delta/\varepsilon_F)^2(x^2+v_s^2\tau^2)^{-\alpha/2} \tag{11}$$

where

$$\alpha = 2(\pi K v_s)^{-1} \tag{12}$$

At $\alpha < 2$ large distances $R \sim v_s/T$ are essential in integral (5) and for the transition temperature we get formula (6).

It is natural to believe that $G(x,\tau)$ diminishes more rapidly for the dirty case than for the pure one. There are two possibilities for the law of decreasing of $G(x,\tau)$. The first possibility is that the impurity depresses the long-wave sound excitations. In this case the correlator $G(x,\tau)$ drops exponentially at large distances, the integral in expression (5) is finite even at the zero temperature, so that the superconducting transition is not observed at small W and at not very small concentration of impurites.

In the second case the long-wave sound excitations exist even in the presence of impurities which change the parameter α in formula (10). Indeed, the coefficient at $(\partial\varphi/\partial x)^2$ in the free energy (10) is proportional to the superfluid density, i.e., to the correlator of the electron velocities. Even for the three-dimensional case this correlator decreases with an increase in the impurity concentration. Below we shall show that for the one-dimensional case its decrease proceeds more rapidly.

To find the velocity correlator let us use the method proposed by Berezinskii (4) for calculating conductivity in the one-dimensional system. The velocity correlator should be expanded into a series in the interaction with impurities and each term of this series should be written in a coordinate representation. The Green's function in a mean field approximation has the form

$$G(x,\varepsilon) = v_F^{-1}\left[\hat{A}e^{ip_F|x|} + \hat{B}e^{-ip_F|x|}\right]\exp\left(-|x|\sqrt{\varepsilon^2+\Delta^2}/v_F\right) \tag{13}$$

where $\varepsilon = (2n+1)\pi T$

$$\begin{pmatrix}\hat{A}\\ \hat{B}\end{pmatrix} = \frac{1}{2\sqrt{\varepsilon^2+\Delta^2}}\begin{pmatrix}-i\left(\varepsilon\pm\sqrt{\varepsilon^2+\Delta^2}\right) & -\Delta \\ \Delta^* & i\left(\varepsilon\mp\sqrt{\varepsilon^2+\Delta^2}\right)\end{pmatrix}; \tag{14}$$

Hence, $\hat{A}\hat{B} = \hat{B}\hat{A} = 0$. Therefore, as in paper (4) the correlator is presented by a loop consisting of two lines. One of them contains the matrices $\hat{B}$. Since $(i\hat{A})^n = i\hat{A}$, and $(-i\hat{B})^n = -i\hat{B}$, the product of matrices in the correlator gives the factor

$$S_p\,\hat{A}\sigma_z\hat{B}\sigma_z = \frac{\Delta^2}{\varepsilon^2+\Delta^2} \tag{15}$$

Factors $\exp(-x\sqrt{\varepsilon^2+\Delta^2})$ may be attributed to the impurity vertices, so that each graph differs from the corresponding one in paper (4) by the substitution of $\omega/2$ by $i\sqrt{\varepsilon^2+\Delta^2}$. Having performed such a substitution in the expression for conductivity (4) in the region $\tau\varepsilon \ll 1$ we get*

$$K v_s^2 = \int_0^\beta d\tau \int dx \langle v(\tau,x) v(0,0)\rangle \approx 32\,\zeta(3)\pi^{-1} v_F (\Delta\tau)^2 \ln(1/\Delta\tau) \quad (16)$$

This formula is valid in the limit $\Delta\tau \ll 1$. The effect of impurities upon the density of states is negligible. Substituting (16) into (12) we obtain

$$\alpha = \frac{\alpha_0}{\Delta\tau\sqrt{32\,\zeta(3)\ln(1/\Delta\tau)}}, \quad (17)$$

where α_0 is the value of α in the absence of impurites, ζ is the Rieman function. When interaction is weak α_0 is close to unity.

Thus, in the dirty limit $\Delta\tau \ll 1$ the parameter α is large, consequently, for $\tau < \tau_{crit} \sim \Delta^{-1}$ the value of α exceeds 2. In this case the integral in formula (5) is finite even at the zero temperature, thus, Eq. (5) at small W has no solution and the superconducting transition is absent.

The occurrence of the dielectric transition in quasi-one-dimensional conductors means that the electron interaction on different threads (V in formula (1)) is larger than the amplitude of the pair jump (W in formula (5)). The dielectric transition can be suppressed by introducing a sufficiently large amount of impurities, but the superconducting transition in such a case will inevitably be suppressed too. And only exotic impurities may save the situation which either do not reverse the electron momentum, or favour the probability of the electron jump from one thread to another.

The quantitative comparison with the experiment is hampered by the difficulty of the experimental determination of free paths in the expressions obtained. Scattering on the quasi-classical impurities does not at all affect conductivity, while the scattering with the reverse of momentum leads to

*The numerical coefficient in this formula is written with regard to the results of A. A. Gogolin et al. (5).

localization of the electron and to the absence of conductivity at low temperatures.

Some quasi-one-dimensional conductors (KCP and the quinolinium and acridizinium TCNQ salts) display an intrinsic disorder. Apparently such a disorder does not result in a short free path, since in all given compounds there is a dielectric transition (6). The electron scattering on the impurities seems to explain the absence of the dielectric transition in the quasi-one-dimensional inorganic polymer $(SN)_x$.

REFERENCES

(1) A. Zawadowski, ZhETF, 54, 1429, 1968.

(2) K. B. Efetov, A. I. Larkin, ZhETF, 69, 764, 1975.

(3) L. N. Bulaevskii, Uspekhi Fiz. Nauk 115, 263, 1975.

(4) V. L. Berezinskii, ZhETF, 65, 1251, 1973.

(5) A. A. Gogolin, V. I. Mel'nikov, E. I. Rashba, ZhETF, 69, 327, 1975.

(6) F. F. Igoshin, A. P. Kir'yanov, V. N. Topnikov, I. F. Shchegolev, ZhETF 68, 1203, 1975; V. N. Topnikov, I. F. Shchegolev, Pisma v ZhETF 20, 404, 1974.

Conference on Organic Conductors and Semiconductors, Siófok, Hungary 1976

SINGULARITY OF THE ELECTRONIC DENSITY OF STATES DUE TO IMPURITIES IN THE ID-SYSTEMS WITH THE HALF-FILLED ZONE

L.P. GORKOV

L.D. Landau Institute for Theoretical Physics, Moscow

/ Title only/

Conference on Organic Conductors and Semiconductors, Siófok, Hungary 1976

CONDUCTIVITY OF QUASI-ONE-DIMENSIONAL METAL SYSTEMS WITH RANDOM IMPURITIES AT T=O

A.A. ABRIKOSOV, I.A. RYZHKIN

L.D. Landau Institute for Theoretical Physics, Moscow, USSR

1. Introduction

One of the important problems of the physics of quazi one-dimensional systems is the interaction with random impurities which in this case play a much more substantial role than in three-dimensional systems. As well Known in three-dimensional metals the impurities cause a finite residual resistance at T=O.

For possible electronic phase transitions the ordinary nonmagnetic impurities at small concentrations have a small influence on the superconducting transition (see e.g. [1]) but can considerably suppress the excitonic (Peierls) transition (see e.g. [2]).

In a purely one-dimensional case an infinitesimal impurity can cause as first stated by Mott and Twose [3] , a localization of electrons and an infinite resistance at T=O. The electronic phase transitions cannot happen in a purely one-dimensional system because of the fluctuations of the order parameter.

In the quasi one-dimensional systems these fluctuations are suppressed to some extent by the interchain hopping of electrons (see e.g. [4]) and this makes the transitions possible. However the role of impurities even in this case will be essential. The mean field approach disregarding the fluctuations leads in the case of attraction to the possibility of both types of transitions [5] . The impurities act differently on the superconducting and

Peierls transitions and this produces a principle way to change the properties of materials. The influence of impurities on the interchain hopping probability is also essential since the hopping is necessary to suppress the fluctuations of the order parameter [6].

A rigorous calculation of the conductivity of a one-dimensional metal was done by V.B.Berezinski [7], who developped a special diagram technique. Unfortunately, this technique is very complicated and does not permit a straightforward generalization to quazi one-dimensional systems which are in fact of physical interest. Therefore we have worked out another technique which is more automatic and makes it possible to take into account the interchain electron hopping. This technique was applied to particular physical systems. The main stress was on the conductivity.

2. Conductivity of a One-Dimensional Metal at T=0

We shall start by the outline of the method using as example the problem already solved in [7] not only because it is the simplest one, but also because the calculation of properties of quazi one-dimensional systems will result in calculation functions of a purely one-dimensional system.

Consider a model of electrons in a linear chain with a spectrum

$$\varepsilon = p^2/2m \tag{1}$$

at T=0. The Fermi momentum is connected with the electron density (spinless electrons) by the relation

$$n_e = 2p_0/2\pi \tag{2}$$

Since only electrons close to the Fermi surface are essential so we introduce the energy

$$\xi = p^2/2m - p_0^2/2m \approx v(|p|-p_0), \quad v = p_0/m$$

If the electron has its momentum close to $|p| = p_0$ so at $p = p_0 + k$, $\xi \approx v k$, if however $p = -p_0 + k$, $\xi \approx -v k$. We shall introduce a matrix description denoting by the index 1 the vicinity of p_0 and by 2 the vicinity of $-p_0$. In this case

$$\xi = v k \sigma_3 \tag{3}$$

where σ_3 is the Pauli matrix.

The interaction with impurities we shall consider in the Born approximation. Since only such scattering processes are essentail in which the electron remains close to the Fermi surface we must introduce two amplitudes, namely $U_1 = U(k=0)$ and $U_2 = U(k=2p_0)$ and the corresponding collision times

$$1/\tau_1 = \pi N_i |U_1|^2/v, \quad 1/\tau_2 = \pi N_i |U_2|^2/v \tag{4}$$

N_i being the impurity concentration. We shall assume $1/\tau_1$, $1/\tau_2 \ll \varepsilon_F$.

The interaction with random impurities can be described in terms of a random potential. Taking into account two types of scattering we introduce two potentials η and ζ, and take the total potential in the form

$$\hat{\varphi}(z) = \eta(z) + \zeta(z)\sigma^{(+)}/2 + \zeta^*(z)\sigma^{(-)}/2 \tag{5}$$

where $\sigma^{(\pm)} = \sigma_1 \pm i\sigma_2$ and σ_1, σ_2 are the Pauli matrices. We assume that the fields η and ζ are averaged with Gaussian functionals

$$\Phi_1[\eta] = \exp[-\int \eta^2(z) dz\, \tau_1/v] \tag{6}$$

$$\Phi_2[\zeta, \zeta^*] = \exp[-\int |\zeta(z)|^2 dz\, \tau_2/v] \tag{7}$$

Hence we have

$$\langle \eta \rangle = 0, \quad \langle \eta(z)\eta(z')\rangle = \delta(z-z')v/\tau_1$$
$$\langle \zeta \rangle = 0, \quad \langle \zeta(z)\zeta(z')\rangle = 0, \quad \langle \zeta(z)\zeta^*(z')\rangle = \delta(z-z')v/\tau_2 \tag{8}$$

The complex potential ζ describes the back-scattering, $p_0 \to -p_0$ as reflected in (5). The ζ-averageing procedure corresponds to the fact that an electron in the p_0 region can only lose a momentum $2p_0$ and the one in the $-p_0$ region is able to gain only $2p_0$.

It can be easily demonstrated that the problem with the random potential $\hat{\varphi}$ is equivalent to the Born approximation. Indeed let us pass from the integrals over z to discrete sums introducing

$$\eta_i = \eta(z_i)\Delta/v, \quad \zeta_i = -i\zeta(z_i)\Delta/v \tag{9}$$

where Δ is the interval between two adjacent points; the limit $\Delta \to 0$ will be taken in the final result. Then instead of the formulas (6) and (8) we get

$$\Phi_1(\eta) = \exp[-\sum \eta_i^2 \tau_1 v/\Delta]$$

$$\langle \eta_i \eta_k \rangle = \delta_{ik}\Delta/v\tau_1$$

The Born approximation corresponds to some expression which contains $\sum_i \eta_i^2$. Averaging this we get

$$(\Delta/v\tau_1)\sum_i \to \int dz_i/v\tau_1 \qquad \text{at } \Delta \to 0$$

If we consider however a fourth order term we get

$$\sum_i \langle \eta_i^4 \rangle \to (\Delta^2/2v^2\tau_1^2)\sum_i \to (\Delta/2v\tau_1)\int dz_i \to 0$$

This proves at the same time that one should not take into account terms containing η_i at one point of higher order than η_i^2. For the ζ_i fields the essential terms are those containing ζ_i, ζ_i^* and $|\zeta_i|^2$. Hence using only these terms we get an approximate solution of the problem which however becomes exact in the limit $\Delta \to 0$.

The retarded Green function $G_{\omega R}(z)$ in the zeroth approximation is

$$G^{(0)}_{\omega R \alpha\beta}(z) = \int \frac{\exp(ipz)\, dp/2\pi}{\omega - vp\sigma_3 + i\delta} = -\frac{i}{v} \exp(i\omega|z|/v) \begin{pmatrix} \theta(z) & 0 \\ 0 & \theta(-z) \end{pmatrix} \quad (10)$$

In case of only the $\eta(z)$ field present the $G(zz')$ acquires the factor

$$\exp\left(-i\sigma_3 \int_{z'}^{z} \eta(z_1)\, dz_1/v\right)$$

In the case of both fields ζ and η acting simultaneously one can pass to the interaction representation with respect to η and then the operators resulting from the field ζ become

$$\sigma^{(+)}\tilde{\zeta}(z) = \exp\left(i\sigma_3 \int^{z} \eta(z_1)\, dz_1/v\right) \sigma^{(+)} \zeta(z) \exp\left(-i\sigma_3 \int^{z} \eta(z_1)\, dz_1/v\right) =$$

$$= \sigma^{(+)} \exp\left(2i \int^{z} \eta(z_1)/v\right) \zeta(z)$$

(11)

In the averaging over ζ the phase factors of $\zeta(z)$ and $\zeta^*(z)$ cancel. Hence only the outer "envelops" of the type $\exp\left(\pm i \int^{z} \eta(z_1)\, dz_1/v\right)$ remain. But if the expressions under consideration are of the form of closed loops (and this is the case of physical quantities) so even these factors cancel. That proves that the field η does not influence the conductivity and therefore one can put from the very beginning $\eta = 0$.

The equation for the Green function in the field ζ has the form:

$$\left(iv\sigma_3\frac{\partial}{\partial z}+\omega-\frac{1}{2}\sigma^{(+)}\zeta-\frac{1}{2}\sigma^{(-)}\zeta^*\right)G_{\omega R}(zz')=\delta(z-z') \tag{12}$$

This equation can be transformed to another form. Let z_1 be an arbitrary point. Then it follows from (12) that

$$G_{\omega R}(zz')=S_\omega(zz_1)G_{\omega R}(z_1z')-\frac{i}{v}S_\omega(zz')\,\theta(z-z')\,\theta(z_1-z') \tag{13}$$

where the S-matrix is

$$S_\omega(zz')=T_z\exp\left[i\int_{z'}^{z}\left[\sigma_3\omega-\frac{1}{2}\sigma^{(+)}\zeta(z_1)+\frac{1}{2}\sigma^{(-)}\zeta^*(z_1)\right]dz_1/v\right] \tag{14}$$

To find the Green function we must know the boundary conditions. They are obtained by the following reasoning. Consider e.g. $G_{\omega R11}(zz')$. The terms of the ζ, ζ^* expansion are of the form

$$\int G_{11}^{(0)}(z-z_1)\zeta(z_1)G_{22}^{(0)}(z_1-z_2)\zeta^*(z_2)\ldots G_{11}^{(0)}(z_n-z')dz_1\ldots dz_n$$

According to (10) $G_{11}^{(0)}\sim\theta(z-z_1)$ and hence the z_1 -integral is limited by $-\infty<z_1<z$ and the z_n - integral by $z'<z_n<\infty$. Hence at $z\to-\infty$ or $z'\to+\infty$, $G_{\omega R11}\to 0$.

In the same way the boundary conditions for other components of $G_{\omega R}$ are obtained. Using these one can put in (13) $z_1\to-\infty$ and then define $G_{\omega R2\beta}(-\infty z')$ from the same equation. One gets

$$G_{\omega R\alpha\beta}(zz')=(i/v)\Big\{S_{\omega\alpha2}(z,-\infty)\left[S_\omega(\infty,z)\sigma_3\right]_{2\beta}/S_{\omega22}(\infty,-\infty)-$$
$$-\theta(z-z')\left[S_\omega(zz')\sigma_3\right]_{\alpha\beta}\Big\} \tag{15}$$

So $G_{\omega R\alpha\beta}$ is expressed via the components of the matrix $S_{\omega\alpha\beta}$ (14). In this matrix we can pass to the interaction representation with respect to ω

$$S_{\omega\alpha\beta}(zz')=\exp\left[i\omega(\sigma_{3\alpha\alpha}z-\sigma_{3\beta\beta}z')/v\right]\tilde{S}_{\alpha\beta}(zz') \tag{16}$$

where the matrix $\tilde{S}$ has formally the same form as (14) with $\omega=0$ but the fields ς, ς^* are changed to

$$\tilde{\varsigma} = \varsigma(z)\exp(-2i\omega z/v), \quad \tilde{\varsigma}^* = \varsigma(z)\exp(2i\omega z/v) \tag{17}$$

Now let us pass from integrals to the discrete sums (9) and remember what was said about keeping terms of the order ς_i, ς_i^* and $|\varsigma_i|^2$ at one point. With this accuracy one can obtain the following form of $\tilde{S}_{11}$

$$\tilde{S}_{11}(zz') = \prod_{i\in(zz')}(1+|\varsigma_i|^2/2)\Big(1+\sum_{z>k>l>z'}\tilde{\varsigma}_k\tilde{\varsigma}_l^* + \sum_{z>k>l>m>n>z'}\tilde{\varsigma}_k\tilde{\varsigma}_l^*\tilde{\varsigma}_m\tilde{\varsigma}_n^* + \ldots\Big) \tag{18}$$

and correspondingly for the other components of $\tilde{S}_{\alpha\beta}$. These expressions permit us to derive the following connections for components of the fall $S_{\omega\alpha\beta}$ (they are true with the same accuracy)

$$\begin{aligned} S^{-1}_{\omega 11} &= S_{\omega 22}, \quad S^{-1}_{\omega 12} = -S_{\omega 12} \\ S^{-1}_{\omega 22} &= S_{\omega 11}, \quad S^{-1}_{\omega 21} = -S_{\omega 21} \end{aligned} \tag{19}$$

It follows that

$$S_{\omega 11}(zz')S_{\omega 22}(zz') - S_{\omega 12}(zz')S_{\omega 21}(zz') = 1 \tag{20}$$

and further

$$\begin{aligned} G_{\omega R11}(zz') &= G_{\omega R22}(z'z) \\ G_{\omega R12}(zz') &= G_{\omega R12}(z'z) \\ G_{\omega R21}(zz') &= G_{\omega R21}(z'z) \end{aligned} \tag{21}$$

These relations help to get rid of the second term in (15).

In the same way one can define the $G_{\omega A\alpha\beta}$ which appears to be

$$G_{\omega\alpha\beta}(zz') = (i/v)\{S_{\omega\alpha 1}(z,-\infty)[S_\omega(\infty,z')\sigma_3]_{1\beta}/S_{\omega 11}(\infty,-\infty) - \Theta(z-z')[S_\omega(zz')\sigma_3]_{\alpha\beta}\} \tag{22}$$

Now from the expressions of the type of (18) one can find the relations resulting from the interchange $\zeta \rightleftarrows \zeta^*$; namely one has simultaneously to interchange $1 \rightleftarrows 2$ and $\omega \to -\omega$. This makes it possible to establish some important connections between G_R and G_A ; namely the interchange $\zeta \rightleftarrows \zeta^*$ leads to $G_R \to -G_A$, $1 \rightleftarrows 2$, $\omega \to -\omega$. Since the ζ -averaging is taken with a functional depending only on $|\zeta|^2$ one can interchange $\zeta \rightleftarrows \zeta^*$ during the averaging and this causes considerable reduction.

By means of (15),(22) and (19) one can show that after averaging of the functions G_R and G_A they acquire in comparison with $G_R^{(0)}$ and $G_A^{(0)}$ the factor $exp(-|z-z'|/2l_2)$ where $l_2 = v\tau_2$. It follows hence that the introduction of completely random impurities does not change the state density although the states themselves are no more free as it will be seen from the conductivity.

Let us pass to the calculation of the conductivity. For the longitudinal current we have the expression

$$j_z(\omega_0 z) = \int_{-\infty}^{\infty} Q_R(\omega_0 z z_1) A(\omega_0 z_1) dz_1 \tag{23}$$

where
$$Q_R(\omega_0) = \begin{cases} Q(\omega_0), & \omega_0 > 0 \\ Q^*(\omega_0), & \omega_0 < 0 \end{cases} \tag{24}$$

$$Q(\omega_0) = -e^2 n_e/mc - (ie^2/4m^2c)\left(\frac{\partial}{\partial z} - \frac{\partial}{\partial z'}\right)_{z'\to z}\left(\frac{\partial}{\partial z_1} - \frac{\partial}{\partial z_1'}\right)_{z_1'\to z_1} \cdot \int G_{\omega+\omega_0\alpha\beta}(zz_1')G_{\omega\beta\alpha}(z_1 z')dz_1 d\omega/2\pi \tag{25}$$

Substracting and adding the same $Q(\omega_0)$ but without impurities one gets for the difference an integral where only the vicinity of $\omega = 0$ and $|p| = p_0$ is essential. This makes it possible to use the expressions (15),(22). Finally we get

$$Q = \frac{ie^2 p_0^2}{m^2 c} \int_{-\omega_0}^{0} \frac{d\omega}{2\pi} \int_{-\infty}^{\infty} dz_1 \, Sp[\sigma_3 G_{R\,\omega+\omega_0}(z z_1) \sigma_3 G_{A\omega}(z_1 z)] \tag{26}$$

After substitution of (15),(22) with regard of (21) we obtain

$$Q = (ie^2\omega_0/2\pi c)\Big\{ \int_{-\infty}^{z} [S_{\omega_0 22}(z_1,-\infty) S_{11}(z_1,-\infty) - S_{\omega_0 12}(z_1,-\infty) S_{21}(z_1,-\infty)] \cdot$$

$$\cdot [S_{\omega_0 22}(\infty,z) S_{11}(\infty,z) - S_{\omega_0 21}(\infty,z) S_{12}(\infty,z)] dz_1 \cdot$$

$$\cdot [S_{\omega_0 22}(\infty,-\infty) S_{11}(\infty,-\infty)]^{-1} + \int_{z}^{\infty} \ldots \Big\} \tag{27}$$

We do not write the second integral explicitly since after averaging it gives the same result as the first one, i.e., one must simply take the double contribution of the $\int_{-\infty}^{z} dz_1$.

Let us first consider the case $\omega_0 = 0$. Then according to (20) the first two brackets in (27) reduce to 1. Let us introduce instead of an infinite chain a finite sample ($0 \div L$).Taking into account that $Q = i\omega_0 \sigma / c$ we get

$$\sigma(0) = \frac{e^2}{2\pi} L F_0(L) \tag{28}$$

where

$$F_n(z) = \left\langle \frac{[S_{12}(z,0) S_{21}(z,0)]^n}{[S_{22}(z,0) S_{11}(z,0)]^{n+1}} \right\rangle \tag{29}$$

Now we pass from a continuous to discrete z and single out the ζ_1, ζ_1^* operators corresponding to the largest z . So we get for example

$$S_{12}(z,0) = (1 + |\zeta_1|^2/2)[S_{12}(z-\Delta,0) + \zeta_1 S_{22}(z-\Delta,0)]$$

Performimg this with all the $S_{\alpha\beta}$ in (29), developping in $\varsigma_1, \varsigma_1^*$ and leading terms only of the necessary order we can average over ς_1 and ς_1^* since these ς's do not enter in the $S_{\alpha\beta}(z-\Delta)$. These results in a system of difference equations connecting $F_n(z)$ with different $F_n(z-\Delta)$ which in the limit $\Delta \to 0$ become differential equations

$$\frac{\partial F_n}{\partial t} = n^2 F_{n-1} + (n+1)^2 F_{n+1} - [n^2+(n+1)^2] F_n \tag{30}$$

where $t = z/l_2$.

Making a Laplace transformation

$$F_n(t) = \int_0^\infty F_{ns} e^{-st} ds \tag{31}$$

and introducing the generating function

$$F_s(x) = \sum_{n=0}^{\infty} F_{ns} x^n \tag{32}$$

we come to an equation for $F_s(x)$:

$$x(1-x)^2 \frac{d^2 F_s}{dx^2} + (1-x)(1-3x)\frac{dF_s}{dx} + [s-(1-x)] F_s = 0 \tag{33}$$

with the boundary condition

$$F(t=0, x) = \int F_s(x) ds = 1 \tag{34}$$

Solving (33) we get

$$F(x,t) = \int_0^\infty d\lambda \frac{2\pi\lambda \operatorname{sh}\pi\lambda}{\operatorname{ch}^2\pi\lambda} \exp[-(\tfrac{1}{4}+\lambda^2)t] \cdot F(\tfrac{1}{2}+i\lambda, \tfrac{1}{2}-i\lambda, 1; \tfrac{-x}{1-x}) \tag{35}$$

$F(\alpha\beta\gamma, z)$ being the hypergeometric function. Putting $x=0$ $t = L/l_2 \gg 1$ and inserting in (28) we obtain:

$$\sigma(0) = \frac{\pi}{4}^{3/2} e^2 l_2^{3/2} L^{-1/2} \exp(-L/4l_2) \tag{36}$$

If $L \to \infty$, $\sigma(0) \to 0$; this means localization.

To calculate $Q(\omega_0)$ we must average the expression (27). The method applied for calculation of $\sigma(0)$, i.e., the singling out of ζ corresponding to the edge point is unapplicable here since one can show easily that it works only for expressions constructed according to Fig.1, where every line $(z_i\ z_k)$ corresponds to all $S(z_i\ z_k)$. The expression in (27) is constructed in a different way, (see Fig.2). Therefore we transform this expression.

One easily obtains

$$[S_{11}(L,0)]^{-1} = [S_{11}(L,z)S_{11}(z,0) + S_{12}(L,z)S_{21}(z,0)]^{-1} =$$

$$= \sum_n (-1)^n \frac{[S_{12}(L,z)S_{21}(z,0)]^n}{[S_{11}(L,z)\,S_{11}(z,0)]^{n+1}}$$

The same we do with $S_{22\omega_0}(L,0)$. Now taking into account that $S_{12}^n(L,z)$ contains n "unpaired" ζ which have to be compensated by the same number of ζ^* belonging to the same interval we get $[S_{11}(L,0)S_{22\omega_0}(L,0)]^{-1} \rightarrow$

$$\rightarrow \sum_n \frac{[S_{12}(L,z)S_{21\omega_0}(L,z)]^n}{[S_{11}(L,z)\,S_{22\omega_0}(L,z)]^{n+1}} \frac{[S_{21}(z,0)S_{12\omega_0}(z,0)]^n}{[S_{11}(z,0)\,S_{22\omega_0}(z,0)]^{n+1}}$$

Inserting this into (27) one can see that now every term of $\sum_n$ has the structure presented at Fig.3, i.e., we have cut off the piece $(L\ z)$. Since this piece is averaged independently of the rest and on the other hand the average in the interval $(0 \div z_1 \div z)$ has just the form of Fig.1, so we can apply again the method used for $\sigma(0)$.

Omitting the details of this calculation we shall write only the final expressions for the dielectric constant and conductivity

$$\varepsilon(0) = 4\pi c \, \mathrm{Re}\, Q / \omega_0^2 = 16\,\zeta(3) e^2 v \tau_2^2 \tag{37}$$

$$\sigma(\omega_0 \ll 1/\tau_2) = \frac{c}{\omega_0} \,\mathrm{Im}\, Q = \frac{4}{\pi} e^2 v \omega_0^2 \tau_2^3 \ln \frac{1}{\omega \tau_2} \tag{38}$$

For $\omega_0 \gg 1/\tau_2$ we get

$$\varepsilon = -\frac{4\pi n_e e^2}{m \omega_0^2}, \quad \sigma = \frac{2 n_e e^2 \tau_2}{m (\omega_0 \tau_2)^2} \tag{39}$$

The results (37),(38) have been obtained already by Berezinski [7] and the results (39) can be found by perturbation theory for small ζ, ζ^*. But the aim of this section was only a brief description of the method.

3. The influence of impurities on pairing [8]

In a one-dimensional metal the formation of Cooper pairs is accompanied by the Peierls dielectric pairing [5]. This leads to a summation of "parquet" diagrammes instead of a "ladder" occuring in the three-dimensional case. The "parquet" corresponds to a logarithmic approximation and the calculation of the next approximations encounters considerable difficulties.. In the logarithmic approximation one obtains that in case of attraction of electrons both types of pairing appear simultaneously. It is not clear however whether this result is reliable and cannot it happen really that the dielectric coupling prevents the formation of Cooper pairs

The exprerimental study of quazi one-dimensional systems based on TCNQ reveals just the latter situation. There are however some substances of this kind which having several equivalent positions for certain complexes possess an intrinsic disorder and it is known that in such substances the dielectric transition is either not observed or happens at very low temperatures [10] . So one can suspect that the scattring by impurities or random inhomogeneities can suppress the Peierls pairing and thus help the superconducting transition.

An anlysis of this question meets complications since in a purely one-dimensional substance the fluctuations exclude the existence of an order parameter (i.e., a phase transition) at finite temperatures. The fluctuations can be suppressed by means of the intercahin hopping but the impurities may prevent this process. The study of the influence of a smooth random potential (the potential $\eta(z)$ in our notations) in the work [6] shows that what concerns fluctuations both types of transitions are suppressed by impurities, i.e., there is no gain on this way. One must have in mind however, that the fluctuation are imposed on the purely one-dimensional pairing and the absence of the latter means definitely no pairing even with hopping.

Zawadowski [11] stidied the influence of the potential $\eta(z)$ on the purely one-dimensional pairing and he found that there is no influence at all in any of the channels. A natural question appears what is the influence of the potential ζ corresponding to the back-scattering $p_0 \to -p_0$.

The major element defining the Cooper pairing is the loop presented at Fig.4. In the absence of scattering it is proportional to $\ln[\mathcal{D}/(-i\omega_0)]$. This results finally in an appearance of an imaginary pole in the fall vertex part for attractive interaction and this means an instability of the Fermi spectrum and a superconducting transition. In the presence of impurities one can expect that for $\omega_0 \ll 1/\tau_2$ this log will be changed to $\ln[\mathcal{D}\tau_2]$

and hence the absence of a pole with respect to ω_0.

To analyze this question one must calculate the integral

$$\Pi_1 = \int_{-\infty}^{\infty} d\omega \int_{-\infty}^{\infty} dz_1 [G_{11}(\omega_0 - \omega; z, z_1) G_{22}(\omega; z, z_1) + \\ + G_{21}(\omega_0 - \omega; z, z_1) G_{12}(\omega; z, z_1)] / 2\pi \tag{40}$$

After the substraction of the corresponding expression without impurities the difference contains only the vicinities of $p = \pm p_0$ i.e., the expressions derived in Section 2 are applicable. A calculation by means of the method described in Sec.2 shows that the diagramme Π_1 remains proportional to $\ln[2\mathcal{D}/(-i\omega_0)]$ whatever the relation between ω_0 and τ_2^{-1} may be.

The main element defining the Peierls instability is the loop at Fig.5 which means

$$\Pi_2 = \int_{-\infty}^{\infty} \frac{d\omega}{2\pi} \int_{-\infty}^{\infty} dz_1 G_{11}(\omega + \omega_0, z, z_1) G_{22}^{2}(\omega; z_1, z)$$

The calculation of this diagramme gives

$$\Pi_2 \approx -\frac{i}{2\pi v} \ln \frac{2\mathcal{D}}{\tau_2^{-1} - i\omega_0} \tag{41}$$

It follows that if $\omega_0 \tau_2 \ll 1$ these log's become unimportant, i.e., the parquet is transformed to a ladder. So there are some reasons to hope that the impurities can help in suppressing the Peierls instability to a larger extent than the superconducting one.

Unfortunately the real quazi one-dimensional systems known at present have either an internal disorder which produces such a strong scattering that both transitions are suppressed or they have a "strong" Peierls transition at temperatures higher than the Debye frequency of phonons that cannot be handled in the framework of the usual scheme of weak interaction. Our reasoning cannot be applied of course to such a material.

4. Conductivity of a quazi one-dimensional metal at T=0 [12]

Consider now a quazi one-dimensional metal with a spectrum

$$\xi(\vec{p}) = \mathcal{E}(\vec{p}) - \mu = v(|p_z| - p_0) + \alpha(p_x, p_y) \tag{42}$$

where $\int \alpha \, dp_x \, dp_y$ (the integral is taken over the (xy) section of the Brillouin zone), $|\alpha| \ll p_0 v$. We sometimes will use as α the strong coupling formula for a rectangular cell:

$$\alpha = \alpha_x \cos(p_x a_x) + \alpha_y \cos(p_y a_y) \tag{43}$$

It is clear that for

$$\alpha l_i / v \gg 1 \tag{44}$$

the metal will practically become three-dimensional and the kinetic equation can be used. The interacting case is however $\alpha l_2 / v \ll 1$ (this corresponds to substances with intrinsic disorder) where the localization effects must play an essential role.

A rough estimate of the conductivity in this case can be obtained from a diffusional consideration. First of all the collisions must lead to a diffusion in the (xy) plane. The corresponding diffusion coefficient will be equal to

$$\mathcal{D}_\perp \sim v_\perp^2 \tau$$

where $\tau = (\tau_1^{-1} + \tau_2^{-1})^{-1}$ and $v_\perp \sim \frac{\partial \alpha}{\partial p_\perp} \sim \alpha a$, "a" being the lattice period in the (xy) plane. The distance at which the electron moves in the transverse direction during the time t is $\rho \sim \sqrt{\mathcal{D}_\perp t}$ If the impurity potential is short ranged it suffices for the electron to jump on the neighbouring chain to get into a completely new localizing potential. Since the localization radius is of the order of l_i the electron can move at a distance of the order of l_z in the z direction. It follows hence that $\rho \sim a$, $t \sim \frac{a^2}{\mathcal{D}_\perp}$

$$\mathcal{D}_{\parallel} \sim l_z^2/t \sim (l_z/a)^2 \mathcal{D}_\perp$$

The conductivity is connected with the diffusion coefficient by the relation

$$\sigma = e^2 \nu(\mu) \mathcal{D}$$

$\nu(\mu)$ being the state density which in our case is

$$\nu(\mu) = (\pi v S)^{-1}$$

S - is the area of the (xy) section of the elementary cell $(S = a_x a_y)$. This all gives

$$\sigma_\perp \sim e^2 \alpha^2 a^2 (l_1^{-1} + l_2^{-1})^{-1} / \pi v^2 S \tag{45}$$

$$\sigma_{\parallel} \sim e^2 \alpha^2 l_z^2 (l_1^{-1} + l_2^{-1})^{-1} / \pi v^2 S \tag{46}$$

If the impurity potential has a long range $r_0 \gg a$ one must take $\rho \sim r_0$. Then it follows

$$\sigma_{\parallel} \sim e^2 \alpha^2 (a/r_0)^2 l_z^2 (l_1^{-1} + l_2^{-1})^{-1} / \pi v^2 S \tag{47}$$

The exact calculation shows that the formula (45) is in accordance with the true result, but the formula (46) is a little oversimplified (it does not contain a factor $\ln^2\left(\frac{v}{\alpha l_z}\right)$ resulting from the exact calculation).

Now we pass to the exact calculation we shall use the representation $G(z,z',\vec{p},\vec{p}',\omega)$, $\vec{p}$ being the transverse momentum. The equation for G has the form

$$\left[\omega + i v \sigma_3 \frac{\partial}{\partial z} - \alpha(\vec{p})\right] G(z,z',\vec{p},\vec{p}',\omega) = (2\pi)^2 \delta(\vec{p}-\vec{p}')\,\delta(z-z') + \\ + \int \hat{\varphi}(\vec{k},z)\, G(z,z',\vec{p}-\vec{k},\vec{p}',\omega)\, d^2\vec{k}/(2\pi)^2 \tag{48}$$

The potential $\hat{\varphi}(\vec{k},z)$ has formally the same shape (5) but now the correlators are

$$\langle \eta(\vec{k},z)\,\eta(\vec{k}',z')\rangle = (2\pi)^2\,\delta(\vec{k}+\vec{k}')\,\delta(z-z')\,\mathcal{D}_1(\vec{k}) \tag{49}$$

$$\mathcal{D}_1(\vec{k}) = \mathcal{N}_i\cdot|\mathcal{U}(k_z=0,\vec{k})|^2$$

and correspondingly for ζ.

The equation (48) can be solved formally in the same way as it was done in Sec.2. But one must observe now that the potential $\hat{\varphi}$ acts not only on the "spin" variables but produces also a momentum displacement $\vec{p}\to\vec{p}-\vec{k}$. Therefore we get e.g.

$$G_{R\alpha\beta}(z<z') = \frac{i}{v}S_{\alpha 2}(z,-\infty)[S_{22}(\infty,-\infty)]^{-1}[S(\infty,z)\sigma_3]_{2\beta} \tag{50}$$

All the $\hat{\varphi}$ entering S are operators acting on the $\vec{p}$ variable and therefore the order in (50) is essential. By means of (50) and the formula for G_A one can derive finally the following formula for the longitudinal conductivity of a sample of the length L in the z direction

$$\sigma_{zz} = \frac{e^2}{2\pi}L\,Sp_{\vec{p}}\{[S_{11}^{-1}(L,0)]^{-1}[S_{11}(L,0)]^{-1}\} \tag{51}$$

Here every field operator makes a displacement $\vec{p}\to\vec{p}-\vec{k}$. The trace means that in the bracket $\{\;\}$ these displacements compensate each other and an integral over the transverse Brillouin zone section is taken.

If we pass to the interaction representation with respect to α the operator ζ for example will take the form

$$\tilde{\zeta}(z)\to\int\zeta(\vec{k},z)\exp\{i[\alpha(\vec{p})+\alpha(\vec{p}-\vec{k})]z/v\}P_{\vec{k}}\,d^2\vec{k}/(2\pi)^2$$

where $P_{\vec{k}}$ is the momentum displacement operator. Mention that in averaging according to (49) the $P_{\vec{k}}$ operators in ζ^* and ζ

are compensated. Now as before we single out the $\zeta(z)$ and $\zeta^*(z)$ belonging to the edge point. In the general case one gets the following phase factor

$$\exp\{i[\alpha(\vec{p})+\alpha(\vec{p}-\vec{k})-\alpha(\vec{p}_1)-\alpha(\vec{p}_1-\vec{k})]z/v\}$$

This factor would become unity if $\vec{p}=\vec{p}_1$. But this demands that between ζ and ζ^* there should not be any unpaired ζ's, i.e., this should be a diagramme with nonintersecting impurity lines (every $\langle\zeta\ldots\zeta^*\rangle$ can be represented by a line). If one takes into account only such diagrammes the calculation of σ_{zz} becomes very simple and one gets

$$\sigma_{zz}=\frac{e^2}{2\pi S}\,l_2\,\frac{t}{t+t_0} \tag{52}$$

where $t=L/l_2$, $t_0\sim 1$. In case $t\gg 1$ one gets the result of the one-dimensional kinetic equation.

The omitted terms with intersections contain phase factors of the type

$$\exp[i\alpha(\vec{p})l_2/v]$$

if $|\alpha|l_2/v\gg 1$ then averaging over $\vec{p}$ in an interval $\Delta p\ll 1/a$ we can turn such a factor to zero, i.e., the criterion (44) corresponds indeed to the applicability of the kinetic equation.

In case $\alpha l_2/v\ll 1$ it is possible to find the first term of the expansion of σ_{zz} in powers of $\overline{\alpha^2}$. However for an infinite sample σ_{zz} is not an anlytic function of $\overline{\alpha^2}$ in the vicinity of $\overline{\alpha^2}=0$. Therefore we take a sample of finite length L where the expansion is possible for α being small enough. We shall omit the details of this calculation and mention only that in case of a short ranged impurity potential one has to take into account only terms where the paired ζ and ζ^* operators are not divided by a single α . Indeed in the latter case we get

$$\hat{\zeta}\alpha(\vec{p})\hat{\zeta}^*\alpha(\vec{p}) \sim \int \alpha(\vec{p})\alpha(\vec{p}+\vec{k})\mathcal{D}_2(\vec{k})d^2k/(2\pi)^2$$

The function $\mathcal{D}_2(\vec{k})$ does not change effectively at momenta of the order of π/a . But since $\bar{\alpha} = 0$ where the average is taken over a cell with periods „ a " the expression above turns to zero.

The result of the calculation has the form (we have written also the zeroth order term in α) which corresponds to the static conductivity of a purely one-dimensional metal, see (36)).

$$\sigma_{zz} \approx \frac{\pi^{3/2}}{4S} e^2 l_2^{3/2} L^{-1/2} \exp\left(-\frac{L}{4l_2}\right) + \frac{2e^2\overline{\alpha^2}}{\pi v^2 S} \cdot \frac{l_1 l_2}{l_1 + l_2} L^2 \tag{53}$$

Now we apply such a reasoning. Let us suppose that the theory contains a "correlation length", i.e., such a value of L where all expansion terms become of the same order of magnitude. Then at larger L the sample is practically infinite and its conductivity becomes independent of L. Comparing both terms in (53) we obtain

$$L_c = 4l_2 \ln(v^2/l_2^2\overline{\alpha^2}) \tag{54}$$

Inserting that in (53) we obtain

$$\sigma_{zz} \sim \frac{16e^2}{\pi S}\overline{\alpha^2}(l_2/v)^2[l_1 l_2/(l_1+l_2)] \ln^2(v^2/\overline{\alpha^2}l_2^2) \tag{55}$$

Up to the factor $\ln^2\left(\frac{v^2}{\overline{\alpha^2}l_2^2}\right)$ this formula corresponds to the diffusion estimation (46). It could be compared also with the formula

(38). They are very much alike, α playing the role of ω_0 .

The transverse conductivity is given by the formula

$$\sigma_{\perp ik} = \frac{e^2}{2\pi} Sp_{\bar{p}\sigma} \int_{-\infty}^{\infty} dz_1 \left[G_R(zz_1) \frac{\partial \alpha}{\partial p_i} G_A(z_1 z) \frac{\partial \alpha}{\partial p_k} \right] \tag{56}$$

In case of a short ranged potential as already mentioned two paired operators η or ζ cannot be separated by a single α . It follows that in the averaging in (56) G_R and G_A are averaged independently. This permits to calculate easily the $\sigma_{\perp ik}$ Having any relation between l and v/α we get

$$\sigma_{\perp ik} = \frac{e^2}{2\pi} \frac{1}{S v^2} \overline{\left(\frac{\partial \alpha}{\partial p_i} \frac{\partial \alpha}{\partial p_k} \right)} \frac{l_1 l_2}{l_1 + l_2} \tag{57}$$

this corresponds exactly to the estimate (45). So the relation between the longitudinal and transverse conductivity in case of a short ranged potential is

$$\frac{\sigma_\perp}{\sigma_\parallel} \sim \left(\frac{a}{l_2} \right)^2 \ln^{-2} \left(\frac{v^2}{\alpha^2 l_2^2} \right) \tag{58}$$

5. Longitudinal conductivity of a semimetal in a very strong magnetic field [13]

Another example of a quazi one-dimensional situation is a semimetal in a very strong magnetic field when all the carriers occupy only one Landau level. The motion in this case is finite in the (xy) plane $(H \parallel z)$ and in absence of impurities it is free along z . The radius of the state in the (xy) -plane is $\lambda = \sqrt{c/eH}$ which is called the magnetic length. When $H \to \infty$, $\lambda \to 0$ i.e., the problem becomes purely one-dimensional.

In case of a finite λ there are some essential distinctions from the previous problem. There the localization was due to impurities and the transverse motion to the band width $\alpha(\bar{p})$

which did not depend on the impurity concentration. The characteristic parameter defining the role of the one-dimensional localization was $\alpha \ell_z / v$ which could change with impurity concentration. In the problem under consideration the transverse motion i.e., the hopping of the electron from one "spiral" to another is due to the impurity scattering. Therefore as we shall see the characteristic parameter of this problem will not depend on the impurity concentration and is defined only by the range of the potential p_0 and λ

Let us consider for simplicity a model of a semimetal having carriers only of one type with an isotropic spectrum

$$\mathcal{E} = \vec{p}^2/2m \tag{59}$$

If all the carriers are concentrated in one Landau band then the Fermi momentum will be connected with the electron density by the relation

$$p_0 = 2\pi^2 n_e \lambda^2 \tag{60}$$

(the electrons are considered as spinless; this corresponds to a large spin splitting).

What concerns the impurity potential, so a real semimetal can contain neutral impurities of atomic range as well as charged impurities with a screened Coulomb potential. The Fourier component of the latter has the form

$$\mathcal{U}(\vec{k}) = \frac{4\pi e^2 Z/\varepsilon}{k_x^2 + k_y^2 + k_z^2 + \varkappa^2} \tag{61}$$

where ε is the dielectric constant and $\varkappa$ - the reciprocal Debye radius. In the following we shall need only the case $k_z = 2p_0$. Assuming $p_0/\varkappa \gg 1$ we can omit $\varkappa$ in (61).

The unperturbed Green function in our problem in the

(x, p_y, p_z, ω) representation has the form

$$G^{(0)}(x,x';p_y,p_z;\omega) = \frac{\psi_0(x-\lambda^2 p_y)\psi_0(x'-\lambda^2 p_y)}{\omega - \xi(p_z) + i\delta\, \mathrm{sign}\,\omega} \tag{62}$$

where $\psi_0(x) = (\pi\lambda)^{-1/4} \exp(-x^2/2\lambda^2)$ is the wave function of the lowest Landau level, $\xi(p_z) = v(|p_z| - p_0)$.

The Dyson equation has the form

$$G(\vec{r},\vec{r}') = G^{(0)}(\vec{r},\vec{r}') + \int G^{(0)}(\vec{r},\vec{r}_1)\hat{\varphi}(\vec{r}_1)G(\vec{r}_1,\vec{r}')\,d\vec{r}_1 \tag{63}$$

We use now the (x, p_y, z, ω) representation and suppose that G has the form

$$G(x,x';p_y,p_y';z,z';\omega) = \psi_0(x-\lambda^2 p_y)\psi_0(x'-\lambda^2 p_y')\mathcal{G}(p_y,p_y';z,z';\omega) \tag{64}$$

The substitution into (63) confirms that $\mathcal{G}$ is in fact independent of x, x' and it is defined by the equation

$$\left(\omega + iv\sigma_3 \frac{\partial}{\partial z}\right)\mathcal{G}(p_y,p_y';z,z';\omega) = 2\pi\delta(z-z')\delta(p_y-p_y') +$$

$$+\int \hat{\varphi}(k_x,k_y,z)\,d(k_x,k_y)\,e^{i\lambda^2 p_y k_x}\mathcal{G}(p_y-k_y,p_y';z,z';\omega)\frac{dk_x\,dk_y}{(2\pi)^2} \tag{65}$$

where $d = \exp[-\lambda^2(k_x^2+k_y^2+2ik_xk_y)/4]$

So in this case as in the previous one the operator $\hat{\varphi}$ acts not only on the "spin" variables but also on the momentum variables and multiplies the function by $d(k_x k_y)\,e^{i\lambda^2 p_y k_x}$. The method of the further solution of (65) and derivation of the static conductivity is in analogy with the previous section. Finally we get

$$\sigma_{zz} = (e/2\pi\lambda)^2 L\, \mathrm{Sp}\{[S_{11}^{-1}(L,0)]^{-1}[S_{11}(L,0)]^{-1}\} \tag{66}$$

here the averaging of the impurity operators is performed according to the following rule

$$\langle \zeta_i^* \ldots \zeta_i \rangle = N_i \int |U(k_x, k_y, 2p_0)|^2 \exp[-\lambda^2(k_x^2 + k_y^2)/2] \cdot$$
$$\cdot \exp[i\lambda^2 k_x \sum_n k_{yn}] dk_x dk_y / (2\pi)^2 \qquad (67)$$

where k_{yn} are the momenta of the ζ's between ζ_i^* and ζ_i.

If one takes into account only the "nonintersecting" diagrammes then these exponents are equal to unity. One comes easily to the result of the kinetic equation

$$\sigma = (e/2\pi\lambda)^2 \, l_2 \qquad (68)$$

The estimation of the omitted terms can be done by a "diffusion" approach. The sum $\sum k_{yn}$ can be expressed as $\sqrt{\mathcal{D} z}$ where z plays the role of "time". Then $\mathcal{D} \sim \overline{k_y^2}/l_2$ plays the role of the diffusion coefficient. But in our problem the $z \sim l_2$ are really essential. That means that the exponent contains a quantity of the order of

$$\left(\lambda^4 \overline{k_x^2}\, \overline{k_y^2}\right)^{1/2} \qquad (69)$$

If this quantity is large then an averaging of such a phase factor over an interval $\Delta k \ll k$ makes it zero i.e., the intersecting diagrammes can be neglected. In the opposite case they must be taken into account. So (69) plays the same role as $d l_2/v$ in the previous problem. As already mentioned before this quantity does not depend on the free path length.

From (67) it follows that the "effective" $|U|^2$ is in fact $|U(k_x, k_y, 2p_0)|^2 \exp[-\lambda^2(k_x^2 + k_y^2)/2]$. When the potential is short ranged, i.e., $r_0 \ll \lambda$, $U(\vec{k})$ changes essentially only at $\vec{k} \sim 1/r_0$. Hence this potential can be practically reduced to $U(k_x = k_y = 0)$. In case of a Coulomb potential and $p_0 \lambda \ll 1$ it is possible on the contrary to assume $\exp[-\lambda^2(k_x^2 + k_y^2)/2] \approx 1$.

Now let us consider the quantity (69). The averaging with an effective potential shows that in case of a short ranged potential it is of the order of unity and in the case of Coulomb potential and $p_0 \lambda \ll 1$ it is much smaller. That means that in no case the neglection of intersecting diagrammes is justified and so the result (68) corresponding to the kinetic equation is always incorrect.

Nevertheless in case of a short ranged potential the truth is not too far from the formula (68). Indeed the calculation of the first intersecting diagrammes shows that due to the exponential factors $exp[i\lambda^2 k_x \sum_n k_{yn}]$ such diagrammes after the averaging over k_i with the effective potential acquire in comparison with the purely one-dimensional case some numerical factors less than unity (1/4 for the first diagramme and smaller factors for the higher orders). But if in the absence of such Factors the intersecting diagrammes compensated entirely the contribution of the nonintersecting ones which resulted into a zero static conductivity so in the present case such a compensation is no more possible. Therefore the total result will be of the form

$$\sigma = \mathcal{A}(e/2\pi\lambda)^2 l_2 \tag{70}$$

where $\mathcal{A} \leq 1$ - is a numerical coefficient (according to the estimation of the first diagrammes $\mathcal{A} \sim 0{,}7 \div 0{,}8$).

The situation is different in the case of a long ranged potential. The quantity (69) is in this case small and so as in the previous section we can expand in powers of the "rate of non one-dimensionality". Such a quantity is represented by

$$\gamma = l_2^2 \langle (\zeta_i \zeta_i^* \zeta_k \zeta_k^* - \zeta_i \zeta_k \zeta_i^* \zeta_k^*) \rangle \tag{71}$$

As before the conductivity of an infinite sample is nonanalytic in the vicinity $\gamma=0$. For a sample of finite length along z we obtain (taking into account the zeroth approximation)

$$\sigma \approx \frac{e^2}{(2\pi\lambda)^2}\left[\frac{\pi^{5/2}}{2} L^{-1/2} l_2^{3/2} e^{-L/4l_2} + \frac{5\pi^3}{256}\gamma L e^{3L/4l_2}\right] \tag{72}$$

In Section 4 we assume that with increasing L the conductivity reaches a limiting value which was defined from the condition that both terms in (72) become equal in order of magnitude. The same idea we use in the present case. But one can suppose that this procedure is justified only to the accuracy of the exponential factors in (72) since it seems probable that there exists a scaling with respect to the Mott's "level splitting" $\sim e^{-L/4l_2}$ and the degree of non one-dimensionality. If this is so the conductivity must have the form

$$\sigma = \text{const}\, q\, f(\gamma/q^{\nu}) \tag{73}$$

where $q=\exp(-L/4l_2)$, $f(0)=1$, $f(x\gg 1)\sim x^{1/\nu}$. Defining ν from the first expansion term in (72) we get $\nu=4$. If the scaling region reaches $\gamma \approx \gamma_0 \sim 1$ and γ_0 is the value for a short ranged potential so the matching with (70) gives

$$\sigma = A\,\frac{e^2 l_2}{(2\pi\lambda)^2}\left(\frac{\gamma}{\gamma_0}\right)^{1/4} \tag{74}$$

From formula (71) we get

$$\gamma = 1 - \overline{\exp[i\lambda^2(k_{x1}k_{y2} - k_{x2}k_{y1})]} \tag{75}$$

This quantity is calculated according to the rule

$$\overline{f(\vec{k}_\perp)} = \frac{\int |U(\vec{k}_\perp, 2p_0)|^2 \exp[-\lambda^2(k_x^2+k_y^2)/2]\, f(\vec{k}_\perp)\, d^2k_\perp/(2\pi)^2}{\int |U(\vec{k}_\perp, 2p_0)|^2 \exp[-\lambda^2(k_x^2+k_y^2)/2]\, d^2k_\perp/(2\pi)^2} \tag{76}$$

In case of the potential (61) we can expand the exponent in γ and in the case of a short ranged potential we can as already mentioned change $\mathcal{U}(\vec{k}_\perp)$ to $\mathcal{U}(0)$ i.e., just cancel $|\mathcal{U}|^2$ in (76). So the result will be

$$\gamma = 4(p_0\lambda)^4 \ln^2[1/(2p_0\lambda)^2], \quad p_0\lambda \ll 1$$

$$\gamma = 1/2 \equiv \gamma_0 \qquad \text{-short ranged potential.} \tag{77}$$

The quantity l_2 itself is given by

$$1/l_2 = N_i \int |\mathcal{U}(\vec{k}_\perp, 2p_0)|^2 d^2k_\perp/(2\pi)^2 = \pi N_i \left(\frac{me^2 Z}{p_0^2 \varepsilon}\right)^2 \tag{78}$$

Inserting (77) and (78) into (74) we obtain

$$\sigma = \frac{1}{2^{3/4}\pi} \cdot \frac{p_0^5 \varepsilon}{N_i m^2 e^2 \lambda Z^4} \ln^{1/2}(1/2p_0\lambda)^2 \tag{79}$$

Unfortunately the practical prediction of the H dependence from this formula and its comparison with experiment is difficult since in the real semimetals the energy spectrum itself can vary considerably with magnetic field. If the number of elelctrons is fixed i.e., they are entirely due to impurities so $p_0 \sim n_e \lambda^2$ and $\sigma \sim H^{-9/2}$. If however there are two intersecting bands and degree of intersection increases with field then $p_0 \sim 1/\lambda$ and $\sigma \sim H^3$. So one sees that the range of possible types of the $\sigma(H)$ dependence is very wide.

6. Conductivity in presence of magnetic impurities [14]

In Sections 4,5 we used the scaling hypothesis to obtain the longitudinal conductivity. To verify this assumption in the framework of the method we use is very complicated since already the calculation of the first expansion term is not simple. Unfortunately it is also difficult to compare the predictions of the previous section with experiment as it was already mentioned. There exists however another way which is much simpler. Let us imagine a quazi one-dimensional metal with a very small α and at $T=0$. The conductivity will be very small in this case. Now let the metal contain magnetic impurities having a localized spin. Then there exists a spin dependent interaction of the electrons with impurities

$$H = \sum_i H_i = -J \sum_i (\vec{\sigma}\vec{S}_i) \tag{80}$$

Here it is essential that due to the noncommutativity of the components of $\vec{\sigma}$ the operators H_i of different impurities do not commute. Supposing free spins S_i we get

$$(1/2) Sp_\sigma [(\vec{\sigma}\vec{S}_i)(\vec{\sigma}\vec{S}_k)(\vec{\sigma}\vec{S}_i)(\vec{\sigma}\vec{S}_k) - (\vec{\sigma}\vec{S}_i)(\vec{\sigma}\vec{S}_i)(\vec{\sigma}\vec{S}_k)(\vec{\sigma}\vec{S}_k)] =$$

$$= -(4/3)[S(S+1)]^2 \tag{81}$$

Since the magnetic interaction J is weaker than the ordinary potential interaction so even at large concentration of the magnetic impurity $l_M \gg l_2$ where

$$l_M^{-1} = N_M \left[\int |J(k_\perp, 2p_0)|^2 d^2k_\perp/(2\pi)^2 \right] S(S+1)/v^2 \tag{82}$$

But in this case we return to the same situation as in Sec.5. Namely the ξ_i operators at different points do not commute and the commutator is relatively small. This permits directly to use the formula (72) where $\gamma = (4 l_2^2/3 l_M^2)$. Taking into account two projections of the spin $\vec{\sigma}$ we get

$$\sigma = \frac{e^2}{\pi S}\left[\frac{\pi}{2}^{5/2} L^{-1/2} l_2^{3/2} e^{-L/4l_2} + \frac{5\pi^3}{192}\frac{l_2}{l_M^2} L e^{3L/4l_2}\right] \tag{83}$$

The critical length obtained from comparison of both terms in (83) is

$$L_c \approx 2 l_2 \ln (l_\mu / l_2) \tag{84}$$

Inserting this into (83) we get for an infinite sample

$$\sigma \sim \frac{e^2}{\pi S} l_2^{3/2} l_\mu^{-1/2} \tag{85}$$

Hence $\sigma \sim \sqrt{N_\mu}$. The verification of this expression would be the best demonstration of the correctness of our scaling hypothesis.

It can happen however that it will be difficult to control the N_μ . In this case use can be done of the fact that in a strong magnetic field $\mu H \gg T$ all the impurity spins are polarized by the magnetic field and $(\vec{s}\vec{S}_i) \to s_z S_{zi}$. But now the different H_i commute and the effect disappears. The region $\mu H \sim T$ cannot be treated by our method since in this region the spin-flip scattering becomes inelastic (at $T \gg \mu H$ it can be considered as elastic and at $T \ll \mu H$ it is suppressed) But from general considerations it follows that $\gamma = \gamma_0 f(\frac{\mu H}{T}, S)$ where $f \approx 1 - a(\frac{\mu H}{T})^2$ for $\mu H \ll T$ and $f \approx b \exp(-\frac{\mu H}{T})$ for $\mu H \gg T$ where $a, b \sim 1$. Since $\sigma \sim \sqrt{\gamma}$ we get

$$\sigma \approx \sigma_0 [1 - \frac{a}{2}(\frac{\mu H}{T})^2], \quad \mu H \ll T$$
$$\sigma \approx \sigma_0 \sqrt{b} \exp(-\mu H / 2T), \quad \mu H \gg T \tag{86}$$

If $\mu = \mu_B g$ is known for the localized spins then the second formula of (86) can be used for verification of the scaling hypothesis.

In all this calculation it is of course assumed that T is above the ordering temperature of the impurity spins.

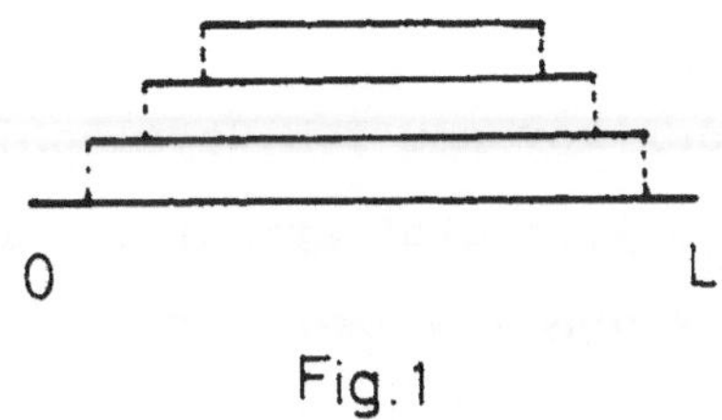

Fig. 1

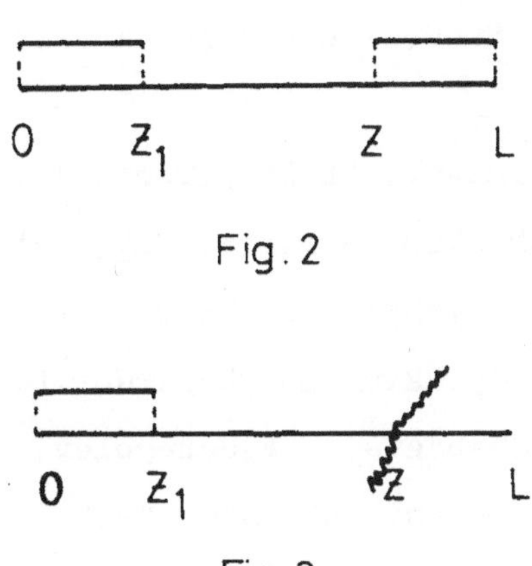

Fig. 2

Fig. 3

Fig. 4

Fig. 5

References

1. A.A.Abrikosov, L.P.Gor'kov and I.E.Dzyaloshinskii. "Methods of Quantum Field Theory in Statistical Physics"(Prentice-Hall, Englewood Cliffs, N.Y.1963).
2. A.A.Abrikosov. Journ.Low Temp.Phys., 10, 3, 1973.
3. N.F.Mott, W.D.Twose. Adv.Phys., 10, 107, 1961.
4. K.V.Efetov, A.I.Larkin. Zh.Eksp.Teor.Fiz., 66, 2290, 1974.
5. Yu.Bychkov, L.P.Gor'kov, I.E.Dzyaloshinskii. Zh.Eksp.Teor.Fyz., 50, 738, 1966.
6. A.I.Larkin, V.I.Melnikov. Zh.Eksp.Teor.Fiz., to be published.
7. V.L.Berezinskii. Zh.Eksp.Teor.Fiz., 65, 1251, 1973.
8. A.A.Abrikososv, I.A.Ryzhkin. Zh.Eksp.Teor.Fiz.,
9. A.A.Abrikosov, I.A.Ryzhkin, Zh.Eksp.Teor.Fiz.,
10. M.L.Hidekel, R.P.Shybaeva, I.F.Schegolev, E.B.Yagubskii. ~~Zh.Eksp.Teor.Fiz~~ Vestnik AN SSSR, 1975, N°11.
11. A.Zawadowskii. Zh.Eksp.Teor.Fiz.,54, 1429, 1968.
12. A.A.Abrikosov, I.A.Ryzhkin. Zh.Eksp.Teor.Fiz., to be published.
13. A.A.Abrikosov, I.A.Ryzhkin, to be published.
14. A.A.Abrikosov. Zh.Eksp.Teor.Fiz.,Pis'ma, to be published.

Conference on Organic Conductors and Semiconductors, Siófok, Hungary 1976

IMPURITY PINNING OF CHARGE DENSITY WAVE IN THE PEIERLS-FROHLICH STATE

H. FUKUYAMA[+]

Department of Physics Tohoku University, Sendai 980, Japan

P.A. LEE

Bell Laboratories Murray Hill, New Jersey 07974, USA

Abstract

The dynamics of impurity pinning of the charge density wave and the frequency dependence of conductivity are investigated in the one-dimensional Peierls-Frohlich state.

§1. Model

In this paper we investigate the dynamics of impurity pinning of the charge density wave (CDW) and the frequency dependence of the conductivity in the one-dimensional Peierls-Frohlich (PF)[1] state at low temperatures. As has been demonstrated by Lee, Rice and Anderson[2] low lying excitations are due to the oscillations of the phase of CDW, ϕ, and the Hamiltoni to describe the dynamics of ϕ in the presence of impurity potential is given by[3]

$$\mathcal{H} = \pi \int dx[v'p(x)^2 + \frac{v_F}{4\pi^2}(\nabla\phi)^2] + V_0\rho_0 \sum_i \cos(Qx_i+\phi(x_i)), \tag{1.1}$$

where $Q=2k_F$, k_F being the Fermi momentum and $[p(x), \phi(x')] = -i\delta(x-x')$, $v'=v^2/v_F$ and v_F and v are the Fermi velocity and the velocity of the collective mode of phase respectively. For convenience we choose $V_0>0$. In eq.(1.1) it is assumed that the impurity potential is of short range with its strength V_0.

The dynamical properties of the phase are conveniently described by the phonon Green function defined by

$$\mathcal{D}(q,q') = \int_{-\beta}^{\beta} d\tau\, e^{i\omega_n\tau} \langle T\tau\phi_q(\tau)\phi_{-q'}\rangle , \tag{1.2}$$

where $\phi(x)=L^{-1/2} \sum_q e^{iqx}\phi_q$ and L is the linear dimension of the system. By use of (1.2) the conductivity is given by[3,4]

$$\sigma(\omega) = \frac{i\omega}{2}\left(\frac{e}{\pi}\right)^2 L\,\mathcal{D}(0,0)\Big|_{i\omega_n\to\omega-i0} \tag{1.3}$$

In order to determine $\mathcal{D}(q,q')$ we derive an equation of motion, which is as follows.

$$\mathcal{D}(q,q') = \delta_{q,q'}\mathcal{D}_0(q) + \frac{V_0\rho_0}{2}\ \mathcal{D}_0(q)\ \sum_i \frac{e^{iqx_i}}{\sqrt{L}}$$

$$\times \int_{-\beta}^{\beta} d\tau\ e^{i\omega_n\tau} \langle T_\tau\ \sin(Qx_i+\phi(x_i,\tau))\phi_{-q'}\rangle\ , \tag{1.4}$$

where $\mathcal{D}_0(q)=4\pi v'[\omega_n^2+(vq)^2]^{-1}$ is the phase phonon Green function in perfect crystals. As is seen the phase obeys a highly non= linear equation.

The characteristic parameter in the present problem, eq. (1.1), is the ratio of the strength of impurity potential and elastic energy per impurity, i.e. $\varepsilon=V_0\rho_0/n_i v_F$ where n_i is the impurity concentration per unit length. According to values of ε we can classify cases as weak pinning ($\varepsilon \ll 1$) and strong pinning (or dilute limit) ($\varepsilon \gg 1$). These cases will be treated separately. In each of these cases there exist characteristic length and then characteristic energy. In the case of weak pinning the phase in the ground state will vary in a length of the order of L_0 which satisfies

$$\frac{1}{2\pi} v_F L_0^{-2} \simeq \frac{1}{2} V_0\rho_0\sqrt{n_i/L_0}\ . \tag{1.5}$$

In eq.(1.5) the left hand side represents the loss of elastic energy due to the spatial variation, whereas the right hand side the gain of energy by impurity potentials compared to the case where the phase does not vary spatially. On the other hand in the case of strong pinning the phase at an impurity site, x_i, will be always $\cos(Qx_i+\phi(x_i))\simeq-1$. This implies that the characteristic length for the spatial variation of the phase is n_i^{-1}. Corresponding to these lengths characteristic energies for the phase phonon dynamics are vL_0^{-1} in the weak pinning and vn_i in the strong pinning respectively.

§2. Frequency dependence of the conductivity

In ref.3, eq.(1.4) was solved under the following approximation.

$$\sin(Qx_i+\phi(x_i,\tau)) \to \phi(x_i,\tau)\cos Qx_i \tag{2.1}$$

After averaging over the random configuration of impurities we obtain the phonon Green function as follows.

$$\mathscr{D}(q,q') = \delta_{q,q'}\,\mathscr{D}(q) = \delta_{q,q'}[\mathscr{D}_0^{-1}(q)-\Gamma(q)]^{-1} \tag{2.2}$$

In eq.(2.2) Γ is the self-energy of phase phonon which was evaluated for the cases of $\varepsilon \ll 1$ and $\varepsilon \gg 1$ separately.

1) $\varepsilon \ll 1$

In this case the impurity scattering would be treated within the Born approximation. However, as has been pointed out in ref.2, $\Gamma\propto|\omega_n|^{-1}$ in this scheme. This indicates the necessity to take higher order terms into consideration. The simplest of such approach is to renormalize the phonon Green function in the intermediate state by the one to be determined self-consistently. In this self-consistent Born approximation the conductivity is given by

$$\sigma_{SB}(\omega) = -\frac{Ne^2}{m^*\gamma}\frac{ix}{x^2+G} \tag{2.3}$$

where m^* being the mass of collective mode[2] $x=\omega/\gamma\,(\gamma=vL_0^{-1})$ and G is a function of x to be determined by

$$G^2(G+x^2)+1=0 \tag{2.4}$$

From eq.(2.4) we see that $\mathrm{Im}G\neq0$ at $x=0$ for the physical solution, i.e. $\mathrm{Re}\,\sigma_{SB}\propto\omega$ as $\omega\to0$. $\mathrm{Re}\,\sigma_{SB}$ is found to have characteristic peaks around $\omega\simeq\gamma$.

ii) $\varepsilon\gg1$

In order to treat this case the perturbation series are summed up to infinite orders in V_0 under the assumption that the scattering from one impurity at a time is most important, i.e. within the framework of self-consistent t-matrix approximation. The conductivity in this case, $\sigma_t(\omega)$, is shown to have a peak around $\omega\simeq\omega_0=vn_i$. In this scheme it is also found $\mathrm{Re}\,\sigma_t(\omega)\propto\omega$ as $\omega\to0$.

These are results under the assumption of eq.(2.1). In either case the result obtained has a difficulty that σ_{SB} and σ_t lead to $\mathrm{Im}\varepsilon(\omega)\neq0$ at $\omega=0$ where $\varepsilon(\omega)$ is the dielectric constant. This difficulty originate from the assumption, eq.(2.1), which will be valid for $\omega\gtrsim\gamma$ ($\varepsilon\ll1$) or $\omega\gtrsim\omega_0$ ($\varepsilon\gg1$) since for the lower frequency region the spatial variation of the phase in the ground state should be taken into account. This has been investigated in detail in ref.5.

§3. Refined treatment of $\sigma(\omega)$

In order to treat low frequency region properly we decompose phase ϕ into sum of that in the ground state, ϕ_0, and the fluctuations above ϕ_0, φ, i.e.

$$\phi = \phi_0 + \varphi \tag{3.1}$$

The spatial variation of ϕ_0 is determined by

$$\frac{\partial E[\phi_0]}{\partial \phi_0} = 0 \ , \tag{3.2}$$

$$E[\phi_0] = \frac{v_F}{4\pi}\int dx(\nabla\phi_0)^2 + V_0\rho_0 \sum_i \cos(Qx_i+\phi_0(x_i)) \ . \tag{3.3}$$

The phonon Green function, eq.(1.2), can be considered as that for φ in this case and the equation of motion, eq.(1.4), will be solved by rewriting $\sin(Qx_i+\phi(x_i,\tau))$ as

$$\varphi(x_i,\tau)\ \cos(Qx_i+\phi_0(x_i)) \ . \tag{3.4}$$

The existence of ϕ_0 in eq.(3.4) changes the property of the perturbation series, and essentially removes the abovementioned difficulty in both cases of $\varepsilon \ll 1$ and $\varepsilon \gg 1$. However the feature obtained in ref.3 of the existence of peaks of $\mathrm{Re}\sigma(\omega)$ around $\omega \simeq \gamma$ or $\omega \simeq \omega_0$ is not changed by ϕ_0.

§4. Finite temperatures

So far discussions are confined to T=0. At finite temperatures the anharmonicity of the fluctuations of φ should be taken into account. Approximate treatment of this effect is to treat the anharmonicity within the self-consistent harmonic approximation[6] and to replace eq.(3.4) by

$$\varphi(x_i,\tau)\, e^{-\langle\varphi(x_i)^2\rangle/2} \cos(Qx_i+\phi_0(x_i)),$$

where <> is the thermal average and $\langle\phi(x_i)\rangle^2$ is to be determined self-consistently. Similar considerations were made by Bruesch et al[7] and Pietronero et al[8] for the case of commensurability pinning in perfect crystals. By ignoring the fact that $\langle\phi(x_i)^2\rangle$ is that at an impurity site and replacing this by that averaged over impurity distributions we find that depinning of the CDW occurs in the case of $\varepsilon \ll 1$ at a temperature $T_c=0.55\sqrt{m^*/m}\,\gamma_0$ where γ_0 is the pinning energy at T=0. In the temperature region of $T \gtrsim T_c$ the metallic behavior is expected in the dielectric constant.

§5. Conclusions

In this paper the impurity pinning of the CDW is investigated in a one-dimensional model system. The effect of three-dimensional Coulomb forces and its consequence on the inelastic neutron scattering and the dielectric properties in realistic systems will be discussed elsewhere[5].

References

+ Visitor to Bell Laboratories in summer 1976.

1. For a review Low-Dimensional Cooperative Phenomena ed. by H.J. Keller (Plenum Press, New York, 1975).
2. P.A. Lee, T.M. Rice and P.W. Anderson, Solid State Commun. 14 703 (1974).
3. H. Fukuyama, J. Phys. Soc. Japan 41 5/3 (1976).
4. M.J. Rice in ref.1.
5. H. Fukuyama and P. A. Lee, in preparation.
6. Y. Okabe and H. Fukuyama, Solid State Commun. (in press).
7. P. Brüesch, S. Strässler and H.R. Seller, Phys. Rev. B12 219 (1975).
8. L. Pietronero, S. Strässler and G.A. Toombs, Phys. Rev. B12 5213 (1975).

Conference on Organic Conductors and Semiconductors, Siófok, Hungary 1976

DISORDERED QUASI-ONE-DIMENSIONAL MATERIALS

MORREL H. COHEN

The James Franck Institute, The University of Chicago,
Chicago, Illinois 60637, USA

A brief review is given of recent progress in several areas of the theory of quasi-one-dimensional materials: Shante's theory of hopping conduction in highly anistropic materials; Theodorou's study of the disordered, one-dimensional Hubbard model with application to NMP-TCNQ; the electron-phonon interaction as dynamic disorder; spin and charge ordering in donor $(TCNQ)_2$ salts; and surfaces of organic charge-transfer salts.

I. INTRODUCTION

I have been collaborating in a series of investigations of the theory of quasi-one-dimensional materials the results of which are as yet largely unpublished. Accordingly, I shall review several of these for the present conference:

Shante's theory of hopping conduction in highly anisotropic disordered materials;[1]

Theodorou's theory of the disordered, one-dimensional Hubbard model with application to the magnetic properties of NMP-TCNQ;[2-6]

The electron-phonon interaction as dynamic disorder;[7,8]

Spin and charge ordering in donor-$(TCNQ)_2$ salts; and

Surfaces of organic charge transfer salts.

II. HOPPING CONDUCTION IN DISORDERED ANISOTROPIC CONDUCTORS

Consider a disordered material such that all states within an energy range much greater than kT of the Fermi level are localized and that these localized states are well separated in space. Electron transport then occurs by phonon-assisted hopping from one localized state to another. Mott showed that because the hopping was of variable range, the electrical conductivity had the form[9,10]

$$\sigma = \sigma_0 e^{-(T_0/T)^{1/4}} \tag{1}$$

Shante has recently generalized Mott's result to highly anisotropic materials with interesting and somewhat surprising consequences.[11]

Let Γ_{ij} be the phonon-assisted hopping rate from localized state i to localized state j. Miller and Abrahams pointed out [12] that the problem of calculating the d.c. conductivity in terms of the Γ_{ij} is equivalent to that of a resistor network with conductances

$$G_{ij} = G_0 e^{-f_{ij}} \propto \Gamma_{ij} \tag{2}$$

where

$$f_{ij} = 2\alpha R_{ij} + \frac{1}{2kT}\left\{|E_i| + |E_j| + |E_j - E_i|\right\}, \tag{3}$$

according to the simplifications introduced by Ambegaokar, Halperin, and Langer (AHL).[13] Here E_i is the energy of the localized state centered at R_i, R_{ij} is the distance between the centers of states i and j, and α is the decay constant of these states. Shante followed Kirkpatrick's extension [14] of the percolation [15] theory introduced by AHL and by Pollak.[16] Electron-electron interactions are neglected in all of these developments.

The argument proceeds by choosing a cutoff G and removing from the network all conductances G_{ij} less than G. Considering a conductance G_{ij} as a bond between sites i and j in the network, the average number of bonds B left per site is

$$B = \frac{\pi}{40} n(\epsilon_F) kT f^4, \tag{4}$$

for the three-dimensional, isotropic case, where $n(E_F)$ is the density of states at the Fermi energy and

$$f = \ln G/G_0 \tag{5}$$

We now take recourse to percolation theory,[15] which states that the conductivity vanishes if B is less than a cririical value B_c below which continuous paths extending to infinity through the network do not occur. That is

$$\sigma(G) = 0, \quad G < G_c, \quad f > f_c, \quad \text{or } B < B_c. \tag{6a}$$

From the work of Pike and Seager[17], we know that $B_c = 2.80$ for isotropic long-range bonds in three dimensions. On the other hand, when percolation can occur, the conductivity is finite and given by

$$\sigma(G) = A \langle G/R \rangle (B - B_c)^{\zeta}, \quad G > G_c, \quad f < f_c, \text{or } B > B_c. \tag{6b}$$

For nearest neighbor hops we know that $\zeta = 1.13$,[18] and this may approach the value 1.5 for long-range hops.[1] In (6b) A is a constant, approximately known.[18] $\langle G/R \rangle$ is an appropriate average of G_{ij}/R_{ij} along the percolation channels[15] where $G_{ij} > G$. The result[1] is

$$\sigma(B) = \sigma_1 (B - B_c)^{\zeta} e^{-\frac{f(B)}{B}} \tag{7}$$

Following Kirkpatrick,[14] Shante obtained a best lower bound to σ by maximizing (7) with respect to B, having left out of σ the contributions to σ of all $G_{ij} < G$ and underestimated the contributions of those G_{ij} kept. The optimal value of B is

$$B = B_c + \zeta / f'(B_c), \tag{8}$$

which gives

$$\sigma = \sigma_0(T) e^{-(T_0/T)^{1/4}}$$

$$\sigma_0(T) = K T^{-1+\zeta/4}$$

$$T_0 = \frac{40 B_c \alpha^3}{\pi n(E_F) k} \tag{9}$$

for isotropic Mott hopping in three dimensions. Similar results were obtained by

Ambegaokar, Cochran, and Kurkijarvi by scaling arguments.[19]

An important result of the above analysis is that hops with

$$R_{ij} > R_m = f_c/2\alpha \propto T^{-1/4}$$

$$\text{or} \quad |E_i| > E_m = kT f_c \propto T^{3/4} \tag{10}$$

lie off the percolation channel and do not contribute significantly to σ. This feature has important consequences in Shante's generalization of the above analysis to anisotropic materials. There is as yet no theory of the shape of localized states in disordered anisotropic materials. Accordingly, Shante was forced to fix the spatial dependence of f_{ij} by analogy with other cases of anisotropic exponential decay:

$$\alpha R_{ij} \longrightarrow [(\alpha_R R_{ij})^2 + (\alpha_X X_{ij})^2 + (\alpha_Y Y_{ij})^2]^{1/2} \tag{11}$$

Here it has been supposed that we are dealing with a quasi-one-dimensional material consisting of stacks of planar molecules with R a distance along the stacking axis and X and Y perpendicular distances. Bloch, Weisman, and Varma[20] were the first to point out the significance of disorder for the transport properties of such materials, introducing the so-called "disorder model" and interpreting the temperature-dependence of σ in terms of hopping conduction. Consider as an example a simple square lattice of molecular stacks with $\alpha_X = \alpha_Y = \alpha_S$ and $X^2+Y^2=S^2$. Let S_o be the minimum distance between stacks and R_o the minimum distance between molecules along a stack. Suppose the system to be highly anisotropic in the sense that $\alpha_S S_o >> \alpha_R R_o$ and that the states are strongly localized perpendicularly to the stacks, i.e. $\alpha_S S_o > 1$. We can assume that $R_o << R_m$ as for the isotropic case, but Shante has shown that we cannot assume that $S_o << S_m$ $(S_m = f_c/2\alpha_S)$. As a consequence, the discrete nature of the crystal structure in the perpendicular direction has an important effect on the temperature dependence of σ.

Shante finds that

$$\left.\begin{array}{l} \sigma(T) = \sigma_m(T)\, e^{-(T_m/T)^{1/m}} \\ T_{um} > T > T_{\ell m} \end{array}\right\} \qquad (12)$$

where $\sigma_m(T)$ is a slowly varying preexponential as $\sigma_o(T)$ is in (9). At $T_{\ell m}$, S_m becomes less than a lattice distance in the transverse direction, and the value of m changes. Above a certain temperature, $S_m < S_o$ holds and the percolation path is apparently one-dimensional, being restricted to a single stack. However, there is no percolation path in one dimension, and the conductivity has a simple activated behavior, according to Kurkijarvi.[21] This activated behavior persists until the temperature is so high that the electron-phonon interaction $\mathcal{H}_{ep}$ cannot be considered small compared to the random potential responsible for localizing the electronic states. The hopping picture breaks down therefore above a temperature near the point of inflection of σ vs. T.

Table I gives Shante's results for the values of m in the different temperature ranges in which percolation can occur. The value of 2.91 for m in the range 1 - 2 of S_m/S_o is indistinguishable on most $\log\sigma$ vs. $T^{-1/m}$ plots from 3. Moreover, the values for m for $S_m/S_o > \sqrt{2}$ are indistinguishable from 4. We thus obtain from Shante's analysis of the square lattice of stacks three temperature ranges of hopping conduction:

$$
\left.\begin{aligned}
\sigma &= \sigma_0\, e^{-T_0/T}, & T_i > T > T_a \\
&= \sigma_1\, e^{-(T_1/T)^{1/2.91}}, & T_a > T > T_b \\
&= \sigma_2\, e^{-(T_2/T)^{1/3.87}}, & T_b > T > 0.
\end{aligned}\right\} \tag{13}
$$

$$
\left.\begin{aligned}
T_a/T_b &= 2.74 \\
T_a &= T_1/(2\alpha_s S_0)^{2.91} \\
T_2 &= T_1\,(2\alpha_s S_0)^{0.962}/0.717 \\
T_1 &= 2\alpha_R B_c\,(2\alpha_s S_0)^{0.905}/k\, n_1(E_F)
\end{aligned}\right\} \tag{14}
$$

In (14) $n_1(E_F)$ is the one-dimensional density of states for a single stack.

Eqs. (13) and (14) provide an interesting basis for data analysis. There are five parameters upon which the conductivity depends sensitively, T_o, T_1, T_2, T_a, and T_b. There are two constraints on these embodied in Eq. (14), leaving three fitting parameters. Coleman <u>et al.</u>[22] have reported data for σ for NMP-TCNQ in the relevant temperature range. They fitted their data by using m = 1 only in the range 18-70 K. Since 18 K was the lowest temperature for which data were taken, this consituted a two parameter fit in the above sense. The r.m.s. error was 0.36. Shante used the scheme of Eqs. (13) and (14) in the temperature range 18-140 K, where the latter is the point of inflection T_i. He varied all five parameters and obtained a fit with an r.m.s. error of 0.057. The constraint on T_a/T_b was satisfied to 4%, and the constraint on T_2 to 20%. Thus, by the addition of a single parameter, he reduced the error by a factor of 6.

When there is significant transverse anisotropy in α, i.e. $\alpha_X \neq \alpha_Y$, and/or in the crystal structure, Shante finds that more ranges are required with somewhat different indices. The scheme of Eqs. (13) and (14) should still provide a rough basis for fitting the data.

Provided $\alpha_S S_o > 1$, $\alpha_S S_o$ should be independent of the degree of disorder. α_R on the other hand, being an inverse localization length, is proportional to the mean square random potential.[23] In a large group of disordered TCNQ salts, the disorder arises from random taking up by the donor ions of two equivalent positions in which their dipole moments are reversed. Thus T_1, T_2, T_a, and T_b are all proportional to the square of the dipole moment. G. Grüner has pointed out to me that a good way to search for the predicted multiplicity of temperature ranges is to plot $\ln \sigma$ vs. $T^{-1/4}$ and look for changes of slope. He and his collaborators have analyzed σ data taken on a homologous series of salts in this way, and have found evidence for a T_b which does scale roughly as the square of the dipole moment.[24]

The theory of hopping conduction described above differs greatly in its fundamental structure from typical theories of conduction which start with electrons in extended states and work out the effect of scattering processes on the transport. The usual starting point for such theories is the Kubo relation,[25] in which the dc conductivity is proportional to the zero wave-vector and zero frequency limits of the current-current correlation function

$$\sigma_{DC} \propto \lim_{k,\omega\to 0} \int_0^\infty dt\, e^{-i\omega t} \langle\langle j_k(t), j_{-k}(0) \rangle\rangle, \tag{15}$$

or its equivalent. Such an approach is completely inadequate for the present problem, in which we deal with strongly disordered materials. The reason is that the strong spatial inhomogeneity of the density of the conduction electrons, assumed at the onset to be in weakly overlapping states, generates important local-field corrections to the uniform external field. Local field effects are not included in (15); the full microscopic generalization of the Kubo relation is required instead, i.e.

$$\vec{j}(\vec{r}) = \int \overleftrightarrow{\sigma}(\vec{r},\vec{r}') \cdot \vec{E}(\vec{r}')\, d^3r' \tag{16}$$

with $\vec{j}(\vec{r})$ the local current density, $\vec{E}(\vec{r})$ the local field, and a generalization of (15) holding for the microscopic conductivity $\overleftrightarrow{\sigma}(\vec{r},\vec{r}')$. The Miller-Abrahams impedance-network approach, on the other hand, takes the local field effects fully into account.

The situation is analogous to that of local field corrections in the dielectric constant ϵ. These arise, as pointed out by Wiser,[26] from a spatially inhomogeneous distribution of polarizability α on a microscopic scale. There are two limiting cases: 1.) The Lorentz-Lorenz case in which the material is so microscopically inhomogeneous that the spatial distribution of α is disjoint. The local-field correction is then L P, P being the macroscopic polarization and L being $4\pi/3$ for a cubic or isotropic material, and ϵ is $[1+\frac{8\pi}{3}N\alpha]/[1-\frac{4\pi}{3}N\alpha]$ for such a material, N being the number density of polarizable entities. 2.) The Sellmeier case in which the distribution of α is uniform. There is no local-field correction, and $\epsilon = 1 + 4\pi N\alpha$.

Local-field effects are similarly important in conductivity problems. Consider a simple-cubic network of identical resistors from which a fraction 1 - p has been removed. The conductivity has the form

$$\sigma = \sigma_0 (p - p_c)^s \tag{17}$$

for p above but near p_c, the percolation threshold. If local field effects were ignored in calculating σ, one would obtain a value of about 2/3 for s, whereas the correct value is 1.14.[18]

Percolation is clearly a case of extreme inhomogeneity, and one cannot then ignore local-field effects. However, it is frequently difficult enough to carry out one's calculations without the local-field effects, and a quantitative measure of inhomogeneity would be very convenient. For $\xi > 1$, there would be inhomogeneity and local-field corrections, and for $\xi < 1$ there would be no significant inhomogeneity or local-field corrections. For phonon-assisted hopping between localized states in disordered quasi-one-dimensional materials, ξ can be taken as the ratio of the mean spacing between states participating in transport and the extent of these states. The states participating in transport have energies in the range $2E_m$ centered at E_F so that their spacing is $[n_1(E_F)\,2E_m]^{-1}$. Their spatial extent is about twice the localization length, or $2/\alpha_R$. We thus have

$$\xi = \alpha_R / 4 n_1(E_F) E_m \quad . \tag{18}$$

This has the value of 30 at T_a (= 100 K) from Shante's analysis[1] of NMP-TCNQ. T_a is the highest temperature for which the Miller-Abrahams impedance network is used, and it is well justified there. ξ increases as the temperature decreases.

III. THE MAGNETIC PROPERTIES OF DISORDERED QUASI-ONE-DIMENSIONAL MATERIALS

Bloch, Weisman, and Varma[20] called attention to the importance of disorder for the electronic properties of quasi-one-dimensional materials, pointed out that there existed x-ray evidence for disorder in a number of such materials, and interpreted the conductivity in terms of Mott hopping in one dimension. Among the materials they considered are NMP-TCNQ, Q-$(TCNQ)_2$, and Ad-$(TCNQ)_2$. Bulaevskii et al.[27] have measured the magnetic susceptibility of these materials and found them to be strongly paramagnetic at low temperatures with a temperature dependence of the form

$$\chi = A/T^{\alpha} \tag{19}$$

for T between 0.1 and 10 K. They found α values of 0.58 for NMP-TCNQ, 0.73 for Q-$(TCNQ)_2$, and 0.74 for Ad-$(TCNQ)_2$. In addition the specific heat was found to have a magnetic contribution[28] with

$$C_v(H=0) = B\,T^{1-\alpha} + C\,T^3 \tag{20}$$

the α in (20) being the same as in (19) and $C_v(H)$ being Schottky-like. The full magnetization curve was found to be strikingly nonlinear[27] with

$$M(H) \propto H^{1-\alpha} \tag{21}$$

at high fields, the α in (21) now being the same as that in (19) and (20).

We shall concentrate primarily on NMP-TCNQ here. We shall assume that charge transfer from NMP to TCNQ is complete. A small deviation of $\sim$5% from complete charge transfer as suggested by Butler, Wudl, and Soos[29] would not affect the arguments reported below. The NMP^+ molecule ion has a dipole moment pointing from its center to its methyl group. Fritchie [30] and Morosin [31] have found the orientation of the methyl group to be random in the plane of the NMP molecule in the samples analyzed by them with x-rays. Kobayashi[32] has found ordering of the methyl groups within a b planes in the samples studied by him. We[2,5] have found striking variations in susceptibility data taken on different samples and in different laboratories, suggesting as do the x-ray studies [30-32] a variable degree of disorder in these materials. We confine our remarks to strongly disordered materials in a sense to be clarified below and suppose the samples of refs. 26 and 27 to be in that category.

Bulaevskii *et al.*,[27] have constructed a very interesting theory of the low temperature properties of NMP-TCNQ based on the following assumptions:

1. The TCNQ one-dimensional conduction band is half-filled. This must be nearly, if not precisely, correct.

2. There is one localized, unpaired spin per TCNQ molecule. This presumably follows from 1. if the disorder is sufficiently great as to give complete localization of the one-electron states to a single site or if one has a Mott-Hubbard metal to insulator transition and is in the strong-coupling limit. However, as we shall see, one does not necessari have one unpaired spin per site when the disorder potential and interaction are comparable

3. The system becomes a Heisenberg antiferromagnet at low temperatures.

4. There is a random exchange J between nearest neighbor spins.

5. The probability distribution $P(J)$ is nonsingular as $J \rightarrow 0$. We believe this to be wrong.

6. There is a one-to-one correspondence between the Heisenberg antiferromagnet and a set of interacting spinless Fermions. This is an exact statement and not an assumption but is listed here as an essential step in the argument.

7. Fermi liquid theory is valid for the set of Fermions. The validity of this is an open question in a disordered system; however, Fermi liquid theory does not adequately describe all of the low-lying excitations of the periodic Heisenberg antiferromagnet in one dimension.[33]

8. The quasiparticle density of states of the Fermi liquid has the form

$$n(\epsilon) \propto \frac{1}{|\epsilon|^{\alpha}}, \quad \alpha < 1. \tag{22}$$

Once the Landau Fermi-liquid theory is set up and the assumption (22) made about the density of states, explicit calculations of M(H) and $C_v(H)$ can be made. Quantitative agreement is found with all of the experimental results including not only (19), (20), and (21), but the full field dependence of M(H) and $C_v(H)$ once α is fixed. These are striking successes, and one would like to have a deeper microscopic understanding of what is going on than is afforded by the assumptions 7. and 8.

Theodorou and myself have attempted to reach such an understanding by study of a more explicit microscopic model. We have made the following four assumptions:

1. The TCNQ one-dimensional conduction band is half-filled (complete charge transfer).

2. The principal effects of the electron-electron interaction can be embodied in a disordered Hubbard model containing three parameters: t, the nearest-neighbor electron transfer matrix element assumed constant; U, the Coulomb repulsion between electrons of

opposite spin on the same TCNQ molecule; and σ , the r.m.s. value of the disorder potential taken to have a Gaussian distribution.[34]

3. t is taken as significantly smaller than U and σ , which are comparable.

4. The electron-phonon interaction can be ignored. The electrical conductivity data [22] suggests this to be the case below 140 K.

These assumptions lead deductively to the following results by treating t as a perturbation on U and σ :

The TCNQ molecular orbitals corresponding to the conduction band go over into localized states because of the disorder. These states are doubly occupied, singly occupied, or empty. The singly occupied states correspond to localized spins which are randomly positioned along the TCNQ stack. There is a probability $p \leq 1$ that a localized state be singly occupied. p turns out to be about 1/3 when the theory is fitted to the χ data of di Salvo.

2. The system behaves as a Heisenberg antiferromagnet at low temperatures.

3. There is a random exchange interaction J between spins which can be limited to nearest neighbors in the spin sequence. These need not be on nearest-neighbor sites along the stack.

4. P(J) is singular at J=0, the dominant dependence for $J \rightarrow 0$ being $J^{-\alpha}$ with $\alpha < 1$. This follows solely from the random spacing between spins and the exponential fall of J with spacing between nearest neighbor spins.

5. The results for χ, C_v, $C_v(H)$ and M(H) observed at low T are obtained.

6. χ (T) can be fitted to the data for all T.

The results 5. can be understood in terms of a very simple cluster model. Divide all exchange interactions into strong, $J > kT$, and weak, $J < kT$. Strong interactions are regarded as bonds linking spins into clusters. Weak interactions are ignored. Internal excitations of the clusters are ignored; they are treated as though they were in their ground states. Then the even-spin clusters do not contribute to χ, and each odd-spin cluster contributes a Curie susceptibility. Thus χ is proportional to T^{-1} times the number of odd-spin clusters. The latter is easily shown to be proportional to $T^{1-\alpha}$ when $P(J) \propto J^{-\alpha}$. It follows that $\chi \propto T^{-\alpha}$. Similar arguments can be made for C_v, $C_v(H)$, and $M(H)$. The cluster arguments are unsuitable for the higher temperatures. The result 6. was obtained by bounding χ from above by the χ of the classical Heisenberg model and bounding it from below by a construct related to the periodic case. As the difference was 10%, Theodorou used the classical Heisenberg model for his calculations. The full P(J) is shown in Fig. 1; the fit to di Salvo's data in Fig. 2.

It should be mentioned that although Bulaevskii *et al*. find that (19) holds for the χ of NMP-TCNQ at least down to about 0.15 K, Azevedo[35] finds that (20) is not followed for C_v below about 1K. It seems clear from the structure of NMP-TCNQ that interstack interactions ultimately have to become important at low temperatures. Perhaps they are responsible for the upturn in C_v obtained by Azevedo. If so, NMP-TCNQ would be a very interesting three-dimensional spin glass of a novel type.

With regard to Q-$(TCNQ)_2$ and Ad-$(TCNQ)_2$, Bulaevskii *et al*. have pointed out that the observed nonmonotonic temperature dependence of their susceptibilities suggests complete charge transfer and a quarter-filled TCNQ conduction band in those materials. Theodorou (unpublished) has extended a simpler version of the above theory to arbitrary filling and has demonstrated that the exponent α increases as the filling decreases, as is observed for the sequence NMP-TCNQ, Ad-$(TCNQ)_2$, and Q-$(TCNQ)_2$.

IV. THE ELECTRON-PHONON INTERACTION AS DYNAMIC DISORDER

We now turn to materials like TTF-TCNQ which have no structural disorder and which are metallic or highly conducting, at least above some phase transition temperature.[36] Where well characterized, these phase transitions have turned out to be complex Peierls transitions, and we wish to confine our attention at first to temperatures above the corresponding mean-field transition temperature, T_p . Moreover, we shall suppose that those temperatures are well above any significant phonon frequency so that we may treat the phonons as classical.

Cohen <u>et al</u>.[7] pointed out that in the adiabatic approximation, the electrons experience a disorder potential for each instantaneous nuclear configuration. In the one-dimensional case, the electronic states associated with each instantaneous nuclear configuration are all localized. The region of localization is roughly of extent L(E) , the localization length for electrons of energy E. It changes its position randomly on a time scale $\tau_c(E)$. This diffusion arises from the variation with time of the random potential V_r and from nonadiabatic transitions, $\tau_c(E)$ being the shorter of the two corresponding times. The former is given by the phonon phase-memory time, which is of order the phonon period for phonons with dispersion but is infinite in the dispersionless case. Thus, as pointed out by Rashba,[37] it is important to distinguish between the dispersionless case, where τ_c is given below, and the case of dispersion where $\tau_c^{-1} \cong \omega_{ph}$, a phonon frequency.

On the basis of this physical picture, the d.c. conductivity is

$$\sigma = n(E_F)\, e^2 D(E_F) \tag{23}$$

where n(E) is the density of states per unit volume. D(E) is the diffusion coefficient, given approximately by

$$D(\epsilon) = \frac{1}{2} \tau_c^{-1}(\epsilon) L^2(\epsilon), \tag{24}$$

and E_F is the Fermi energy. Bush[38] has shown that over a wide range of conditions

$$L(\epsilon) \propto \langle V_r^2 \rangle^{-1}. \tag{25}$$

For V_r arising from first-order coupling between electrons and classical phonons,

$$\langle V_r^2 \rangle \propto T \tag{26}$$

so that $L(E) \propto T^{-2}$. For both phonons with dispersion ($\tau_c^{-1} = \omega_{ph}$ and dispersionless phonons,[8] τ_c^{-1} is temperature independent. Thus we have the remarkable result that the ideal phonon resistivity is proportional to T^2 at high temperatures.

Madhukar and I have considered the case of dispersionless phonons in detail.[8] We have used the techniques of Berezinskii,[39] of Gogolin et al.,[40] and Keldysh[41] and have found that, as expected,[37] the adiabatic effects do not enter and nonadiabatic effects lead to a reduction of $\tau_c^{-1}(E_F)$ below ω_{ph} by the factor

$$\frac{\omega_{ph}}{E_F} \frac{1}{2} \left(\lambda(0) + \lambda(2k_F)\right)$$

where the λ's are dimensionless phonon coupling constants. The conductivity is

$$\sigma = \frac{8 \zeta(3) e^2}{N \pi v_F} \frac{\lambda(0) + \lambda(2k_F)}{\lambda^2(2k_F)} \left(\frac{\omega_{ph}}{T}\right)^2, \tag{27}$$

where $\zeta(3)$ is a zeta-function and v_F is the Fermi velocity.

Shante[42] has taken over this theory of dynamic localization of electrons and has derived the corresponding thermoelectric power. He has generalized the theory to cases where both the donor and acceptor chains are conducting, as in the TTF-TCNQ family of materials, and has explored the extent to which the observed conductivities and thermopowers of those materials are quantitatively consistent with the present picture. He finds excellent fits to the available data for parameters consistent with our present understanding of these materials.

Finally, we should note that this picture of dynamical localization has built into it a consistency condition. For localization to occur, an electron must undergo many collisions before it escapes from its region of localization. We require therefore that

$$\frac{L(\epsilon_F)}{v_F} << \tau_c(\epsilon_F) \tag{28}$$

for dynamic localization to occur. If (28) is violated, the electronic motion reverts to that characteristic of higher dimensionality, and the ideal phonon resistivity becomes proportional to T. Such can be expected to occur at lower temperatures, when it is also not possible to neglect the coherent effects of the electron-phonon interactions which ultimately give rise to the Peierls transitions.

V. SPIN AND CHARGE ORDERING IN SOME (DONOR) $(TCNQ)_2$ SALTS

With A. Zawadowski I have developed a simple model of certain nonconducting complex TCNQ salts.[43] In these salts there are two acceptor (TCNQ) stacks for each donor stack. One electron is transferred per donor molecule. The donor ions have dipole moments which alternate. This leads to a dimerization of the TCNQ molecules with one excess electron per dimer. These structural features are summarized in Fig. 3.

The simple model, called the box model, encloses each dimer within an imaginary box and assigns the transfer matrix elements and Coulumb interactions within and between boxes shown in Fig. 4a. We suppose that we are in the limit that

$$U_0, U_2 >> t, t_1, U, \tag{29}$$

so that there is only one electron per box. We also suppose that

$$|t_1| << t, U \tag{30}$$

so that charge transfer between boxes can be ignored and the primary effect of t_1 is to introduce an antiferromagnetic exchange $J > 0$ between boxes, as indicated in Fig. 4.

Each box has two spin states and two charge states. It is convenient to use an isospin representation of the charge states, as shown in Fig. 5. The Hamiltonian of the system can now be written as

$$\mathcal{H} = -2t \sum_i I_i^x - \sum_i (U + J \vec{S}_i \cdot \vec{S}_{i+1})(I_i^z + \tfrac{1}{2})(I_{i+1}^z - \tfrac{1}{2}). \tag{31}$$

The ground states of (31) are readily obtained for the three limiting cases in which one of the three terms in $\mathcal{H}$ is dominant. These are listed in Table II together with the nature of the long-range order possessed in charge and in spin.

Referring to Table II, a passage from one domain of the t, U, J parameter space to another, e.g. from 1. to 2., entails a change in long-range order and is thus a phase transition. However in one dimension long-range order and phase transitions can exist only at T = 0 K. For phase transitions to occur at finite temperature, coupling with three-dimensional phonons, or interchain interactions (Coulomb or magnetic), or both must occu Examples of such phase transitions as displaved by the magnetic susceptibility χ are show in Figs. 6a and 6b. The phase transition shown in Fig. 6a corresponds to that observed in MEM$(TCNQ)_2$ by the Groningen group [44] and that shown in Fig. 6b to that observed in PY$(TCNQ)_2$ by the Budapest group.[45]

The possibility raised by these observations that a phase transition between two different kinds of ordering can go either way with temperature deserves further study. Accordingly we have pursued the box model further. In doing so our goals have been 1.) to establish the phase diagram at T=0, 2.) study the effects of phonons, and 3.) extend the phase diagram to finite temperatures. We have reached 1.), but have only preliminary results for 2.), and no results for 3.). We report here our results for 1.), the phase diagram.

We treat the spin-spin interactions and the isospin-isospin interactions exactly, but treat the isospin-spin interactions in mean-field theory. The problem then separates into two parts, for which the self-consistent-field Hamiltonians are

$$\mathcal{H}_s^I = -2t\sum_i I_i^x - \sum_i \left(U + J\langle \vec{S}_i \cdot \vec{S}_{i+1}\rangle\right)\left(I_i^z + \tfrac{1}{2}\right)\left(I_{i+1}^z - \tfrac{1}{2}\right) \tag{32}$$

$$\mathcal{H}_s^S = -J\sum_i \left\langle \left(I_i^z + \tfrac{1}{2}\right)\left(I_{i+1}^z - \tfrac{1}{2}\right)\right\rangle \vec{S}_i \cdot \vec{S}_{i+1} \tag{33}$$

At this point, it is convenient to introduce the following notation for the various charge ordered states:

I F (Ising ferromagnetic) for ferroelectric

I A (Ising antiferromagnetic) for antiferroelectric

I P (Ising paramagnetic) for paraelectric

In addition we distinguish two paraelectric states, I P(F) with ferroelectric short-range order and I P(A) with antiferroelectric short-range order. For the I F, I P(F), and I P(A) types of charge ordering in the ground state, the self-consistent solution of (32) and (33) for the spin ordering gives

$$\langle \vec{S}_i \cdot \vec{S}_{i+1} \rangle = -m^2 \tag{34}$$

Thus, $\mathcal{H}_s^I$ becomes

$$\mathcal{H}_s^I = -2t \sum_i I_i^x - (U - Jm^2)\left(\sum_i I_i^z I_{i+1}^z - \tfrac{1}{4}N\right), \tag{35}$$

an I F Hamiltonian ($U > Jm^2$) in a transverse field $2t$ or the corresponding I A Hamiltonian ($U < Jm^2$). This problem was solved by Pfeuty[46] for the I F case. His solution is trivially extended to the I A case but can only be used in the I P(A) domain because Eq. (34) and therefore Eq. (35) no longer hold when there is long-range order of I A type.

The principal result is that charge ordering, i.e. long-range order in I_i^z ($\langle I_i^z \rangle \neq 0$), occurs for $t < \frac{1}{4}|U - Jm^2|$. The details of the charge ordering are

$$\left.\begin{aligned} \langle I_i^z \rangle &= \gamma, \quad U > Jm^2, \quad IF \\ \gamma &= \tfrac{1}{2}(1-\mu^2)^{1/2}, \quad \mu < 1, \quad IF \\ &= 0 \quad\quad, \mu \geq 1, \quad IP(F),\ IP(A) \\ \mu &\equiv t / \tfrac{1}{4}|U - Jm^2| \end{aligned}\right\} \tag{36}$$

A second-order phase transition occurs between IF and IP(F) or IA and IP(A) when $\mu = 1$. The IP(F) state changes over continuously into the IP(A) state across the surface $U = Jm^2$ where $\mu = \infty$. Other properties of Pfeuty's solution of interest are the short-range order,

$$
\begin{aligned}
\langle I_i^z I_{i+1}^z \rangle &= n^2 , \quad IF , \; IP(F) \\
&= -n^2 , \quad IP(A)
\end{aligned}
\tag{37}
$$

where n^2 is a smooth, monotonically decreasing function of μ, $n^2 = \frac{1}{4}$ for $\mu = 0$, and

$$
\langle I_i^x \rangle = \xi , \tag{38}
$$

where ξ is a smooth monotonically increasing function of μ.

We are now in a position to do the spin ordering for the IF, IP(F), and IP(A) cases. The results are collected in Table III. In that table, the acronym AFM stands for antiferromagnetic. Thus we see that the property (34) fed initially into the cycle of self-consistency is obtained once again at its end; the spin Hamiltonians in Table III are all Heisenberg AFM with spin-ordering in the ground state consistent with (34). From the Bethe solution of the linear Heisenberg AFM,[47] the value of m^2 is 0.443.

We now turn to the IA case. We suppose to begin with that

$$
\langle I_i^z I_{i+1}^z \rangle = -n^2 , \quad IA \tag{39a}
$$

$$
\langle I_i^z \rangle = (-1)^i \nu , \quad IA \tag{39b}
$$

The spin Hamiltonian obtained under the assumption (39) is listed in Table III. It is that of an alternating Heisenberg AFM. As $|\mu|$ decreases from one to zero, ν and $\frac{1}{4} + n^2$

each monotonically approach $\frac{1}{2}$. When $\mu = 0$, the exchange alternates between 2J and zero. The nearest-neighbor spin correlation function thus has the form

$$\langle \vec{s}_i \cdot \vec{s}_{i+1} \rangle = - m^2 (1 + (-1)^i \eta) . \tag{40}$$

η increases monotonically from 0 to 1 as $|\mu|$ decreases from 1 to 0. Simultaneously, m^2 decreases from 0.443, the Bethe value,to 3/8 = 0.375, the value appropriate to singlet pairs.

Substituting (40) into (32) leads to

$$\mathcal{H}_s^I = -\frac{1}{4}(Jm^2 - U)N - 2t\sum_i I_i^x - Jm^2\eta\sum_i (-1)^i I_i^z + (Jm^2 - U)\sum_i (1 + (-1)^i \gamma) I_i^z I_{i+1}^z . \tag{41a}$$

where

$$\gamma = \eta / (Jm^2 - U) \tag{41b}$$

for the isospin Hamiltonian. In contrast to the previous forms of $\mathcal{H}_s^I$, (41) cannot be diagonalized by standard methods, and its ground state is unknown. It can be diagonalized when $t = 0$. In that case $\langle I_i^z I_{i+1}^z \rangle$ and $\langle I_i^z \rangle$ are given by (39) with $n^2 = \frac{1}{4}$ and $\nu = \frac{1}{2}$. The term in t can now be treated as a perturbation. Because it is translationally invariant, it cannot modify the spatial dependence of $\langle I_i^z I_{i+1}^z \rangle$ and $\langle I_i^z \rangle$, only their magnitudes. Thus as long as perturbation theory in t converges, the initial supposition (39) is derived from $\mathcal{H}_s^I$ and is therefore self-consistent. Unless there is an intervening novel phase transition, perturbation theory should converge up to the boundary of the IA phase, $|\mu| = 1$. We cannot rule out such a possibility in the absence of an exact ground state for all $-1 \leq \mu \leq 0$. Accordingly, we have obtained what should be regarded as a minimal phase diagram, which is shown in Figs. 7 and 8.

VI. THE SURFACES OF QUASI-ONE-DIMENSIONAL CHARGE-TRANSFER SALTS: A SPECULATION

Substantial ohmic contact resistance ($\sim 20\,\Omega$ for mm^2 contacts) is characteristic of quasi-one-dimensional charge-transfer salts. Mappings of surface equipotentials on TTF-TCNQ crystals have led to a proposed nonconducting surface layer.[48] X-ray photoelectron spectroscopy at both normal and grazing emergence angles has led to the suggestion of a neutral layer of molecules at the surface of TTF-TCNQ crystals.[49] These three bits of evidence suggest that there may be no charge transfer at the surface of TTF-TCNQ. In the present section, we speculate that an absence of charge transfer may occur quite at the surfaces of quasi-one-dimensional charge-transfer salts and give a discussion of why it might be so.

Let z_o be a measure of the charge transfer which actually occurs in a crystal, $0 \le z_o \le 2$. In simple salts, it is the electron transfer from each donor to each acceptor molecule. In complex salts, it can be taken as the mean electron transfer per acceptor molecule. Suppose that z_o is merely the bulk equilibrium value of a quantity z which can vary continuously throughout the allowed range of z_o, $0 \le z \le 2$. However, as z varies, no change is allowed in the molecular structure of the crystal; the structure is fixed to be that of the equilibrium salt, z_o. $z = 0$ corresponds, therefore, to a system of neutral molecules in the structure of the charge transfer salt.

Let $f(z)$ be the free-energy density of a material in which the charge transfer is uniformly z, and $\mathfrak{f}(z)$ the same quantity measured relative to the neutral crystal

$$\mathfrak{f}(z) = f(z) - f(0). \tag{42}$$

At sufficiently low temperatures, f (z) is close to the cohesive energy and accordingly has the dominant contributions,

$$f(z) = E_{MAD} + E_{POL} + E_{VDW} + E_{SR} + E_{BS} + E_{CORR} + I - A, \tag{43}$$

and, correspondingly,

$$f(0) = E'_{MAD} + E'_{POL} + E'_{VDW} + E'_{SR} . \tag{44}$$

In (43), E_{MAD} is the Madelung energy, E_{POL} the polarization energy, E_{VDW} the Van der Waals energy, E_{SR} the short-range repulsion, E_{BS} the band-structure energy associated with the finite width of the donor molecules' valence bands and of the acceptor molecules' conduction bands, E_{CORR} the corresponding correlation energy, A the acceptor electron affinity and I donor ionization energy. When $z = 0$, the molecules are neutral overall, but E'_{MAD} and E'_{POL} arises from the charge variations within the molecules. These are both small, and we neglect them for simplicity. Moreover, we suppose that E_{VDW} and E_{SR} depend only weakly on z. Consequently, we have that

$$f(z) \cong E_{MAD}(z) + E_{POL}(z) + E_{BS}(z) + E_{CORR}(z) + I(z) - A(z). \tag{45}$$

Existing studies of the cohesive energy of salts with segregated stacks indicate that $E_{MAD}(z)$ is not nearly enough to overcome the large positive value of $I(z) - A(z)$. While both $E_{BS}(z)$ and $E_{CORR}(z)$ are undoubtedly important, especially the latter, we expect that E_{POL} is important. For such structures and with such polarizabilities, we estimate that it falls in the range -0.5 to -1.5 eV. Accordingly, we wish to raise the possibility that the rest of $f(z)$ is positive, at least for some materials.

Let us now examine the effect of a surface upon the various contributions to the free-energy $f(z)$. Let x be the distance to the surface and $F(z,x)$ the local free-energy density. We suppose that the same decomposition can be made for $F(z,x)$ as was made for $f(z)$ in (45). Of all the terms in that decomposition, only $E_{POL}(z,x)$ does not return to its bulk value $E_{POL}(z)$ within a few molecular planes of the surfaces. Consider a point charge q at the center of a spherical cavity in a polarizable medium a distance x from the surface. The energy of polarization of the medium by q, $E_{POL}(x)$ has the distance dependence

$$E_{POL}(x) = E_{POL}(\infty) + K/x \,, \tag{46}$$

where the second term is an image potential. We therefore expect that the surface correction to $E_{POL}(z,x)$ is a slowly varying function of x.

This makes possible the introduction of a Ginzburg-Landau (GL) formalism. We use the simplest possible form of GL free-energy functional

$$F(z,x) = -\frac{1}{2} b z_0^2 + \frac{1}{2} b (z-z_0)^2 + \frac{1}{2} c \left(\frac{dz}{dx}\right)^2 \tag{47}$$

In (47) b and c are constant, but z_0 is a function of x of the kind given in Fig. 9,

$$Z_0 = Z_0(x) \tag{48}$$

$z_0(x)$ ceases to exist as a minimum in $F(z,x)$ for $x < x_0$ because there the image potential has reduced $|E_{POL}(z,x)|$ below the minimum necessary to decrease $F(z,x)$ below zero for $z > 0$. Thus, if it were not for the $(dz/dx)^2$ term in $F(z,x)$, $z(x)$ would equal $z_0(x)$, and there would be a neutral layer at the surface. Instead z(x) is a solution of the GL equation

$$-c\nabla^2 z + b(z - z_o) = 0, \tag{49}$$

which requires that z(x) vary on a distance scale given by

$$\xi = (c/b)^{1/2}. \tag{50}$$

Thus a neutral layer can exist at the surface if 1.) x_o exists and is at least several molecular separations in size and 2.)

$$\xi < x_o . \tag{51}$$

We have not examined the absolute stability of the surface, but only the relative stability of the surface with respect to charge transfer and without surface reconstruction. If charge transfer is unstable at the surface, the neutral layer at the surface can in general be expected to undergo reconstruction. One can expect both structure and composition changes at the surface consequent to the loss of charge transfer.

For those salts with open shell donors such as NMP, Q, Ad, etc. return of electrons from acceptors to donors would lead to striking modifications of the magnetic properties of the surfaces. Miljak *et al.*[50] have reported very interesting experimental results which may bear on this point.

VII. ACKNOWLEDGMENT

I am deeply indebted to V. K. Shante, G. Theodorou, A. Madhukar, F. Zawadowski, and J. Lill for their collaboration and to G. Grüner for helpful and informative conversations. I am grateful to the Central Research Institute for Physics of the Hungarian Academy of Sciences for hospitality while the work reported in § V was carried out. The remainder of the work was supported in part by NSF DMR75-13343 and the Materials Research Laboratory of the National Science Foundation at The University of Chicago.

TABLE I: m-Values for Square Lattice of Stacks.

Range of S_m/S_o	m	$\frac{T_{um}}{T_{\ell m}}$	approximate m
$1 - \sqrt{2}$	2.91	2.74	2.91 or 3
$\sqrt{2} - 2$	3.87	3.82	3.87 or 4
$2 - \sqrt{5}$	3.91	1.53	⋮
$\sqrt{5} - \sqrt{8}$	3.96	2.54	⋮
$\sqrt{8} - 3$	3.97	1.26	⋮
⋮	⋮	⋮	⋮
$\sqrt{13} - 4$	4.00	1.52	⋮

TABLE II: Ground States in the Box Model.

LIMITING CASES	ORDER CHARGE	ORDER SPIN	OBSERVED
1. $\frac{t}{U}, \frac{J}{U}$ small	Ferroelectric  ∣○ ●∣○ ●∣	Heisenberg Antiferromagnetic	Maybe
2. $\frac{t}{J}, \frac{U}{J}$ small	Antiferroelectric  ∣○●∣●○∣	Alternating Heisenberg Antiferromagnetic (Singlet-Triplet or Spin Peierls)	Yes
3. $\frac{J}{t}, \frac{U}{t}$ small	Paraelectric  ∣• •∣• •∣ (Bonding Level Occupied in Dimer)	Heisenberg Antiferromagnetic	Maybe

TABLE III: Spin Ordering of the Various Charge Order Ground States.

CHARGE ORDERING	PHASE DOMAIN	$\mathcal{H}_s^s$	SPIN ORDERING
IF	$\mu < 1$ $U > Jm^2$	$(\frac{1}{4} - n^2) J \sum_i \vec{S}_i \cdot \vec{S}_{i+1}$	Heisenberg AFM
IP(F)	$\mu > 1$ $U > Jm^2$	$(\frac{1}{4} - n^2) J \sum_i \vec{S}_i \cdot \vec{S}_{i+1}$	Heisenberg AFM
IP(A)	$\|\mu\| > 1$ $U < Jm^2$	$(\frac{1}{4} + n^2) J \sum_i \vec{S}_i \cdot \vec{S}_{i+1}$	Heisenberg AFM
IA	$\|\mu\| < 1$ $U < Jm^2$	$(\frac{1}{4} + n^2) J \sum_i \left(1 + (-1)^i \frac{\nu}{\frac{1}{4} + n^2}\right) \vec{S}_i \cdot \vec{S}_{i+1}$	Alternating Heisenberg AFM

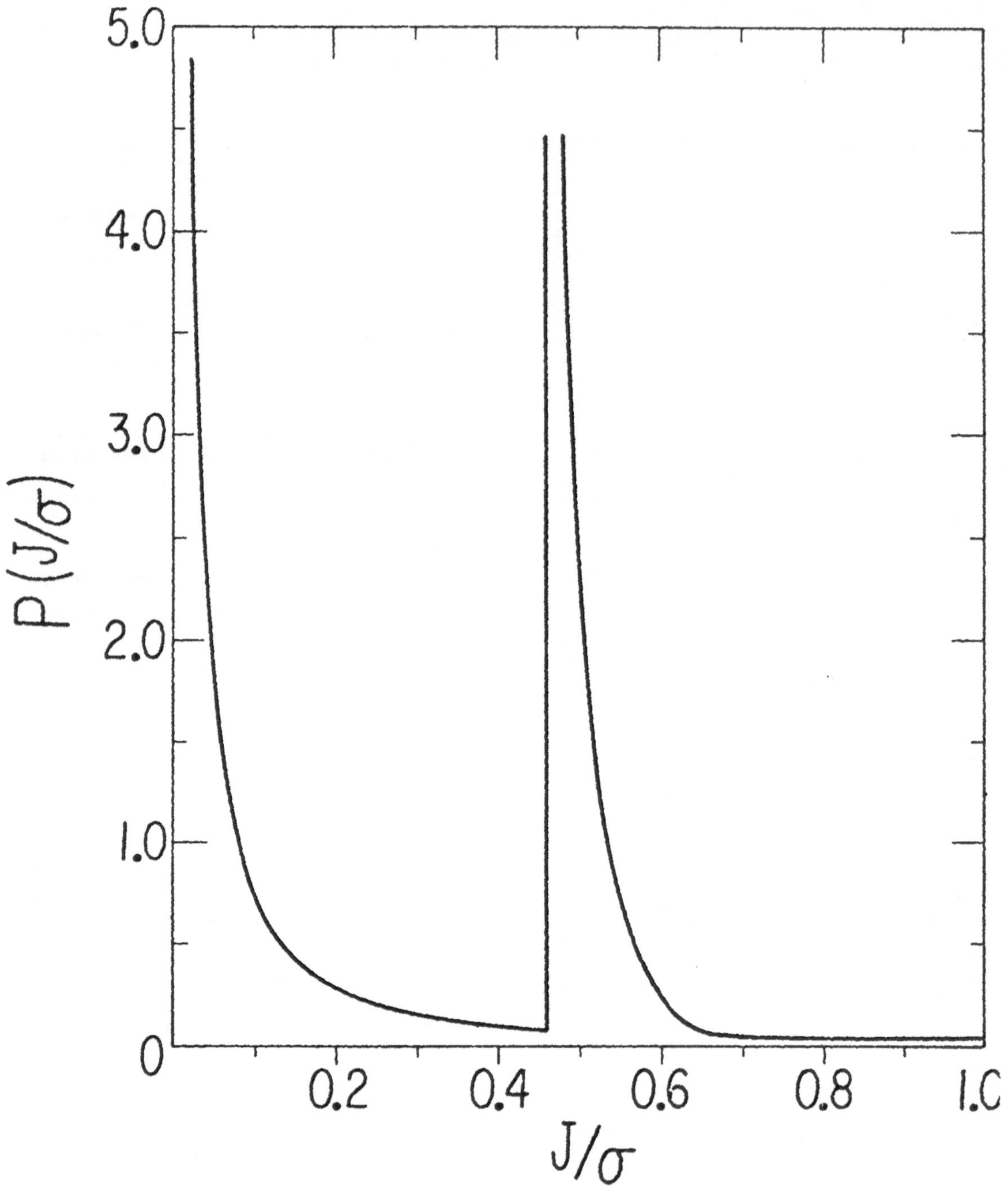

Fig. 1. Probability distribution of J/σ as a function of J/σ for $\sigma = 0.136$ eV, $t = 0.055$ eV, and $U = 0.130$ eV. (after refs. 2 and 4).

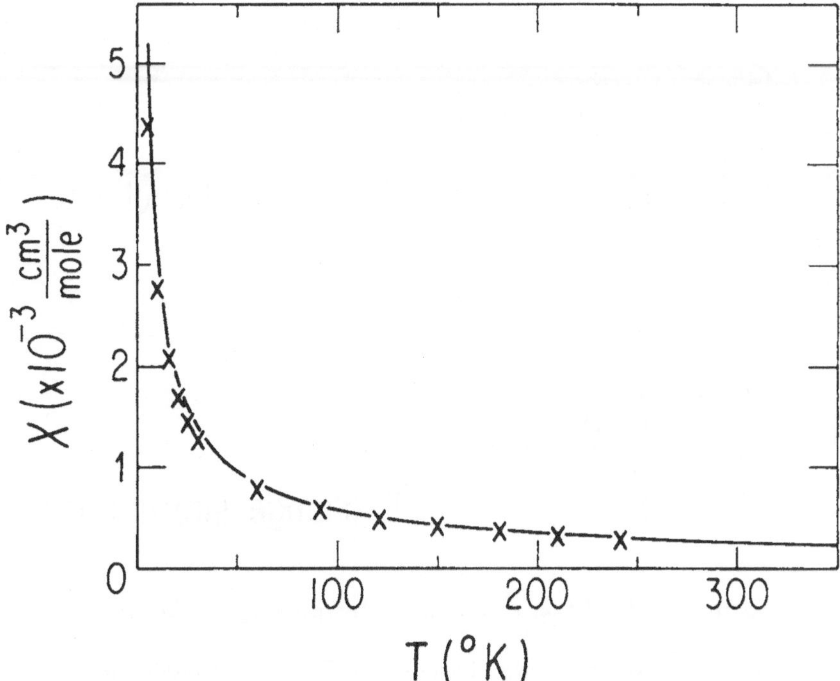

Fig. 2. Spin susceptibility of NMP-TCNQ. Crosses are DiSalvo's data; solid curve is theory of refs. 2 and 5 with σ', t, and U as in Fig. 1. (after refs. 2 and 5).

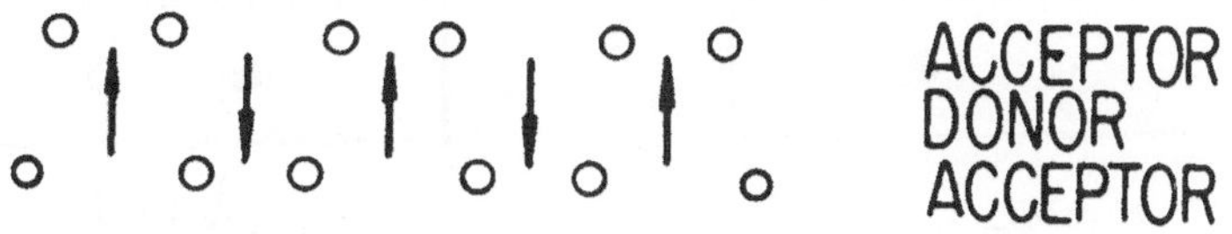

ACCEPTOR
DONOR
ACCEPTOR

Fig. 3. Schematic rendering of the prototypic structural elements of ordered salts of (Donor) $(TCNQ)_2$ with alternation donor-ion dipole moments. The donor ions are symmetrically disposed between the acceptor molecules, causing an alternation of acceptor molecule spacing, i.e. dimerization.

a.

$-t$ t_1 Transfer matrix elements

U_0 U_2 U Coulomb interactions

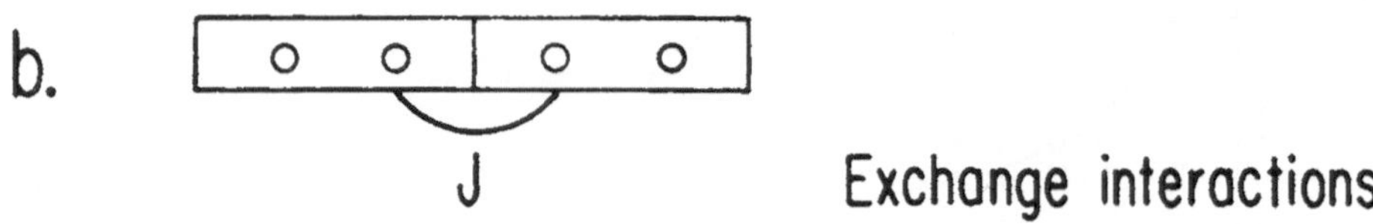

Fig. 4. Interactions within boxes and between nearest neighbor boxes:

(a) Electron transfer matrix elements and Coulomb interactions;

(b) Exchange interaction.

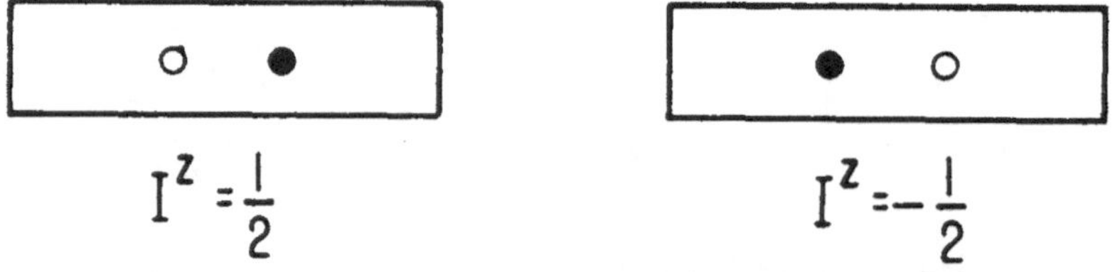

Fig. 5. Isospin representation of charge state of a box. Solid circle represents an occupied site; open an unoccupied site.

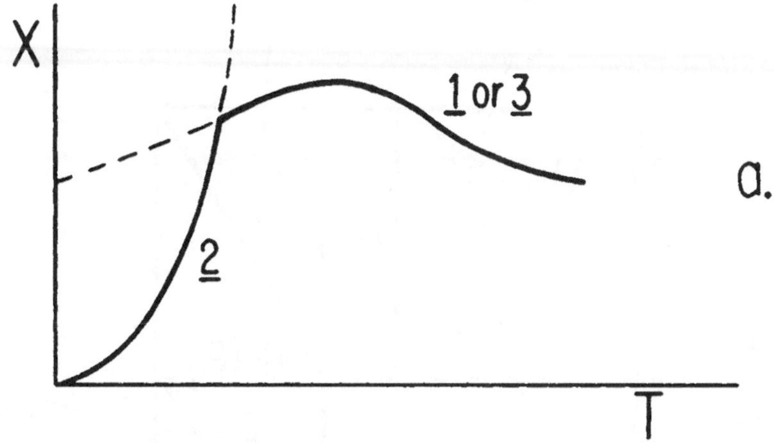

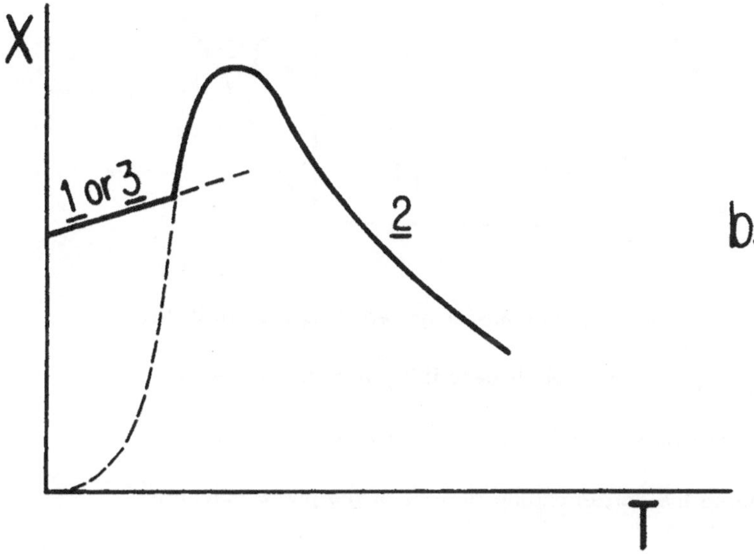

Fig. 6. Magnetic susceptibility versus temperature showing a) a transition from ordering of type 2. in Table II to ordering of type 1. or 3. , and b) the reverse.

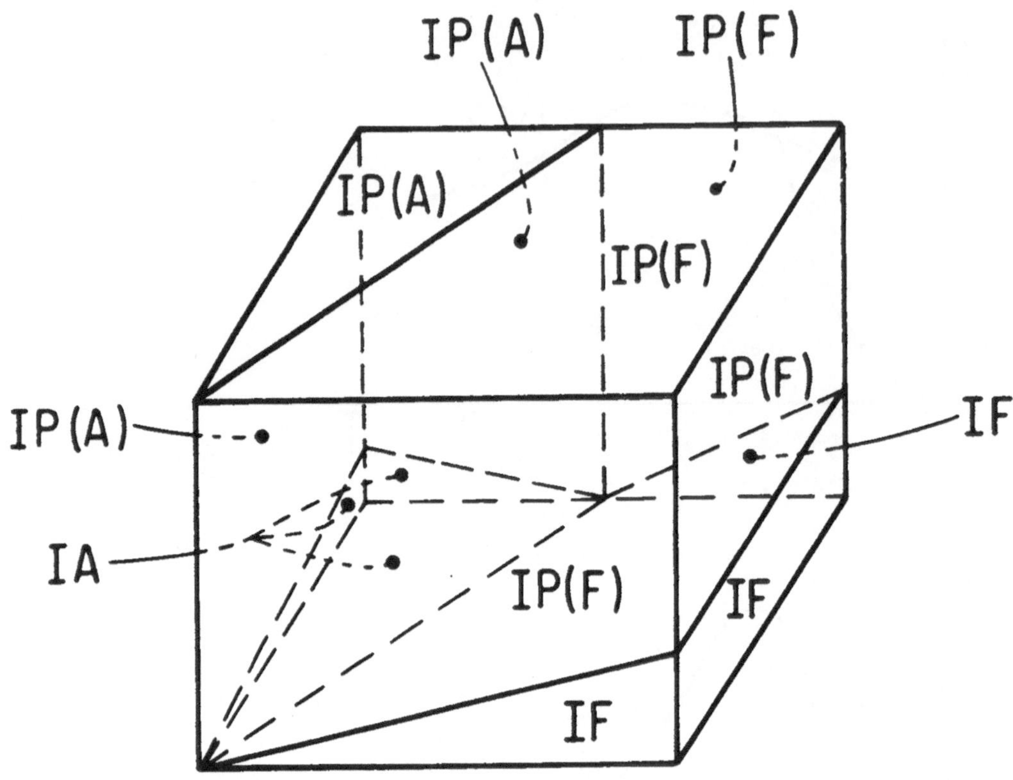

Fig. 7. Phase diagram of the box model at zero temperature in the t, U, J space. All labels describing the charge ordering in the ground state refer to planes of constant t, U, or J bounding the corresponding regions of the phase volume.

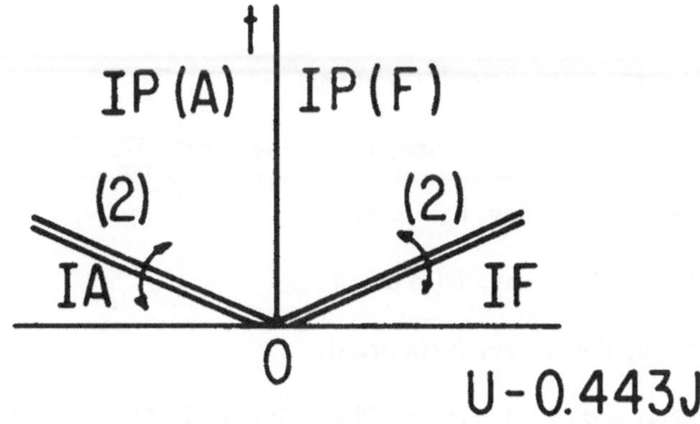

Fig. 8. Projection of the (t, U, J) phase volume of Fig. 7 onto the t, U - 0.443J phase plane. The numbers in parentheses indicate the order of the phase transition at T = 0 K across each indicated boundary.

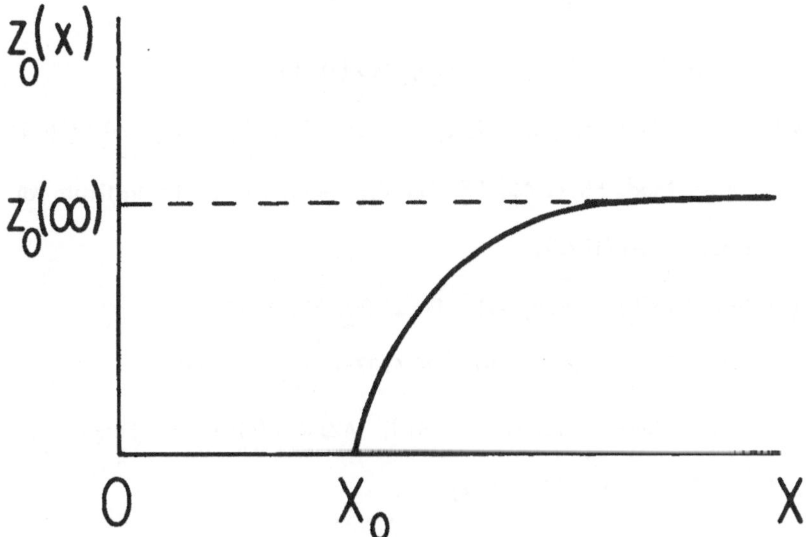

Fig. 9. The local equilibrium value of the charge transfer $z_o(x)$ as a function of distance from the surface x .

REFERENCES

1. V.K.S. Shante, Phys. Rev. B (in press).
2. G. Theodorou and M. H. Cohen, Phys. Rev. Lett. 37, 1014 (1976).
3. G. Theodorou, Phys. Rev. B (in press).
4. G. Theodorou, Phys. Rev. B (in press).
5. G. Theodorou, Phys. Rev. B (in press).
6. G. Theodorou and M. H. Cohen, Phys. Rev. B13, 4597 (1976).
7. M. H. Cohen, J. A. Hertz, P. M. Horn, and V.K.S. Shante, Int. J. Quant. Chem. 8, 491 (1974).
8. A. M. Madhukar and M. H. Cohen, Phys. Rev. Lett. (in press).
9. N. F. Mott, J. Non-Cryst. Solids 1, 1 (1968).
10. N. F. Mott, Phil. Mag. 19, 835 (1969).
11. V.K.S. Shante, Phys. Rev. B (in press).
12. A. Miller and E. Abrahams, Phys. Rev. 120, 745 (1960).
13. V. Ambegaokar, B. I. Halperin, and J. S. Langer, Phys. Rev. B4, 2612 (1971).
14. S. Kirkpatrick, Rev. Mod. Phys. 45, 574 (1973); Report B15 of the Institute for Solid State Physics, Tokyo (1973).
15. V.K.S. Shante and S. Kirkpatrick, Adv. Phys. 20, 325 (1971).
16. M. Pollak , J. Non-Cryst. Solids 8-10, 486 (1974); 11, 1 (1971).
17. G. E. Pike and C. H. Seager, Phys. Rev. B10, 1421 (1974); C. H. Seager and G. E. Pike, Phys. Rev. B10, 1435 (1974).
18. I. Webman, J. Jortner, and M. H. Cohen, Phys. Rev. B (in press).
19. V. Ambegaokar, S. Cochran, and J. Kurkijarvi, Phys. Rev. 138, 3682 (1973); J. Kurkijarvi, Phys. Rev. B9, 770 (1974).

20. A. N. Bloch, R. B. Weisman and C. M. Varma, Phys. Rev. Lett. 28, 753 (1972).

21. J. Kurkijarvi, Phys. Rev. B8, 922 (1973).

22. L. B. Coleman, J. A. Cohen, A. F. Garito and A. J. Heeger, Phys. Rev. B7, 2122 (1973).

23. G. Grüner, private communication.

24. R. L. Bush, Phys. Rev. B6, 1182 (1972).

25. R. Kubo, J. Phys. Soc. Japan 12, 570 (1957).

26. N. Wiser, Phys. Rev. 129, 62 (1963).

27. L. N. Bulaevskii, A. V. Zvarykina, Yu. S. Karimov, R. B. Lynbovskii and I. F. Shchegolev, Sov. Phys. JETP 35, 384 (1972).

28. L. N. Bulaevskii, A. A. Gusseinov, O. N. Evemenko, V. N. Topnikov and I. F. Shchegolev, Sov. Phys. Solid State 17, 498 (1975)

29. M. A. Butler, F. Wudl and Z. G. Soos, Phys. Rev. B12, 4708 (1975).

30. C. J. Fritchie, Jr., Acta Cryst. 20, 892 (1966).

31. B. Morosin, Phys. Lett. 53A, 455 (1975).

32. H. Kobayashi, Bull. Chem. Soc. Japan 48, 1373 (1975).

33. J. des Cloiseaux, J. Math. Phys. 7, 2136 (1966).
I am grateful to Dr. P. Fazekas for calling this paper to my attention.

34. C. Papatriantafillou and M. H. Cohen, unpublished.

35. L. Azevedo, Ph.D. thesis, Department of Physics, University of California (Los Angeles), 1975.

36. For a survey cf, e.g., A. N. Bloch, this volume.

37. E. Rashba, this volume.

38. P. L. Bush, Phys. Rev. B6, 1182 (1972).

39. V. L. Berezinskii, Sov. Phys. JETP 38, 620 (1974).

40. A. A. Gogolin, V. I. Mel'nikov and E. I. Rashba, Zh ETF 69, 327 (1975).

41. L. V. Keldysh, Sov. Phys. JETP 20, 1018 (1965).

42. V. K. S. Shante, unpublished.

43. A. Zawadowski and M. H. Cohen, Phys. Rev. B (in press).

44. P. I. Kuindersma, Ph.D. thesis, University of Groningen, 1975.

45. K. Holczer et al. this volume.

46. P. Pfeuty, Ann. Phys. 57, 79 (1970).

47. H. Bethe, Z. Physik 71, 205 (1931); L. Hulthén, Arkiv. Mat. Astron. Fysik 26A , No. 11 (1938).

48. L. R. Bickford and K. K. Kanazawa, J. Phys. Chem. Solids 37, 839 (1976).

49. R. S. Swingle et al., Phys. Rev. Lett. 35, 452 (1975).

50. M. Miljak, J. Cooper, K. Holczer and G. Grüner, this volume and to be published.

Conference on Organic Conductors and Semiconductors, Siófok, Hungary 1976

THE EFFECT OF PHONONS ON LOCALIZATION AND ELECTROCONDUCTIVITY IN 1d CONDUCTORS

E.I. RASHBA, A.A. GOGOLIN, V.I. MEL'NIKOV

L.D. Landau Institute for Theoretical Physics, Academy of Sciences of the USSR, Moscow, USSR

The effect of electron-phonon interaction on the localization and conductivity in 1d conductors is investigated. This interaction is shown to result both in the destroying of localization produced by impurities and in its strengthening depending on the frequency of phonons, their dispersion and so on. The first process leads to an increase in conductivity, and the second to its decrease with increasing temperature. The results are applied to interpret the maxima in the temperature dependence of conductivity in some 1d conductors.

It is well known that in a 1d disordered system all quantum states are localized. This fact was first pointed out by Mott and Twose /1/ and then proved for various systems in a number of papers.

A most significant consequence of this localization is

$$\sigma(\omega) \to 0 \quad \text{at} \quad \omega \to 0$$

that is, the low frequency electrical conductivity vanishes /2/. As a result, the diffusion in such systems is absent. The laws which obey the real and imaginary parts of σ in the low frequency range were established by Berezinsky /3/ for the weak scattering case: $\sigma_1 \sim \omega, \sigma_R \sim \omega^2 \ln^2 \omega$.

In this paper we discuss the effect of the electron-phonon interaction on the localization and conductivity in 1d conductors. We show that the effect of phonons is very diversified, the phonons may destroy the localization as well as reinforce it depending on the mean frequency of phonons, their dispersion, and the values of some other parameters. We also investigate the influence of these factors on conductivity and dielectric permeability.

1. General Approach

To find the conductivity it is convenient to calculate the polarization loop. We shall suppose that the impurity scattering is sufficiently weak. Under these conditions in the 3d case ladder diagrams of the type in Fig. 1a dominate, and the contribution of the diagrams with intersections /Fig. 1b and 1c/ may be neglected. As a result, the usual kinetic equation is valid. On the contrary, in the 1d case all these diagrams give comparable contributions.

A large contribution from the diagrams with intersections arises owing to specific conditions for interference of scattered waves in the 1d case. To see this it is convenient to use the coordinate representation. For example, in the 1d case the path length and, hence, the arising phase difference for the diagram shown in Fig. 2a does not depend at all on the position x_1 of the first scatterer, if only $x_1 < x_2$. So, all the scattered waves strengthen one another and, thus, the averaging over x_1 /for $x_1 < x_2$ only!/ does not influence the contribution of this diagram. On the contrary, in the 3d case the path length for the analogous diagram /Fig. 2b/ does depend on $\vec{r}_1$; the contribution strongly oscillates as a function of $\vec{r}_1$, and, as a result, the averaging over $\vec{r}_1$ nearly cancels this contribution due to interference quenching of scattered waves. Namely, this difference in the behaviour of the diagrams in Fig. 2a and 2b causes the strong difference in the magnitude of vertex correction /Fig. 1b/ in 1d and in 3d cases.

Thus, it is necessary, using the coordinate representation, to select all the space ordered diagrams, which give a large contribution, and to sum them up. A convenient method for such a summation was proposed by Berezinsky /3/.

In particular, this method allows us to find the correlator of charge densities. Its asymptotical behaviour /4/

$$\langle \rho(0,0)\rho(x,t)\rangle \backsim \frac{1}{|x|^{3/2}}\, e^{-|x|/4\ell_i} \qquad \begin{matrix} |x| \to \infty \\ t \to \infty \end{matrix} \qquad /1/$$

clearly manifests the localization and shows the form-factor of localized states at $|x| \gg \ell_i$. Here ℓ_i is the mean free path of electrons for the impurity back scattering. It is seen that the localization length $\ell_{loc} = 4\ell_i$. We suppose everywhere that $p_F \ell_i \gg 1$, p_F being the Fermi momentum.

The zero-frequency dielectric permeability equals /4/

$$\mathcal{E} = \frac{32\,\zeta(3)\,e^2}{hSv_F}\,\ell_i^2 \qquad /2/$$

where $\zeta(3) = 1.208...$, v_F is the Fermi velocity, and S is the cross-section per one chain. The dispersion of $\mathcal{E}$ arise at frequencies $\omega\tau_i \sim 1$, $\tau_i = \ell_i/v_F$.

When the electron-phonon interaction is taken into account side by side with the impurity scattering, various phonon lines must be inserted in the diagrams of Fig. 1a, b, c. As a result, the summation of diagrams becomes much more complicated since the diagrams now include the electronic lines with different energies. Thus, the solution of the problem is possible only in some limiting cases.

2. Phonon Induced Conductivity

Let us begin with the case of high frequency, strongly dispersive phonons. And namely we suppose that

$$\bar{\omega}\tau_i \gg 1 \quad /3a/ \qquad \Delta\tau_i \gg 1 \quad /3b/$$

$\bar{\omega}$ being the mean frequency of phonons, and Δ their dispersion.

Under these conditions all the interference effects in phonon scattering are strongly suppressed and, thus, the phonon skeleton for the polarization loop /Fig.3a/ is the same as in the kinetic equation. But it must be completed by the impurity lines. It can be shown that the diagrams of the Fig. 3b type introduce a large contribution and must be taken into account. As for the diagrams with intersections of phonon lines with impurity lines, some of these diagrams /for example, the diagrams shown in Fig. 3c and d/ exactly cancel one another, and the other diagrams are small in the parameters $(\bar{\omega}\tau_i)^{-1}, (\Delta\tau_i)^{-1} \ll 1$.

The summation of all essential diagrams may be carried out under the assumption $\tau_{ph} \gg \tau_i$, τ_{ph} being the characteristic time corresponding to phonon induced electron jumps between localized states. This criterion correspond to the regime when the electronic localization is only slightly destroyed by the electron-phonon interaction. The dc conductivity equals /4/

$$\sigma_{dc} = \frac{8\zeta(3)\, e^2}{\pi \hbar S} \cdot \frac{l_i^2}{l_{ph}} \,, \qquad l_{ph} = v_F \tau_{ph} \qquad /4/$$

The corresponding diffusion coefficient is

$$D = 4\zeta(3)\, l_i^2 / \tau_{ph} \qquad /5/$$

This result for $\mathcal{D}$ has a simple physical meaning. The mechanism of conductivity is of a diffusive type: the electrons make jumps at the distances of the order of the localization length l_{loc} separated by time intervals of the order τ_{ph} The diffusion is produced by both the forward and back scattering of electrons by phonons.

In the opposite limiting case $\tau_{ph} \ll \tau_i$ the localization is totally destroyed and the usual Drude law holds for conductivity. The temperature dependence of σ_{dc} is shown in Fig. 4a; it is essential that the maximum corresponds to $l_{ph}(T) \sim l_i$. At high temperatures $(T \gtrsim \bar{\omega})$ $\tau_{ph} \backsim T^{-1}$ and $\sigma_{dc} \backsim T^{-1}$

As for the frequency dependence of conductivity, the localization influences it only at low frequencies where at $T \approx 0$ it follows the law /3/

$$\sigma_R = \frac{8}{\pi} \frac{e^2 l_i}{\hbar S} (\omega \tau_i)^2 \ln^2 (\omega \tau_i) \qquad /6/$$

The frequency τ_i^{-1} corresponding to the ballistic motion of electron between two successive scatterings by impurities is here the characteristic one. At $\omega\tau_i \gg 1$ the usual Drude theory holds. Figure 4b illustrates the $\sigma_R(\omega)$ dependence.

3. Dispersionless Phonons

Let us now see what happens when the above restrictions are removed. At first we consider the case when the electron interacts with one branch of dispersionless phonons $\omega(\vec{q}) = \omega_0 =$ const., then $\Delta = 0$. We suppose that

$$\omega_0 \tau_i \gg 1 \quad , \quad \omega_0 \gg T \qquad /7/$$

This case is of physical interest especially in connection with the interaction of electrons with intramolecular phonons.

When dispersion is absent, the energy of the electron returns precisely to its initial value after some phonons are absorbed and then the same number of phonons are emitted: $N_{abs} = N_{em}$. As a result, some interference conditions are restored and many new phonon diagrams, which are analogous to interference impurity diagrams, must be taken into account. One such diagram is shown in Fig. 5a. The close pairs of dashed lines correspond to phonons successively absorbed and reemitted by the electron, and every short section of electronic lines between them, shown in Fig. 5 by hachured lines, corresponds to a hot electron or a hot Fermi hole. These hot carriers have the energy $\sim \omega_0$, therefore their phonon mean free path $l_h \sim N(\omega_0)\, l_{ph} \ll l_{ph}$ is much less than that l_{ph} for thermal carriers, $N(\omega_0)$ are the Planck's mean occupation numbers. From every "hot" section the long "tongues" with the length $\sim l_{ph}$ formed by "cold" lines,may be thrown out by the multiple processes of successive emission and reabsorption of phonons. This is illustated by Fig. 5b and 5c. The diagram shown in Fig. 5c may be considered as a compound multitail vertex.

We have carried out the exact summation of all diagrams, which are essential under the conditions /7/, $p_F l_h \gg 1, \varepsilon_F/\omega_0$ and $l_i \gg l_h$; ε_F is the Fermi energy. The main results are as follows /5/.

In the perfect system the contributions of all non-ladder diagrams dramatically cancel. As a result, the conductivity may be described by usual kinetic equation, i.e. it is of the Drude type.

In the disordered system conductivity $\sigma_{dc} = 0$, i.e. the localization perists. It is of special interest that the localization length and the form-factor of localized states are not influenced by the phonon scattering at all. But the frequency dependence of σ_R is strongly changed; when $\tau_{ph} \ll \tau_i$, the new characteristic frequency τ_{ph}/τ_i^2 arises. τ_i^2/τ_{ph} is the time of diffusive motion at the distance $\sim l_i$ with diffusion coefficient $D \sim v_F^2 \tau_{ph}$ controlled by phonon scattering. We stress that this diffusion is purely a quantum one, this is seen from the fact that it does not destroy the localization.

In the low-frequency region $(\omega \ll \tau_{ph}/\tau_i^2)$ the conductivity is described by the formula

$$\sigma_R = \frac{8}{3\pi} \frac{e^2}{\hbar S} \frac{\ell_i^2}{\ell_{ph}} (\omega \tau_i)^2 \ln^3 (\tau_{ph}/\omega \tau_i^2) \qquad /8/$$

and at $\omega \gg \tau_{ph}/\tau_i^2$ it follows the Drude law, as in the perfect crystal. This behaviour is sketched in Fig. 6. The wide plateau corresponds to

$$\sigma_R \approx \frac{2}{\pi} \frac{e^2}{\hbar S} \ell_{ph}(T) \qquad /9/$$

When the concentration of impurities decreases, the region of frequencies - where the impurity scattering influences the conductivity - diminishes.

4. Low Frequency Phonons. Adiabatic Regime

Let us now remove the criterion /3a/ and consider low frequency highly dispersive phonons. We begin with the case of a disordered system /6/ and suppose $\bar{\omega}\tau_i \ll 1$. This criterion means that the phonon frequency is much less than the characteristic frequency of localized electron τ_i^{-1} . Thus the electrons "see" the quasistatic instantaneous field of phonons. As a result, the phonons shorten the localization length but produce hardly any real transitions between the localized states. At high temperatures $(T \gtrsim \bar{\omega})$ where the phonons may be described classically, the effective localization length ℓ_{eff} is determined by

$$\frac{1}{\ell_{eff}} = \frac{1}{\ell_i} + \frac{1}{\ell'_{ph}(T)} \quad , \quad \ell'_{ph}(T) \sim T^{-1} \qquad /1o/$$

where l_{ph} is the usual electron mean free path for quasi-elastic phonon scattering. Due to $\bar{\omega}\tau_i \ll 1$ the strong inequality $l'_{ph} \ll l_{ph}$ may be expected. So, under adiabatic conditions the phonons shorten the localization length and thus reinforce localization.

σ_{dc} and D are determined by formulae /4/ and /5/ with l_i substituted by l_{eff}. When both $l_{ph}, l'_{ph} \backsim T^{-1}$ then at high temperatures $\sigma_{dc} \backsim T^{-1}$. The dependence of σ_{dc} vesus T is illustrated by Fig. 7. Here the maximum corresponds to $l_i \ll l_{ph}$ and is caused by the strengthening of localization rather than by destroying it.

In the perfect crystal the adiabatic regime may arise too, but only when the electron-phonon coupling is sufficiently strong and the temperature sufficiently high /5/. Indeed, when $\bar{\omega} \ll v_F / l'_{ph}$, the frequency v_F / l'_{ph} of the electron motion in the instantaneous field of phonons strongly exceeds the phonon frequencies, and thus the criterion of adiabaticity is fulfilled. Under these conditions momentary localization arises: the electron is trapped inside the region with the size $\sim l'_{ph}$ during the time when the field of phonons still remains stationary. For highly dispersive phonons /i.e. $\Delta \sim \bar{\omega}$ / the phonon potential is destroyed for the time $\sim \bar{\omega}^{-1}$, so the diffusion coefficient may be estimated as

$$D \sim (l'_{ph})^2 \, \bar{\omega} \sim D_{Dr} \bar{\omega} l'_{ph} / v_F \qquad /11/$$

where $D_{Dr} \sim v_F \, l'_{ph}$ is the usual diffusion coefficient /7/. It is seen that in the adiabatic region $D \ll D_{Dr}$, while at $\bar{\omega} > v_F / l'_{ph}$ the usual diffusion with $D \sim D_{Dr}$ is maintained. For dispersionless phonons the potential is exactly restored after the time $2\pi / \omega_0$. Thus, the diffusion may arise due to nonadiabaticity only, and thus D may be expected to be even less.

5. Non -linear Electron-Phonon Interaction

A high temperature decrease in σ /see Fig. 7/ caused by shortening of the localization length may arise also due to non-linear elecrron-phonon coupling /6/. The major part of the intramolecular vibrations belongs to the non-totally-symmetric /NTS/ type. For such vibrations the linear intra-molecular electron-phonon interaction is absent, but only if the molecule is not of the Jahn-Teller type. The Hamiltonian is thus of the type $\frac{1}{2}\Delta\omega a_n^+ a_n (b_n^+ + b_n)^2$, where a_n and b_n are the electron and the phonon operators. The term $\Delta\omega a_n^+ a_n b_n^+ b_n$ corresponds to the purely elastic scattering of electron by phonons. Analogously to this impurity scattering it produces localization. With T increasing, this localization becomes stronger owing to an increase in $\overline{b_n^+ b_n}$. The terms with $(b_n^+)^2$ and b_n^2 describing the non-elastic processes are small at low temperatures. In accordance with Section 3 they do not destroy at all the localization produced by the elastic term when only the dispersion of phonon may be neglected. These statements are valid when $\Delta\omega/\varepsilon_F \ll 1$.

6. Discussion of Experimental Data

The ideas described above were applied in Ref. 6 to the analysis of experimental data on complex conductivity of TCNQ salts with asymmetric cations - quinolinium /Qn/ and acridizinium /Adz/. It was supposed that the structural disorder caused by randomly oriented cations produced the localization, and the temperature dependence of σ was attributed to the effect of phonons on the localization. The data of Schegolev and Zolotukhin on conductivity of Adz $/TCNQ/_2$ are shown in Fig. 8. Their data on dielectric permeability are shown in Fig. 9. For Adz$/TCNQ/_2$ the low temperature localization length $\ell_i \approx 17$ Å has been found in Ref. /6/, using the data of Fig. 9, formula /2/ and the value $v_F \approx 6.4 \cdot 10^7$ cm/sec. Then from Eq. /4/ and the data of Fig. 8 it follows that $l_{ph} \sim 1500$ Å in the temperature region near the maximum of σ_{dc}. Since $l_{ph} \gg l_i$ .we attribute the maximum in σ_{dc} to the phonon induced shortening of localization length by the mechanisms discussed in Sects. 4 and 5. The theoretical curve of Fig. 8 is drawn for the model with a two-branch phonon spectrum: the phonons of the first branch destroy the localization, and those of the second branch strengthen it according to mechanism described in Sect. 5. The values of parameters were chosen so that the theory and the experiment match as well as possible. The theoretical curve for $\mathcal{E}(T)$ drawn using the same values of parameters is shown in Fig. 9; the analogous curve for $Qn/TCNQ/_2$ is shown too. The agreement of the experimental and theoretical data is seen to be quite satisfactory. Unfortunately, it follows from the numerical values of parameters found in Ref. 6 that the weak scattering approximation used everywhere above turns out to be rather rough. We believe, nevertheless, that the general agreement of the theoretical data with experimental evidence is in favour of the general picture of the effect of phonons on the localization outlined above.

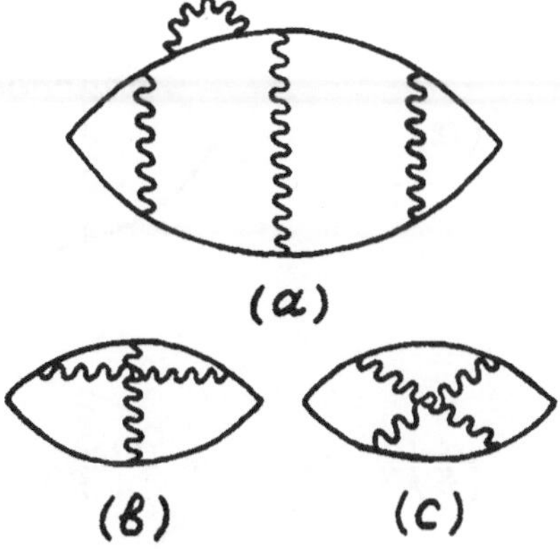

Fig. 1 Diagrams contributing to conductivity of disordered 1d conductors

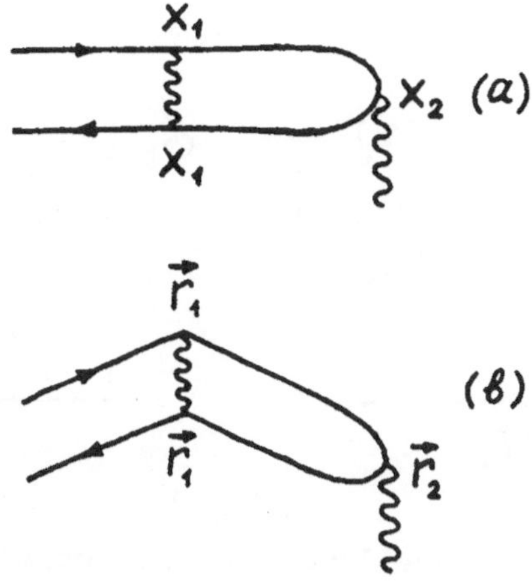

Fig. 2 Diagrams manifesting the difference in the interference conditions for the 1d case /a/ and 3d case /b/

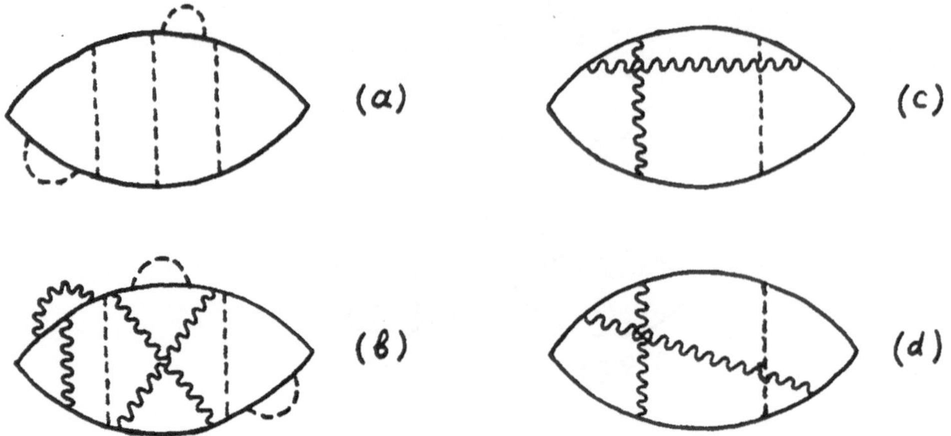

Fig. 3 Diagrams with impurity and phonon lines contributing to conductivity of 1d conductors

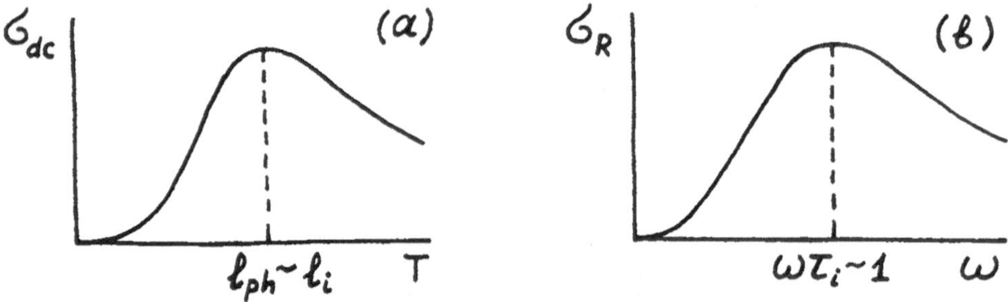

Fig. 4 T and ω dependence of conductivity /a/ Temperature dependence of dc conductivity. In high temperature region the localization is destroyed by electron-phonon interaction, and conductivity follows the Drude law; /b/ Frequency dependence of conductivity

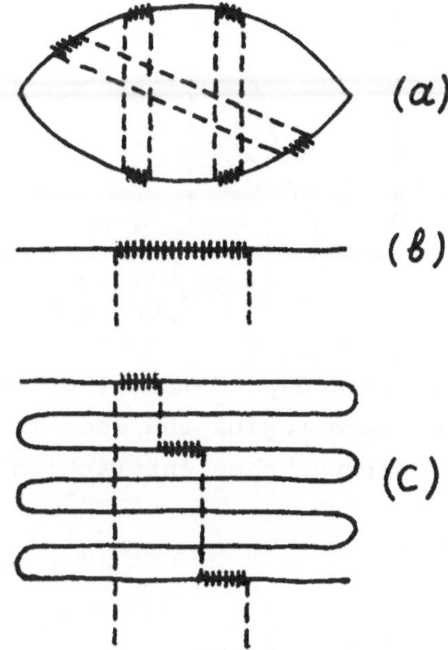

Fig. 5 Diagrams contributing to conductivity for the case of dispersionless phonons

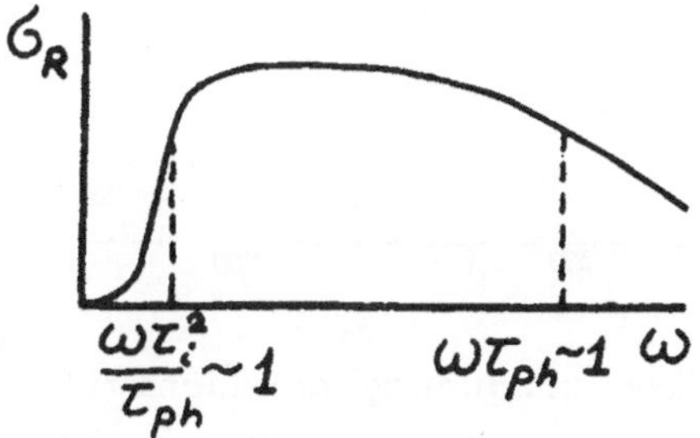

Fig.6 Frequency dependence of conductivity for disordered system. Electrons are scattered by high-frequency dispersionless phonons

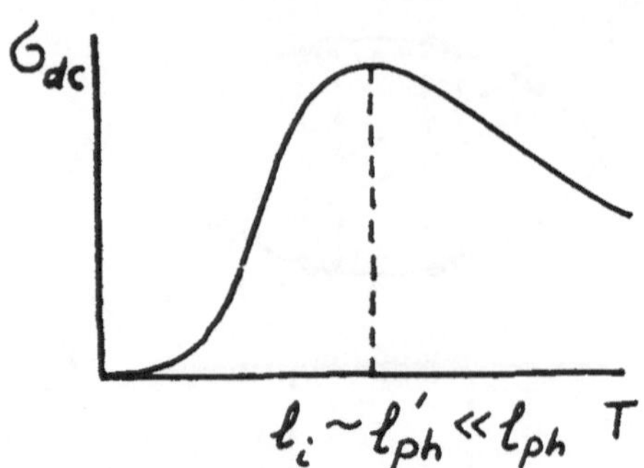

Fig. 7 Temperature dependence of dc conductivity. In high temperature region the localization is strengthened by electron-phonon interaction

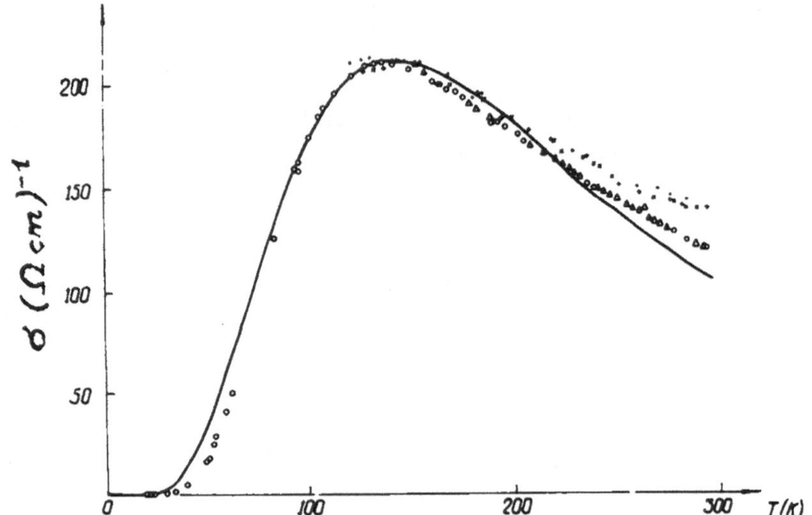

Fig. 8 dc and low frequency conductivity of Adz $/TCNQ/_2$ according to Ref. /6/. The solid curve represents theory

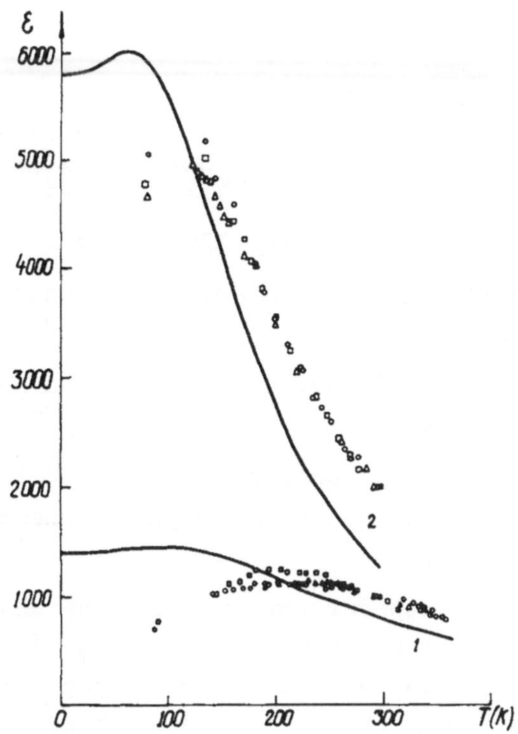

Fig. 9 Dielectric permeability of $Qn/TCNQ/_2$ and $Adz/TCNQ/_2$ /curves 1 and 2/; the solid curves-theory

REFERENCES

/1/ N. F. Mott, W. D. Twose, Adv. Phys. 10, 107 /1961/.

/2/ B. I. Halperin, Adv. Chem. Phys. 13, 123 /1967/.

/3/ V. L. Berezinsky, ZhETF 65, 1251 /1973/.

/4/ A. A. Gogolin, V. I. Mel'nikov, E. I. Rashba,ZhETF 69 327 /1975/.

/5/ A. A. Gogolin, V. I. Mel'nikov, E. I. Rashba, ZhETF 71, N° 6 /1976/.

/6/ A. A. Gogolin, S. P. Zolotukhin, V. I. Mel'nikov, E. I. Rashba, I. F. Schegolev, Prisma v ZhETF 22, 564 /1975/.

/7/ L. P. Gor'kov has informed us that the analogous estimation for D has been obtained by J. Hertz and M. H. Cohen in an unpublished paper. We particularly stress here the crucial importance of the adiabaticity criterion for the validity of this result.

IV

PHASE TRANSITIONS IN TTF-TCNQ AND RELATED COMPOUNDS

Conference on Organic Conductors and Semiconductors, Siófok, Hungary 1976

SYMMETRY CHANGES AND PHASE TRANSITIONS IN TTF - TCNQ

E. ABRAHAMS

Department of Physics, Rutgers University, New Brunswick, New Jersey, USA

J. SÓLYOM and F. WOYNAROVICH

Central Research Institute for Physics, Budapest, Hungary

As it is well known from recent experiments (1, 2), TTF-TCNQ has several subsequent phase transitions at low temperatures. Below 54 K, there is a distortion with wave vector $k = [(1/2 - q)\, a^*,\ 0.295\, b^*,\ 0]$ where q varies between zero at 54 K and 1/4 at 38 K. Bak (3, 4) and Emery (3) have interpreted the data in terms of three consecutive phase transitions involving charge density waves (CDW's) on the various molecular chains, as follows: The first transition at 54 K has $q = 0$ and involves only one set of chains (either TTF or TCNQ); at the second, near 47 K, the other chains order and q varies continuously from 0 toward 1/4 as the temperature is lowered. At 38 K, there is a first order transition involving locking of the distortion at $q = 1/4$. While this theory is a phenomenological one, Šaub, Barišić and Friedel (5) proposed a mechanism to explain the phase sliding.

In our calculation we have analysed the situation within the framework of the Landau theory (6), taking proper account of the symmetry properties of TTF - TCNQ, without specifying the interactions driving the phase transitions. We argue that the symmetry requirements are not properly taken into account in the Bak-Emery theory and therefore at the first transition, in contrast to Refs. 3, 4, both sets of chains order and the second transition is not related to the ordering of the second set of chains. Our explanation for the third transition is different, too. Similar calculations have been performed by Bjeliš and Barišić independently (7).

In the Landau theory (6), the change $\delta\rho$ in the actual charge density as one moves away from the phase transition point is expanded in terms of a complete set of functions which are basis functions of the irreducible representations (IR's) of the space group of the high temperature phase

$$\delta\rho(r, p, T) = \sum_{\alpha} c_{\alpha}(p, T)\, \varphi_{\alpha}(r) \tag{1}$$

where $\varphi_{\alpha}(r)$ is a basis function of an IR, α labels the IR and the row, p is the pressure and T is the temperature. If the transition is of second order, or weakly first order, then $\delta\rho$ and hence the c_{α} are small near the transition point and the free energy may be expanded in terms of the c_{α}(up to fourth order, say). Only those combinations of the c_{α} which are invariant under the symmetry operations of the space group may enter. The coefficients of the various invariants in the expansion of the free energy are functions of p and T, but their functional form is not determined by the Landau theory. The values of the c_{α} are determined by minimizing this expression for the free energy and these values determine the symmetry of $\delta\rho$ in the low temperature phase.

The chemical unit cell of TTF-TCNQ contains two TTF and two TCNQ molecules. The space group is C_{2h}^{5} ($P2_1/b$). The symmetry elements are the translations, the identity E, the inversion I, a two-fold screw axis C_2^{*} parallel to b and a glide plane (the ac plane) $\sigma^{*} = C_2^{*} I$.

To study the 54 K transition with q = 0, we consider the IR's corresponding to $k_0 = 1/2 a^{*} + \mu b^{*}$. The star of k_0 then contains only k_0 and $-k_0$, and the small group of k_0 has two one-dimensional representations. The IR's of the space group are generated in the usual manner (8). There are two two-dimensional IR's D_+ and D_- with basis function of Bloch type

$$\begin{aligned} \varphi_1^{\pm}(r) &= u^{\pm}(r)\, e^{ik_0 r} \pm u^{\pm}(C_2^{*-1} r)\, e^{ik_0' r}, \\ \varphi_2^{\pm}(r) &= u^{\pm}(-r)\, e^{-ik_0 r} \pm u^{\pm}(-C_2^{*-1} r)\, e^{-ik_0' r}, \end{aligned} \tag{2}$$

where $k_0' = k_0 - a^{*}$. Expanding $\delta\rho$ in terms of these functions with coefficients $c_i^{\pm}$, it is easy to construct the second and fourth order invariants which enter the free energy expansion:

$$\begin{aligned} F = F_0 &+ A_+ c_1^+ c_2^+ + A_- c_1^- c_2^- \\ &+ B_+ (c_1^+ c_2^+)^2 + B_- (c_1^- c_2^-)^2 \\ &+ C(c_1^+ c_2^+ c_1^- c_2^-) + D(c_1^+ c_2^- + c_1^- c_2^+)^2 + \ldots \end{aligned} \tag{3}$$

In the high temperature phase all the coefficients A, B, C and D are positive (C and D may have small negative values), the transition takes place when one A, say A_+, becomes negative. The minimization of F yields

$$C_1^+ = C_2^{+*} = c\,e^{i\Theta} \quad , \quad c = \sqrt{-\frac{A_+}{2B_+}} \quad , \quad C_1^- = C_2^- = 0 \tag{4}$$

and the free energy is

$$F = F_0 - \frac{A_+^2}{4B_+} \quad . \tag{5}$$

The phase Θ is undetermined, corresponding to the incommensurability in the b direction.

Replacing this solution for $C_i^{\pm}$ into $\delta\rho$, the charge density for the two possibilities ($A_{\pm} < 0$, $A_{\mp} > 0$) may be written as

$$\begin{aligned}\delta\rho^{\pm}(r) = c\,\{ & e^{i\Theta}u^{\pm}(r)e^{ik_0 r} + e^{-i\Theta}u^{\pm}(-r)e^{-ik_0 r} \\ & \pm e^{i\Theta}u^{\pm}(C_2^{*-1}r)e^{ik_0' r} \pm e^{-i\Theta}u^{\pm}(-C_2^{*-1}r)e^{-ik_0' r}\} .\end{aligned} \tag{6}$$

From this, it is possible to find the values of $\delta\rho$ on the various chains. In the elementary cell, there are two inequivalent TTF sites $f = (0,0,0)$ and $f' = (0,1/2,1/2)$ and two TCNQ sites $q = (1/2,0,0)$ and $q' = (1/2,1/2,1/2)$. They generate four distinct chains in the b direction. We note that C_2^* shifts from f or q sites to f' or q' respectively and use the values of k_0 and k_0' given earlier. We find along the various chains

$$\begin{aligned}
\delta\rho_f^{\pm}(y) &= c\left[e^{i(\Theta+\mu y)}u_F^{\pm}(y) + e^{-i(\Theta+\mu y)}u_F^{\pm}(-y)\right], \\
\delta\rho_{f'}^{\pm}(y) &= \pm c\left[e^{i(\Theta+\mu y)}u_F^{\pm}(y-1/2) + e^{-i(\Theta+\mu y)}u_F^{\pm}(-y+1/2)\right], \\
\delta\rho_q^{\pm}(y) &= ic\left[e^{i(\Theta+\mu y)}u_Q^{\pm}(y) - e^{-i(\Theta+\mu y)}u_Q^{\pm}(-y)\right], \\
\delta\rho_{q'}^{\pm}(y) &= \mp ic\left[e^{i(\Theta+\mu y)}u_Q^{\pm}(y-1/2) - e^{-i(\Theta+\mu y)}u_Q^{\pm}(-y+1/2)\right],
\end{aligned} \tag{7}$$

where y is the coordinate along b in units of the lattice spacing and the $u_{F,Q}$ are periodic, e.g. $u_F^{\pm}(y) = u^{\pm}(0,y,0) \pm u^{\pm}(0,y-1/2,1/2)$. We see that D_+ and

D_- each describe non-vanishing CDW's on both the TTF and TCNQ chains. With D_+, the CDW's on the inequivalent TTF chains have similar amplitude and are essentially in phase while the CDW's on the TCNQ chains are essentially out of phase. With D_-, the relative phase relations are reversed. These statements are the more precise the smoother the $u^{\pm}$ or the smaller the μ.

We now study the possibility of changing the CDW phase along a in which case the wave vector of the distortion becomes $k_1 = (1/2 - q)\, a^* + \mu b^*$. k_1 belongs to a single four-dimensional IR of the space group, the other members of its star being $k_2 = -k_1$, $k_3 = -k_4 = -(1/2 - q)\, a^* + \mu b^*$. The basis functions are

$$\chi_1 = u(r)\, e^{ik_1 r}, \qquad \chi_2 = u(-r)\, e^{ik_2 r},$$
$$\chi_3 = u(C_2^{*-1} r)\, e^{ik_3 r}\, e^{-i\mu/2}, \qquad \chi_4 = u(-C_2^{*-1} r)\, e^{ik_4 r}\, e^{i\mu/2}. \tag{8}$$

The basis functions can be chosen in such a way that the χ's go over to basis functions of D_+ or D_- as $q \to 0$

$$\chi_1 \to \varphi_1^+, \quad \chi_2 \to \varphi_2^+, \quad \chi_3 \to e^{-i\mu/2} \varphi_1^+, \quad \chi_4 \to e^{i\mu/2} \varphi_2^+, \tag{9}$$

or

$$\chi_1 \to \varphi_1^-, \quad \chi_2 \to \varphi_2^-, \quad \chi_3 \to e^{-i\mu/2} \varphi_1^-, \quad \chi_4 \to e^{i\mu/2} \varphi_2^-.$$

If the structure below 54 K corresponds to D_+, we have to consider the first case. The functions χ_i enter the expansion of the charge density with coefficients d_i, say, and the second and fourth order invariants involving the d_i (for a particular q) give an additional contribution to the free energy to be added to Eq. (3)

$$F_q = A_q\,(d_1 d_2 + d_3 d_4) + B_q\,(d_1 d_2 + d_3 d_4)^2 + C_q\, d_1 d_2 d_3 d_4 + \ldots \tag{10}$$

There are also mixed invariants involving the d_i and the c's which also have to be added

$$F_q' = D_q^+\, c_1^+ c_2^+ (d_1 d_2 + d_3 d_4) + D_q^-\, c_1^- c_2^- (d_1 d_2 + d_3 d_4) + E_q (c_1^+ c_2^- + c_1^- c_2^+)(d_1 d_2 - d_3 d_4) + \ldots \tag{11}$$

The effective coefficient of the bilinear term $(d_1 d_2 + d_3 d_4)$ is $A_q + D_q^+ c_1^+ c_2^+ + D_q^- c_1^- c_2^- + \ldots$ which remains positive even if A_q should become negative at some temperature below the q = 0 transition. Therefore in a finite temperature range below 54 K the CDW belongs to q = 0.

It follows from the continuity of the basis functions as a function of q that the Landau coefficients (9) are continuous functions of q and

$$A_q \to A_+ , \quad B_q \to B_+ , \quad C_q \to 2B_+ \quad \text{as } q \to 0. \tag{12}$$

There is the possibility that $A_q(T)$ has a behaviour such that above the transition it is smallest near q = 0 and at some lower temperature its minimum shifts to a finite value of q. In this case it may be that the CDW state with wave vector $k = (1/2 - q)\, a^* + \mu b^*$ has lower energy than the q = 0 state. To investigate this, we minimize F_q for fixed q. There are two types of solutions:

1/ $$d_1 = d_2^* = d e^{i\Theta} , \quad d_3 = d_4 = 0$$
or
$$d_1 = d_2 = 0 , \quad d_3 = d_4^* = d e^{i\Theta} \qquad d = \sqrt{-\frac{A_q}{2B_q}} , \tag{13}$$

2/ $$d_1 = d_2^* = d e^{i\Theta} , \quad d_3 = d_4^* = d e^{i\Theta'} , \quad d = \sqrt{-\frac{A_q}{4B_q + C_q}} . \tag{14}$$

The excess free energy due to the ordering is

1/ $$\Delta F = - A_q^2 / 4B_q , \tag{15}$$

and

2/ $$\Delta F = - A_q^2 / (4B_q + C_q) . \tag{16}$$

The first solution has lower free energy, since $C_q > 0$ for small values of q. For this first solution $\Delta F_q \to \Delta F$ as $q \to 0$. The sliding of q from q = 0 takes place when $|\Delta F_q|$ becomes larger than ΔF from Eq. (5) . This can happen either for small values of q or at a finite q and therefore the transition can be either continuous or discontinuous.

Finally, we remark on the phase locking at q = 1/4. This situation is again commensurate in the <u>a</u> direction and the free energy has a new fourth order invariant

$$(d_1 d_4)^2 e^{i\mu} + (d_2 d_3)^2 e^{-i\mu} . \tag{17}$$

This term can lower the free energy if its coefficient is sufficiently large. The minimization leads to a solution like the second solution for the general q case with a fixed relative phase $\Theta - \Theta' = -\mu/2$ or $\Theta - \Theta' = \pi/2 - \mu/2$ and there is a finite free energy difference between the cases $q = 1/4$ and $q \rightarrow 1/4$; it is the commensurability energy and is due to the presence of the new invariant. This explains the locking of the phase at $q = 1/4$ with a jump in the value of q.

We want to stress that this discontinuity is not in contradiction with the continuity argument we used in the $q \rightarrow 0$ case. There is a fourth order invariant

$$
\begin{aligned}
& e^{-i\mu} \chi_1(1/4+q')\,\chi_1(1/4-q')\,\chi_4(1/4+q')\,\chi_4(1/4-q') \\
& + e^{i\mu} \chi_2(1/4+q')\,\chi_2(1/4-q')\,\chi_3(1/4+q')\,\chi_3(1/4-q')
\end{aligned}
\tag{18}
$$

which couples basis functions belonging to different stars. This invariant plays no role for general q but is of great importance in the $q' = 0$ ($q = 1/4$) situation.

SUMMARY

No symmetry argument requires that at 54 K only one set of chains should order. The type of ordering and the amplitude of the CDW is different for the two types of molecules. The amplitude may be much smaller on TTF than on TCNQ, but it is finite. In the Bak-Emery theory the same type of ordering is assumed for both TTF and TCNQ and that is the reason that they find no coupling.

When the phase starts to shift from $q = 0$, this shift can be either continuous or discontinuous. The description is different from that given by Bak and Emery, since the coefficients of the expansion of the free energy should be even functions of q.

The 38 K transition is related to a fourth order Umklapp invariant and not to the term proposed by Bak and Emery. This result is in agreement with Dzyaloshinsky's (10) theorem, which states that any commensurate situation will always have lower energy than nearby non-commensurate ones, since in the commensurate situation there are always new invariants due to the extra translational symmetry. By going to sufficiently high order in the Landau treatment we could see the commensurability energies between $q = 0$ and $q = 1/4$ but they are probably very small and experimentally inaccessible.

Finally we mention that there is no symmetry reason to suppose that q moves along the a^* direction. The wave vector of the CDW may have a small component parallel to c^* as well.

REFERENCES

(1) F. Denoyer, R. Comés, A. F. Garito and A. J. Heeger, Phys. Rev. Lett. 35, 445 (1975); S. Kagoshima, H. Anzai, K. Kajimura and T. Ishiguro, J. Phys. Soc. Japan 39, 1143 (1975).

(2) R. Comés, S. M. Shapiro, G. Shirane, A. F. Garito and A. J. Heeger, Phys. Rev. Lett. 35, 1518 (1975).

(3) P. Bak and V. J. Emery, Phys. Rev. Lett. 36, 978 (1976).

(4) P. Bak, Phys. Rev. Lett. 37, 1071 (1976).

(5) K. Šaub, S. Barišić and J. Friedel, Phys. Letters 56A, 302 (1976).

(6) L. D. Landau, E. M. Lifsitz, Statistical Physics, Pergamon Press, London, 1958; G. Ya. Lyubarskii, The Application of Group Theory in Physics, Pergamon Press, Oxford, 1960.

(7) A. Bjeliš and S. Barišić, see the paper in this volume.

(8) O. V. Kovalev, Irreducible Representations of the Space Groups, Gordon and Breach Science Publisher, New York, 1963.

(9) The details of this argument will be published elsewhere.

(10) I. E. Dzyaloshinsky, in Discussion following the paper by J. R. Schrieffer, Proceedings of Nobel Symposium (24), Stockholm 1973.

Conference on Organic Conductors and Semiconductors, Siófok, Hungary 1976

PERPENDICULAR PHASE DEPENDENCE IN (TTF) (TCNQ)

A. BJELIŠ and S. BARIŠIĆ

Institute of Physics of the University Zagreb, Croatia, Yugoslavia

The types of ordering with q_b=0.295b* in (TTF)(TCNQ) below 54K are discussed within the simple model of coupled one-dimensional charge density waves. It is argued that the bilinear interchain coupling is essential for the ordering above 38K while the anharmonic interchain coupling governs the first-order transition to the commensurate structure at 38K.

A rather complex picture of structural anomalies in (TTF)(TCNQ) was recently established by X-ray[1] and neutron scattering[2] measurements. In this contribution we discuss the three-dimensional ordering observed from 54K to 38K and at 38K using a simple Ginzburg-Landau model for coupled one-dimensional Peierls distorted (q_b=0.295b*) chains[3]. We argue that the ordering of TCNQ chains (at 54K) and TTF chains (at 49K), as well as the temperature dependence of q_a below 49K, can be explained by means of the bilinear interchain coupling. In contrast, the first order transition at 38K and the lock-in of q_a at a*/4 depend strongly upon the anharmonic (fourth-order) interchain coupling, as has already been shown elsewhere[4].

First we shall briefly describe the type of reasoning

which led to the conclusions concerning the 54K and 49K transitions, first put forward in Ref.5. This reasoning is in many respects similar to the phenomenological model of Bak and Emery[6], with the additional effort to estimate the relevant parameters within the Coulomb coupling model. Finally we propose a phenomenological model of the commensurability transition at 38K.

We shall assume that the dominant part of the bilinear interchain coupling comes from the Coulomb interaction of CDWs on neighboring chains. Here we neglect the finite size and the tilt of molecules in the chains and associate a purely one-dimensional CDW to each chain. Then for all interchain distances $d_\perp$ in (TTF)(TCNQ) $q_b d_\perp \gtrsim 1$, so that the Coulomb coupling constant reduces to the asymptotic form[5]

$$V_{d_\perp} \simeq \frac{e^2}{d_\perp} \frac{1}{(q_b d_\perp)^{1/2}} \exp(-q_b d_\perp) . \tag{1}$$

The characteristic values for neighboring pairs of chains are thus found to be: $V_{c/2} \simeq 7\text{meV}$, $V_{a/2} \simeq 60\text{meV}$, $V_a \simeq 1\text{meV}$. Within this model the simplest way to include the full symmetry of (TTF)(TCNQ) is to consider two one-dimensional CDWs on the wings of the TCNQ molecules. The basic results obtained within the above model, i.e. equation (1), are not altered by this generalization[4], which is consistent with general symmetry considerations[7].

Depending on the strength of the interchain couplings, we have to distinguish the situations in which the unrenormalized transverse correlation lengths are either larger or smaller than the corresponding interchain distances.

In the former case there tends to be compensation between the interchain and intrachain Coulomb contributions to the effective coupling constant which determines the critical temperature of the CDW instability. The total coupling is therefore still taken as dominated by the phonon mediated electron-electron coupling. This is important for the applicability of the Ginzburg--Landau picture[3]. The Peierls critical fluctuations are then of the three-dimensional type (i.e. occur very close to the critical temperatures) and, although enhanced, can be neglected for our purposes: therefore in this limit a mean-field calculation is adequate.

The opposite limit of weak interchain coupling is characterized by strong one-dimensional n=2 fluctuations below the highest mean-field critical temperature of a single chain. As before[8], the transverse fluctuations below the cross-over to the three-dimensional fluctuation regime will be treated by the mean-field approximation within the transfer matrix method. For a single family of chains this approximation yields the exact cross-over index[9]. Close to the cross-over temperature the dominant intrachain fluctuations entering in the deformation energy expansion are fluctuations of the phase of the order parameter. The additional approximation made here is that of neglecting the phase dependence in all, interchain as well as intrachain,fourth-order terms. This point is perhaps worthy of further investigation since even the anharmonic terms which are local in the displacements representation are phase dependent in terms of the order parameter for the Peierls unstable system with $q_b \neq 0, b^*/2$ (Ref.4). This non-locality is the consequence of the inherent non-locality of the order parameter involved in the Ginzburg-Landau expansion.

For both above cases the quadratic part of the deformation energy appropriately averaged in the longitudinal direction is given by

$$F = \sum_{q_a, q_c} \{ (a_Q+\alpha_Q)\rho_Q^2 + (a_F+\alpha_Q)\rho_F^2 + 2\alpha_{QF}\rho_Q\rho_F\cos(\phi_Q-\phi_F) \} \qquad (2)$$

where

$$\alpha_Q = 2V_a\cos(q_a a) + 2V_{c/2}\cos(q_c c/2), \qquad (3a)$$

$$\alpha_{QF} = 2V_{a/2}\cos(q_a a/2). \qquad (3b)$$

$\rho_{Q,F}\exp(i\phi_{Q,F})$ are Fourier components of CDW densities in TCNQ and TTF chains. The form of parameters $a_{Q,F}$ for a given family of chains depends on which of the above mentioned limit of interchain coupling is under consideration. For the strong interchain coupling case

$$a \cong \frac{\lambda^{-1} - \log(2.28E_F/T)}{n_F\log^2(2.28E_F/T)} + U_0. \qquad (4)$$

Here n_F and E_F are the bare density of states and the Fermi energy, λ is the electron-phonon coupling constant and U_0 is the intrachain Coulomb interaction. When the interchain coupling is weak

$$a \cong \frac{\lambda^2 T_{mf}^2}{8n_F^3E_F^2\Delta^4} T^2, \qquad (5)$$

where T_{mf} is the mean-field critical temperature for a single chain and Δ the value of the (pseudo-)gap[10] in the electron spectrum. The criterion[8,9] for taking into account only fluctuations of phase in eqs.(2,5) is $T \ll n_F E_F \Delta^2/T_{mf}$.

For both limits of interchain coupling we proceed by picking out the three-dimensional periodic configurations for which the integrand in eq.(2) has a minimum and we neglect all other non-equilibrium configurations. The configurations so determined are characterized by the star of wave vectors in the Brillouin zone. In our model the wave number of ordering in z-direction is equal to c* at all temperatures, in agreement with the experimental results[1,2]. This means that the star can have either four points $(\pm q_a, \pm q_b)$, if q_a is different from a*/2, or two points $(a^*/2, \pm q_b)$, if $q_a = a^*/2$. In the former case the three-dimensional ordered periodic configuration can be amplitude or phase modulated, i.e. either the two diagonal or all four Fourier components in the star are activated (Fig.1). The type of modulation which actually occurs depends on the nature of anharmonic fourth-order coupling. As long as this coupling is dominated by the local (i.e. intra-chain) contribution, the phase modulation is energetically more favourable for all q_a except $q_a = a^*/4$. Thus the ordering in (TTF)(TCNQ) above 38K is very probably phase modulated. The same conclusion follows from the experimental observation[1,2] that there is no discontinuity in the type of ordering at 49K, the temperature at which the two-point star ($q_a = a^*/2$) transforms into the four-point one[4].

When $q_a = a^*/4$ and with purely local anharmonic terms, the amplitude modulation becomes as favourable as the phase one due to the switching in of the Umklapp terms. Then even a small interchain anharmonic coupling stabilizes the phase modulation if it is repulsive, or leads to the lock-in of commensurate amplitude modulated configuration if it is attractive[4]. We have argued previously[4] that the latter situation explains the observed lock-in below 38K in (TTF)(TCNQ). The commensurate $q_a = a^*/4$ configuration is stabilized here by the Umklapp terms involving only the four points of the star (Fig.1). This contrasts with the alternative explanation of Ref.6, where the coupling involving the Fourier component with $q_b = -3 \cdot 0.295 b^*$ was invoked.

Finally we discuss the temperature behavior of three-dimensional ordered CDWs. The expression for the deformation energy obtained by the mean-field treatment of the interchain coupling has to be minimized in particular with respect to the relative phase between TCNQ and TTF CDWs, which gives

$$\cos(\phi_Q-\phi_F) = -1$$

at all temperatures. In order to facilitate the presentation of all possible types of the temperature variation of q_a and the CDW amplitudes $\rho_{Q,F}$, we introduce four characteristic temperatures: T_Q and T_F are defined by

$$a_{Q,F}(T_{Q,F}) = 2V_{c/2} + 2V_a \tag{6}$$

and T_Q', T_F' by

$$a_{Q,F}(T'_{Q,F}) = 2V_{c/2} + 2V_a + V^2_{a/2}/V_a. \tag{7}$$

Physically speaking the temperatures $T_{Q,F}$ would be the critical temperatures for the ordering of one family of chains in the absence of (the interaction with) the other family. In contrast to this, $T'_{Q,F}$ depend on the coupling between different chains. It can be easily seen that $T_Q'>T_Q$, $T_F'>T_F$ in both limits of strongly and weakly coupled CDWs. Our reasoning is valid even if among the temperatures (6,7) only those characterizing the actual structural instabilities are consistent with the appropriate limit of the interchain coupling.

Depending on the relative position of $T_{Q,F}$, $T'_{Q,F}$, three various types of temperature behavior of q_a are possible:

Two of them occur when $(T_Q', T_F')>(T_Q, T_F)$. Then both families of chains order simultaneously at a temperature between $\max(T_Q, T_F)$ and $\min(T_Q', T_F')$. Depending on the parameters, the value of q_a at this temperature is either finite or equal to zero (periodicity a). The former case occurs whenever

$$\left(a_Q(T_c)-2V_{c/2}-2V_a\right)^{1/4}\left(a_F(T_c)-2V_{c/2}-2V_a\right)^{1/4}/2V_a^{1/2} < 1, \qquad 8)$$

where the critical temperature T_c is the zero of the equation

$$\left(a_Q(T_c)-2V_{c/2}-2V_a\right)^{1/2} + \left(a_F(T_c)-2V_{c/2}-2V_a\right)^{1/2} = \frac{V_{a/2}}{V_a^{1/2}} \qquad (9)$$

$\cos(q_a a/2)$ at $T=T_c$ is then given by the left-hand side of the inequality (8), and $q_a(T<T_c)$ changes with temperature. If the inequality (8) is not satisfied, the instability occurs at the temperature given by

$$(a_Q-2V_{c/2}+2V_a)(a_F-2V_{c/2}+2V_a) = 4V_{a/2}^2, \qquad (10)$$

and the ordered phase is characterized by $q_a=0$ at all temperatures.

The third possibility corresponds in our model to the behavior observed in (TTF)(TCNQ), and is realized when $T_Q>T_F'$. Then the TCNQ chains become ordered at $T=T_Q$ ($\simeq$54K for (TTF)(TCNQ)), while TTF chains remain disordered. In the temperature range $T_Q>T>T_F'$, q_a remains at a*/2. At $T=T_F'$ ($\simeq$49K) TTF chains also order. With ρ_F finite $\cos(q_a a/2)$ becomes also finite being given by

$$\cos\frac{q_a a}{2} \simeq \frac{V_{a/2}}{2V_a}\cdot\frac{\rho_F}{\rho_Q} \qquad (11)$$

in agreement with the statement of Ref.6. The temperature dependence of q_a below T_F' is governed by ρ_F. For T just below T_F', $(a^*/2-q_a)\sim(T_F'-T)^{1/2}$. This agrees with the result of Ref.6 and the experimental findings[1,2]. When q_a reaches the value a*/4, the already mentioned commensurability lock-in may be expected.

In summary, the following conclusions concerning the ordering in (TTF)(TCNQ) can be drawn from the present model: The temperature variation of q_a in the range 49K>T>38K is possible only if both families of chains are ordered in this temperature range (Eq.11) (we neglect the "symmetry" coupling arising from the finite size of molecules!). The ratio of CDW (or phonon) amplitudes on TTF and TCNQ chains is however very small, since it is essentially determined by the ratio $V_a/V_{a/2}$ ($\sim 10^{-1}$ within the Coulomb coupling model). The first-order phase transition at 38K is characterized by the change from a phase modulated incommensurate to an amplitude modulated commensurate three-dimensional CDW configuration.

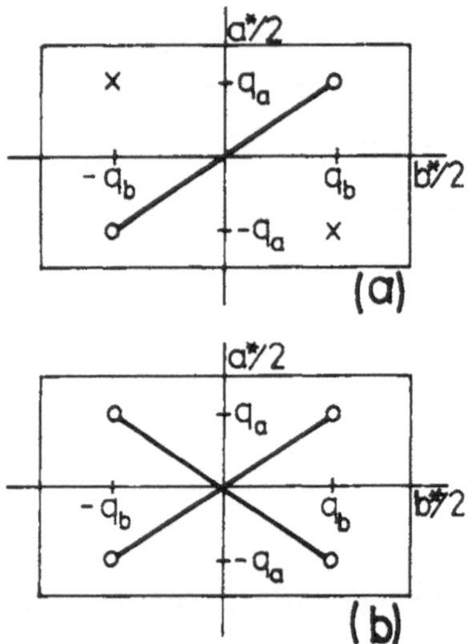

Fig.1. The phase modulated (a) and the amplitude modulated (b) star in the Brillouin zone with q_c=c*.

1) F. Denoyer, R. Comès, A. F. Garito and A. J. Heeger, Phys. Rev. Lett. 35, 445 (1975); S. Kagoshima, H. Anzai, K. Kajimura and T. Ishiguro, J. Proc. Soc. Japan 39, 1143 (1975); J. P. Pouget, S. Khanna and R. Comès, Proceedings of this conference.

2) R. Comès, S. M. Shapiro, G. Shirane, A. F. Garito and A. J. Heeger, Phys. Rev. Lett. 35, 1518 (1975); W. D. Ellenson, R. Comès, S. M. Shapiro, G. Shirane, A. F. Garito and A. J. Heeger, Solid State Commun. 20, 53 (1976). G. Shirane, Proceedings of this conference.

3) S. Barišić, Fizika 8, 181 (1976), and Proceedings of this conference.

4) A. Bjeliš and S. Barišić, to be published.

5) K. Šaub, S. Barišić and J. Friedel, Phys. Lett. 56A, 302 (1976).

6) Per Bak and V. J. Emery, Phys. Rev. Lett. 36, 978 (1976).

7) E. Abrahams, J. Sólyom and F. Woynarovich, Proceedings of this conference.

8) K. B. Yefetov and A. I. Larkin, ZhETF 66, 2290 (1974); D. J. Scalapino, Y. Imry and P. Pincus, Phys. Rev. B11, 2042 (1975); P. Manneville, J. Physique 36, 701 (1975).

9) S. Barišić and K. Uzelac, J. Physique 36, 1267 (1975).

10) P. A. Lee, T. M. Rice and P. W. Anderson, Phys. Rev. Lett. 31, 462 (1973); A. Bjeliš and S. Barišić, J. Physique Lettres 36, L-169 (1975).

Conference on Organic Conductors and Semiconductors, Siófok, Hungary 1976

A MODEL FOR THE METAL-INSULATOR TRANSITION IN TSeF-TCNQ

T.D. SCHULTZ

IBM Watson Research Center, Yorktown Heights, New York, USA

TSeF-TCNQ differs from its isostructural analog, TTF-TCNQ, in several important ways: 1) It has only one phase transition, not three or more; 2) The transition is at a much lower temperature; 3) Both chains seem to develop three-dimensional order in this transition, not just one chain; 4) The component q_a of the superlattice vector along the a* direction is a*/2, at least at all temperatures down to 15 K, in contrast to TTF where q_a varies with temperature; 5) The transition temperature is increased markedly, not decreased slightly, by small percentages of doping with the "other" donor molecule; 6) the width of the transition is also increased much more significantly by small percentages of such donor-doping.

A Ginzburg-Landau model for the coupling of the distortions on the donor and acceptor chains is introduced in which the essential difference from TTF-TCNQ is the presence in TSeF-TCNQ of larger values for the mean square distortions and longer correlation lengths on the donor stacks. This means that the donor Peierls fluctuations tend to screen the acceptor interactions, lowering T_c. Small concentrations of impurities can impede this screening, thereby raising the transition temperature while setting up random fields that broaden the transition. The larger values of the donor order parameter can also make fourth-order terms in the order parameters important and thereby explain why both the donor and acceptor subsystems can have transitions at the same temperature, despite the decoupling of these subsystems implied by $q_a = a^*/2$. The fourth order terms can also explain why q_a remains at a*/2 as the temperature is lowered.

Conference on Organic Conductors and Semiconductors, Siófok, Hungary 1976

CHIRAL CHARGE DENSITY WAVES IN QUASI ONE-DIMENSIONAL ORGANIC CONDUCTORS

H. MORAWITZ

IBM Research Laboratory San Jose, California

Abstract

We discuss the interaction of a partially filled electronic conduction band in a segregated donor-acceptor stack system with librational modes of the solid. The orientational Peierls instability predicted by us earlier leads to the formation of chiral charge density waves, which interact and phase-lock below the metal-insulator transition T_c via the Coulomb interaction. The effect of the resulting order on the physical properties of the system and the implications for the understanding of the recent neutron scattering data for the occurrence of several transitions in TTF-TCNQ will be discussed.

The intense theoretical and experimental activity in the organic charge transfer crystals exemplified by TTF-TCNQ has led to several intriguing new results in the past year. In particular, the observation of the Kohn anomaly and Peierls instability in both x-ray and neutron scattering studies[1-3] and the $4k_F$ instability[4] provide the impetus for a more detailed reconsideration of the interaction between the condensing charge density wave structures on the TTF and TCNQ stacks.

We propose in this paper that the observed Peierls instability in TTF-TCNQ and the richness of structural transitions arises from the interaction of the π electron and hole system on separate TCNQ and TTF chains, respectively, with orientational modes (librons) of the solid[5]. We suggest, that the chiral charge density waves (CCDW), which result from the orientational distortion of the TCNQ and TTF stacks account for the observed continuous increase in the unit cell dimension from $a' = 2a$ at 54° to $a' = 4a$ at $38^{\circ}K$.

We note at the outset our reasons for believing that orientational modes (librons) couple strongly to the π electron conduction band:

(I) In a molecular lattice, the usual set of 3 center of mass coordinates per atom has to be extended to 6, three to determine the center of mass position, three Euler angles to determine the orientation of the molecule in the crystal[6].

(II) The diagonalization of the resulting larger dynamical matrix leads to a coupling of translational ("phonon") and rotational ("libron") degrees of freedom within the first Brillouin zone.

(III) The arrangement of molecules within the crystal is determined by steric considerations and hence deviations from the preferred steric arrangement are intrinsically interacting.

(IV) The π electron wavefunction of the conduction band in planar organic molecules like TCNQ is large in regions far from the center of mass and small amplitude angular displacements lead to large displacements in space at the ends of the molecule.

(V) For staggered or eclipsed configurations small rotations of the molecules around an axis perpendicular to the molecular plane bring about a large change in the electronic overlap integral between 2 molecules. This leads to large electron-libron coupling.

We also add that there are definite indications from recent x-ray work on TSeF-TCNQ[7], $(TTF)_7\ I_5$[8] that rigid body hindered rotation is involved in the metal-insulator transition in these materials.

We consider a specific example of a 1/3 filled band and have accordingly for the Fermi wavevector $k_F = \pi/3a$, where a is the stacking axis lattice spacing. The corresponding soft libron has wavevector $Q^L_{soft} = 2k_F = 2\pi/3a$. In figure (1a) and (1b), we show the resulting <u>static</u> orientational lattice distortion and its effect on the electronic band structure and libron spectrum above, at and below the structural transition.

We next pursue the consequences of the resulting charge distribution residing on the orientationally distorted stack on the transverse order in the a direction of the crystal. It is well established from the work of Barisic [9], that for a system of single chains the Coulomb interaction will lead to a simple doubling of the period in both transverse directions. In TTF-TCNQ the observed soft mode shows no component in the c-direction, but an original doubling of the unit cell in the a-direction at 54°K. We, therefore, assume following earlier work by Etemad and Schultz[10], that only the TCNQ chains are ordered at 54°K by the Coulomb interaction between them. In contrast to all work published to date we want to consider an electronic CDW, which is not only compressional (corresponding to a soft longitudinal acoustic phonon triggering the Peierls instability), but <u>torsional</u>[5] as well. We note that the additional transverse degress of freedom of a CCDW wave, which we model for simplicity by a helix allow more varied ordering arrangements. We are well aware that other forms of transverse displacements may lead to similar effects, but will stick in this paper to the model of helical CDW's interacting with each other by the Coulomb interaction. In addition we will briefly treat the case of a helical CDW interacting with a compressional (longitudinal) CDW.

We consider the Coulomb potential due to a helical charge distribution at some distance d larger than the molecular dimensions of the assumed chiral charge distribution: (see Fig. 2)

$$\rho_{CDW}^{chiral} = \rho_o \cos(2k_F z + q_x a \cos pz + q_y b \sin pz)$$

$$\Phi(d,z_o) = \rho_o \int_{-\infty}^{\infty} \frac{\cos(2k_F z' + q_x a \cos pz' + q_y b \sin pz')dz'}{[(z'-z_o)^2 + (d-a \cos pz')^2 + b^2\sin^2 pz']^{1/2}} . \tag{1}$$

We can reexpress (1) in the form

$$\Phi(d,z_o) = \frac{\rho_o}{2} \cos 2k_F z_o \sum_{n=-\infty}^{\infty} \int_{-\infty}^{\infty} \frac{dz' \cos(2k_F z')[e^{in(p(z'+z_o)-\Phi)} J_n(D) +}{[z'^2+d^2+a^2-2da \cos p(z'+z_o)]^{1/2}}$$

$$+ e^{-in(p(z'+z_o)-\Phi)} J_n(-D)] \tag{1'}$$

where

$$D = \sqrt{q_x^2 + q_y^2}\, a << 1 \quad \text{(we have taken } a = b \text{ in 1)}$$

$$tg\ \Phi = \frac{q_y}{q_x} .$$

Since $D << 1$ we will keep only the 2 lowest order terms in the Bessel-function expansion, i.e. only take $n = 0, \pm 1$. After some algebra, we obtain for the Coulomb potential at $x = d$, $z = z_o$

$$\Phi(d,z_o) = \rho_o \{J_o(aq\ )\ [\cos 2k_F z_o K_o(2k_F(d^2+a^2)^{1/2}) + \frac{ad2k_F}{(d^2+a^2)^{1/2}} \cos 4k_F z_o$$
$$K_1(4k_F(d^2+a^2)^{1/2})] + J_1(aq\ )[K_o(4\ k_F(d^2+a^2)^{1/2}) - \cos 4k_F z_o L_1]\} \tag{2}$$

where $L_1 = \int_{-\infty}^{\infty} \frac{dz}{(z^2+d^2+a^2)^{1/2}}$ is a logarithmically divergent integral, and

K_o, K_1 are modified Bessel functions.

The interaction between CCDW's on adjacent like stacks (e.g. TCNQ stacks separated by a_o=12.3Å in the crystallographic a direction) is given by

$$H_{int}^{0-1} = \rho_o(\ x=a_o+a\cos pz_o, y=a\sin pz_o, z_o)\ \Phi\ (a_o, z_o) \tag{3}$$

In the absence of long range order on the TTF chains located at $x_1 = 1\frac{a_o}{2}$, the phasedifference on successive CCDW should be Π, corresponding to a doubling of the a unit cell dimension. Writing a Ginzburg-Landau expression for the charge density wave structures in a single (a,b) sheet for the crystal, we have for the free energy density

$$F_Q^{G-L} = \sum_{j=1}^{3} \sum_{i=1}^{N}{}' \{a_i^Q|\rho_i|^2 + b_i^Q|\frac{\delta\rho_i}{\delta x_j}|^2 + c_i^Q|\rho_i|^4 + H_{int}^{i-i+1}\} \qquad (4)$$

Minimizing equation (4) with respect to the transverse phase $\phi_x = q_x a_o$ leads to $\phi_x = \Pi$, i.e. $q_x = \frac{\pi}{a_o}$ as expected, corresponding to the transition at 54°K.

The interest in the TTF-TCNQ system lies, however, in the multiple transitions observed[1-3], presumably corresponding to the subsequent build-up of long range order on the TTF chains in the field of the condensed CCDW on the TCNQ chains. We have to augment the free energy (4) by additional terms corresponding to the CCDW's on TTF chains and include a coupling term for CCDW's on adjacent TTF-TCNQ chains with <u>opposite</u> sign.

$$F^{G-L} = \sum_{\substack{j=1 \\ \alpha=Q,F}}^{3} \sum_{i=1}^{N}{}' \{a_i^\alpha|\rho_i^\alpha|^2 + b_i^\alpha|\frac{\delta\rho_i^\alpha}{\delta x_j}|^2 + c_i^\alpha|\rho_i^\alpha|^4 + H^{i\ i+1}_{int_{\alpha\alpha}} + H^{i\ i+1/2}_{int_{\alpha\beta}}\} \qquad (5)$$

We observe that the form of the Coulomb potential (equation 4) due to CCDW introduces <u>higher</u> harmonics of $2k_F$ in the interaction terms, which is of interest in view of the $4k_F$ instability[4]. The departure of the transverse phase from $\phi_x = \Pi$ is obtained directly by including the additional terms in (5) in the variational equation, explaining the increase in the unit cell dimension a'=2a - 4a below 54°K. The experimentally observed locking at a'=4a, corresponding to the first order transition at T=38°K is presumably due to an Umklapp term as suggested by Bak and Emery[11].

In conclusion, we note that the consideration of librational motion coupling to the Π electron and hole bands allows inclusion of <u>multipolar</u> interaction effects (higher harmonics) in the Coulomb interaction energy and may lead to a better quantitative understanding of the metal-insulator transition and CCDW ordering in quasi 1D organic solids.

A more detailed description of this work is in preparation[12].

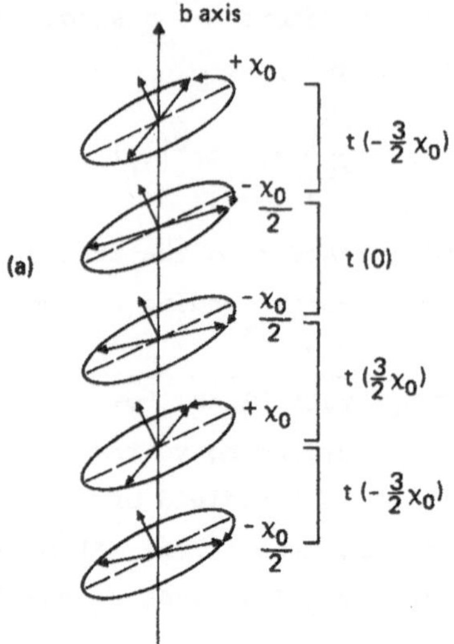

Figure 1a: Orientationally distorted stack: $\chi = \chi_o \cos 2k_f z$

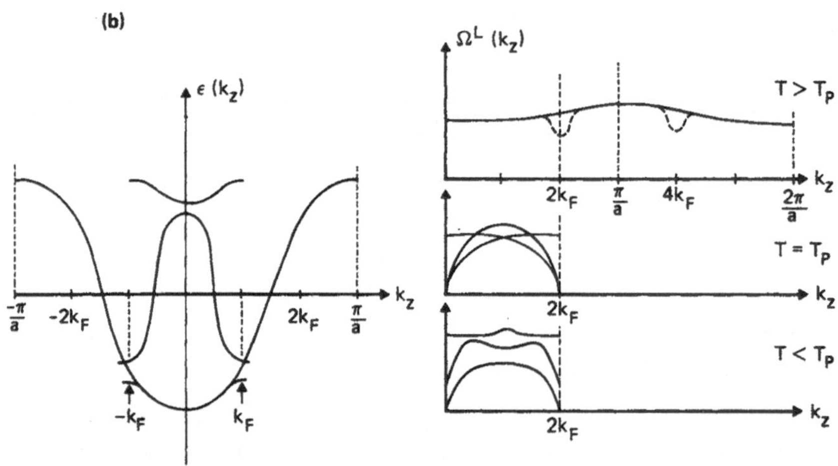

Figure 1b: Modified electronic and librational dispersion relations $\varepsilon(k_z)$, $\Omega^L(k_z)$.

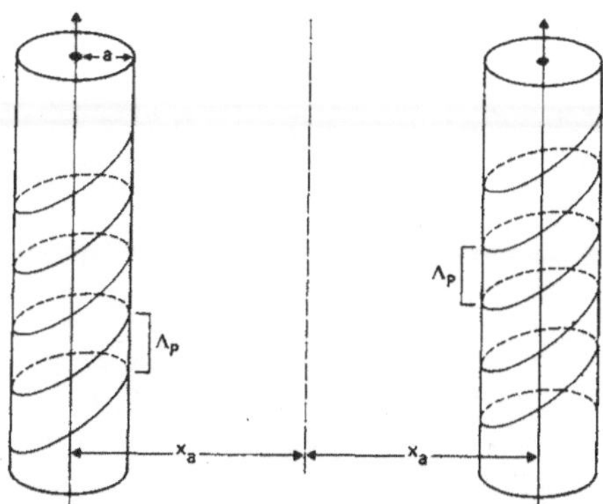

Figure 2: Two helical charge density waves $\rho=\rho_o \cos(2k_f z + k_x a \cos \frac{2\Pi}{\Lambda p}x + k_y a \sin \frac{2\pi}{\Lambda p}y)$.

References

(1) S. Kagoshima, H. Anzai, K. Kajimura, and T. Ishiguro, J. Phys. Soc. Jpn. 39, 1143 (1975);

(2) R. Comès, S. M. Shapiro, G. Shirane, A. F. Garito, and A. J. Heeger, Phys. Rev. Lett. 35, 1518 (1975);

(3) F. Denoyer, G. Comès, A. F. Garito, and A. J. Heeger, Phys. Rev. Lett. 35, 445 (1975);

(4) J. P. Pouget, S. K. Khanna, F. Denoyer, R. Comes, A. F. Garito, and A. J. Heeger, Phys. Rev. Lett. 37, 437 (1976);

(5) H. Morawitz, Phys. Rev. Lett. 34, 1096 (1975);

(6) A. I. Kitaigorodsky, "Molecular Crystals and Molecules", Chapt. III, Academic Press, New York and London (1973);

(7) C. Weyl, E. M. Engler, S. Etemad, J. Jehanno, and K. Bechgaard, Solid State Comm. 19, 925 (1976);

(8) C. K. Johnson and C. R. Watson, J. Chem. Phys. 64, 2271 (1976);

(9) S. Barisic, Phys. Rev. B5, 941 (1972);

(10) T. D. Schultz and S. Etemad, Phys. Rev. B13, 4928 (1976);

(11) P. Bak and V. J. Emery, Phys. Rev. Lett. 36, 978 (1976);

(12) H. Morawitz and J. Lajzerowicz (to be published);

V

EXPERIMENTAL INVESTIGATIONS ON TTF-TCNQ AND ITS DERIVATIVES

Conference on Organic Conductors and Semiconductors, Siófok, Hungary 1976

THE PEIERLS INSTABILITY IN THE I D METAL TTF-TCNQ

A.J. HEEGER

Department of Physics and Laboratory for Research on the Structure of Matter
University of Pennsylvania Philadelphia, PA 19174, USA

Abstract

The electronic properties of TTF-TCNQ and related 1 D conductors are reviewed in the context of the structural data which has established the incommensurate charge density wave ground state and soft-mode fluctuations associated with the Peierls instability. The frequency-dependent complex conductivity, the distribution of electronic oscillator strength, and the magnetic properties are discussed in terms of the Peierls-Fröhlich model with complex order parameter for $T > 54K$ which becomes pinned below 54K. Far infrared studies together with measurements of the magnitude and temperature dependence of the microwave conductivity and dielectric constant provide insight into the origin of the dc conductivity in the conducting regime and the large dielectric constant in the insulating regime in this class of 1 D conductors. (A detailed account of the results presented will be published in the Proceedings of the NATO Summer School on the Physics and Chemistry of 1D Conductors, Ed. by H. J. Keller, to be published by Plenum Press, 1976.)

Conference on Organic Conductors and Semiconductors, Siófok, Hungary 1976

ORGANIC CONDUCTORS DERIVED FROM SUBSTITUTED TETRASELENAFULVALENES

K. BECHGAARD and C. BERG

H.C. Ørsted Institutet, Copenhagen,

J.R. ANDERSEN,

Kemiafdelingen, Risø

C.S. JACOBSEN,

Fys. Lab. III, DTH, Denmark

A summary of some important molecular properties of several tetraselenafulvalenes(Photoelectron spectra,Electrochemistry,Mass spectra and optical absorbtion spectra) will be presented and the results compared with those obtained for the corresponding tetrathiafulvalenes.

The tetraselenafulvalenes in question form highly conducting charge-transfer salts with TCNQ and analogous compounds such as MTCNQ(Methyl-TCNQ) and DMTCNQ(Dimethyl-TCNQ), but often, depending on the conditions of crystallization, also insulating,diamagnetic (1:1) compounds can be isolated.

HMTSF-TCNQ exhibit unusual properties as a result of a more 3-dimensional character of the electronic states at low temperature(ref. 1 and 2). The 3-dimensional character is broken if small amounts of MTCNQ or DMTCNQ are introduced in the lattice probably due to the increase in size of the dopants,when compared to TCNQ. Also the low temperature properties of HMTSF-TCNQ depend somewhat on the conditions of crystallization.

A new low symmetric donor DEDMTSF(DiethylDimethyltetraselenafulvalene) has been prepared. DEDMTSF-TCNQ (1:1) can be obtained in both a highly conducting and an insulating form. Preliminary results of the temperature dependent conductivity of single crystals of the conducting form will be reported.

References:

1) HMTSF-TCNQ (Di-2,3:6,7-Trimethylene-tetraselenafulvalene-tetracyanoquirodimethane)
T.E.Phillips,T.J.Kistenmacher,A.N.Bloch and D.O. Cowan, J.C.S.Chem.Commun.,334(1976).

2) G.Soda,D.Jerome,M.Weger,K.Bechgaard and E.Pedersen. Solid State Commun., to be published.

Conference on Organic Conductors and Semiconductors, Siófok, Hungary 1976

CHEMICAL TRENDS IN ORGANIC CONDUCTORS: STABILIZATION OF THE NEARLY ONE-DIMENSIONAL METALLIC STATE

A.N. BLOCH

Department of Chemistry, The Johns Hopkins University, Baltimore, Maryland, USA

I. INTRODUCTION

High electrical conductivity can occur in a purely organic system only under special chemical conditions.[1] Among the most significant is the constraint that a sizeable population of carriers can be energetically accessible at ordinary temperatures only if the basis states for conduction consist of extended π-molecular orbitals. At least two important consequences follow.

First, because π-π interactions are relatively weak, the carriers are confined to comparatively narrow bands. In fact, for the charge-transfer salts considered here, the dominant π-π overlaps are out-of-plane and the corresponding tight-binding bandwidth parameter (charge-transfer integral) $t_{||}$ is but of order 0.1 eV.[2,3] This is no larger than typical room-temperature scattering rates and intramolecular optical phonon frequencies, and is probably smaller than typical unscreened intramolecular coulomb correlation energies. Under these conditions, the electrical,[4] magnetic,[5] and optical[6] response need not resemble those of a simple metal, and their interpretation may require special care.

Second, because of the geometry of the π-orbitals, the bandwidths and hence the electronic properties are inherently anisotropic. Nowhere is this effect more dramatic than in the conducting charge-transfer salts. These contain planar, open-shell molecular ions, stacked so as to form a periodic array of parallel conducting chains.[7] Insofar as the interchain charge-transfer integral $t_{\perp}$ is very much smaller than $t_{||}$, the propagation of carriers is confined effectively to one dimension.

Now, it is well known that physics in one dimension is pathological. As the one-dimensional limit is approached, we should expect the metallic state to become unstable[8] and fluctuation effects, static[9] and dynamic,[10] to become severe. Hence the electronic properties in this regime ought to be especially sensitive to any chemical modifications which relax or tighten the one-dimensional constraints -- that is to say, which alter the coupling between conducting chains. To illustrate, we survey here the series of compounds progressing in this manner from TTF-TCNQ[11] to HMTSF-TCNQ,[12] the first organic whose conductivity remains metallic in magnitude as $T \to 0$. We shall find that the trends and distinctions within this sequence afford physical perspectives not readily apparent from the study of any single material, and suggest approaches toward optimization of the organic metallic state.

Fortunately, TTF-TCNQ is a particularly suitable prototype, because of the amenability of the TTF donor molecule to chemical modification. Some of the possibilities[1f] are illustrated in Figure 1. Any or all of the sulfur heteroatoms can be replaced with selenium, and the terminal protons can be replaced with any of a wide variety of substituent groups. To study the role of interchain coupling, we seek to hold the molecular electronic structure and intrachain stacking patterns as nearly constant as possible, while introducing steric factors chosen so as to force changes in the crystal structure. In this respect our work complements some by the IBM-Yorktown Heights group, which has studied systematically the weaker but significant effects of adjusting molecular parameters within an isostructural series.[13]

II. STRUCTURES AND BAND STRUCTURES

Let us examine the interchain coupling problem in more detail. Of particular interest is the persistence in these materials of the instabilities of the one-dimensional metal toward formation of charge- or spin-density waves. Here the interchain coupling has two effects. When the flat sheets of Fermi surface at $\pm k_F$ are sufficiently curved due to the non-zero $t_\perp$, the divergences in the response functions at $q = 2k_F$ are alleviated and the instabilities suppressed.[14] On the other hand, we recall that since long-range order at finite temperatures cannot develop in a strictly one-dimensional system,[10] a phase transition can occur only as a result of the finite interchain coupling. Hence the actual transition will be observed at a temperature T_c not necessarily equal to its mean-field value T_p, but dependent upon both T_p and the strength of the relevant interchain coupling.[15] In the case of a Peierls transition the latter probably consists largely of the coulomb coupling between charge-density waves on neighboring chains.[16] Hence it may be difficult to predict the effect of a given structural modification upon T_c. An increase in interchain separation, for example, may tend to reduce T_c by reducing the interchain coulomb interaction, but it may also tend to enhance T_c by reducing $t_\perp$, thereby enhancing T_p. Some flexibility is available, however, insofar as $t_\perp$ depends upon both interatomic distances and relative molecular orientations, whereas the coulomb coupling depends only upon the former.

To appreciate the role of $t_\perp$, we consider the band structure predicted by simple one-electron theory.[2,3] The essential features of our results were anticipated by the speculations of Cohen et al,[17] and independently by those of Bernstein et al,[18] though both invoked incorrect molecular physics.

The crystal structure of TTF-TCNQ,[17] shown in Figure 2, consists of parallel conducting chains of TTF (Figure 1a) and TCNQ (Figure 1d) molecular ions, separately stacked along the crystallographic b-axis. In the limit of zero interchain bandwidth, the one-electron conduction band structure[2,3] would consist of one-dimensional donor and acceptor bands chemically constrained to cross at the Fermi level. The interchain coupling lifts this degeneracy over most of the Brillouin zone, leaving isolated pockets of Fermi surface to form a semimetal.

Figure 3 presents the density of states N(E) for undistorted TTF-TCNQ, calculated[2] in the tight-binding approximation using Slater atomic orbitals and a Hartree potential adjusted to the experimental[20] value of the charge transfer. Experience argues that because of the artifically rapid decay of the Slater orbitals, the absolute energies in the figure may be too small by a factor of 2, but that the relative values should be fairly trustworthy. Our best estimate is that the average $t_{||} : t_\perp \sim 25:1$, with $t_{||} \sim 0.1$ eV, in fair agreement with the rough calculations of Berlinsky et al.[3] .

Since $t_\perp < T$ above $T_c \sim 53K$, it is doubtful that covalent bonding between chains maintains any long-range interchain coherence, or that the calculated structure in N(E) near E_F is meaningful, in the high-temperature metallic state.[17] Nevertheless, our results do anticipate that states near the Fermi level should be especially sensitive to somewhat larger values of $t_\perp$; that, other factors being equal, an increase in $t_\perp$ by less than an order of magnitude should be sufficient to suppress the Peierls transition; and that at least at low temperatures the resulting material should behave as a semimetal. As we shall see, in large measure this prospect appears to be realized in HMTSF-TCNQ.[21]

To see how an adjustment of $t_\perp$ can be effected, we consider the interchain coupling on a molecular level. The perspective of Figure 2 emphasizes that in TTF-TCNQ, the a-axis coupling occurs largely through the four short S•••N contacts between TTF and TCNQ molecules separated by $\pm \frac{1}{2}$ a $\pm \frac{1}{2}$ b. As Table I indicates, the S•••N separations of 3.20 and 3.25A are substantially shorter than the nominal van der Waals separations, and comparable with intrachain interplanar stacking distances. Even so, $t_\perp$ remains small because of the geometry of the π-molecular orbitals and the relative tilts of the TTF and TCNQ molecular planes.[19]

The c-axis coupling can better be appreciated from the perspective of Figure 4a, which views the structure along the stacking b-axis. The strongest interaction occurs between TCNQ molecules separated by $\pm \frac{1}{2}$ b $\pm \frac{1}{2}$ c and is mediated by the short contacts between cyano groups (Table I).

The situation is drastically altered when the TTF protons are replaced by electronically inert but physically bulky substituent groups. As Figure 4b demonstrates, even the tetramethyl derivative TMTTF[22] is simply too large to be accommodated by the crystal structure of TTF-TCNQ. The methylation does increase the stacking distance[23] (and probably the charge transfer)[24] slightly, but the principal effect is to force a rotation of each stack about its axis.

The change in the pattern of interchain contacts (Table I) from Figure 4a to Figure 4b is striking. The strongest interchain coupling in TTF-TCNQ, the c-axis N•••N interaction, has altogether disappeared: each TCNQ cyano group now faces the inert methyl substituents of the cation. The remaining contacts are at least as distant as the van der Waals separation.

Thus, its crystallography suggests that TMTTF-TCNQ is considerably more "one-dimensional" than its TTF parent compound. We shall find, in fact, that its interchain coupling is the weakest of any material in the series we have studied.[25]

When the sulfur heteroatoms in TMTTF are replaced with selenium to form TMTSF[26] (Figure 1b), the arrangement of interchain contacts in the TCNQ salt (Figure 5a) is qualitatively unchanged.[27] However, because of the greater spatial extent of Se and the shorter Se•••N separations (Table I), $t_\perp$ is likely to be considerably larger. The TMTSF-TCNQ structure[47] is also the simplest in the series in that it is the only one containing but one cation-anion molecular pair per unit cell.

TABLE I

Electronically Significant Interchain Contacts

	Contact	Number	Distance, A	van der Waals Distance, A	Ref.
TTF-TCNQ	S•••N (a)	2	3.20	3.35	19
	S•••N (a)	2	3.25	3.35	
	C•••N (c)	2	3.29	3.10	
TMTTF-TCNQ	S•••N	2	3.45	3.35	23
	S•••S	2	3.66	3.70	
TMTSF-TCNQ	Se•••N	2	3.36	3.50	32
	Se•••Se	2	4.00	4.00	
HMTSF-TCNQ	Se•••N	4	3.10	3.50	26

This material is of special interest for another reason. By any spectroscopic or electrochemical measure,[12,28] the TMTSF molecule is electronically identical with its hexamethylene analog HMTSF,[12] Figure 1c. Further, the intrachain molecular overlaps and stacking distances in their TCNQ salts [20,29] are practically indistinguishable. Hence, TMTSF-TCNQ and HMTSF-TCNQ can differ only in their respective interchain couplings, and comparison of the two materials promises a sharp differentiation of interchain from intrachain and molecular effects.

The contrast in interchain coupling is emphasized in Figure 5. The larger HMTSF molecule cannot enter the structure of Figure 5a, and HMTSF-TCNQ adopts that of Figure 5b instead.[29] In the process the number of short Se•••N contacts per molecule is increased from 2 to 4, and their length is shortened to a remarkable 3.10A, considerably shorter than even the S•••N distances in TTF-TCNQ (Table I). Further, the familiar herringbone alternation between the stacks in TTF-TCNQ[19] (Figure 2), preserved throughout the rest of the series,[23,25] is replaced in HMTSF-TCNQ by the geometry of Figure 6. This arrangement is far more conducive to direct Se•••N π-bonding. The clear inference is that in HMTSF-TCNQ, the (b-axis) interchain coupling is substantially larger than the corresponding parameter for any other material in the series.

Along the a-axis, on the other hand, the coupling must be exceptionally weak. Here unlike molecules again alternate, but the electron-rich TCNQ cyano groups face only the saturated outer rings of the HMTSF cations. As Figures 5 and 6 indicate, these rings form crude electrical insulators between the π-electron systems of adjacent b-c layers, and also increase their spacing so that even the interlayer coulomb interaction is diminished. Indeed, so weak is the a-axis coupling that the structure is substantially disordered in this direction, as strong diffuse streaking in X-ray photographs attests.[29] Electron micrographs[30] show evidence of macroscopic layering in the same direction. The ordered structure of Figures 5b and 6 corresponds, in fact, to only 80% of the total heavy-atom electron density in the crystal, the remainder corresponding to some b-c layers displaced along the stacking c-axis, and some composed of molecules tilted in the direction opposite the one shown.

Their crystallography strongly suggests, then, that HMTSF-TCNQ is the most "two-dimensional" and TMTTF-TCNQ the most "one-dimensional," of the conductors in this series. Based upon available information,[13,19,23,29,31,32] we should expect on chemical and crystallographic grounds that in effective "dimensionality" the rest should rank as follows:

$$\text{HMTSF-} > \text{TSF-} > \text{TMTSF-TCNQ} \qquad (1a)$$

and

$$\text{HMTTF-} > \text{TTF-} > \text{TMTTF-TCNQ} \qquad (1b)$$

with each selenium compound substantially higher than its sulfur counterpart. (Further information is required to compare, say, TMTSF-TCNQ and HMTTF-TCNQ.) In the next section we inquire to what extent these trends are reflected in the electronic properties of the materials.

III. EXPERIMENTAL RESULTS

Of the substantial body of experimental data now available for these materials, we concentrate upon those aspects most directly pertinent to the trends discussed above. Some of this information in summarized in Table II.

A. Spin-Resonance Linewidths

In contrast with other materials of high conductivity, organic conductors in this class typically display sharp, well-defined spin-resonance signals.[33] In collaboration with Tomkiewicz and Schultz, we have shown that the linewidth probes the effective dimensionality of a conductor.[34,35]

In an isotropic conductor, the electronic spin-lattice relaxation time T_1 is usually governed by the Overhauser[36] and Elliott[37] processes, second-order in the electron-phonon interaction and in the spin-orbit coupling. Yafet[38] has shown that the two mechanisms in combination produce a linewidth $\Delta H \sim T_1^{-1} \sim (\delta g)^2 \tau_{ph}^{-1}$, where δg is the average g-shift and τ_{ph}^{-1} the scattering rate characterizing the phonon part of the resistivity.

Our point[35,39] is that in the one-dimensional limit, these mechanisms are ineffective. Here the only possible scattering wavevectors lie near $q = 0$ or $q = 2k_F$. In the former case the matrix element[38] is never larger than order q^2. Spin relaxation via backscattering through $q = 2k_F$, on the other hand, is forbidden by time-reversal symmetry,[34,38] and the matrix element is proportional to $|q - 2k_F|$. Hence the spin resonance line for a one-dimensional metal can be sharp even where that for an isotropic metal of the same τ_{ph} would be too broad to observe.

As $t_\perp$ is increased, so is the phase space available for spin-flip scattering and hence the linewidth. The most important processes involve interchain scattering with $q_{||} \sim 2k_F$. A rough calculation, taking account of the phase-space cutoffs imposed by conservation of energy, yields in the high-temperature limit:

$$\Delta H \sim (\delta g)^2 \tau_{ph}^{-1} \left(\frac{t_\perp}{t_{||}}\right)^2 \left(\frac{\max[t_\perp, \omega_{ph}]}{t_\perp}\right), \qquad T \gg t_\perp, \omega_{ph}, T_c \tag{2}$$

where ω_{ph} is the frequency of a phonon at $q = 2k_F$. At lower temperatures the problem becomes considerably more complicated, particularly when phonon softening is taken into account. Nevertheless, at room temperature the dependence on $(\delta g)^2$ is experimentally established,[35] and the linewidths listed in Table II suggest precisely the trends anticipated in Equation (1). These are reinforced if the room temperature conductivities σ_{RT}, also listed in the table, are a measure of τ_{ph}.

Particularly interesting is the result that HMTSF-TCNQ, which we have taken to possess the highest dimensionality in the series, is the only member of the class in which no spin-resonance signal is observed. In our experiment this implies a linewidth substantially larger than 4000 G, at least throughout the temperature range 4-300 K. Our interpretation of both the technique and the material is buttressed by a comparison with TMTSF-TCNQ, the compound

Table II

Properties of 1:1 TCNQ Salts of:

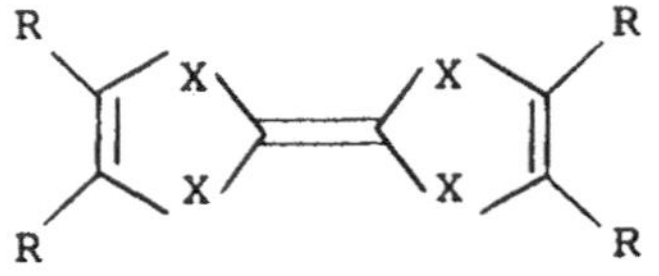

	R	X	$\sigma_{RT}(cm^{-1}\Omega^{-1})$	σ_{max}/σ_{RT}	T_{max}	T_c	$\Delta g\times10^4$ (300K)	ΔH (300K),G	
TTF	-H	S	500	20	59	53,47,38	20,40,-2	6	JHU
DMTTF	2-H,2-CH_3	S	50	25	~50	~35 (broad)	37,2	5	JHU/Penn
TMTTF	-CH_3	S	350	15	60	34	37,37,1	3-4	JHU
HMTTF	-$CH_2CH_2CH_2$-	S	500	4	80	50,40	∿40	11	IBM(SJ)
DTDSF	-H	2S,2Se	500	7	64	~45 (broad)	∿100	250	IBM(Y)
TSF	-H	Se	800	12	40	28	∿100	500-650	IBM(Y)
TMTSF	-CH_3	Se	1,200	6	61	57	88,-30	100	JHU
HMTSF	-$CH_2CH_2CH_2$-	Se	2,000	3.5	--	(32,12?)	Not Observable		JHU

which differs from HMTSF-TCNQ only in interchain coupling. The appearance of a well-defined (100 G) line in TMTSF-TCNQ is strong evidence that the extra broadening in the HMTSF salt is not an intramolecular or intrachain effect. Fruther, we note that observable spin resonance is also absent in the semimetallic normal state of the polymer $(SN)_X$.

B. Static Magnetic Susceptibilities

Qualitatively, the temperature-dependent magnetic susceptibility $\chi(T)$ for each member of the series has same the general shape as the plots in Figure 7, falling almost linearly with decreasing temperature above the metal-to-insulator transition temperature T_C, and quickly dropping and flattening below. We have verified by explicit calculation[2] that the undistorted band structure alone cannot account for the temperature dependence of $\Delta\chi \equiv \chi(T) - \chi(0)$, which is not fully understood. We emphasize that a correct explanation must be adequate to explain the nearly identical behavior of $\Delta\chi(T)/\Delta\chi(300K)$ above T_C for all members of the class.[40-42] In the context of discussions of the magnitude of the effective electron-electron interaction,[43,44,45] it is interesting that this behavior is shared by TTF-TCNQ, which exhibits[20] a $4k_F$ phonon softening from room temperature down, and an onset of $2k_F$ softening near 150 K; by TSF-TCNQ, which exhibits[46] $2k_F$ softening below 230 K and no $4k_F$ softening; and by HMTSF-TCNQ, which exhibits[46] $2k_F$ softening even at room temperature, but no anomaly at $4k_F$.

Despite the ubiquity of the temperature dependence, it is apparent from the figure and the table that the magnitude of $\chi(T)$ in HMTSF-TCNQ is anomalous. Here $\chi(0)$ is nearly twice the molecular core value, and the material remains diamagnetic even above room temperature.[42,47] Comparison with TMTSF-TCNQ (Figure 7) is again instructive: since the molecular and intrachain electronic structures of the two materials are identical, and the excess diamagnetism in the HMTSF salt must arise from the difference in interchain coupling. Consistent with the discussion of Section III, we assign it to formation of the large intermolecular coherent orbits characteristic of a semimetal.[42,47]

Figure 7 also reveals a sharp break in the slope of $\chi(T)$ for HMTSF-TCNQ near 32 K, and heat capacity measurements[42] confirm a phase transition at this temperature. The transition obviously does not lead to an insulating state, but it is reflected as a sharp change in the slope of the temperature-dependent conductivity.[48] Hence studies of the conductivity under pressure[49,50] can be interpreted as indicating that the transition is continuously suppressed to lower temperatures with increasing P, and disappears entirely below 2Kbar.

C. Conductivities and Excess Noise

Along the stacking axis, the room temperature d.c. conductivities of compounds in this class (Table II) are near the lower threshold of the metallic regime: they correspond to nominal mean free paths[51,52] ranging from about one lattice constant in the sulfur compounds to about five in HMTSF-TCNQ. With cooling, the conductivity rises rapidly, then typically peaks at a temperature T_{max} and decreases perceptibly before the onset of single or multiple phase transitions to a low-temperature semiconducting state. The sole exception in this series[53] is HMTSF-TCNQ, in which the maximum is extremely broad, the metal-to-insulator transition does not occur (notwithstanding the unexplained phase transitions mentioned in the previous section), and the metallic magnitude of the conductivity persists at least to 6mK.[54]

The temperature-dependent resistivities of the four materials we have singled out for extended discussion are compared in Figure 8. For convenience we consider the various temperature regions separately.

1. High-Temperature Conductivities $T > T_{max}$. It has been widely observed[55,56] that above T_{max}, the resistivity of TTF-TCNQ can be fitted to the empirical formula

$$\rho(T) = \rho_o + b\, T^{\gamma} \tag{3}$$

where ρ_o is a sample-dependent constant that presumably arises from crystal imperfections. We find this fit to be universal for the series,[12,13,41] with the value of b varying among different compounds but the constant γ common to them all. The dependence on ρ_o and b can be eliminated from (3) by normalizing $\rho(T) - \rho_o$ to its value at 300 K, as in Figure 8b. Plotted in this way, the high-temperature resistivities of the four compounds are practically indistinguishable from one another, with $\gamma = 2.3 \pm 0.1$.

We emphasize that apart from the importance of ρ_o,[55,56] no direct physical significance need be attached to the precise form of the empirical expression (3): it is but the simplest of several functional forms that fit the data equally well.[48] Nevertheless, the universality of the fit does imply that at high temperature all of these materials conduct by the same mechanism. It follows that the temperature dependence of the conductivity, like that of the magnetic susceptibility,[41-48] is sensitive neither to the strength of the interchain coupling nor to the development of $2k_F$ or $4k_F$ phonon anomalies.[20,46]

This last observation, together with the short coherence lengths observed in diffuse X-ray scattering,[20,46] the modest apparent mean free paths,[52] and the experimental validity of the Wiedemann-Franz Law for TTF-TCNQ,[57] militates against the hypothesis,[56,57] that a current-carrying collective mode contributes appreciably to conduction at these temperatures. Much of the original impetus for this hypothesis arose, of course, from early reports[31,59] of unusually large conductivities in occasional samples. We have verified experimentally[60,52] the suggestion by Shafer and Thomas[56,52] that identical results can arise artificially from inhomogeneous current distributions in small, highly anisotropic crystals measured by the standard four-probe technique.

To avoid these difficulties, we measured some time ago[61] the microwave conductivity of TTF-TCNQ using the standard Buravov-Schegolev technique.[62] Our results were in substantial agreement with "ordinary" d.c. conductivities, of magnitude consistent with simple metallic transport.[52]

The Buravov-Schegolev analysis is correct, however, only for specimens thin compared with the skin depth, so that the field is uniform and parallel to the conducting axis throughout the volume of the sample.[62] As Cohen et al have observed,[63] this would no longer be the case if the conductivity were to become so large that the sample thickness exceeds twice the classical skin depth $\delta_0 = c/(2\pi\sigma\omega)^{\frac{1}{2}}$. For ω = 10 GHz, δ_0 is smaller than 5 μ for σ in excess of $10^4\ \Omega^{-1}\ \text{cm}^{-1}$. Here the analysis of the microwave response for ordinary-sized crystals requires special care.

The complexity of this problem has not always been fully appreciated. In particular, widespread interest has surrounded claims[63] of exceptionally high microwave conductivities in TTF-TCNQ based upon a surface impedance analysis valid for isotropic conductors. We have devoted considerable effort to theoretical and experimental studies of the microwave response of small, strongly anisotropic conductors under skin-depth limited conditions. Our conclusion is that the isotropic analysis does not apply, and that the reported measurements[63] bear no simple relationship to the true microwave conductivity of TTF-TCNQ.

When the skin depth is much less than the simple thickness, the loss for a small isotropic conductor is proportional to the surface impedance $Z = (1/\sigma)\text{Re}K$, where $K = (1 + i)/\delta_0$ is the wavevector parallel to the surface inside the body. We find that in the anisotropic case the components of Z retain this form, but K is severely modified. For a rectangular parallelopiped of biaxial TTF-TCNQ with dimensions $b \gg a \gg c^*$ and $E \| b$, we find that the leading Fourier component of the loss occurs with:

$$K = \frac{1 + i}{\delta_b} \{1 - \frac{\lambda^2}{\varepsilon_a} [\frac{1}{b^2} + \frac{\varepsilon_{c*}}{\varepsilon_b} \frac{1}{c^2}]\}^{\frac{1}{2}} \tag{4}$$

Depending upon the sample dimensions and the complex dielectric tensor $\underset{\approx}{\varepsilon}$, ReK can become very small or even vanish. This is a new effect, related to the Clogston effect in laminated transmission lines[64] but somewhat more analogous to total reflection at a dielectric interface above the critical angle. Under these circumstances, naive use of the isotropic formulae[63] can lead to gross overestimates of the conductivity.

To demonstrate, we present in Figure 9 experimental results on a batch of crystals of TMTSF-TCNQ. In all specimens from this batch, the d.c. conductivity (solid line) rises with cooling to a maximum near 70K of about 5.8 times its room-temperature value of ca. 1500 Ω^{-1} cm^{-1}. For <u>thin</u> (< 10 μ) samples, the microwave loss could be analyzed in the usual manner, as in our previous work,[61] and yielded conductivities in quantitative agreement with the d.c. results. In contrast the loss in thick (> 100 μ) samples was clearly skin-depth limited: analysis of this loss using the isotropic formulae of Ref. 63 led to apparent peak conductivities 100 times larger than d.c. Such artifacts are just as readily produced in TTF-TCNQ.

The precise mechanism for the high-temperature resistivity is still in question. A strong temperature dependence can arise from the strong energy dependence of N(E) near E_F implied by the band-structure calculations.[17,18,11,2] A T^2 contribution is also predicted for one dimension by calculations[45] of carrier propagation in the dynamically disordered environment created by incoherent multiphonon excitations at high temperatures. Optical experiments[66] indicate substantial coupling of the electrons to intramolecular optical modes; their contribution to the d.c. resistivity is strongly temperature-dependent in any dimension. Nor can the possibility[62] of a T^2 term due to strong electron-hole scattering be lightly dismissed. At present our work does not distinguish unambiguously among such effects, but studies of the variation of b [Equation (3)] with chemical trends in $t_{||}$, ionic masses, and charge transfer are in progress.

2. <u>Conductivities in the Region of Phase Transitions</u>. The data suggest that collective effects precursive to the phase transition may assume importance as the temperature is reduced to the vicinity of T_{max}. In the case of TTF-TCNQ, for example, the resistivity begins to rise at T_{max} ~ 59 K and has increased by about 50% before a second-order phase transition occurs at T ~ 53 K.[68,20] With diminishing sample quality, T_c is unchanged but the increase in resistivity

commences at higher temperatures and lower conductivities;[59] conversely, in highly purified samples[69] we have observed T_{max} as low as ~56 K. Such behavior is consistent with the theoretical result[70,71] that the resistivity, and in particular the impurity scattering, can be dynamically enhanced through the divergence of the electronic polarizability at $q = 2k_F$ as $T \to T_C$. The greater the impurity concentration, the larger would be this fluctuation contribution to the resistivity, and the higher the temperature at which it rises above the contribution (3) to form the conductivity maximum.

The development of this part of the resistivity with temperature is apparent in the derivative plot[72] of Figure 10, where the rate of rise of $\partial(\ell n\ \rho)/\partial(T^{-1})$ with cooling begins to accelerate as early as 70 K. Horn and co-workers[73] were the first to demonstrate experimentally that the enhanced scattering rate leads to a critical divergence in $\partial\rho/\partial T$ as $T \to T_c$ from above; this behavior is also evident in the figure.

It is now well known that TTF-TCNQ undergoes at least three closely spaced phase transitions. On the basis of neutron scattering studies, Bak and Emery[74] have plausibly identified these as the three-dimensional ordering of one set of chains at ~53 K; the ordering of the other set at ~47 K, along with the onset of a continuous shearing of the superlattice so as to lower the interchain coulomb energy; and a first-order locking of the shear with transverse period 4a near 38 K. Etemad[13c] is responsible for the first convincing evidence that the 38 K transition is reflected in the conductivity. Figure 10 reproduces this result, and reveals[75] a weak anomaly at 47 K as well.

We have emphasized the complexity of the metal-to-insulator transformation in TTF-TCNQ so as to contrast it with the other materials in our series. We have already remarked in Section II that one should expect to find no detailed systematic variation of T_c (Table II) with our qualitative estimates of dimensionality, because of the competition between the tendencies of the interchain coupling to suppress T_p and to mediate the three-dimensional ordering at T_c. More interesting is the variety of behavior within the series once the region of T_c is reached.

For example, in TMTTF-TCNQ, whose structure and spin resonance suggest the weakest interchain coupling in the series, a critical divergence in $\partial\rho/\partial T$ (Figure 10) does not appear at all. The position of T_{max} 65 K is typical for the series, but the maximum (Figure 8) is broad and the region of what we have called fluctuation resistivity is greatly extended, as in a system more one-dimensional[70] than TTF-TCNQ. Only near 34 K does the resistivity show any sign of a phase transition, and this takes the form of a weak discontinuity in $\partial\rho/\partial T$, surprisingly reminiscent of the 47 K resistance anomaly in TTF-TCNQ (Figure 10). Indeed, the entire TMTTF-TCNQ derivative curve strongly suggests the TTF-TCNQ and TMTSF-TCNQ curves with their critical divergences subtracted. Apparently the growth of one-dimensional fluctuations with cooling below T_{max} in this material does not lead to three-dimensional ordering as directly as in its more strongly coupled analogs. More precise conclusions await studies of the diffuse X-ray scattering and excess noise.

A critical divergence does occur in $(\partial\rho/\partial T)$ for TMTSF-TCNQ, whose interchain coupling we have taken to be of intermediate strength. This, the simplest structure[22] in the series with but one molecule pair per unit cell and one pair of short Se···N contacts, exhibits only one phase transition, with $T_c = 57$ K. It is apparent from Figure 10 and especially from Figure 8 that the divergence of $\partial\rho/\partial T$ as $T \to T_{c+}$ is stronger than that in TTF-TCNQ, but that the development of the gap below T_c is very substantially slower. In this respect TMTSF-TCNQ differs from its unsubstituted analog TSF-TCNQ, in which the divergence is sharper than in the TTF salt on both sides of the transition.[73] The differences in critical behavior among the three materials appear difficult to rationalize fully in terms of the simple theory of Horn et al.[73]

In the higher-dimensional analog HMTSF-TCNQ,[12] the situation is of course entirely different. Here the Peierls transition is suppressed and the resistivity merely flattens below 120 K, remaining essentially featureless on the scale of Figures 8 and 10 throughout this intermediate temperature region. Even at the 32 K phase transition[42] there is only a weak discontinuity in $\partial\rho/\partial T$ (which becomes apparent when Figure 10 is expanded by a factor of ten). The coincidence of the gradual drop in $\partial\rho/\partial T$ below 120 K with an increase[42] in $\partial\chi/\partial T$ (Figure 7) and in the magnetoresistance[49,50] suggests that these effects may arise from establishment of the full semimetallic band structure with cooling.[17,47,50,42]

3. <u>Direct Measurement of Resistance Fluctuations: The Excess Noise</u>. The phase transitions and their influence on the conduction can be studied in another way. One of us (T.F.C.) has observed that when an order parameter is coupled to the resistivity, its fluctuations should be manifested as fluctuations in the resistance at constant current, or excess noise.[76,77]

Experimentally, the excess noise is the difference between the mean-square voltage fluctuations $\langle|\delta V|^2\rangle$ measured in the presence and absence of a quiet d.c. voltage $V_o = I_oR$. We determine it using a four-probe technique designed to eliminate the relatively small excess contact noise.

For TTF-TCNQ, we find that the spectral dependence of the excess noise is 1/f at least from 1 to 10,000 Hz at all temperatures. Of more immediate interest here, however, is the temperature dependence of the broadband excess noise power, shown in Figures 11 and 12 for two samples of TTF-TCNQ.

Above T_{max}, the excess noise power varies little despite the strong temperature dependence of $\rho(T)$ and $\partial\rho/\partial T$. In contrast, it is evident from Figure 10 that the rise in resistivity below $T_{max} \sim 59$ K is accompanied by a strong buildup in the mean-square resistance fluctuations, $<|\delta R|^2> = <|\delta V|^2>I_o^2$. Indeed, the noise power begins to diverge critically as the second order phase transition near 53K is approached. This divergence is even stronger than that in $(\partial\rho/\partial T)^2$.

Analysis[48] reveals that for a parallel set of independently conducting chains, the excess noise power is proportional to A_c, the cross-sectional coherence area. Hence, it may be that critical exponents in the noise power directly reflect the growth of the transverse coherence length at the three-dimensional ordering temperature. Our data, however, are not yet of sufficient quality to evaluate this proposition quantitatively.

Below 53 K, $<|\delta R|^2>$ eventually becomes nearly proportional to R^2, as in a semiconductor.[70] In particular, the discontinuity in $\rho(T)$ at the first-order 38 K transition also is reflected in $<|\delta R|^2>$, Figure 11. But in the normalized noise power $\delta P/P = <|\delta R|^2>/R^2$ (Figure 12), only the 53 K structure appears. In other words, critical resistance fluctuations appear at the second-order,[70,73] but not the first-order transition. There is also some indication of weak structure in Figure 12 near 47 K, but because of the scatter in the data this observation remains tentative.

4. Low Temperatures. At the lowest temperatures, the plots of Figure 8 and especially Figure 10 reveal at least two remarkable features. First, as described earlier, the conductivity of HMTSF-TCNQ[12] never becomes activated, but falls slowly toward a T = 0 intercept of order $10^3\ \Omega^{-1}\ cm^{-1}$; this behavior persists at least to 6 mK.[54] Unlike the results at higher temperatures, this drop in conductivity is sample-dependent,[12] and becomes very mild indeed in the best specimens.[70] Whether this temperature dependence is really due to fluctuations effects as we originally speculated[12] (based in part upon our misunderstanding of the crystal structure[29]), or whether the energy dependence of the semimetallic density of states,[12,2,49] is not at present clear. The situation is complicated by the identification of the 32 K phase transition[42,48] and by indications in the heat capacity,[42] thermoelectric power,[12] conductivity,[48] and excess noise[48] of a possible second, even weaker phase transition near 12 K.

The second important low-temperature feature of Figure 10 is the puzzling similarity among the apparent activation energies in the semiconducting states of TTF-, TMTTF-, and TMTSF-TCNQ. The low-temperature values of $\partial \ell n\rho/\partial(T^{-1}) \sim$ 100 - 120 K for these materials are also in the range of Etemad's[13c] results for TSF- and DTDSF-TCNQ. It appears, then, that the activation energy is little influenced by significant variations in the parameters -- $t_{||}$, $zt_{\perp}$, E_F, λ, etc. - that normally govern a Peierls transition. This persistence is particularly surprising in light of differences in the low-temperature magnetic g-values, whicl indicate that the states near the band edge are cation-like in some of the materi (such as TTF-TCNQ[33] and TMTTF-TCNQ[30,34]) but anion-like in others (such as TMTSF-TCNQ[30,34]).

At present the insensitivity of the apparent activation energy to otherwise well-established chemical trends is unexplained. It is true that our findings fo TTF-TCNQ contradict those of Etemad,[13c] but they do agree with measurements by Horn[79] on samples collected from several different laboratories.

The complete absence of low-temperature Curie behavior in our TTF-TCNQ sampl invites further study of the energy gaps. Our data for $\chi(T)$ below the 38 K phase transition can be expressed as the sum of two activated terms, one of activation energy Δ_1 = 104 K and prefactor $T^{-\frac{1}{2}}$ consistent with an intrinsic gap, and the other of activation energy Δ_2 = 10.2 K and prefactor T^{-1} consistent with about 50 ppm non-magnetic defects or impurities. When published[80,81] static[40] and spin-resonance[81,80] susceptibility data are treated in the same way, Δ_2 is uncert because of the Curie contribution, but Δ_1 is in excellent agreement with our resu

These same activation energies describe our conductivity data for these samples. Indeed, we can go farther. We have also measured[82] the current-voltage characteristics of semiconducting TTF-TCNQ at higher fields, using a pulsed technique designed to eliminate sample heating effects. The results do not include a negative differential resistance region as found in d.c. measurements,[83,84,82] but are strongly non-ohmic. To describe the data, we assume that a steady state is established and adapt standard hot-electron theory[85] to the case of a small-gap, nearly one-dimensional semiconductor. We obtain excellent fits to the entire set of I-V curves, from 4 K to 38 K and from V = O to field saturation by using the values for Δ_1, Δ_2, and the impurity concentration that were obtained from the magnetic susceptibility. The fits require that momentum relaxation occur via elastic collisions with impurities or defects, but that energy relaxation occur principally via low-lying optical phonon modes. Since we take the phonon frequencies from neutron data[86] and the bandwidth parameters are fairly well established,[2] the only adjustable parameter in the theory is the power of T in the low-field mobility

μ_o. The best fit to the conductivity in the low-field limit gives $\mu_o \propto T^{3/2}$ almost exactly; the family of high-field curves are then fitted with no adjustable parameters whatever. The average low-field mobility ranges from ~ 40 to ~ 10^3 cm^2/V-sec between 4 and 38 K; these values are not overly large considering that $\Delta_1/t_{||}$ is small and that the effective mass near the band edge is $m^* \sim \hbar^2\Delta_1 / 4b^2t_{||}^2 \sim 0.1$ m.

Taken together, these results bear important implications for current discussions of the low-temperature character of TTF-TCNQ and presumably of the other materials as well. First, the smallness of the implied gap $2\Delta_1$ ~ 208 K suggests that the mean-field transition temperature T_p may not be so far above T_c as some theoretical treatments[16] might predict. Second, the success of the magnetic activation energies in fitting the whole temperature-field conductivity surface is exceptionally strong evidence that the magnetic and semiconducting gaps in TTF-TCNQ are identical. Hence our results flatly contradict the assertion,[43] based upon the data of Ref. 13c, that a set of spin excitations lies below the semiconducting gap. Finally, we find no persuasive evidence that phase soliton excitations [87,84,85] play any appreciable role in low-temperature transport. The combined magnetic and electric data would require that the soliton carry a spin, and in any case the non-ohmic conductivities are well described by standard semiconductor theory. Indeed, at low temperatures we find nothing to distinguish TTF-TCNQ from a very ordinary, if very anisotropic, small-gap semiconductor.

D. Some Comments on the Optical Conductivity of TTF-TCNQ

The relatively small intrinsic low-temperature gaps deduced in the last section from d.c. conductivity and magnetic data invite comparison with the structure in published optical data for TTF-TCNQ.[88] The far-infrared optical conductivity shows a threshold near 300 cm^{-1} and rises to a peak at ~ 1000 cm^{-1}, and the latter figure is sometimes interpreted as the Peierls gap.

We disagree. On physical grounds, it is inconceivable that a gap of order 0.1 eV could arise simply from the weak Peierls modulation of $t_{||}$ which is itself of order 0.1 eV. On spectroscopic grounds, it is more appropriate to assign the "gap" to the absorption threshold at low temperatures than to the absorption peak.

Even then, the apparent optical gap of ~ 300 cm^{-1} is nearly twice as large as our estimates. This may, of course, simply reflect the difference between a direct and an indirect gap. There is evidence, however, that the 300 cm^{-1} absorption "edge" in the optical conductivity is not a simple semi-conducting gap. First, the edge remains practically temperature-independent even up to the room-temperature metallic state.[88] Second, Torrance and Bloch[66] have demonstrated, both experimentally and theoretically, the presence at higher frequencies of strong Fano[89] anti-resonances in the optical conductivity due to the coupling to the electronic continuum of discrete, totally symmetric intramolecular optical phonon modes. Such modes also occur in the far infrared, and the calculated couplings suggest a strong, nearly temperature-independent dip in the optical conductivity very close to the apparent 300 cm^{-1} absorption edge. The coupling to these low-frequency phonons must also renormalize the electronic ground state,[90] so that the conductivity resembles that of a polaronic system. In fact, simple polaron theory is remarkably successful in describing the optical conductivity of TTF-TCNQ.[91]

At least one other aspect of the high-frequency response of TTF-TCNQ bears upon some of the discussion in this volume. The plasmon spectrum has been experimentally determined,[92] and extends without Landau damping across the entire Brillouin zone. The dispersion is very well described within the random phase approximation.[93] However, when a modest amount of exchange is included in such calculations for nearly one-dimensional systems, the dispersion is severely modified and Landau damping is introduced at large wavevectors.[94] Hence rationalization of the plasmon spectrum may be a significant burden for descriptions of TTF-TCNQ[43,50,45,44] in which the short-range electron-electron interaction is taken to be large.

IV. CONCLUDING REMARKS

We summarize our survey as follows. The extraordinary instabilities and fluctuation effects characteristic of a hypothetical one-dimensional metal are quenched rapidly as the effective dimensionality is increased. Hence we should expect that the physics of a nearly one-dimensional conductor should be especially sensitive to the effective interchain coupling. This is particularly true in two-band organic systems based upon the prototype TTF-TCNQ, where the one-electron band structure in the undistorted state is nominally semimetallic, and the shape of the Fermi surface and density of states at the Fermi level are dominated by interchain charge-transfer integrals.

These observations suggest a means of systematically adjusting the electronic properties. By adding saturated substituent groups to the molecular components of such systems, we are able to control molecular size and interchain spacing without appreciably affecting the electronic structures of the molecule, the interchain stacking patterns and distances, or the effective radii of the conducting strands. We find that while the details of the interchain coupling apparently have little effect upon the temperature dependences of the high-temperature conductivity and magnetic susceptibility, or upon the low-temperature semiconducting gap, they do dominate the evolution of the metal-to-insulator transition. As the dimensionality is lowered, for example, the multiple phase transitions of TTF-TCNQ give way to a single very weak transition (TMTTF-TCNQ), with indications that the width of the critical region may approach the magnitude of the transition temperature itself. Conversely, with increased interchain coupling the metal-to-insulator transition is suppressed and the semimetallic band structure fully established. The result is HMTSF-TCNQ, whose electronic properties appear midway between those of TTF-TCNQ and a conventional semimetal. The crucial role of dimensionality in this system is emphasized by contrast with the more conventional TMTSF-TCNQ, which differs from HMTSF-TCNQ only in interchain coupling.

The development of HMTSF-TCNQ has not only underscored the physical importance of higher dimensionality, but illuminated the chemical means for attaining it. For example, preliminary studies indicate that the analogous series of compounds based upon the electron acceptor TNAP (Figure 1e) displays a wider range of electrical behavior than does the TCNQ sequence. It may not be overly optimistic to look forward to the emergence of a class of higher-dimensional organic conductors which is much broader and electrically more versatile than the materials we have so far developed.

V. ACKNOWLEDGEMENTS

Much of this paper is based upon work supported by a grant from the Materials Science Office, Advanced Research Projects Agency, Department of Defense, and by the Alfred P. Sloan Foundation (A.N.B.). We are very grateful to F. J. Di Salvo, to T. J. Kistenmacher, and to V. K. S. Shante for permission to present Figures 7, 6, and 3, respectively, prior to publication elsewhere.

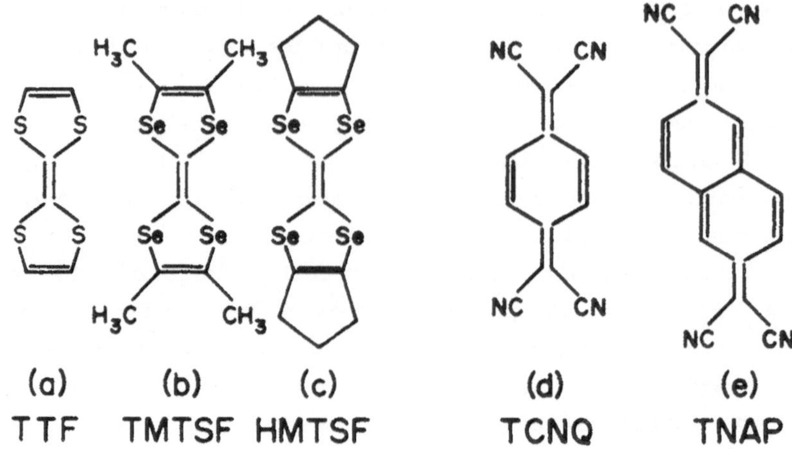

FIGURE 1: Molecular structures of TTF, TCNQ, and some of their analogs.

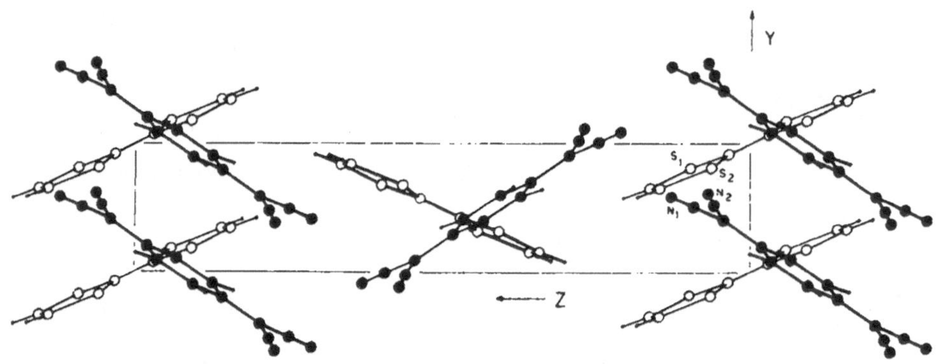

FIGURE 2: Crystal structure of TTF-TCNQ projected along the a-axis.

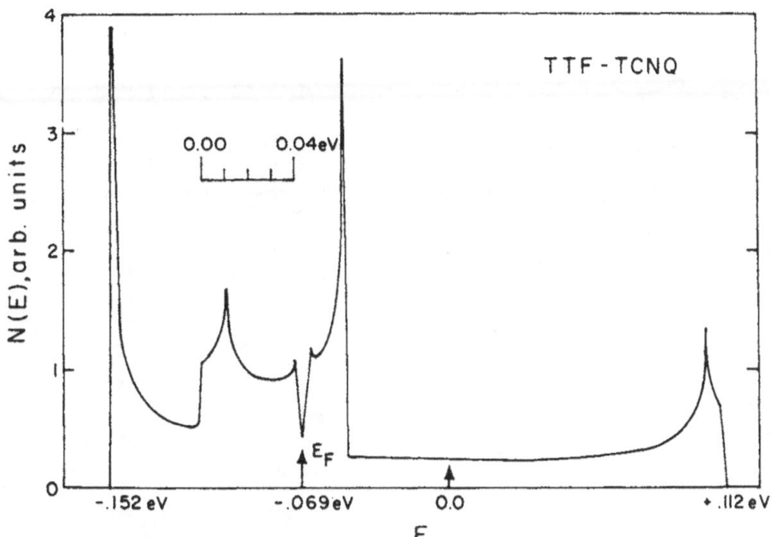

FIGURE 3: Density of states for undistorted TTF-TCNQ, calculated in one-electron tight-binding theory at T = 0.

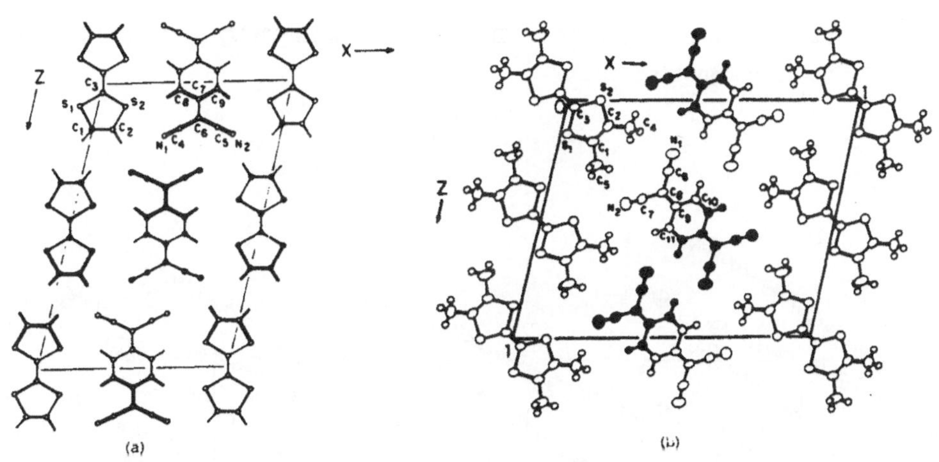

FIGURE 4: Comparison of the crystal structures of (a) TTF-TCNQ and (b) TMTTF-TCNQ, projected along their conducting axes.

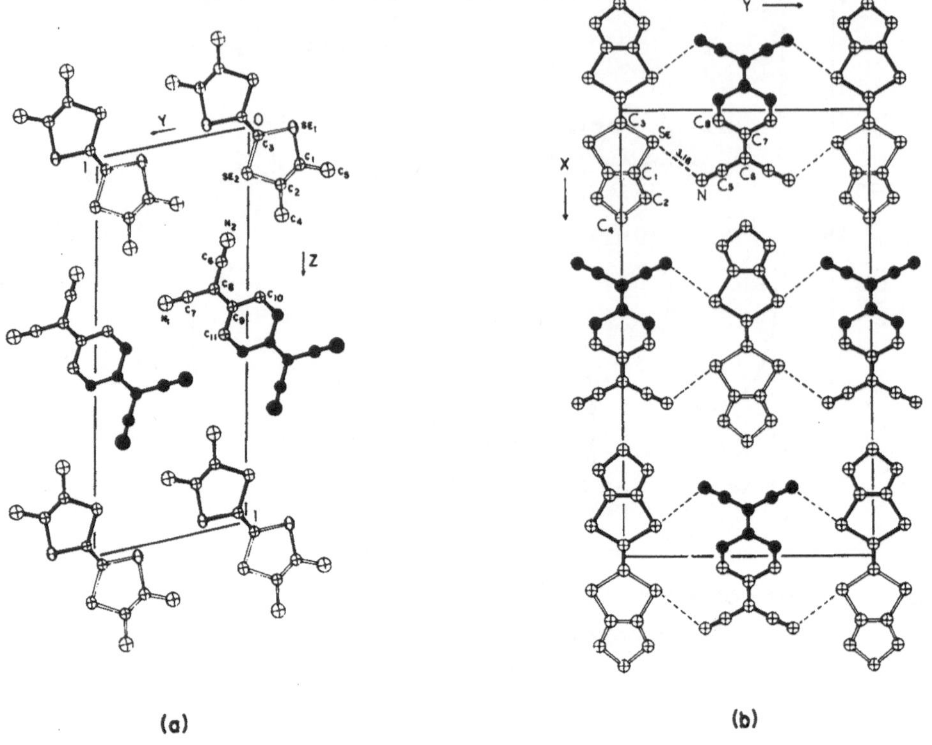

FIGURE 5: Structures of (a) TMTSF-TCNQ and (b) HMTSF-TCNQ, projected along their conducting axes.

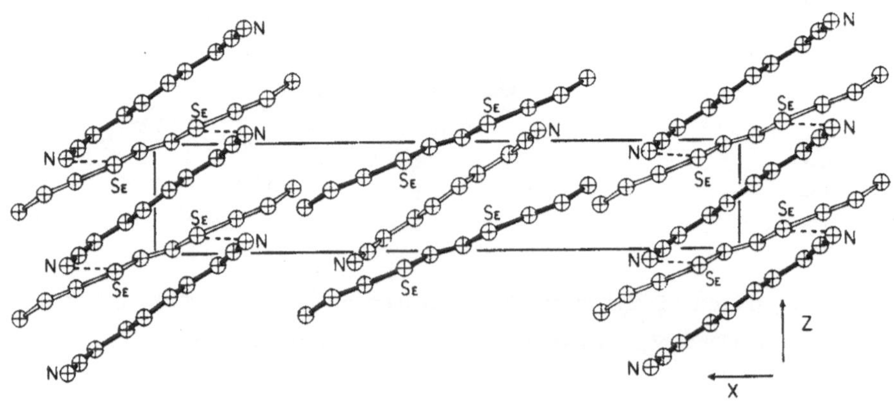

FIGURE 6: Crystal structure of HMTSF-TCNQ, projected along the b-axis.

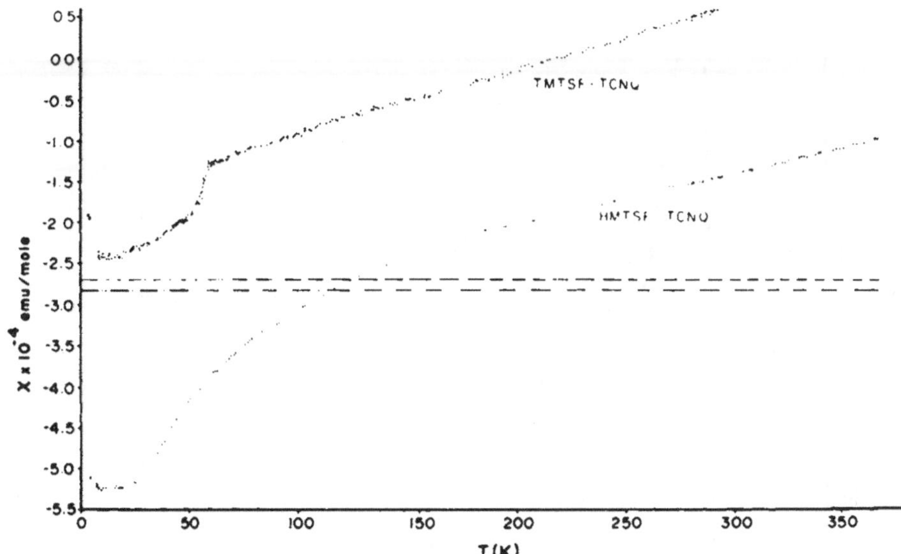

FIGURE 7: Temperature dependence of the static magnetic susceptibilities of TMTSF-TCNQ and HMTSF-TCNQ. The experimental core diamagnetisms of $TMTSF^{O}$ + $TCNQ^{O}$ (– – –) and $HMTSF^{O}$ + $TCNQ^{O}$ (– ·· – ·· –) are shown for comparison.

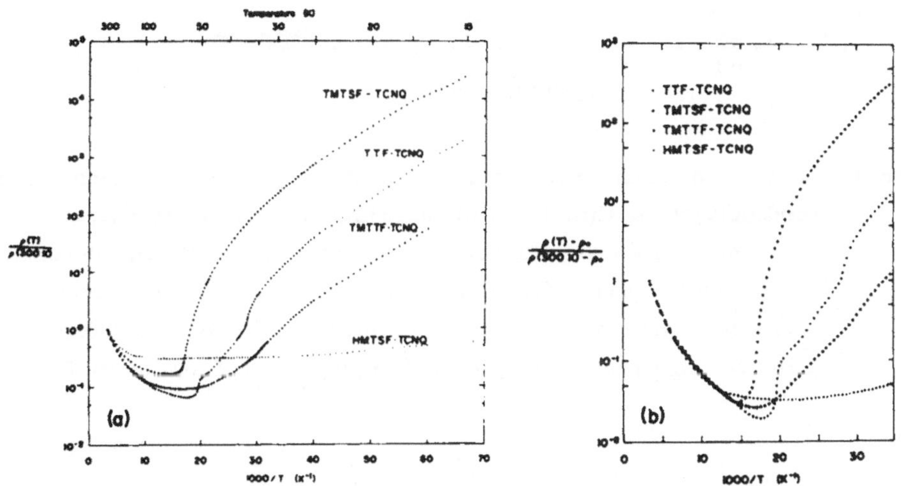

FIGURE 8: (a) Temperature-dependent resistivities of some organic conductors. (b) Data of Figure 8a, corrected for sample-dependent constant ρ_o and normalized to values at 300 K.

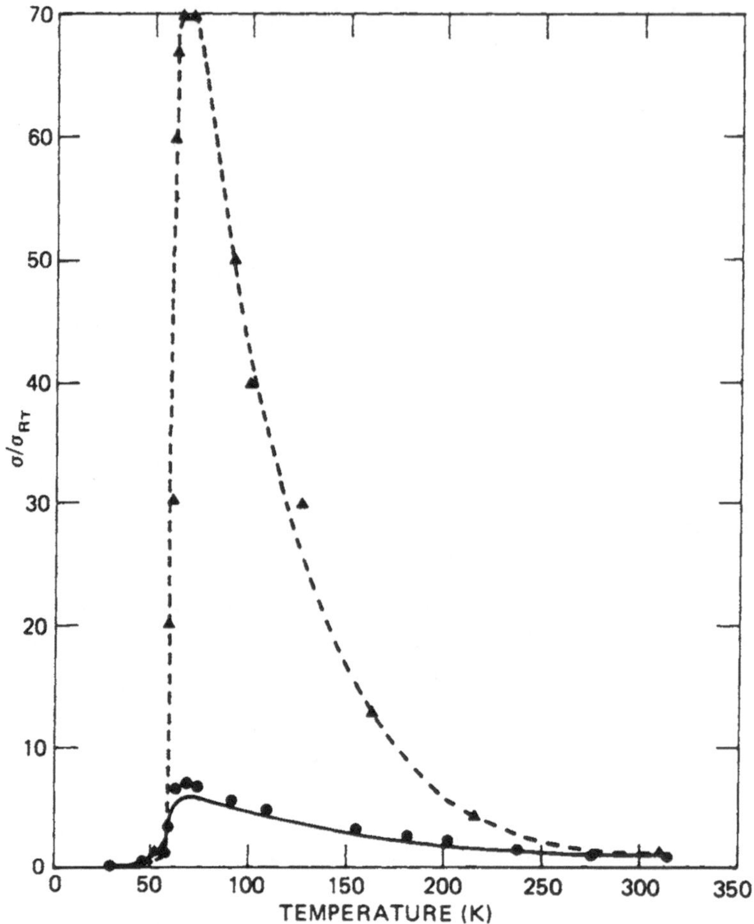

FIGURE 9: Real and apparent conductivities of TMTSF-TCNQ. Solid line: d.c. conductivity of thin (< 10 μ) and thick (> 100 μ) crystals. (•): Microwave conductivity of thin crystals at 10 GHz, determined using the dielectric formulae of Refs. 62 and 61. (▲): Apparent microwave conductivity of thick crystals at 10 GHz, as determined from the incorrect isotropic surface impedance formulae of Ref. 63.

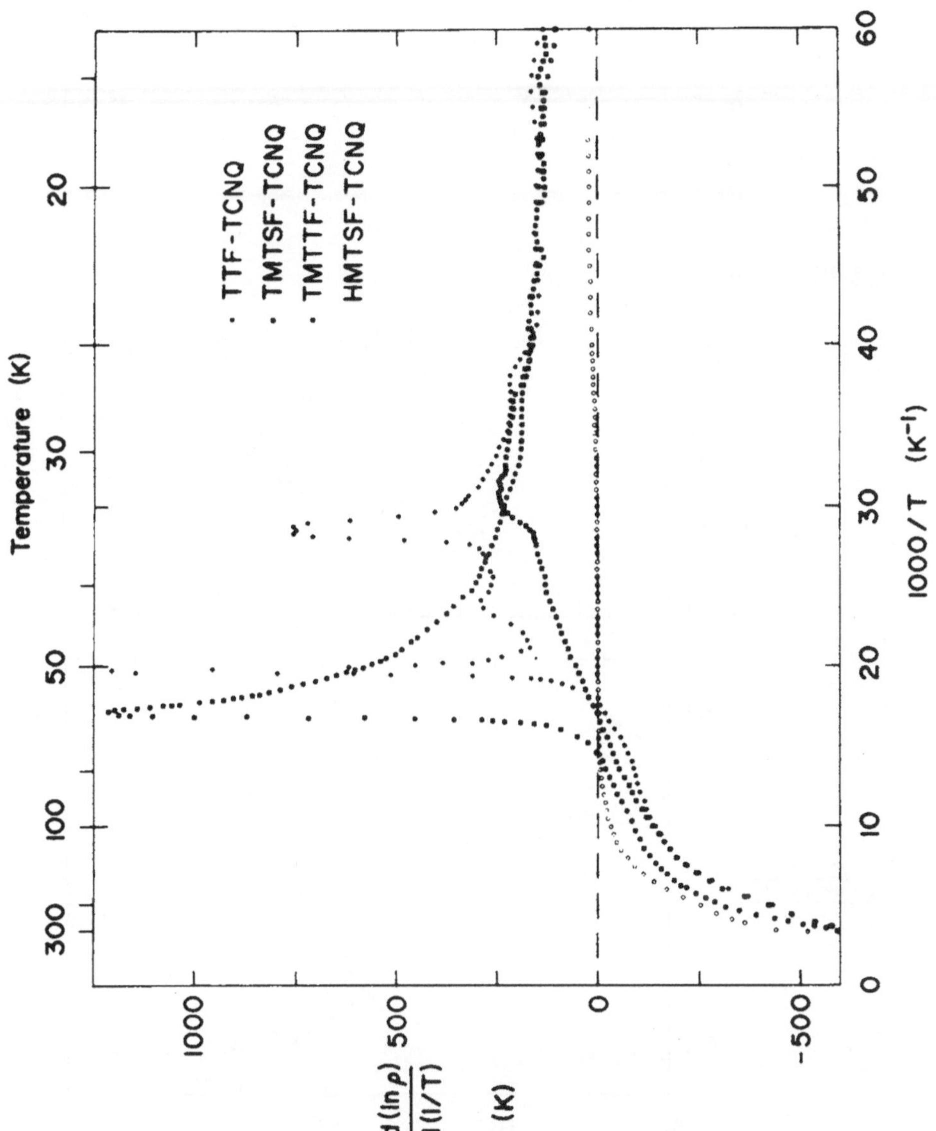

FIGURE 10: Temperature derivatives of the data of Figure 8.

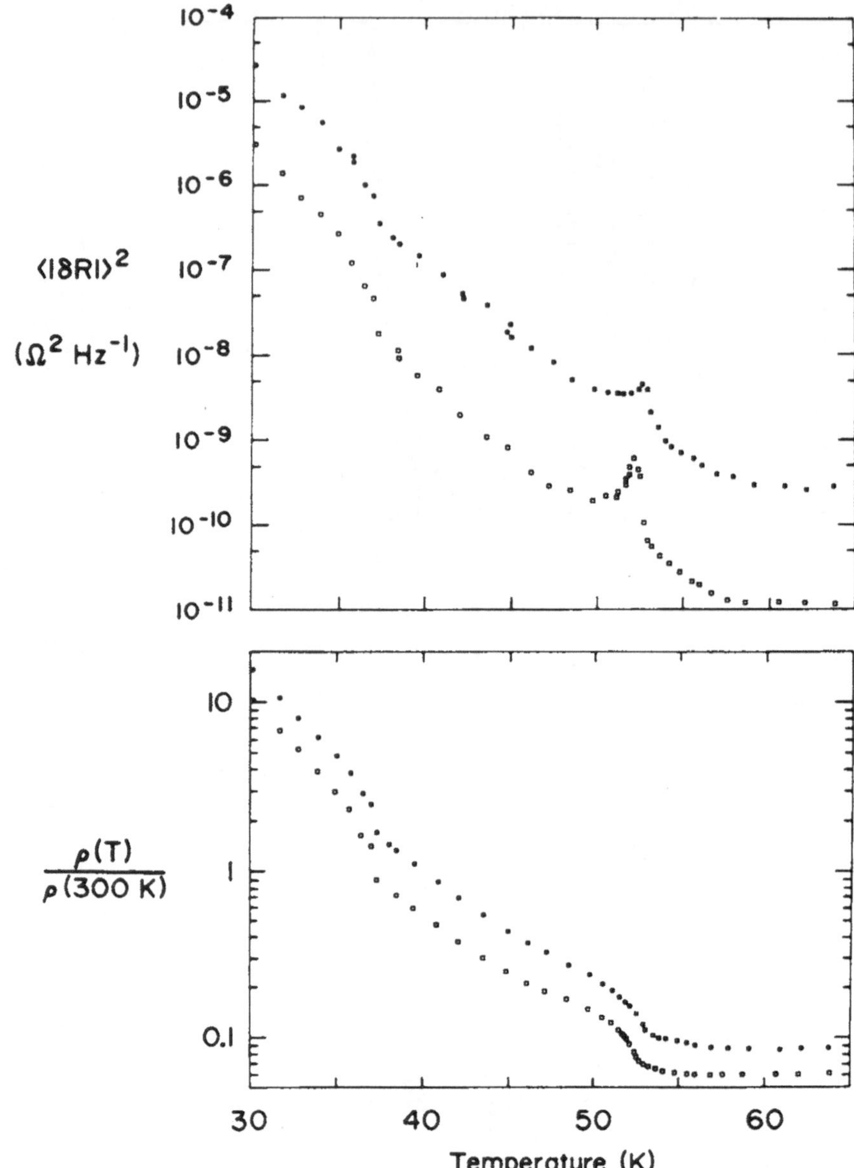

FIGURE 11: Mean-square broadband (1 - 1000 Hz) resistance fluctuations and d.c. resistivity of two samples of TTF-TCNQ.

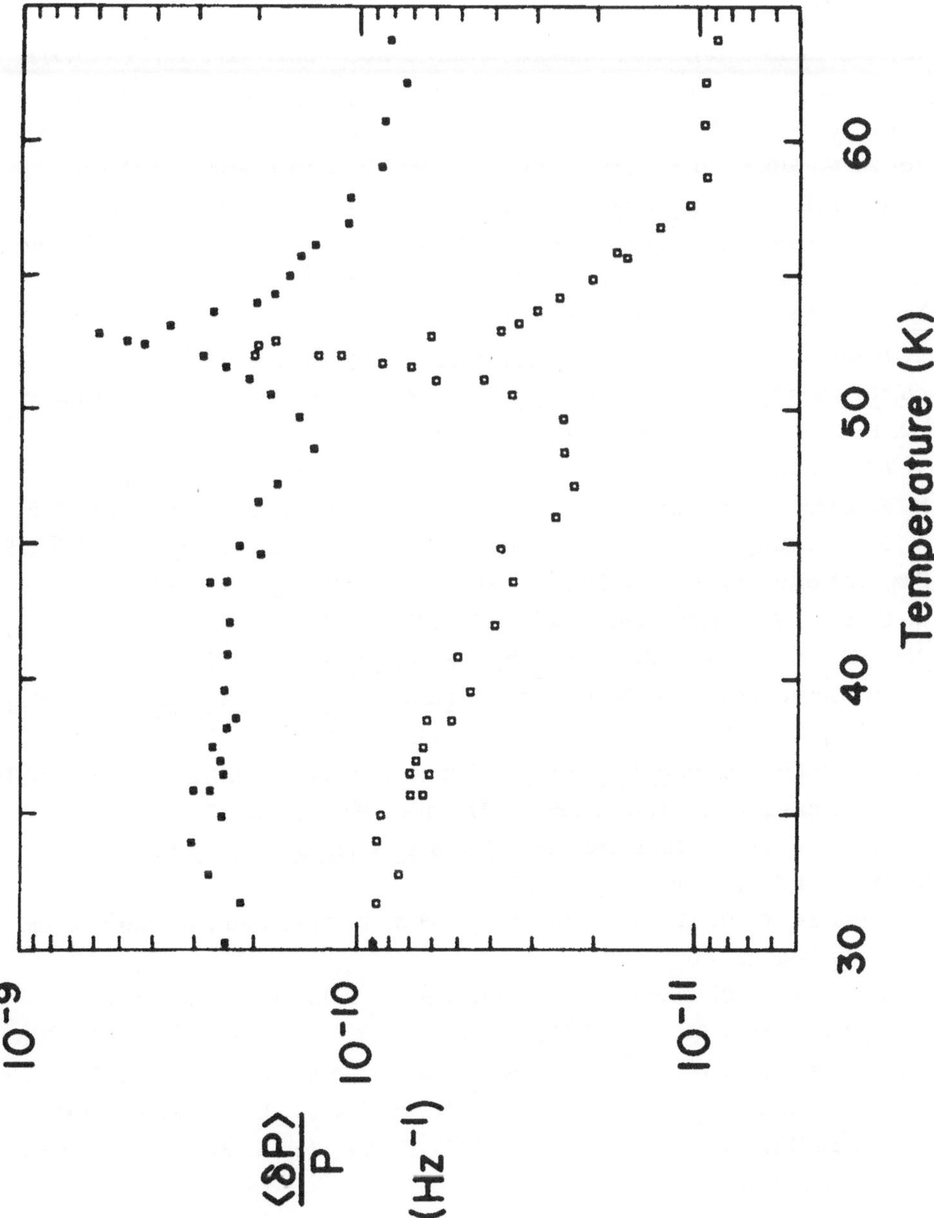

FIGURE 12: Normalized broadband excess noise power $\delta P/P = <|\delta R|^2>/R^2$ for the samples of Figure 9.

VI. REFERENCES

1. Reviews include (a) I. F. Schegolev, Phys. Stat. Sol. (A), 12, 9 (1972); (b) A. N. Bloch, in Energy and Charge Transfer in Organic Semiconductors, edited by K. Masuda and M. Silver (Plenum Press, New York, 1974), p. 159; (c) A. N. Bloch, D. O. Cowan, and T. O. Poehler, in Energy and Charge Transfer in Organic Semiconductors, edited by K. Masuda and M. Silver (Plenum Press, New York, 1974), p. 167; (d) A. F. Garito and A. J. Heeger, Accounts Chem. Res., 7, 232 (1974); (e) A. J. Berlinsky, Contemp. Phys., 17, 331 (1976); (f) D. O. Cowan, P. Shu, C. Hu, W. Krug, T. Carruthers, T. Poehler, and A. Bloch, in Proceedings of the NATO Conference on the Chemistry and Physics of One-Dimensional Metals, Bolzano, Italy, ed. H. J. Keller (Plenum Press, New York, 1977), in press.
2. V. K. S. Shante, A. N. Bloch, D. O. Cowan, W. M. Lee, S. Choi, and M. H. Cohen, Bull. Amer. Phys. Soc., 21, 287 (1976); and to be published.
3. A. J. Berlinsky, J. F. Carolan, and L. Weiler, Sol. St. Comm., 15, 795 (1974).
4. See, for example, M. H. Cohen, J. Non-Cryst. Sol., 2, 432 (1970);
5. Z. G. Soos, Ann. Rev. Phys. Chem., 25, 121 (1974).
6. P. F. Williams and A. N. Bloch, Phys. Rev., B10, 1097 (1974).
7. For a review, see F. H. Herbstein, in Perspectives in Structural Chemistry, edited by J. D. Dunitz and J. A. Ibers (Wiley, New York, 1972), Vol. IV, p. 166
8. R. E. Peierls, Quantum Theory of Solids (Oxford Press, London, 1955), p. 108.
9. See, for example, B. I. Halperin, Adv. Chem. Phys., 13, 123 (1966).
10. L. D. Landau and E. M. Lifschitz, Statistical Physics (Pergamon Press, New York, 1958), p. 482.
11. J. Ferraris, D. O. Cowan, V. Walatka, and J. H. Perlstein, J. Amer. Chem. Soc., 95, 498 (1973).
12. A. N. Bloch, D. O. Cowan, K. Bechgaard, R. E. Pyle, and R. H. Banks, Phys. Rev. Lett., 34, 1561 (1975).
13. (a) S. Etemad, T. Penney, E. M. Engler, B. A. Scott, and P. E. Seiden, Phys. Rev. Lett., 34, 741 (1975); (b) Y. Tomkiewicz, E. M. Engler, and T. D. Schultz, Phys. Rev. Lett., 35, 456 (1975); (c) S. Etemad, Phys. Rev., B13, 2254 (1976).
14. G. Beni, Sol. St. Comm., 15, 269 (1974).
15. D. J. Scalapino, Y. Imry, and P. Pincus, Phys. Rev., B11, 2042 (1975).
16. See, for example, P. A. Lee, T. M. Rice, and P. W. Anderson, Phys. Rev. Lett., 31, 462 (1973).
17. M. H. Cohen, J. A. Hertz, P. M. Horn, and V. K. S. Shante, Int. J. Quant. Chem. Symp., No. 8, 491 (1974).

18. U. Bernstein, P. M. Chaikin, and P. Pincus, Phys. Rev. Lett., 34, 271 (1975).
19. T. J. Kistenmacher, T. E. Phillips, and D. O. Cowan, Acta Cryst.. B30, 763 (1974).
20. F. Denoyer, R. Comes, A. F. Garito, and A. J. Heeger, Phys. Rev. Lett., 35, 445 (1975); R. Comes, S. M. Shapiro, G. Shirane, A. F. Garito, and A. J. Heeger, Phys. Rev. Lett., 35, 1518 (1975); S. Kagoshima, H. Anzai, K. Kajimura, and T. Ishigoro, J. Phys. Soc. Japan, 39, 1143 (1975); R. Comes, in Proceedings of the NATO Conference on the Chemistry and Physics of One-Dimensional Metals, Bolzano, Italy, ed. H. J. Keller (Plenum Press, New York, 1977), in press.
21. A similar interpretation of this material has been conceived independently by M. Weger, Sol. St. Comm., 19, 1149 (1976). See also Refs. 31 and 49.
22. J. P. Ferraris, T. O. Poehler, A. N. Bloch, and D. O. Cowan, Tet. Lett., 27, 2553 (1976).
23. T. E. Phillips, T. J. Kistenmacher, A. N. Bloch, J. P. Ferraris, and D. O. Cowan, Acta. Cryst. (in press).
24. M. A. Butler, J. P. Ferraris, A. N. Bloch, and D. O. Cowan, Chem. Phys. Lett., 24, 600 (1974).
25. Perhaps because these interactions are so weak, several distinct crystalline forms of TMTTF-TCNQ have been observed (Refs. 23 and 32). The one described here, though different from that originally reported in Ref. 22, is by far the most prevalent (Ref. 23).
26. K. Bechgaard, D. O. Cowan, and A. N. Bloch, Chem. Comm., 1974, 937.
27. K. Bechgaard, T. J. Kistenmacher, A. N. Bloch, and D. O. Cowan, Acta. Cryst. (in press).
28. K. Bechgaard, D. O. Cowan, and A. N. Bloch, Mol. Cryst. and Liq. Cryst., 32, 237 (1976).
29. T. E. Phillips, T. J. Kistenmacher, A. N. Bloch, and D. O. Cowan, J. C. S. Chem. Comm., 1976, 334 (1976).
30. T. O. Poehler, unpublished.
31. L. B. Coleman, M. J. Cohen, D. J. Sandman, F. G. Yamagishi, A. F. Garito, and A. J. Heeger, Sol. St. Comm., 12, 1125 (1973).
32. T. J. Kistenmacher, T. E. Phillips, D. O. Cowan, J. P. Ferraris, and A. N. Bloch, Acta. Cryst., B32, 539 (1976).
33. Y. Tomkiewicz, B. A. Scott, L. J. Tao, and R. S. Title, Phys. Rev. Lett., 32, 1363 (1974).
34. T. O. Poehler, J. Bohandy, A. N. Bloch, and D. O. Cowan, Bull. Amer. Phys., Soc., 21, 287 (1976).

35. Y. Tomkiewicz, T. D. Shultz, E. M. Engler, A. R. Taranko, and A. N. Bloch, Bull. Amer. Phys. Soc., 21, 287 (1976). See also Ref. 13b.
36. A. W. Overhauser, Phys. Rev., 89, 689 (1953).
37. R. J. Elliot, Phys. Rev., 96, 266 (1954).
38. Y. Yafet, Sol. St. Phys., 14, 1 (1963).
39. Y. Tomkiewicz, D. Garrod, A. R. Taranko, and A. N. Bloch, preprint.
40. J. C. Scott, A. F. Garito, and A. J. Heeger, Phys. Rev., B10, 3131 (1974).
41. S. Etemad, private communication.
42. F. J. DiSalvo, W. A. Reed, F. Hsu, A. N. Bloch, and D. O. Cowan, unpublished.
43. J. B. Torrance, in Proceedings of the NATO Conference on the Chemistry and Physics of One-Dimensional Metals, Bolzano, Italy, ed. H. J. Keller (Plenum Press, New York, 1977), in press.
44. P. A. Lee, T. M. Rice, and R. A. Klemm, preprint.
45. V. J. Emery, in Proceedings of the NATO Conference on the Chemistry and Physics of One-Dimensional Metals, Bolzano, Italy, ed. H. J. Keller (Plenum Press, New York, 1977), in press.
46. C. Weyl, E. M. Engler, S. Etemad, K. Bechgaard, and G. Jehanno, Sol. St. Comm., 19, 925 (1976).
47. The data for HMTSF-TCNQ (Ref. 42), first presented at the 1975 March meeting of the American Physical Society (Denver, Colo.), have been essentially reproduced by G. Soda, D. Jerome, M. Weger, K. Bechgaard, and E. Pederson, Sol. St. Comm., 19, (in press); see also Ref. 50a. Our results and interpretation were communicated directly to these authors in early 1976.
48. T. Carruthers, unpublished.
49. J. R. Cooper, M. Weger, D. Jerome, D. Lefur, K. Bechgaard, A. N. Bloch, and D. O. Cowan, Sol. St. Comm., 19, 749 (1976).
50. (a) D. Jerome and M. Weger, Proceedings of the NATO Conference on the Chemistr and Physics of One-Dimensional Metals, Bolzano, Italy, ed. H. J. Keller (Plenum Press, New York, 1977), in press; (b) G. Soda, D. Jerome, M. Weger. J. M. Fabre, and L. Giral, Sol. St. Comm., 18, 1417 (1976).
51. A. N. Bloch, R. B. Weisman, and C. M. Varma, Phys. Rev. Lett., 28, 753 (1972).
52. G. A. Thomas, et al., Phys. Rev., B13, 5105 (1976).
53. Substantial residual conductivities as $T \to 0$ have also been observed in HMTSF-TNAP (Ref. 12) and in the halides of tetrathiatetracene, TTT (E. Perez-Albuerne, work presented at the 1976 March meeting of the American Physical Society [Atlanta, Ga.]; I. F. Schegolev, this volume.)
54. R. L. Greene, unpublished.
55. R. P. Groff, A. Suna, and R. E. Merrifield, Phys. Rev. Lett., 33, 418 (1974).

56. D. E. Schafer, F. Wudl, G. A. Thomas, J. P. Ferraris, and D. O. Cowan, Sol. St. Comm., 14, 347 (1974).
57. M. B. Salamon, J. W. Bray, G. DePasquali, R. A. Craven, R. Herman, G. Stucky, and A. Schultz, Phys. Rev., B11, 619 (1975).
58. J. Bardeen, Sol. St. Comm., 13, 357 (1973).
59. M. J. Cohen, L. B. Coleman, A. F. Garito, and A. J. Heeger, Phys. Rev., B10, 1298 (1974).
60. R. V. Gemmer, D. O. Cowan, A. N. Bloch, R. E. Pyle, and R. Banks, Mol. Cryst. and Liq. Cryst., 32, 227 (1976).
61. A. N. Bloch, J. P. Ferraris, D. O. Cowan, and T. O. Poehler, Sol. St. Comm., 13, 753 (1973).
62. L. J. Buravov and I. F. Schegolev, Prib. Tek. Eksp., 2, 171 (1971).
63. M. Cohen, S. K. Khanna, W. J. Gunning, A. F. Garito, and A. J. Heeger, Sol. St. Comm., 17, 367 (1975).
64. A. M. Clogston, Bell System Tech. J., 30, 491 (1951).
65. A. Madhukar and M. H. Cohen, preprint; A. A. Gogolin, V. I. Mel'nikov, and E. I. Rashba, this volume.
66. J. B. Torrance, E. E. Simonyi, and A. N. Bloch, Bull. Amer. Phys. Soc., 20, 497 (1975), and to be published.
67. See, for example, P. E. Seiden and D. Cabib, Phys. Rev., B13, 1846 (1976).
68. R. A. Craven, M. B. Salamon, G. DePasquali, R. M. Herman, G. Stucky, and A. Schultz, Phys. Rev. Lett., 32, 769 (1974).
69. R. V. Gemmer, D. O. Cowan, T. O. Poehler, A. N. Bloch, and R. H. Banks, J. Org. Chem., 40, 3544 (1975).
70. H. Fukuyama, T. M. Rice, and C. M. Varma, Phys. Rev. Lett., 33, 305 (1974).
71. A. Luther and V. J. Emery, Phys. Rev. Lett., 33, 589 (1974).
72. T. Carruthers, A. N. Bloch, and D. O. Cowan, Bull. Amer. Phys. Soc., 21, 313 (1976), and to be published.
73. P. M. Horn and D. Rimai, Phys. Rev. Lett., 36, 809 (1976); and P. M. Horn and D. Guidotti, preprint.
74. P. Bak and V. J. Emery, Phys. Rev. Lett., 36, 978 (1976).
75. The 47 K anomaly in the conductivity, though absent in the data of Ref. 13c, has now been observed independently in several laboratories. To our knowledge it was first reported by us (Ref. 72), and separately by P. Horn and D. Guidotti, at the 1976 March meeting of the American Physical Society (Atlanta, Ga.).
76. See, for example, Aldert van der Ziel, Noise: Sources, Characterization, Measurement (Prentice-Hall, Englewood Cliffs, N. J., 1970).

77. V. K. S. Shante, A. N. Bloch, T. Carruthers, and D. O. Cowan, Bull. Amer. Phys. Soc., 21, 313 (1976), and to be published.

78. K. Bechgaard and B. S. Jensen, this volume.

79. P. Horn, private communication.

80. J. E. Gulley and J. F. Weiler, Phys. Rev. Lett., 34, 1061 (1975).

81. Y. Tomkiewicz, A. R. Taranko, and J. B. Torrance, Phys. Rev. Lett., 36, 751 (1976).

82. T. O. Poehler, R. M. Somers, A. N. Bloch, and D. O. Cowan, preprint.

83. H. Kahlert, Sol. St. Comm., 17, 1161 (1975); K. Seeger, Sol. St. Comm., 19, 245 (1976).

84. M. J. Cohen, P. R. Newman, and A. J. Heeger, preprint.

85. See, for example, K. Seeger, Semiconductor Physics (Springer-Verlag, New York, 1973).

86. H. A. Mook and C. R. Watson, preprint.

87. M. J. Rice, A. R. Bishop, J. A. Krumhansl, and S. E. Trullinger, Phys. Rev. Lett., 36, 432 (1976).

88. D. B. Tanner, C. S. Jacobsen, A. F. Garito, and A. J. Heeger, Phys. Rev. Lett., 32, 1301 (1974); 33, 1559 (1974); Phys. Rev., B13, 3381 (1976).

89. U. Fano, Phys. Rev., 124, 1886 (1961).

90. This effect is distinct from the stabilization of the Peierls gap by the softening of intramolecular optical modes suggested by M. J. Rice, C. B. Duke, and N. O. Lipari [Sol. St. Comm., 17, 1089 (1975)]. In fact, their calculation is flawed in its assumption that the phases of the intramolecular and intermolecular soft modes are arbitrary. The correct theory [A. Madhukar, Chem. Phys. Lett., 27, 606 (1974)] recognizes that the former modes build up charge around the lattice sites and the latter in the bonding regions; hence the two instabilities compete.

91. H. Hinkelmann and H. G. Reik, Sol. St. Comm., 16, 567 (1975).

92. J. J. Ritsko, D. J. Sandman, A. J. Epstein, P. C. Gibbons, S. E. Schnatterly, and J. Fields, Phys. Rev. Lett., 34, 1330 (1975).

93. P. F. Williams and A. N. Bloch, Phys. Rev. Lett., 36, 64 (1976).

94. G. Giulliani, E. Tosatti, and M. P. Tosi, Lett. Nuovo Cimento, 16, 385 (1976).

Conference on Organic Conductors and Semiconductors, Siófok, Hungary 1976

OPTICAL PROPERTIES OF HEXAMETHYLENE-TETRASELENAFULVALINIUM TETRACYANOQUINODI-METHANIDE (HMTSF-TCNQ)

C.S. JACOBSEN

Physics Laboratory III, Technical University of Denmark,
DK-2800 Lyngby, Denmark

K. BECHGAARD

H.C. Ørsted Institute, DK-2100, Denmark

J.R ANDERSEN

Chemistry Department, Risø, DK-4000, Denmark

Polarized single crystal reflectance studies of HMTSF-TCNQ are presented. One reflectance component perpendicular to the highly conducting direction is flat and rather low, the other shows a transition at $\omega = 16000\ cm^{-1}$, much stronger than usually seen in conducting TCNQ salts. The highly conducting direction displays a plasma edge. An analysis yields a plasmon frequency $\omega_p = 7400\ cm^{-1}$, an inverse life time $\Gamma = 1300\ cm^{-1}$, and an optical mass $m^* = 1.9\ m_o$. The low-temperature reflectance gives $\omega_p = 8200\ cm^{-1}$ and a 2.4 times longer lifetime, consistent with the temperature dependence of the dc conductivity.

1. Introduction

In the study of quasi-one-dimensional organic metals, optical measurements have proven to be a valuable tool in probing the electronic structure. From analysis of the plasma edge, inevitably found for the electric field parallel to the conducting direction, information on single particle lifetimes and band masses can be found /1, 2/. The infrared reflectance spectrum yields the optical conductivity which gives clues to the origin of the dc conductivity. For example in TTF-TCNQ an infrared energy gap has been found /3/ suggesting that the dc conductivity is not limited by single particle scattering.

Recently a new organic conductor HMTSF-TCNQ was synthesized /4/. The absence of the usual low temperature transition into an insulating state is one of the remarkable properties of this compound. In order to help clarify why HMTSF-TCNQ behaves drastically different from, for example TTF-TCNQ, we have studied the optical properties in some detail.

2. Experimental

HMTSF-TCNQ crystals of excellent optical quality were obtained by slow cooling of the solvent. Typical dimensions were 3x0.3x0.2 mm^3 frequently with the largest faces in the crystallographic *a*-*c* plane /5/. These crystal faces can be easily recognized, since they have a distinct red shine. The *b*-*c* crystal faces are those of highest optical quality and they show a low, white reflectance in the visible. In order to explore the importance of surface conditions, the reflectance for $E \parallel c$ was measured on both types of planes but no significant difference was found. Absolute values of reflectance were obtained by partly covering the crystals with an aluminium film. Polarization of the light was obtained using a Perkin-Elmer gold wire grid polarized in the range from 500-5000 cm^{-1} and a Glan-Thompson prism polarizer from 5000-25000 cm^{-1}. A description of the reflectance spectrometer will be given elsewhere.

3. Results

In Fig. 1 the three normal incidence reflectance components are shown in the range from the infrared to 25000 cm^{-1} /3.1 eV/. As expected $R(E \| \underline{b})$ is almost frequency independent with a value of 11-12 % corresponding to a background dielectric constant $\varepsilon_b \simeq 4$. This behaviour is similar to the one found in the measured transverse reflectance in TTF-TCNQ /1, 2/. In both cases the electric field vector is perpendicular to the long axis of the molecules. The second transverse component $R(E \| \underline{a})$ has a strong, broad transition centred at 15000-16000 cm^{-1}. The high reflectance on the low-energy side gives rise to the characteristic red shine.

The metallic component $R(E \| \underline{c})$ displays the expected plasma edge with a minimum at 9200 cm^{-1} and high, metallic reflectance in the infrared. Apart from a shift of plasma frequency, no significant difference is found between TTF-TCNQ and HMTSF-TCNQ in the reflectance data. For example, a small dip in the reflectance around 1600 cm^{-1} is seen in both materials.

4. Optical Conductivity

Searching for possible, less pronounced differences, we have performed a Kramers-Kronig analysis of the $R(E \| \underline{a})$ and $R(E \| \underline{c})$ specrta. Due to uncertainties in the high frequency extrapolations, the resulting oscillator strengths in the range above 10000 cm^{-1} should be regarded with some caution. In both directions we assumed a ω^{-2} dependence for $\omega >$ 25000 cm^{-1}. At low frequencies we assumed constant reflectance for $E \| \underline{a}$, $\omega < 1200$ cm^{-1}, and a Hagen-Rubens extrapolation $R = 1-A\omega^{1/2}$ for $E \| \underline{c}$, $\omega < 600$ cm^{-1}.

The resulting optical conductivities as functions of frequency are shown in Fig. 2 /solid and dashed lines/. Let us discuss the high frequency region first, referring to the recent paper by Torrance, Scott and Kaufman /TSK/ /6/. The onsetting transition at the highest measuring frequencies and seen in both components is obviously the D transition of TSK, always present in TCNQ salts. At 15000-16000 cm^{-1} another transition is seen, again visible in both components, but far strongest for $E \parallel \underline{a}$, and corresponding to the C transition of TSK.

The C and D transitions are generally agreed to be intramolecular excitations in the TCNQ /6, 7/. This is consistent with their presence in all TCNQ salts, and with the fact that the transitions seem to be polarized along the long axis of the molecules /7, 8/. Since the molecules in HMTSF-TCNQ are tilted with respect to the crystallographic a-b planes, the transitions are present in both a and c directions. The most remarkable feature is the strength of C for E a. Usually C is weak in conducting salts and strong in insulating salts /for powder absorption spectra, see TSK/. Obviously model calculations of the optical properties from the crystal structures and molecular properties are needed to understand these differences. The possible influence of intermolecular interactions /charge transfer, etc./ should also be considered.

Finally the weak B transition of TSK at 12000 cm^{-1} is seen in both components supporting the view /9/ that, contrary to the suggestion by TSK, this transition is also essentially intramolecular.

Let us next turn to the low frequency behaviour. $\sigma(E \parallel \underline{a})$ shows a weak shoulder, but drops effectively to zero apart from a molecular line at 2200 cm^{-1} /the C-N stretch in the TCNQ/. $\sigma(E \parallel \underline{c})$ is close to simple Drude behaviour in the near IR, but at lower frequencies deviations prevail, as for example the dip at 1500 cm^{-1} and the apparent decrease below 900 cm^{-1}. With reasonable extrapolation procedures for $0 < \omega <$

600 cm^{-1} no real energy gap is found, but certainly the conductivity goes below the simple Drude dependence, which should extrapolate to the dc value of 1400-2200 /Ω^{-1} cm^{-1}/4. The gap in TTF-TCNQ has been linked /3/ to the formation of $2k_F$-charge density waves, suggesting the same origin of the low frequency behaviour in HMTSF-TCNQ. Charge density waves have been seen by diffuse X-ray scattering at room temperature /1o/. The less pronounced gap in HMTSF-TCNQ may indicate a large single particle contribution to the dc conductivity.

5. Analysis

The last part of this paper concerns sum rule calculations, Drude fits and temperature dependence in the near infrared. In Fig. 3 is given a normalized sum rule plot for E || c:

$$\frac{m_o}{m^*} \times N_{eff}(\omega) = \left(\frac{1}{8}\int_0^{\omega} \sigma_c(\omega')\, d\omega'\right) \Big/ \left(\frac{4\pi N_c e^2}{m_o}\right) \qquad /1/$$

where N_c = 3.72x1o^{21} molecules/cm^3 /5/, m^* is the optical band mass and m_o the free electron mass. /1/ is expected to saturate above the intraband region corresponding to N_{eff} = = 0.74, the effective number of electrons per TCNQ /1o/. From Fig. 3 a saturation level /m_o/m^*/N_{eff} = 0.385 is found, yielding m^*=1.9 m_o. The band width /within a single band tight binding model/ is then:

$$W = \frac{\pi \hbar^2}{a^2 m^*} = 0.9 \text{ eV}$$

where a = 3.89 A is the molecular spacing /5/. For comparison W in TTF-TCNQ is 0.5 eV, the difference presumably arising from a change in cation overlap.

The region near the plasma edge in 1d conductors /any gap $\ll \omega <$ interband region/ is known to approximately obey the Drude formula for the dielectric function:

$$\tilde{\varepsilon}(\omega) = \varepsilon_{core}\left(1 - \frac{\omega_p^2}{\omega(\omega + i\Gamma)}\right) \qquad /2/$$

ω_p is the plasmon frequency, which can be taken from a plot of the loss function $-\mathrm{Im}(1/\tilde{\varepsilon})$. The experimental loss function is shown in Fig. 2, giving ω_p = 7400 cm^{-1} at room temperature. Using Γ = 1300 cm^{-1} and ε_{core} = 2.6, a reasonable fit to the room temperature reflectance data is obtained /Fig. 4/.

Also shown in Fig.4 are preliminary low temperature data /T = 20 K/ in the plasmon region. Several interesting features are seen such as clear deviations from Drude behaviour at 6000-8000 cm^{-1}. A tentative Drude fit yield the parameters ω_p = 8200 cm^{-1} and Γ = 540 cm^{-1}, using an unchanged value of ε_{core}. The distinct blue shift of the plasma edge /10 %/ may explain the smearing of the plasma minimum, since the region of interband transitions is approached. The narrow resonance near 7000 cm^{-1} has been seen in several crystals and in various geometries and may be due to disorder induced, optical excitation of plasmons /11/. The ratio of the high and low temperature relaxation rates $\Gamma_{300}/\Gamma_{20K}$ equals 2.4, a number that is not too different from the dc resistivity ratio /4/, ρ_{300}/ρ_{MIN} = 3.4. This feature seems to confirm the single particle nature of the conduction mechanism. More detailed low temperature studies are presently being carried out in our laboratory.

Acknowledgments We would like to thank Dr. H. Soling for identifying the crystallographic directions and H. Axel for valuable technical assistance.

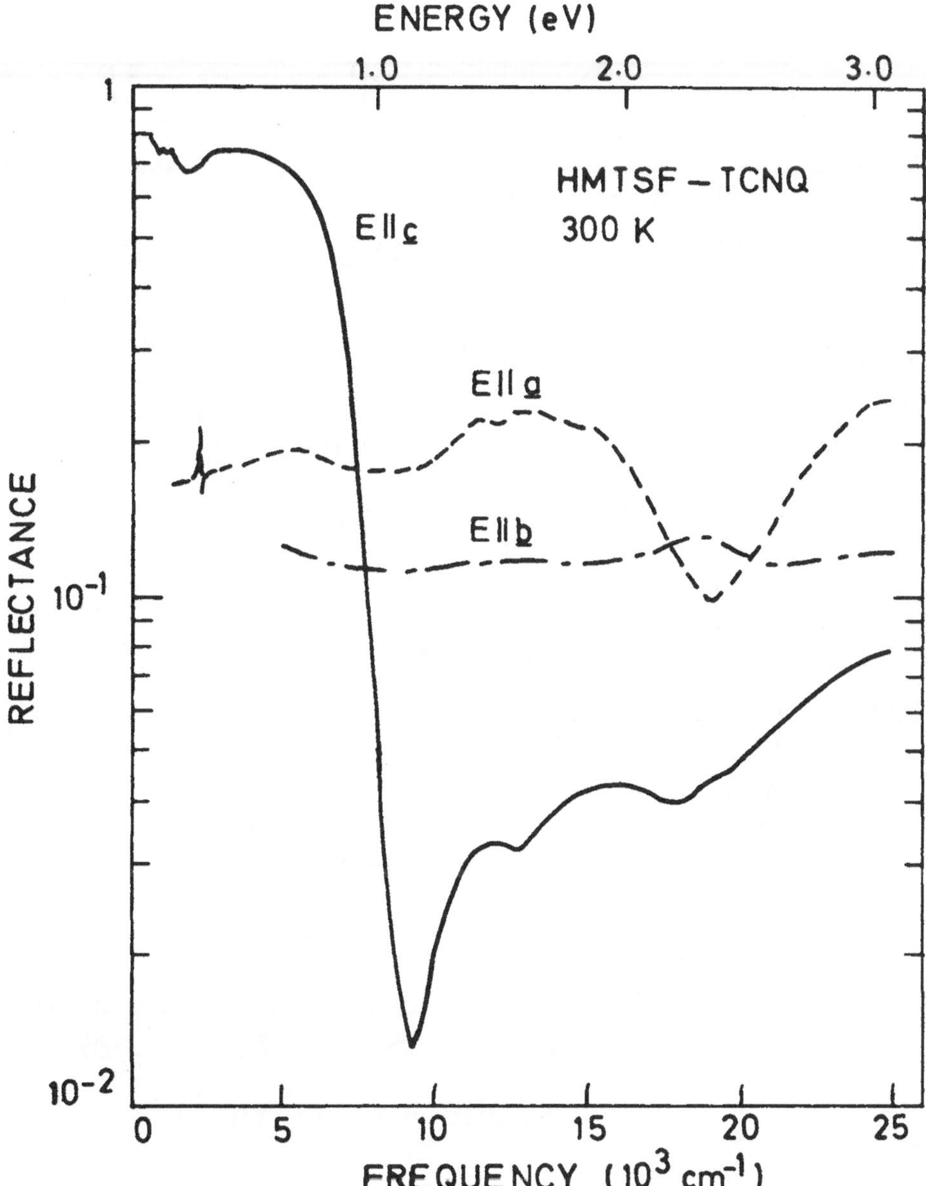

Fig. 1 Polarized reflectance of HMTSF-TCNQ vs. frequency from 600-25000 cm^{-1}. Notice the logarithmic reflectance scale

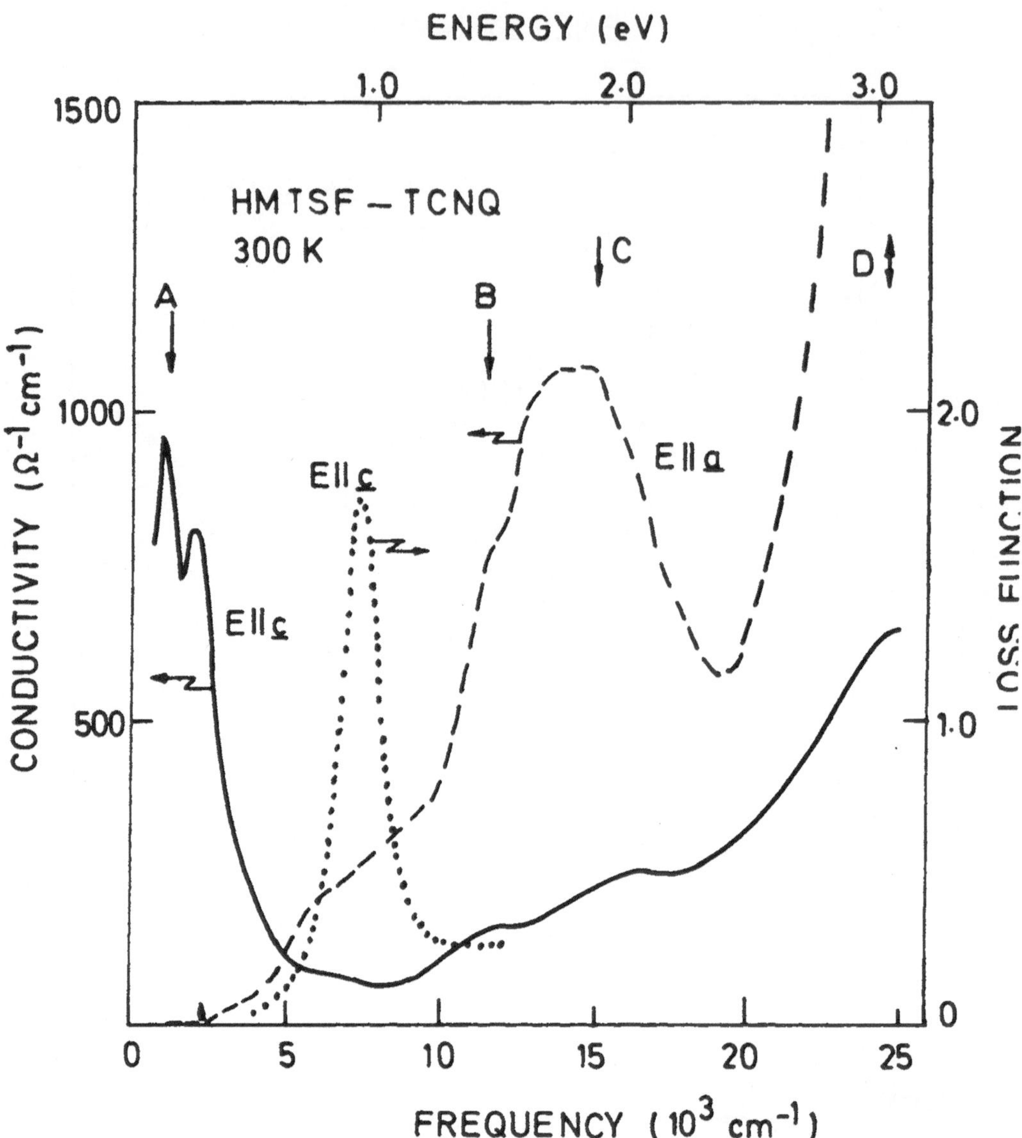

Fig. 2 Optical conductivity of HMTSF-TCNQ for E ‖ c /solid line/ and E ‖ a /dashed line/ vs. frequency from 600-25000 cm^{-1}. Also shown is the loss function $-Im\left(1/\tilde{\varepsilon}\right)$ for E ‖ c /dotted line/. Capital letters A, B, C and D refer to Ref. 6

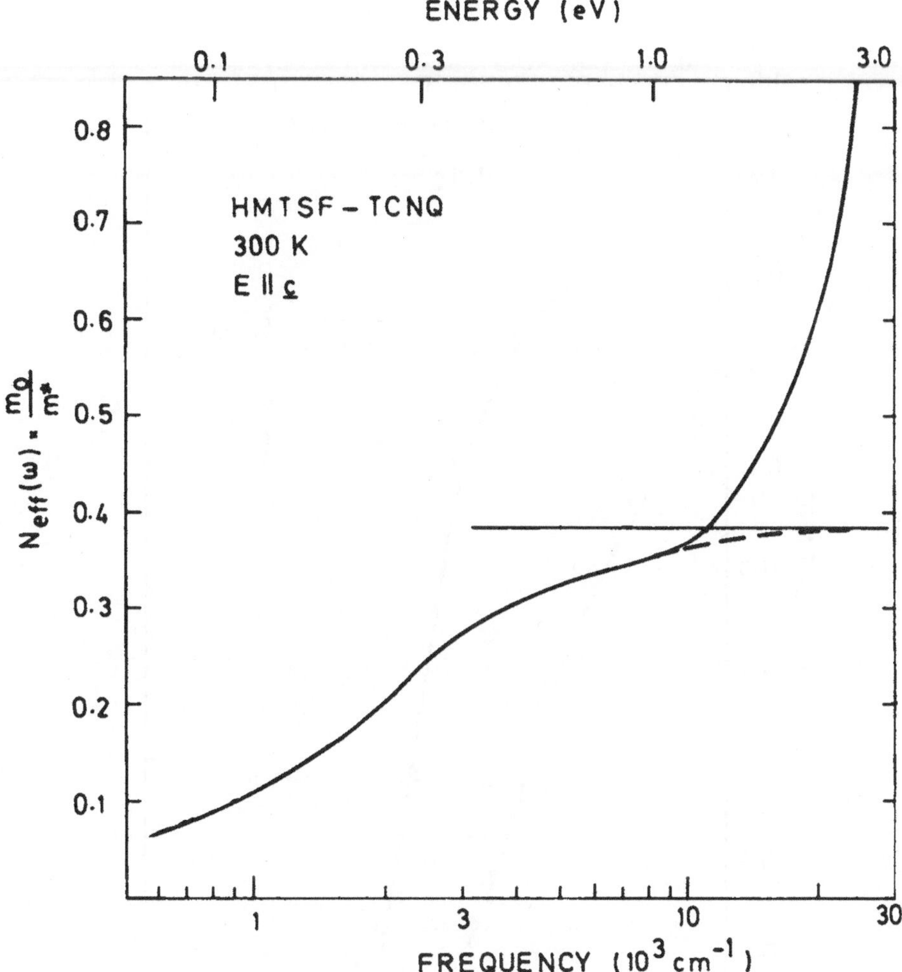

Fig. 3 Number of effective electrons obtained from the oscillator strength sum rule for HMTSF-TCNQ vs. frequency from 600-25000 cm^{-1}. E∥ c. Notice the logarithmic frequency scale

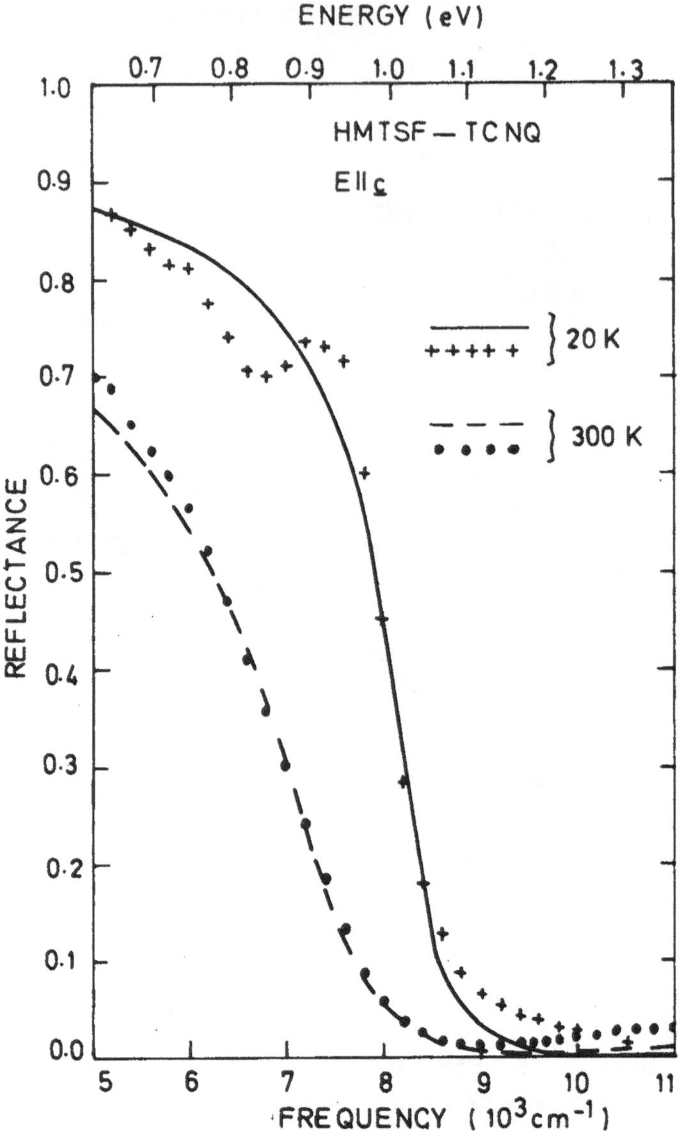

Fig. 4 Reflectance of HMTSF-TCNQ for E ∥ c vs. frequency from 5000-11000 cm^{-1}. Crosses: T = 20 K. Dots: T = 300K. Solid line, Drude fit /T = 20K/: ω_p = = 8200 cm^{-1}, Γ = 540 cm^{-1}; dashed line, Drude fit /T = 300K/: ω_p = 7400 cm^{-1}, Γ = 1300 cm^{-1}

REFERENCES

/1/ A. A. Bright, A. F. Garito,and A. J. Heeger, Solid State Commun. 13, 943 /1973/; Phys. Rev. B1o, 1328 /1974/.

/2/ P. M. Grant, R. L. Greene, G. C. Wrighton, and G. Castro, Phys. Rev. Letters 31, 1311 /1973/.

/3/ D. B. Tanner, C. S. Jacobsen, A. F. Garito, and A. J. Heeger, Phys. Rev. B13, 3381 /1976/.

/4/ A. N. Bloch, D. O. Cowan, K. Bechgaard, R. E. Pyle, R. H. Banks, and T.O. Poehler, Phys. Rev. Letters 34, 1561 /1975/.

/5/ T. E. Phillips, T. J. Kistenmacher, A. N. Bloch, and D. O. Cowan, J. C. S. Chem. Commun. 1976, 334.

/6/ J. B. Torrance, B. A. Scott, and F. B. Kaufman, Solid State Commun. 17, 1369 /1975/.

/7/ R. M. Vlasova, A. I. Gutman, N. F. Kartenko, L. D. Rozenshtein, L. S. Agroskin, G. V. Papayan, L. P. Rautian, and A. I. Sherle, Fiz. Tverd. Tela 17, 3529 /1975/ /Sov. Phys. Solid State 17, 2302 /1976/.

/8/ P. M. Grant, R. L. Greene, and G. Castro, Bull. Am. Phys. Soc. 2o, 496 /1975/.

/9/ D. B. Tanner, C. S. Jacobsen, A. A. Bright, and A. J. Heeger, /to be published/.

/1o/ C. Weyl, E. M. Engler, K. Bechgaard, G. Jehanno, and S. Etemad, Solid State Commun. 19, 925 /1976/.

/11/ P. F. Williams, M. A. Butler, D. L. Rousseau, and A. N. Bloch, Phys. Rev. B1o, 1109 /1974/.

Conference on Organic Conductors and Semiconductors, Siófok, Hungary 1976

X-RAY SCATTERING OF TSeF-TCNQ AND HMTSeF-TCNQ

C. WEYL

Laboratoire de Physique des Solides, Université
Paris-Sud, 91405 Orsay, France,

E.M. ENGLER

IBM Thomas Watson Research Center, Yorktown Heights,
New-York 10598, USA

K. BECHGAARD

H.C. Oersted Institute, Universitetsparken 5, DK-2100
Copenhagen, Denmark,

G. JEHANNO

Service de Physique du Solide et de Résonance Magnétique,
Centre d'Etudes Nucléaires de Saclay, 91190 Gif-sur-Yvette France,

S. ETEMAD

Department of Physics, Aria-Mehr University, P.O. Box 3406, Tehran, Iran

Some results of a X-ray diffuse scattering investigation on the charge transfer compounds TSeF-TCNQ and HMTSeF-TCNQ are presented. In the metallic state the results show the formation of a 1-D distortion or a Kohn anomaly in the phonon spectrum. For TSeF-TCNQ this is observed below T$\sim$238 K and corresponds to a periodicity of (3.15$\pm$ 0.05) $|\vec{b}|$ along the chain axis. For HMTSeF-TCNQ the distortion was detected even at room temperature and has a period of (2.7$\pm$ 0.1) $|\vec{c}|$
In both materials the distortion has both transverse components. In the insulating state of TSeF-TCNQ below 29 K the 1-D distortion on each stack are correlated to form a 3-D super-lattice.

Conference on Organic Conductors and Semiconductors, Siófok, Hungary 1976

OBSERVATION OF A LARGE HALL EFFECT IN HMTSF-TCNQ BELOW ROOM TEMPERATURE

J.R. COOPER

Institute of Physics of the University, POB 304, Zagreb, Yugoslavia

M. WEGER[+], G. DELPLANQUE and D. JÉROME

Laboratoire de Physique des Solides[x], Université de Paris-Sud, Orsay, France

K. BECHGAARD

H.C. Oersted Institute, Universitetsparken 5, DK-2100, Copenhagen, Denmark

ABSTRACT

Hall effect measurements are reported for three single crystals of the charge transfer salt HMTSF-TCNQ in the temperature range 1.4-200 K at ambient pressure and under hydrostatic pressures of approximately 6 Kbars. There is evidence that the high conductivity of this material at low temperatures arises from a small number of electrons with a high mobility and a low degeneracy temperature; as suggested by other experiments and a recent band-structure calculation.

The charge transfer salt HMTSF-TCNQ is exceptional in that it remains a good electrical conductor down to very low temperatures, with a conductivity parallel to the stacking axis, σ_{hc}, of at least 500 $(\Omega\, cm)^{-1}$ down to 0.05 K, and there is good agreement between the DC and microwave results.[1] A recent investigation[2] showed that this "metallic" behaviour was even more pronounced under hydrostatic pressures of 4 kbars or more, where $d\sigma_{hc}/dT$ was often

+ On visit from the Hebrew University, Jerusalem, and the Nuclear Research Centre, Negev, Israel.

x Laboratoire associé au C.N.R.S. - Work at Orsay supported in part by D.G.R.S.T. Contract no. 75-7-0820 negative over the whole temperature range and conductivities as high as $10^4 (\Omega cm)^{-1}$ were obtained at low T. Under pressure there was evidence for a T^2 term in the conductivity between 0.2 and 2 K and also a large positive transverse magnetoresistance below 100 K. These results indicated that, under pressure at least, the conductivity arose from a small number of carriers with a high mobility and a low Fermi energy, i.e. at low T the material behaved like a semi-metal or a degenerate semiconductor.

Shortly after the above work an anomalously large diamagnetic susceptibility, χ_{dia}, was observed below 30 K[3] and ascribed to Landau-Peierls diamagnetism arising from the two or three D character of the electronic states of HMTSF-TCNQ at low T. It is now clear that independent measurements by other groups[4] give the same value for χ_{dia}, and at present this seems to be the strongest experimental evidence in favour of the semi-metallic, as opposed to the degenerate semiconductor picture. (The impurity states or the free electrons in a degenerate semiconductor can also give strong diamagnetism but it should then vary strongly from sample to sample).

A one electron band structure proposed recently [5] for HMTSF-TCNQ accounts very well for the observed diamagnetism and many other electronic properties of this material. In this picture the Fermi surface consists of small electron-like ellipsoids and hole-like cylinders (or vice versa) at $k_b = \pm k_F$ (to be consistent with reference 5 the high conductivity axis is called the b axis here) extending over about 1/3 of the k_a, k_c plane. The band structure parameters were estimated from experimental data, in particular from the low and high temperature susceptibility, and the anisotropy in the conductivity at room temperature. The width of both electron and hole pockets was estimated to be $2\delta k_F^b \simeq 2.10^{-2} k_F^b$ and the effective masses $m_b \simeq 2.10^{-2} m_e$. At high temperatures ($kT \gtrsim$ the parameters t_{split} and t_{shift} which are defined in reference 5 and which were taken to be approximately 10-20 mev) the fine electronic structure is washed out and the effective Fermi surface consists of electron and hole planes associated with the TCNQ and HMTSF molecules respectively.

As a further test of this picture we have measured the Hall effect for three samples in the temperature range 1.4 to 200 K, both at ambient pressure and under applied pressures of 5-7 kbars. More details of the experimental conditions and results are to be published elsewhere[6] The sample geometry is shown in the lower part of figure 1a, current was passed along the long axis, the high conductivity (hc) axis, the magnetic field was directed along the low conductivity (lc) axis and the Hall voltage developed along the intermediate conductivity (ic) axis, which is usually the smallest side of the crystal. Particularly long and thin samples were chosen because we believe that in order to avoid either shorting of the Hall voltage by the current contacts or a non-uniform current distribution, the normal condition for the length/ thickness ratio, $L/T \gtrsim 3$ for isotropic materials has to be replaced by the more stringent one $L/T \gtrsim 3\sqrt{A}$, where A is the anisotropy of the conductivity. For the arrangement shown in figure 1a, A is equal to σ_{hc}/σ_{ic} which is 33 ± 13 at ambient temperature and pressure. Preliminary measurements of the T and P dependence of the anisotropy[6] indicate that the above condition was satisfied, except for samples 1 and 2 below about 10 K at P =0.

Before discussing the results shown in figures 1a and 1b we wish to mention a few standard results for the low field Hall effect and magnetoresistance for different anisotropic Fermi surfaces, in the usual relaxation time approximation. For a single band the kinetic formula for the low field Hall coefficient, $R_H = 1/nec$, is usually valid. For example for a plane or slightly curved F. S. such as that obtained in the tight binding approximation: $R_H = (nec)^{-1} k_F b/\tan k_F b$, where b is the molecular spacing along the hc direction (which is also that of current flow). So the kinetic formula is a good approximation except for a nearly half-filled band.

For an ellipsoidal F.S. the elementary formulae $R_H = (nec)^{-1}$ and $\sigma_i = ne^2\tau/m_i$ are also exact for a constant relaxation time[7] τ. But in all these cases the low field components of the tranverse and longitudinal magnetoresistance vanish, essentially because the Lorentz force and the Hall field cancel for a single band and the carriers continue to propagate as if there were no field.

A two band model is clearly more appropriate for these materials and for a F.S. consisting of electron and hole ellipsoids of equal volume, with principal axes along the x, y, z directions using the standard methods[7] we obtain the following formulae:

$$(1) \quad R_H = \frac{1}{n|e|c} \cdot \frac{(\sigma_x^h \sigma_y^h - \sigma_x^e \sigma_y^e)}{(\sigma_x^h + \sigma_x^e)(\sigma_y^e + \sigma_y^h)} = \frac{1}{n|e|c} \cdot \frac{\sigma_x^h - \sigma_x^e}{\sigma_x^h + \sigma_x^e} \text{ for } \frac{\sigma_x^h}{\sigma_x^e} = \frac{\sigma_y^h}{\sigma_y^e}$$

and for the transverse magnetoresistance,

$$(2)\quad \Delta\rho_x = \frac{H^2}{(nec)^2}\ \frac{\sigma^h_y \cdot \sigma^e_y}{\sigma^h_y + \sigma^e_y}$$

These formulae apply for current flow along the x axis and a magnetic field H along the z axis. $\sigma^{h,e}_{x,y}$ are the zero field conductivities of the hole and electron ellipsoids in the principal directions perpendicular to the field.

The longitudional magnetoresistance is still zero in this model in agreement with experiment.[6] Unfortunately the above formula for R_H implies that it does not always correspond to the sign of the carriers with the higher mobility in the longitudinal direction. For example negative values of R_H also occur for $\sigma^h_x \geq \sigma^e_x$, $\sigma^e_y \geq \sigma^h_y$, so it is also necessary to look at the sign of thermopower Hovever the kinetic formula still gives an upper limit to the carrier concentration n, and if $/R_H\, ec/$ is large n must be small. The absolute value of the magnetoresistance is only dependent on the transverse effective mass, while the relative value $\Delta\rho_x/\rho_x$ goes as $/m_x\, m_y/^{-1}$.

Turning now to the experimental results shown in figures 1a and 1b the main point to notice is the extremely small values of n, especially at low T. R_H is positive and hole-like at high temperatures; in the band picture it starts to increase when the F. S. details become sharper, this is consistent with the diamagnetic susceptibility which starts to develop around 100 K[3]. It can be seen that n ≃ 0.1 at 100 K. For P = 0, R_H reaches a maximum around 50K and changes sign at 32K, becoming large and electron-like at low T. The sign of R_H is consistent with that of the thermopower[1], only the latter changes sign at 50K and shows a large negative peak at 30K. In the region 60-200K, a pressure of 6 Kbars is equivalent to an increase of 1.4 in the temperature scale. Within the band picture it seems that the parameters t_{split} and t_{shift} increase under pressure and the 3-D region becomes larger.

We have also made some preliminary measurements of the Hall effect in TTF-TCNQ with similar sample geometry, namely current along the hc(b) axis, H along the lc(a) axis and Hall field along the ic(c*) axis.[14] At ambient pressure and temperature the four samples have R_H values between -0.4 and -1.2 x 10^{-10} Vcm/Agauss,[++] and R_H decreases at lower T. For one sample under pressure R_H is smaller, -0.25x10^{-10}Vcm/Agauss at 290K, but

++ See reference (14)

increases at lower temperatures, by factors of 2 and 4 at 100 K and 60 K respectively.

Therefore from 60-200 K the two materials show quite different behaviour which may be related to the larger transverse electron overlap in one direction for HMTSF-TCNQ; as deduced from the crystal structure, which shows rather short Se-N distances[8], and the lack of frequency dependence in the proton NMR relaxation rate at room temperature[9,10].

Since R_H is large and negative at low T (Figure 1b) and has the same sign as the thermopower it is reasonable to estimate n from the formula for a single electron band, as is done in Figure 1b. The fall in R_H for samples 1 and 2 below 10K at P=0 is believed to arise from the geometrical effect referred to previously, sample 3 had a much larger L/T ratio ($\simeq 100$) and did not show this effect. Accepting this argument it can be seen that the qualitative behaviour of R_H is not altered by a pressure of 6 Kbars, in contrast to the conductivity. At 4.2 K and below there is good agreement between the value of n derived from figure 1b and the volume of the electron pockets in the proposed F.S.[5,10], the strong T dependence of R_H above about 10 K is also consistent with the estimated degeneracy temperature of 10-20 K for the electron pockets. We can also estimate the electron mean free path ℓ and the scattering time τ from the Hall mobility at low T, using the formula $R_H \sigma_{hc} = e\tau/m_b = e\ell/\hbar\,\delta k_F^b$ and the results are summarised in Table 1. According to reference 5 $\delta k_F^b \simeq 10^{-2} k_F^b$, but there is a serious problem here because even if we take a value a factor of 3 larger as in Table 1 we find $\ell \simeq 50$ Å and $\ell\,\delta k_F^b = 0.2$ which means that the F.S. is not well-defined. It appears that δk_F^b must be at least another factor of two larger for the band picture to be valid at P = 0, or alternatively, as mentioned elsewhere[5] that the measured value of σ gives an underestimate of the true conductivity at P = 0. Under pressure ℓ is 250-1000Å and this problem does not arise. The value of τ under pressure is also roughly consistent with the observed transverse magnetoresistance with H along the 1c axis. If we use equation (2) with m_a^h equal to the free electron mass m_e then we obtain $\omega_c\tau \simeq 0.1$ at $H = 10^3$ Gauss, which is consistent with the magnitude and field dependence of the magnetoresistance. However for H along the ic direction the observed magnetoresistance was only about a factor of two smaller, which indicates that $m_c^h \simeq 2m_a^h$ and thus the approximation of a cylindrical F.S.[3] may be too crude(5).

To summarise, it seems clear that at low T the conductivity of HMTSF-TCNQ arises from small pockets of electrons with a degeneracy temperature of about 20K and there is mounting evidence that these are an intrinsic property of the material. The proposed band structure accounts for many of the experimental observations but a number of problems remain. One interesting experiment is X-ray diffuse scattering at low T. It is known that there is a longitudinally polarised "$2k_F$" soft phonon down to 100 K[11], if this condenses to give a 3D superlattice then the band structure would be modified. In fact it is even possible that a 3D superlattice could be responsible for the electron pockets at low T and the change in sign of R_H. In HMTSF-TCNQ the transverse transfer integral, t_a say, may well be large enough[8] to satisfy the condition $t_a \gtrsim T_P$ [12] for suppressing a $(0, 2k_F, 0)$ superlattice, or the more stringent one for a $(\pi/a, 2k_F, \pi/a)$ superlattice, which is $t_a \gtrsim 3{\cdot}5\sqrt{T_p T_F}$ [13], where T_p is the mean field Peierls transition temperature for $t_a = 0$ and T_F the Fermi temperature. In either case, before the 3D superlattice transition is completely suppressed, semi-metallic behaviour could arise from a small number of states near the boundaries of the new Brillouin zone, which although strongly perturbed by the lattice distortion, remain metallic down to T=0. The number of such states must necessarily be small, substantially less than Δ/T_F per molecule where Δ is the low T energy gap over most of the F.S. However their effective mass would also be small, of order $m_e \Delta/T_F$, so the conductivity could remain large below the 3-D ordering temperature.

Acknowledgments

We wish to thank Dr. G.Soda and Dr. C.Weyl for stimulating discussions and G.Malfait for constant assistance in operating the high-pressure equipment. **The preliminary work on TTF-TCNQ was also done in collaboration with M. Miljak using samples supplied by the group of Professor L. Giral in Montpellier.**

TABLE 1

Parameters estimated from low T Hall coefficient and conductivity of HMTSF-TCNQ.

Pressure (Kbars)	R_H (cm^3/Coul)	σ $(\Omega cm)^{-1}$	n (el/mol)	μ (cm^2/Vsec)	ℓ (2) (A°)	τ (2) (sec)	E_F^* (1,2) (°K)
0	$\simeq 2$	$\simeq 10^3$	$\simeq 0.002$	$\simeq 2\times10^3$	$\simeq 50$	$\simeq 6\times10^{-14}$	14
5–7	1.5–3.5	$0.5-1\times10^4$	0.001 –0.002	$1-4\times10^4$	250–1000	$3-12\times10^{-13}$	14

(1) taking $k_F^b = 3\times10^{-2} k_F^b = 4.5\times10^{-3}$ A°$^{-1}$
(2) taking $m_b = 6\times10^{-2} m_e$

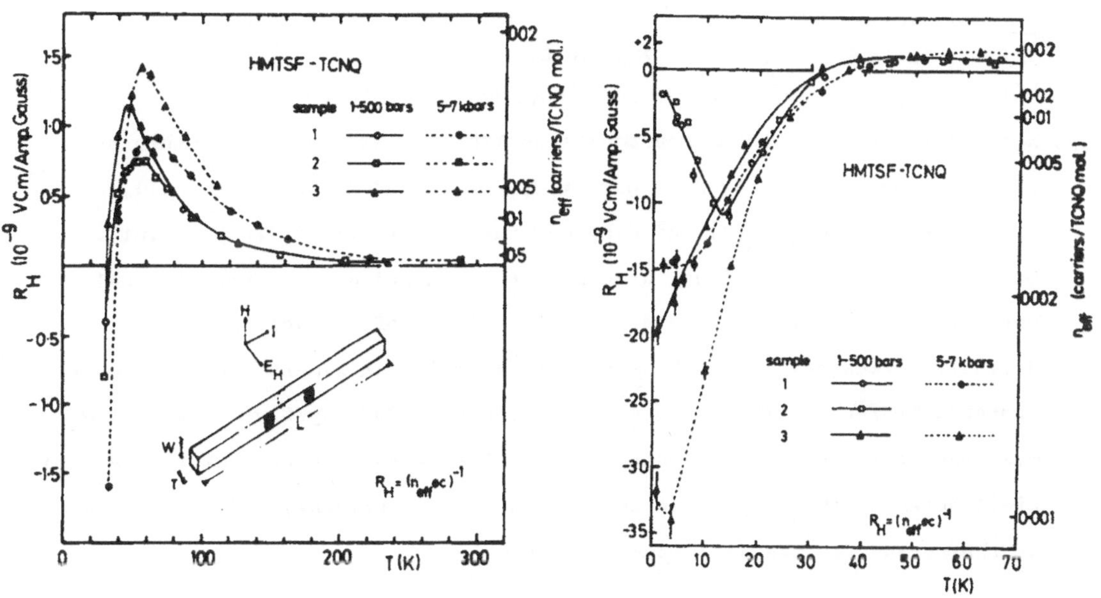

FIGURES 1(a) and 1(b)

Low field Hall coefficient for three samples of HMTSF-TCNQ at P ≃ 0 and P ≃ 6 Kbars

References

1. A.N.Bloch, D.O.Cowan, K.Bechgaard, R.E.Pyle, R.H.Banks and T.O.Poehler, Phys.Rev.Lett.34,1561(1975)
2. J.R.Cooper, M.Weger, D.Jérome, D.Lefur, K.Bechgaard, D.O.Cowan and A.N.Bloch, Solid State Comm.19,749(1976)
3. G.Soda, D.Jérome, M.Weger,K.Bechgaard and E.Pedersen, Solid State Comm.(in press
4. A.N.Bloch et al, this conference, and A.N.Bloch private communication.
5. M.Weger, Solid State Comm.19,1140(1976)
6. J.R.Cooper, M.Weger, G.Delplanque, D.Jérome and K.Bechgaard,Journ.de Physique Lettres, in press.
7. R.G.Chambers, in "The Fermi Surface" Wiley, New York (1960)
8. T.E.Phillips, T.J.Kistenmacher, A.N.Bloch and D.O.Cowan JCS Chem.Comm.334(1976)
9. G.Soda et al, this Conference.
10. D.Jérome and M.Weger, Proceedings of the NATO summer school on One-dimensional Conductors, Bolzano, Italy (August 1976)
11. C.Weyl, E.M.Engler, S.Etemad, K.Bechgaard and G.Jehanno, Solid State Comm. 19 925(1976)
12. G.Beni, Solid State Comm. 15, 269 (1974)
13. B.Horowitz, H.Gutfreund and M.Weger, Phys.Rev.B 12, 3174 (1975)
14. Note added: Hall effect measurements on TTF-TCNQ at room temperature have already been reported by N.P. Ong, A.M. Portis and K. Kanazawa, Bull. Am. Phys. Soc. 20, 465 (1975) and N.P. Ong and A.M. Portis (1976-preprint). Using a microwave technique these authors obtained a positive Hall mobility and were able to interpret it in terms of a model involving diffusive transport along the a direction, which is almost certainly more appropriate for TTF-TCNQ than a band picture. Our negative values of R_H must therefore be viewed with caution, but we wish to note that the magneti field orientation was not the same in the two experiments and this could al be the reason for the discrepancy in the sign.

Conference on Organic Conductors and Semiconductors, Siófok, Hungary 1976

NUCLEAR SPIN RELAXATION STUDIES OF ONE-DIMENSIONAL ORGANIC CONDUCTORS, TTF-TCNQ AND HMTSF-TCNQ

G. SODA[+], D. JÉROME and M. WEGER[++]

Laboratoire de Physique des Solides,[x] Université Paris-Sud, 91405-ORSAY (France)

J.M. FABRE and L. GIRAL

Laboratoire de Chimie Organique Structurale USTL,[xx] 34060-MONTPELLIER (France)

K. BECHGAARD

H.C. Oersted Institute, Universitetsparken 5, DK-2100 COPENHAGEN (Denmark)

In this report we will present the characteristic aspects of nuclear spin relaxation in 1D organic conductors. The diffusive picture of the electron motion in conduction chain (in other words the diffusive character of the response function for small $\underline{q}$ SDW excitation, $\underline{q}\,\lambda \ll 1$, λ being the mean-free-path) and the inter-chain coupling (or the electron hopping between the conduction chains) will be discussed.

---

[x] Work at Orsay supported in part by D.G.R.S.T. contract N° 75-7-0820

[xx] Laboratoire associé au C.N.R.S.

[+] Permanent address : Faculty of Science, Osaka University, Toyonaka, Osaka-560 (Japan).

[++] Permanent address : Racah Institute of Physics, The Hebrew University, Jérusalem (Israël).

§ 1 -. Nuclear spin Relaxation in 1D Organic Conductors

Since a strong positive Overhauser effect has been observed in TTF-TCNQ (the enhancement factor being more than zero)[1], the hyperfine interaction proportional to $\underline{I}.\underline{S}$ is the dominant one for nuclear relaxation. The relaxation rate T_1^{-1} in this case is given by the summation over all the momentum transfer components $\underline{q}$ of the SDW response function[2]

$$T_1^{-1} = \frac{2\gamma^2 kT}{g^2 \mu_B^2} \sum_{\underline{q}} |A\underline{q}|^2 \frac{\chi''_\perp(\underline{q},\omega_N)}{\omega_N} \qquad (1)$$

$$\chi''_\perp(\underline{q},\omega_N) = \pi(f_{k\downarrow} - f_{k+\underline{q}\uparrow})\ \delta(\omega_N - \varepsilon_{k+\underline{q}\uparrow} + \varepsilon_{k\downarrow})$$

where $A\underline{q} = -\frac{8\pi}{3} g\mu_B |U\underline{q}(0)|^2$ and ω_N is the nuclear Larmor frequency. Taking a free electron model or a tight-binding model (ω, $kT \ll E_F$), Eq.(1) reduces to the Korringa relation even in 1D case, and T_1^{-1} is independent of frequency. However, T_1^{-1} observed in TTF-TCNQ at room temperature shows the following frequency dependence (see fig.1) : $T_1 \propto \sqrt{\omega_e}$ for high field side, and T_1^{-1} independent of frequency for low field side.

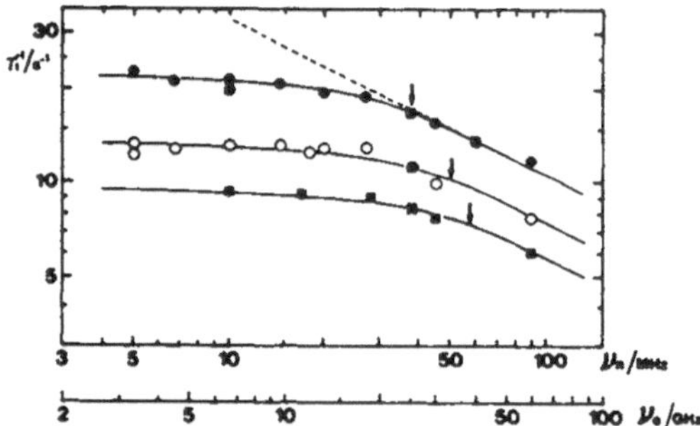

Figure 1. Frequency dependence of the proton relaxation rate at room temperature (1 atm.). The straight line shows $T_1^{-1} \propto \omega_e^{-1/2}$. The solid lines in this figure are the best-fit curves of Equation(4) in the text. Results from reference[3] are included (✖). The arrows mean the position of the respective cut-off frequency. Top TTF-TCNQ(D4), Middle TTF-TCNQ, Bottom TTF(D4)-TCNQ

Since $\hbar\omega_e$, $kT \ll \varepsilon_F$, the components of the SDW response function contributing to T_1^{-1} are only those of $|\underline{q}| \sim 0$ and $|\underline{q}| \sim 2K_F$. In TTF-TCNQ, electrons are running along the conduction chain with the Fermi velocity $v_F \sim 1.8 \times 10^7$ cm/s and the electron relaxation time $\tau_v \sim 3 \times 10^{-15}$s at room temperature (the mean free path being $\lambda = v_F \tau_v \sim 5.8 \times 10^{-8}$ cm)[4]. SDW excitations of $|\underline{q}| \sim 0$, which widely spread along the conduction chain, are heavily influenced by the scattering effects due to phonon, electron-correlation or defects and therefore the $\chi_\perp(\underline{q},\omega)$ for $|\underline{q}| \sim 0$ shows a diffusive character[5],

$$\frac{\chi''_\perp(\underline{q},\omega)}{\omega} = \chi \cdot \frac{D\,\underline{q}^2}{(D\,\underline{q}^2)^2 + \omega^2} \qquad (2)$$

On the other hand the $|\underline{q}| \sim 2k_F$ components of $\chi_\perp(\underline{q},\omega)$ may be treated by a tight-binding picture.

The correlation time τ_c^{-1} associated with the inter-chain hopping is given by the amplitude of the matrix element $t_\perp$ connecting initial and final states of the carriers and the density of states for one spin orientation $n(\varepsilon_F)$ to be

$$\tau_c^{-1} = \frac{2\pi}{\hbar}\, t_\perp^2\, n(\varepsilon_F) \qquad (3)$$

which is order of magnitude $\tau_c^{-1} \sim 10^{12}\ s^{-1}$, same order of ω_e. The auto-correlation function $\phi(t)$ of the electron which gives the power spectrum in Eq.(2), $\phi(t) = (4\pi\, Dt)^{-1/2}$, should be, therefore, replaced by $\phi(t) = (4\pi\, Dt)^{-1/2}\, e^{-t/\tau_c}$.

The final result for the Korringa product is given by

$$\left[T_1\, T\, \chi_s^2 \left(\frac{A}{g\,\mu_B}\right)^2\right]^{-1} = \frac{S}{2}^{-1} \left\{\sqrt{\frac{3\,\tau_c}{2\pi^3\,\tau_v}}\; g(\omega_e)\, K'(\alpha) + K(\alpha)\right\} \qquad (4)$$

$$S = \left(\frac{\gamma_e}{\gamma_n}\right)^2 \frac{\hbar}{4\pi\, k_B} \quad , \quad g(\omega) = \sqrt{\frac{1 + \sqrt{1+\omega^2\tau_c^{\,2}}}{2\,(1+\omega^2\tau_c^{\,2})}}$$

$$\chi_s = \frac{1}{2}\, g^2\, \mu_B^2\; n(\varepsilon_F)\, \frac{1}{1-\alpha}$$

$$K(\alpha) = 2\,(1-\alpha)^2 \left\langle \left[1 - \alpha\, F(\underline{q})\right]^{-2} \right\rangle_{\underline{q} \sim 2K_F}$$

$$K'(\alpha) = 2\,(1-\alpha)^2 \left\langle \left[1 - \alpha\, F(\underline{q})\right]^{-1.5} \right\rangle_{\underline{q} \sim 0} \quad , \quad \alpha = U\, n(\varepsilon_F)$$

The first term comes from the $|\underline{q}| \sim 0$ component and the second from $|\underline{q}| \sim 2K_F$. In the previous publication[6] we treated only the first term which is more effective to T_1^{-1} than the second by a factor of 7 at $H_0 \sim 10^4$ gauss. For $\omega\ \tau_c \ll 1$, T_1^{-1} becomes independent of frequency and for $\omega\ \tau_c \gg 1$, T_1 is proportional to $\omega^{1/2}$ neglecting the $|\underline{q}| \sim 2K_F$ contribution. The treatment of the component $|\underline{q}| \sim 2K_F$ here, in Eq.(4), namely,a coherent picture, may be an over-simplified one. This mode could also become diffusive if we take into account strong electron-electron interactions. This theoretical point is still open for discussions[7].

Solid lines in Fig.(1) are the best-fit curves of the theoretical formula, Eq.(3), and arrows show the respective cut-off frequences.

§ 2 -. Inter-chain coupling in TTF-TCNQ

The cut-off frequencies of the order of $10^{11} s^{-1}$ derived from the breaks in the frequency dependence of T_1^{-1} are different for each complex as will be seen in Fig.1. This means that the escaping time of the carriers from TTF-chain is different with that from TCNQ-chain. This reflects the differences in the transverse transfer integrals between the interacting two chains, t_{QQ}, t_{QF}, t_{FF} and also the difference in the density of states at the Fermi surface of both chains : $n_F/n_Q = 3/2$.

The observed hopping rates of both chains are given by[6]

$$\left.\begin{aligned} \tau_Q^{-1} &= \frac{2\pi}{\hbar} \cdot 2 \left[(2\ C\ t_{QQ})^2 n_Q + (2\ S\ t_{QF})^2 n_F \right] \\ \tau_F^{-1} &= \frac{2\pi}{\hbar} \cdot 2 \left[(2\ S\ t_{QF})^2 n_Q + (2\ C\ t_{FF})^2 n_F \right] \end{aligned}\right\} \qquad (5)$$

If we take a diffusive picture for the electron motion in chain, $C = S = 1$. On the other hand, for the coherent picture we should, instead, put the phase difference factors between the interacting electrons : $S = \sin K_F\ b$, $C = \cos K_F\ b/2$.

By using the results of MO calculation by Berlinsky et al. (t_{QQ} = 1.7 meV , t_{QF} = 1.3 meV , t_{FF} = 0.25 meV)[8] and the observed values of $t_{//} = 0.23$ eV[4,9], $k_F = 0.55\ \pi/2b$[10], we derive the values of τ_Q and τ_F shown in Table I, which is in good agreement with the values determined by the cut-off frequency.

- TABLE I -

Obsd.		Calcd. Coherent	Calcd. Diffusive
τ_Q	4.2×10^{-12}s	6.0×10^{-12}s	8.4×10^{-12}s
τ_F	6.4×10^{-12}s	2.2×10^{-11}s	2.5×10^{-11}s
τ_{av}	4.8×10^{-12}s		

The inter-chain hopping of electrons observed by the NMR experiment could be the same mechanism for the transverse conductivity. Assuming the diffusive mechanism for the transverse conductivity, $\sigma_\perp$; $\sigma_\perp$ is given by

$$\sigma_a = n \frac{\ell_a^2\ \ell^2}{kT\ \tau_a} \quad , \quad \sigma_c = n \frac{\ell_c^2\ \ell^2}{kT\ \tau_c}$$

By using the inter-molecular spacing $\ell_a = 6.15$ A , $\ell_c = 9.2$ A[11], the observed conductivities, $\sigma_a = 0.5 \sim 1\ (\Omega\ \mathrm{cm})^{-1}$, $\sigma_c = 4 \sim 6\ (\Omega\ \mathrm{cm})^{-1}$[12], give the escaping times along the a- and c- axes, τ_a and τ_c ,

$$\tau_a = 5 \times 10^{-12}\mathrm{s} \quad , \quad \tau_c = 2.5 \times 10^{-12}\mathrm{s}$$

which are the same order of magnitude of the escaping time derived from NMR cut-off frequency.

Increasing pressure, σ_a increases by a factor of 3.8 up to 8 kbar[13]. Figure 2 shows the frequency dependence of T_1^{-1} for TTF-TCNQ (D4) under several pressures up to 8 Kbar. Here also the solid lines mean the best-fit curves of the theoretical formula and arrows, the positions of cut-off frequency.

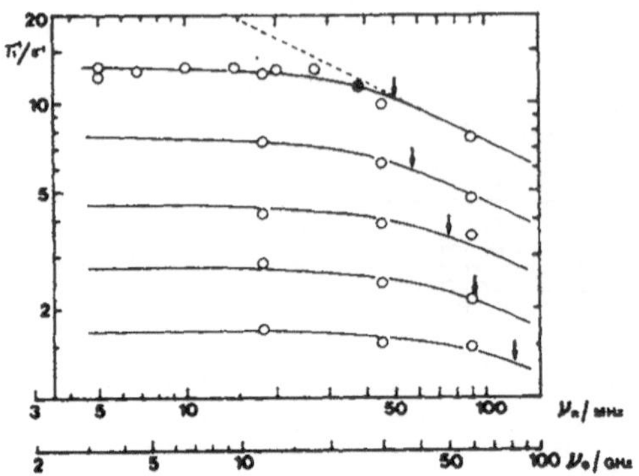

Figure 2. Frequency dependence of the proton relaxation rate in TTF-TCNQ under several pressures at room temperature. The solid lines are the best-fit curves of Equation(4) in the text and the arrows means the positions of the cut-off frequency.
From top to bottom 0, 2, 4, 6, 8 kbar

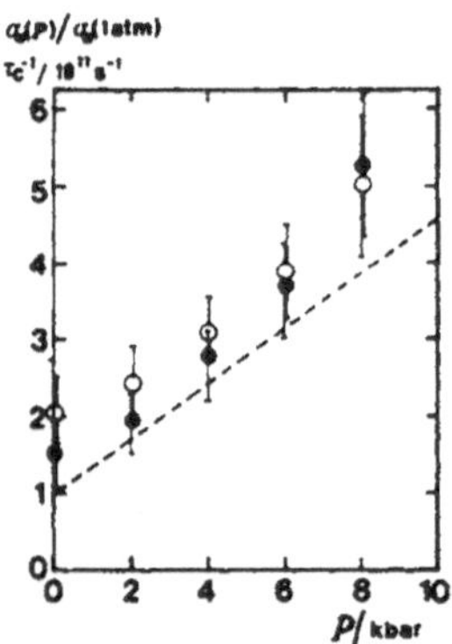

Figure 3. The pressure variation of the cut-off frequency τ_c^{-1} for TTF-TCNQ and TTF-TCNQ(D4). The pressure variation of the normalized transverse conductivity $\sigma_a(P)/\sigma_a(1\ \mathrm{atm})$ is also shown.

Figure 3 shows the pressure variation of the cut-off frequency τ_c^{-1} for TTF-TCNQ(D4) and TTF-TCNQ and that of σ_a normalized by 1 atm value(13). In TTF-TCNQ(D4), τ_c^{-1} increases by a factor of 3.8 and in TTF-TCNQ τ_c^{-1} increases by a factor of 2.7.

In conclusion for this section, the inter-chain hopping derived from the breaks in the frequency dependence of T_1^{-1} is consistent with the MO calculation of transverse transfer integrals and with the transverse conductivity (absolute value and its pressure variation at room temperature). This means that the breaks in the frequency dependence of T_1^{-1} reflects the same process of electron motion, as in the transverse conductivity, and Eq.(3) gives a good estimate for the escaping time. The value of $\tau_{av} = 4.8 \times 10^{-12}$s gives the effective value of the transverse transfer integral $(t_\perp)_{av} = 5$ meV. Frequency dependence of T_1^{-1} in the high field side gives the information about the electron motion in conduction chain : diffusive character of the electron motion. This point will be reported in the other publication[14].

§ 3 -. Inter-chain coupling in HMTSF-TCNQ[15]

HMTSF-TCNQ is the first organic conductor found to be metallic down to very low temperatures[16]. Quite short S_e-N distances[17] and the relatively small anisotropy of conductivity ($\sigma_b/\sigma_a = 33 \pm 13$, $\sigma_b/\sigma_c = 450 \pm 150$ at room temperature)[18] suggest the larger inter-chain coupling in this material comparing with TTF-TCNQ. Figure 4 shows the temperature variation of T_1 at various frequencies. Above 130K the temperature variation resembles that of TTF-TCNQ. T_1 is, however, about 15 times longer and is independent of frequency upto 276 MHz. This indicates a much shorter hopping time at least by a factor of 9, in other words much stronger inter-chain interaction. Then, we can expect the three-dimensional character in electronic state to dominate at low enough temperatures. The upper limit of the hopping time $\tau_c < 4 \times 10^{-13}$s gives the transverse transfer integral of $t_\perp > 20$ meV produce small electron and hole pocket at 0 K with Fermi energy of the order of 10 ∿ 50 K. Thus, at low temperatures the material appears to be a semi-metal, and shows a large Landau-Peierls diamagnetism below 30 K. Figure 5 shows the temperature variation of the susceptibility of HMTSF-TCNQ after substracting the core diamagnetism and the Curie paramagnetism. Above 30 K the excitations of the carriers smear out the three dimensional nature of the Fermi surface and the system becomes to show the one-dimensional character.

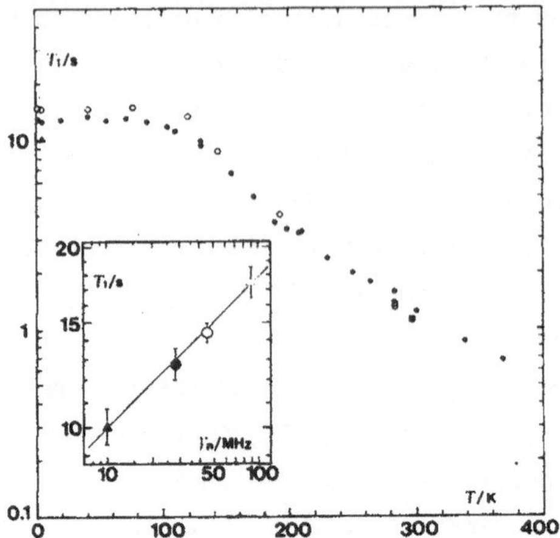

Figure 4. Temperature variation of T_1 in HMTSF-TCNQ at various frequencies: 10 MHz (▲), 28 MHz (●), 45 MHz (O), 90 MHz (Δ) and 276 MHz (■). Inset : Frequency dependence of T_1 at 4.2 K, solid line being $T_1 \propto H^{1/4}$.

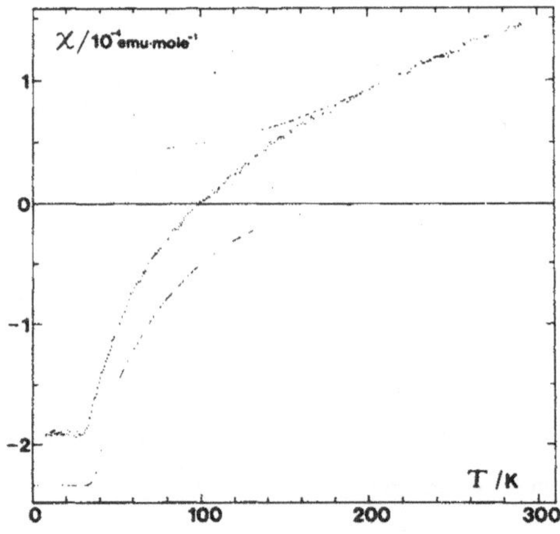

Figure 5. Temperature variation of the residual susceptibility after substracting the core diamagnetism and the Curie paramagnetism. Expected temperature variations of the Pauli and the Landau-Peierls susceptibilities are shown by .... and —·—— – —· respectively.

Acknowledgement : *We are pleased to acknowledge the profitable discussions with Drs. J.R. Cooper and C. Weyl we have had during this work. One of the authors (G.S.) express his sincere thanks to Dr. H. Fukuyama for the illuminating discussions on the diffusive susceptibility. We are grateful to Mrs. G. Delplanque and G. Malfait for the technical assistance.-*

REFERENCES

(1) J. Gallice, J.P. Blanc, H. Robert, J. Alizon: Report at Siofok Conference (1976) and private Communication.

(2) T. Moriya : J. Phys. Soc. Japan, 18, 516 (1963).

(3) E.F. Rybaczewsky, A.F. Garito, A.J. Heeger and E. Ehrenfreund : Phys. Rev. Lett. 34, 524 (1975) ; J.E. Gulley and J.F. Weiher, Bull. Am. Phys. Soc. 19, 222 (1974).

(4) A.A. Bright, A.F. Garito and A.J. Heeger : Solid State Commun. 13, 943 (1973) ; Phys. Rev. B10, 1328 (1974).

(5) P. Fulde and A. Luther : Phys. Rev. 170, 570 (1968).

(6) G. Soda, D. Jérome, M. Weger, J.M. Fabre and L. Giral : Solid State Comm. 18, 1417 (76).

(7) Private Communication from P. Lederer and M.T. Béal-Monod.

(8) A.J. Berlinsly, J.F. Carolan and L. Weiler : Solid State Commun. 15, 795 (1974).

(9) See the foot-Note (20) cited in the paper ref.(15).

(10) F. Denoyer, R. Comès, A.F. Garito and A. J. Heeger : Phys. Rev. Lett. 35, 445 (1975) ; R. Comès, S.M. Shapiro, G. Shirane, A.F. Garito and A.J. Heeger, Phys. Rev. Lett. 35, 1518 (1975) ; S. Kagoshima, H. Anzai K. Kajimura and T.Ishiguro, : J. Phys. Soc. Japan 39, 1143 (1975).

(11) T.E. Phillips, T.J. Kistenmacher, J.P. Ferraris and D.O. Cowan : Chem. Commun. 14, 471 (1973).

(12) M.J. Cohen, L.B. Coleman, A.F. Garito and A.J. Heeger : Phys. Rev. B10, 1298 (1974).

(13) J.R. Cooper, Private Communication.

(14) G. Soda, D. Jérome, M. Weger : To be published.

(15) G. Soda, D. Jérome, M. Weger, K. Bechgaard, E. Pedersen : Solid State Commun. 20, 107 (1976).

(16) A.N. Bloch, D.O. Cowan, K. Bechgaard, R.E. Pyle, R.H. Banks and T.O. Poelher : Phys. Rev. Lett. 34, 1561 (1975).

(17) T.E. Philipps, T.J. Kistenmacher, A.N. Bloch and D.O. Cowan, : J.C.S. Chem. Commun. 334 (1976).

(18) J.R. Cooper, M. Weger, G. Delplanque, D. Jérome, K. Bechgaard,: J. Physique Lett. (To be published).

Conference on Organic Conductors and Semiconductors, Siófok, Hungary 1976

MAGNETIC PROPERTIES OF ONE-DIMENSIONAL CHARGE TRANSFER CONDUCTORS: PRESSURE EFFECTS*

D. JÉROME

Laboratoire de Physique des Solides, Université Paris-Sud, 91405, Orsay, France

L. GIRAL

Laboratoire de Chimie Organique Structurale, USTL, 34060, Montpellier, France

A B S T R A C T

We report a systematic investigation under pressure of spin susceptibility, nuclear relaxation rate and temperature dependence of the resistivity in the charge transfer salts TTF-TCNQ, TMTTF-TCNQ and HMTSF-TCNQ. The very large pressure dependences observed in the TTF-TCNQ family, which cannot be explained solely by the band broadening point to the picture of a magnetism governed by the properties of a one dimensional strongly correlated electron gas. An interpretation of our data in terms of the Hubbard model for one dimensional itinerant magnetism provides an estimation of the ratio electron-electron repulsion to band width $U/4t \approx 1$ in TTF-TCNQ at ambient pressure. The effects of correlations tend to become weaker for the selenium compounds, in particular for HMTSF-TCNQ, or as pressure is applied to the TTF-TCNQ family.

It is suggested that pressure enhances the screening of electron-electron repulsion via the excitonic mechanism.

This work is part of a more general study of the electronic properties of one-dimensional conductors, supported partly by a CNRS ATP no. A206 and partly by a DGRST Contract no. 75-7-0820.

1. Introduction

In the last few years a large amount of work has been devoted to the search for better organic charge transfer conductors. Most of the efforts have been concentrated on the study of the TTF-TCNQ family /TQ/ and of the selenium analogue family, TSeF-TCNQ /TSe-Q/.

We wish to summarize briefly here some of the conclusive results which have been obtained.

/i/ The stacking of planar charge transfer molecules along a unique direction, allowing therefore a large overlap of π-like wave functions between adjacent molecules of the same chain, is a necessary condition for the achievement of a good metallic conduction along the chain axis.

/ii/ The chain-like stacking, however, confers one-dimensional properties to the organic charge transfer conductors and therefore enhances the instability towards a Peierls insulating state.

/iii/ The electron-electron interaction, usually important for a large number of charge transfer compounds, can be significantly reduced by the high polarizability of molecules stacked on neighbouring chains.

/iv/ Application of hydrostatic pressure leads to large effects on the electronic properties of organic conductors /1/:

/a/ An intrachain effect is responsible for the pressure enhancement of the metal insulator transitions at 53K and 29K in TQ /2, 3/ and TSe-Q /4/ respectively, through the lattice stiffening.

/b/ The pressure enhancement of the electron tunnelling matrix element has been observed for KCP /5/ and TQ /1, 6/. Regarding KCP, pressure removes most of the large 1d charge fluctuations observed between the mean field Peierls temperature T_p^{mf} and T_p /5, 7, 8/. It is also believed that pressure stabilizes the conducting phase at low temperature /9/ . This has been observed for HMTSF-TCNQ /HQ/ above 3 kbar /1o/.

/v/ The nature of the low temperature state of CT salts results from a subtle interplay between the electron-electron interaction, the intrachain kinetic energy and the interchain coupling. In particular a small interchain coupling, together with a small e-e interaction and a large kinetic energy situation leads to a low temperature Peierls insulator /11/.

A large e-e interaction with respect to the bandwidth leads to a Mott-Hubbard magnetic insulator at low temperature, irrespective of the interchain coupling /12/. Finally, there is some hope to maintain the conducting state at low temperature whenever the interchain coupling is large and e-e interaction small /1/.

As far as TQ is concerned, it was concluded from early experiments of T_1 /proton relaxation time/ /13/ and recent results of T_1 in C^{13} enriched samples /14/ that Coulomb correlations do not play a dominant role. The verification of the relation $\chi_S^{-2} T_1 T = C^{st}$ between spin susceptibility and T_1 was thought as justifying the neglect of e-e interactions.

The purpose of this report is to reexamine the magnetic properties of the whole TQ family, including the recently investigated case of HQ /15/. In this respect, pressure has proved a very useful tool.

We look at the significance of T_1 in a quasi 1-d conductor in Section 2. We present the pressure dependence of χ_S and T_1 in TQ, TMQ /tetramethyl/ /16/ and HQ in Section 3, together with a comparative study of χ_S , T_1 and resistivity for CT conducting salts. We propose a model of strongly correlated 1d electron gas in Section 4. We shall conclude more generally in Section 5 on the existence of large e-e interactions in TQ but of weaker ones in HQ. The effect of e-e interactions is also attenuated under pressure.

2. Spin-lattice relaxation time in quasi 1-d conductors

It has already been known for some time that one-dimensionality may lead to low frequency divergences in spin correlation functions of magnetic insulators /17, 18/. However, the investigation of the spin correlation functions of 1-d conductors has been up to now the subject of much less work /19-21/. It is the purpose of this section to show that a frequency dependence study of T_1 can provide much information in TQ an d HQ.

The relaxation rate induced by the electron-nuclear coupling is given, in terms of electron and nuclear Larmor frequencies, by a relation /22, 23/

$$1/T_1 = A_+ g^+(\omega_e) + A_z g^z(\omega_n) \qquad /1/$$

where A_{+z} are geometrical factors associated with the scalar /+/ and dipolar /z/ parts of the hyperfine coupling. The function $g(\omega)$ is the Fourier transform of the spin autocorrelation function

$$g(t) = \sum_q \langle S_q^+(t) S_q^-(0) \rangle$$

The existence of a large Overhauser enhancement of the TTF and TCNQ protons magnetization is a first confirmation of the dominant scalar interaction in the hyperfine coupling /24/. The observation of relaxation times, nearly ten times longer in HQ than in TQ, is another indication that the dipolar coupling in /1/ can be discarded /25/.

For 3d metals , the Fourier transform of g/t/ is a Lorentzian curve with a cut-off frequency $E_F/\hbar$ usually much larger than ω_e /26/.

Whenever the electrons are constrained to move on a chain by the one-dimensional character of the structure, the electron-nucleus coupling becomes a very efficient relaxation process. In 1-d systems, there will always be a return of random walking electrons to the initial position. The probability of return to the initial site goes like $t^{-1/2}$, leading to a divergence of the Fourier transform of $g(t) \sim t^{-1/2}$ at low frequency and therefore to a $1/T_1$ enhancement in /1/.

However, the possibility of 1d enhancement is limited by the one-dimensional life time of the electron. Introducing this life time effect into the spin autocorrelation function the relaxation rate

$$1/T_1 \propto \int_0^\infty \cos(\omega_e t)\, t^{-1/2} \exp -\frac{t}{\tau}\, dt$$

becomes

$$1/T_1 \begin{cases} \text{constant} & \text{if} \quad \omega_e \tau < 1 \\ 1/\sqrt{\omega_e} & \text{if} \quad \omega_e \tau > 1 \end{cases} \qquad /2/$$

a cross-over between the field independent regime and the $H_0^{1/2}$ dependence occurs at a magnetic field $H_c = \omega_c / g\mu_B$ such as $2\omega_c \tau = 1$ /23/.

The behaviour preicted by /2/ is very well observed in TQ and tetramethyl TTF-TCNQ /TMQ/, see Fig. 1.

For a general value of $\omega_e \tau$, T_1 is given by

$$1/T_1 \propto \frac{\sqrt{1+\sqrt{1+\omega_e^2\tau^2}}}{\sqrt{1+\omega_e^2\tau^2}}$$

For TQ, the average 1-d life time τ_{Av} can be estimated from a fit of the theory with Fig. 1

$$\nu_c = 1.6 \times 10^{10}\ \text{Hz} \quad \Rightarrow \quad \tau_{Av} = 3.7 \times 10^{-12}\ \text{s}$$

In fact a detailed investigation which has been performed on deuterated samples /TTF/D4/-TCNQ and TTF-TCNQ /D4/ /27/ provides a life time on the TTF chain nearly three times larger than that on the TCNQ chain /6, 23/.

This is in agreement with the calculation of the various overlap integrals between TCNQ-TCNQ chains, TTF-TTF chains or TTF and TCNQ chains /28/. The average 1d life time of the carriers is roughly 1000 times longer than the scattering time derived from the optical experiments /29, 3o/. As a basic approximation, the motion of the carriers along the chains can be described as coherent /a questionable assumption at room temperature since the mean free path is not much larger than the intermolecular spacing/. At high temperature, namely $kT > t_{\perp}$, ($t_{\perp}$ is the interchain wave function overlap) the electron wave functions which can be built from the superposition of incoherent atomic wave functions in each individual chain lead to localized states. Therefore, the motion of the carriers from chain to chain is diffusive, due to a hopping mechanism. The correlation time associated with this hopping is given by the golden rule:

$$1/\tau = \frac{2\pi}{\hbar} t_{\perp}^{2} n(E_F)$$

where $t_{\perp}$ is the matrix element of the kinetic energy term between initial and final states of the carriers /the tunnelling interchain coupling/. With a tight binding model for the density of states, $t_{\parallel} = 0.23$ eV and a charge transfer of 0.55 carriers per molecule /31/, we derive $t_{\perp} = 5.8$ meV, veryfying "a posteriori" the high temperature condition for transverse diffusive motion.

The cut-off frequency has been found strongly pressure dependent /6/ and the large increase observed at low temperature /32/ is correlated with the low temperature increase of the transverse conductivity /23, 33/. The transverse motion progressively changes from a diffusive model at high temperature to a coherent model at low temperature. A detailed experimental analysis of the 1d to 3d change occurring in the electronic properties as a function of the temperature has been performed in HQ /25/.

A plot of T_1 versus $H_0^{1/2}$, Fig. 2, has been performed for TQ with a high field measurement corresponding to $\nu_n = 276$ MHz which deviates significantly from the $T_1 \propto H_0^{1/2}$ line. This deviation cannot be attribuated to the experimental accuracy which is $\pm$ 3 %, or even to the existence of paramagnetic impurities, since under pressure, on the same sample, much longer /x1o/ T_1 have been observed.

As the magnetic field is increased, the electron Larmor period may become as short as the time spent by an electron on every nucleus $(\hbar\omega_e \sim E_F)$. Therefore, in sufficiently high fields, an electron undergoes a single scattering with a given nuclear spin within the Larmor period, and the fact that the electron belongs to a quasi 1-d conductor now becomes irrelevant as far as the 1-d $1/T_1$ multiple-scattering enhancement effect is concerned. The very high field behaviour, Fig. 2, may also be attributed to the dipolar contribution to . In this case the estimate is $(T_1^{-1})_{dip} = 1/3\,(T_1^{-1})_{obs}$ at $H_0 \leqslant 10\,kOe$

In particular, T_1 should become field independent again above a second cut-off frequency. Large correlation effects tend to increase the electron localization time over that $\hbar/E_F$ of a free electron gas. Therefore, the high field regime in a strongly correlated 1-d electron gas may be reached in the 200 kOe range. Admittedly the data of Fig.2 may corroborate the picture of strong correlations, but measurements at higher fields are still needed.

The enhancement of $^1/T_1$ at low field is limited by the interchain coupling. Very crudely, within the Hubbard model /2o/ the low field relaxation rate should be given in terms of a bare band relaxation rate $(1/T_1)_0$ by a relation such as:

$$\left(1/T_1\right)_{\omega_e < \omega_c} \approx \sqrt{\frac{U}{\omega_c}}\left(1/T_1\right)_0$$

In TQ $(1/T_1)/(1/T_1)_0 \approx 200$ /see Section 3/ yielding U=4eV, $(U/4t_\parallel \sim 4)$. This crude estimation of $U/4t_\parallel$ suggests a non negligible contribution of electron-electron interaction in TQ; in particular the temperature variation of $1/T_1$ depends on that of ω_c can $(1/T_1)_0$. Since ω_c has been found temperature dependent, increasing by a factor 3 /at least/ from 300K to 100K /32/, the existence of a relation $\chi_s^2 T_1 T = c^{t}$ for TQ in the same range of temperature does not entail the absence of e-e interactions in this 1-d electron gas, and may be fortutious.

T_1 for HQ has been found field independent up to ν_n =276 MHz with an accuracy of ± 5 % /25/. The very short 1-d life time of the carriers is in agreement with the strong 2-d coupling determined by other measurements /1/. The very long T_1 observed in HQ is also a sign for a low value of V/ω_c, suggesting weak 1-d enhancement and/or possibly weaker electron-electron repulsion than in TQ.

Similar $1/T_1$ enhancement could play a role in the Pt^{195} relaxation of KCP which has been found ~ 6o times faster than the calculated bare value at room temperature /34, 35/. This is corroborated by the observed increase /x2o/ of T_1 at 2o kbar, accompanied by a large increase of the interchain coupling /5,8/. However, for KCP the low frequency fluctuations in the density of states may, below room temperature, appreciably shorten T_1 as well.

3. The pressure dependence of susceptibility and relaxation rate: experiments

The temperature dependence of the spin susceptibility is one of the most intriguing problems of the conducting charge transfer salts. Basing their arguments on T_1 measurements, some authors maintain that the large decrease of χ_s observed between room temperature and the metal-insulator transition temperature is related to the development of a pseudo-gap in the density of states at the Fermi level, as temperature becomes smaller than a mean-field Peierls temperature /3o, 36, 37/. The pressure study at hand was motivated by the following reasons: since pressure /via the increased interchain coupling/ is most efficient in removing the charge fluctuations above a 3d ordering temperature /see for example the work on Pt^{195} Knight shift under pressure in KCP /8/, we may accordingly expect a spectacular change of the temperature dependence of χ_s in TQ between 300 and 53K.

The relatively narrow EPR line in TQ an TMQ has fortunately allowed a direct determination, via low field EPR experiments ($\nu_e = 40$ MHz), of the spin susceptibility as a function of both temperature and pressure. Data regarding χ_s and T_1 under pressure have already been published /38, 39/ and more details should appear in a forthcoming paper /4o/.

The conducting phase of TMQ /16/. Microwave conductivity experiments, performed at low temperature on the samples used for the pressure experiment, have succeeded in showing an increase by a factor 10 from 300 K down to loo K, /41/. The possibility of susceptibility measurement via low field method is limited by the EPR line broadening occurring at low temperature and under pressure. The spin susceptibility was derived from a fit of the EPR absorption line shape with a Lorentzian curve. The proton relaxation time was measured under pressure at low field ($\omega_e < \omega_c$) with a pulse spectrometer. Figures 3 and 4 summarize the pressure and temperature dependence of χ_s in TQ and TMQ.

The electronic contribution to the relaxation rate of protons in TMQ has been arrived at after a careful sustraction of the methyl group rotation contribution, Fig. 5. The pressure dependence of the room temperature T_1 in TQ and TMQ is shown in Fig. 6. A log-log plot of $(T_1 T)^{-1}$ versus χ_s reveals a phenomenological relation $(T_1 T)^{-1} \sim \chi_s^{\alpha}$ with an exponent $\alpha = 2.2 \pm 0.1$ for both materials TQ and TMQ /Fig.7/. In HQ, the pressure dependence of T_1 is weaker than in TQ /a factor 2 at 8 kbar instead of a factor $\sim$ lo in TQ/. No signal has been observed in low field EPR experiment on the selenium family.

The pressure dependences of χ_s, $1/T_1$ and $t_{\parallel}$ are presented in Table 1. The value $\frac{\partial \ln t_{\parallel}}{\partial P}$ can be evaluated either from the knowledge of the phonon dispersion curve, as in TQ /42/ or experimentally, from the pressure induced shift of the plasma frequency, as was reported by the IBM group for

TQ and TSQ /43/. Both estimations give $\frac{\partial \ln t_{\parallel}}{\partial P} \approx + 2$ % $kbar^{-1}$, and we have no reason to believe that this coefficient is significantly different in HQ.

Table 1

	TTF-Q	TMTTF-Q	HMTSF-Q	TSeF-Q
χ_s^{exp} (300K) (10^{-4} emu/mole)	/48/ 6	/48/ 5	/25/ 1.4	/47/ 3.3
$\frac{\partial \ln \chi_s^{exp}}{\partial P}$ (300K) (%. $kbar^{-1}$)	-9±2	-9±1		
B ($10^{-2} \times \mu\Omega cm\, K^{-2}$)	/46/ 2,2		~0.3	1.6
$\frac{\partial \ln B}{\partial P}$ (% $kbar^{-1}$)	-17±1		-12±2	
$\frac{\partial \ln 1/T_1}{\partial P}$ (% $kbar^{-1}$)	-23±2	-19.5±2	-8	
$\frac{\partial \ln t_{\parallel}}{\partial P}$ (% $kbar^{-1}$)	/43/ +2	/43/ +2.2	/+2.2/*	

* No actual measurement has been published for HMTSF-Q. However there is no reason to expect for this compound a pressure broadening of the band very different from that of TTF-Q or TSeF-Q

A comparative study of χ_s, T_1 and ρ in TQ and HQ reveals three clear-cut questions which a proper theory should explain:

/a/ χ_s has been found 4 times smaller in HQ than in TQ /3o, 25/.

/b/ χ_s and $1/T_1$ are both strongly pressure dependent in TQ, but $1/T_1$ is only weakly pressure dependent in HQ /1, 6/.

/c/ The resistivity of all metallic CT salts /44 - 46/ follows a quadratic temperature dependence in the conducting domain which appears to be strongly dependent on pressure /3/.

We shall try for the moment to comment on those three points.

/a/ Although we cannot say much about HQ, we can estimate $t_{\parallel}$ by deriving the effective mass, using the plasma edge frequency for HQ and TQ. Taking into account the incomplete charge transfer of 0.55 elec/molecule and 0.74 elec/molecule /49/ in TQ and HQ respectively, the overlap and effective mass become $m^* = 2.8\ m$, $t_{\parallel} = 0.23$ eV in TQ, and $m^* = 2\ m$, $t_{\parallel} = 0.32$ eV in HQ. The Pauli susceptibility per formula unit is given in a tight binding model by

$$\chi_s^P = 4\mu_B^2\, n(E_F)$$

where $n(E_F) = \dfrac{1}{2\pi t_{\parallel} \sin k_F b}$ yielding $\chi_s^P(TQ) = 1.1 \times 10^{-4}$ emu/mole and $\chi_s^P(HQ) = 0.67 \times 10^{-4}$ emu/mole

The enhancement of the experimental susceptibility is 5.45 in TQ and 2.1 in HQ /25, 48/. Accordingly, the large χ_s in TQ, compared to HQ, may be partly due to the narrower band and the smaller charge transfer, but this is not enogh to explain the ratio 4 between the two experimental values.

/b/ The change of $t_{\parallel}$ under pressure can contribute to the decrease of χ_s by a factor 1.17 in 8 kbar, but not by a factor 2 as it is actually observed.

It was proposed to picture TQ in terms of a Peierls-Fröhlich condensation of the conduction electron with a high mean field temperature /3o/, with electronic properties dominated by CDW fluctuations in the domain $T_P < T < T_P^{mf}$

This picture is probably adequate in KCP for which the behaviour of χ_s under pressure is known /8/, Fig. 8. The application of 2okbar removes the charge fluctuations and consequently shrinks the temperature domain (T_P , T_P^{mf}) .

However, no shrinking is observed in the CT salts, as seen from the results of TMQ, Fig. 9. This fact suggests that the explanation of the temperature dependence of (χ_s) for KCP is not appropriate for TQ.

/c/ A good fit of the resistivity of TQ with a law $\rho = \rho_0 + BT^2$ was obtained for a broad pressure range, see Fig. 10, /3/. At ambient pressure, there is certainly an upward deviation from a T^2 law, noticed both in our measurements and in the literature /50/. We can see from a comparative study of resistivity and susceptibility in TQ, TSeQ and HQ that the magnitude of the T^2 law contribution is connected with the magnitude of χ_s in the same compound, namely, the larger the larger B /Table 1/.

Following the same lines, the initial pressure dependence of B is approximately twice that of χ_s in TQ /Table 1/. In HQ the rise of ρ above 100K is somewhat similar to a T^2 law, with a small value of B; but the decrease of the number of carriers noticed as T decreases makes the analysis of the resistivity more difficult /25, 51, 52/. We have established some connections between resistivity and susceptibility not only among the various members of organic conductors but also as pressure is increased. Consequently, a reasonable explanation for χ_s should at the same time explain the behaviour of ρ .

4. The strongly correlated 1-d electron gas in TTF-Q

Many investigations of 1-d localized spin systems have been carried out, and in particular the study of the 1d antiferromagnetic Heisenberg model. Bonner and Fisher /53/ gave a detailed analysis of the thermodynamic properties at high temperature. This model has been very helpful for the understanding of the 1d antiferromagnets and for VO_2 /54/. Performing machine calculations with the Hubbard Hamiltonian on finite length chains or rings, Shiba and Pincus have extended the calculation of the thermodynamic properties to the situation of an itinerant 1d electron system. The complete calculation of χ_s as a function of the temperature has been performed in the half filled band case /55/, but the effect of partial filling of the band has been taken into account in a zero temperature calculation of the 1d Hubbard model /56/. The question is the following: if we try to fit the experimental results of χ_s in TMQ with a 1d Hubbard model, do we find ratios $U/4t_{\parallel}$ which tally with the other data of the charge transfer compounds, namely resistivity and T_1?

Let us assume for this estimation that the susceptibilities of both chains of the compounds are equal, though we know that the experiments give $0.3 < \frac{\chi_s(TCNQ)}{\chi_s} < 0.4$ /14, 57, 58/. We feel justified in doing so since we are interested, for the moment, in factors 5 to 10 and not in 20% effects or so. The values of the experimental susceptibility are summarized in Table 2, either per molecule of TMTTF /TCNQ/ or per electron at P=0 kbar and P= 8kbar, assuming a 0.55 elec/molecule charge transfer and $t_{\parallel}$=0.23 eV. Assuming that $\chi_s(80K) \sim \chi_s(T=0)$, we derive for N/N_a =0.55 from Table 2 and from an interpolation, Fig. 5 of Ref. /56/ the values of the parameters: $U/t_{\parallel}$ =4.5 at P=0 kbar and $U/t_{\parallel}$=1.4 at 8 kbar, leading to U= 1 eV P=0 kbar and U= 0.4 eV P=8 kbar /taking into account the weak pressure b roadening of the band/.

Table 2

	T = 80 K	T = 300 K
0 kbar	1.8×10^{-28} emu/molecule $1.5 \frac{\mu_B^2}{t_{\parallel}}$ /electron	4.2×10^{-28} emu/molecule $3.45 \frac{\mu_B^2}{t_{\parallel}}$ /electron
8 kbar	1.1×10^{-28} emu/mulecule $0.9 \frac{\mu_B^2}{t_{\parallel}}$ /electron	1.8×10^{-28} emu/molecule $1.5 \frac{\mu_B^2}{t_{\parallel}}$ /electron

Since no high temperature calculation exists yet for the partially filled band case, a direct use of the high temperature value of Table 2 cannot be made. However, since it is predictible that as N/N_a decreases from 1 to 0.55 both the low temperature and the high temperature susceptibility values will nearly double, the experimental results of χ_s are in good agreement with the 1d Hubbard model. Considering the uncertainty in the determination of both $t_{\parallel}$ and the band filling, we can say with some confidence that $1 < V/4t_{\parallel} < 2$ in TQ at P= =O kbar. Moreover, U is found to be strongly pressure dependent, see Fig. 11. The value $V/4t_{\parallel} \gtrsim 1$ in TQ, derived from the susceptibility, seems reasonable indeed. In fact if $V/4t_{\parallel}$ had been found much larger than unity, then the ground state in TQ should have been that of a magnetic insulator, as in NMP-TCNQ where $V/4t_{\parallel} > 2$ /59/. Had $V/4t_{\parallel}$ been much smaller than unity, the Coulomb interactions would have played only a minor tole in the magnetism and the electronic properties. This is probably the case for HQ which eplains:

/i/ the small value of the room temperature susceptibility beind only 2 times larger than the estimated Pauli susceptibility. /In HQ the change of 1d to 3d band model around 200K and the existence of a strong diamagnetic contribution at low temperature makes the interpretation of the susceptibility in terms of the 1d Hubbard model more ambiguous than in TQ./

/ii/ the large value of T_1, its field independence and its weak pressure dependence. Since the ratio $\sqrt{\frac{V}{\omega_c}}$ is probably small in HQ, we expect the experimental T_1 to be closer to the bare band value.

A large value of ω_c for HQ /25/ is a consequence of the particular crystalline structure. Both cation and anion stacks have their molecular planes tilted by the same direction with regard to the high conductivity direction /60/. Consequently, in HQ the Se-N bond /3.1 Å/ is smaller than the S-N bond in TQ /3.25 Å/. The short Se-N bonds may explain both the strong 2d coupling along the intermediate conductivity direction and strong screening of U by the presence of a highly polarizable molecule nearby /61, 62/. The molecular electron-electron repulsion is reduced by the polaron binding energy E_B /63/, which varies roughly like d^{-4} /d=interchain distance/, yielding the effective electron-electron repulsion $V = V_0 - 2E_B$.

Therefore it appears that, starting from values V_0 =2eV and $2E_B$=1 eV, the pressure dependence of U observed in TQ is accounted for by a transverse compressibility of -1 % $kbar^{-1}$ This value of the transverse compressibility agrees very well with the longitudinal compressibility of -/0.6-0,7/% $kbar^{-1}$, and with the finding of considerably more softness along the transverse direction than along the chain axis /64/.

We attempt to propose an explanation for the large T^2 law exhibited by the TQ resistivity and which appears, in pressure dependence, closely related to the spin susceptibility /see Table 2/. We have previously attributed the temperature dependence of the spin susceptibility to the formation of quasi 1-d antiferromagnetic ordering at low temperature. As temperature is raised, disorder in the spin alignment occurs progressively up to the characteristic temperature T_{ex} which is related to the antiferromagnetic exchange interaction /in the Heisenberg model $kT_{ex} \sim J$/. The spin fluctuations can be considered as due to the existence in time of nearly magnetically ordered large effective mass, electrons. One can imagine a Normal inelastic scattering process in which light carriers /$m \sim m_e$/ exchange momentum 2 k_F and ene rgy $\approx kT_{ex}$ with antiferromagnetic spin fluctuations /65 - 67/. The scattering time is given in terms of the J_{s-sf} interaction by /66/, $1/\tau \sim \frac{J^2_{s-sf}}{t_{\parallel}} \left(\frac{T}{T_{ex}}\right)^2$ contributing a T^2 law to the resistivity up to T_{ex} /higher than room temperature in TQ/ varying like $\frac{\chi_s^2}{t_{\parallel}}$ since $\chi_s \sim 1/T_{ex}$ This model for the resistivity introduces the necessary connection between χ_s and ρ . It supports the results of Table 1 and provides a consistent picture of the magnetic properties of TQ, TMQ, TSeQ and HQ, at P=0kbar as well as under pressure.

We have not said much in this report about the EPR properties of the TQ family. However, in the light of the pressure and temperature dependence of the electron 1d life time, we propose a possible explanation. It has been suggested that the EPR line width, which is due to the phonon modulation of the spin-orbit interaction, is greatly reduced by the one dimensionality of the conductor /68/. The increase of the EPR line width occurring in the conducting phase of TQ or TMQ, both as pressure is increased or as temperature is lowered /Fig.12/ corroborates the proposed explanation since NMR relaxation time measurements have shown /see Sect. 2/ an important increase of ω_c occurring under pressure and at low temperature /6, 32/. From EPR line width, ω_c and χ_s measurements under pressure, we may conclude that the 3d coupling increases either between 300K to 100K or by the application of a $\sim$ 10kbar hydrostatic pressure.

5. Summary

That TQ may behave like a strongly correlated 1-d electron gas was suggested by a systematic investigation of T_1, χ_s and resistivity as a function of pressure and crystal structure. This is also suggested by optical experiments /69/. One-dimensional properties dominate the magnetic behaviour, probably because of a very low 3d spin ordering temperature, T_S. The interaction between neighbouring spins on different chains is $J' \ll 10^{-2}$ times the interaction J between spins on the same chain /70, 71/. In the case of TQ, $/\frac{J}{k} \sim 400K/$, T_S is somewhat below 50K. Therefore, for $T > T_P$ in TQ the system behaves magnetically as an assembly of 1d chains. It is suggested that pressure plays an important role, reducing the screened Coulomb interaction whenever it exists, e.g. in TQ or TMQ $\left(V/4t_{\parallel} \sim 1 \text{ at } P=0\right)$. Following the same lines, the carriers in HQ behave as those of a weakly correlated electron gas.

We propose a unified picture for the quasi 1d systems, under pressure, sketched in Fig. 13, in which $t_{\perp}/t_{\parallel}$ and $U/4t_{\parallel}$ are the important parameters. It appears that pressure is a very rewarding instrumental tool since it allows changes not only in $t_{\perp}$ but also in U for the charge transfer salts. Admittedly this new aspect of the organic charge transfer salts needs to be confirmed by more experimental work and a further development of the theory. However, we find it satisfactory to propose an explanation of the magnetic properties of the conducting CT salts which does not treat TQ and HQ independently of other salts such as NMP-TCNQ, Qn-/TCNQ/$_2$ or TEA-/TCNQ/$_2$ /72/.

Acknowledgements. This report is the summary of more detailed articles either already published or to be published in the near future. As such, it is the result of a fruitful cooperation between many of our colleagues. We wish to acknowledge J. R. Cooper, G. Soda and M. Weger for their very efficient collaboration at Orsay. Some early results were obtained by C. Berthier, L. Zuppiroli and C. Weyl. J. M. Fabre, E. Torreilles and P. Calas participated in the chemistry work done at Montpellier. We also had very fruitful exchanges with K. Bechgaard from Copenhagen. Much technical help with the experiments has been provided by G. Delplanque and G. Malfait. We had useful discussions with S. Etemad, J. Friedel and S. Barišić.

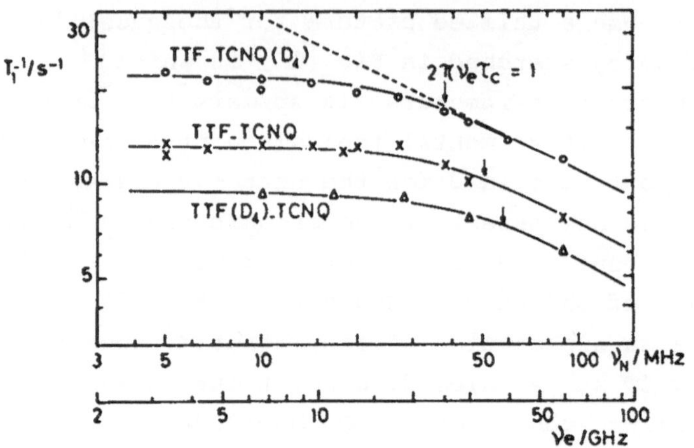

Fig. 1 Magnetic field dependence of $1/T_1$ in TTF-TCNQ

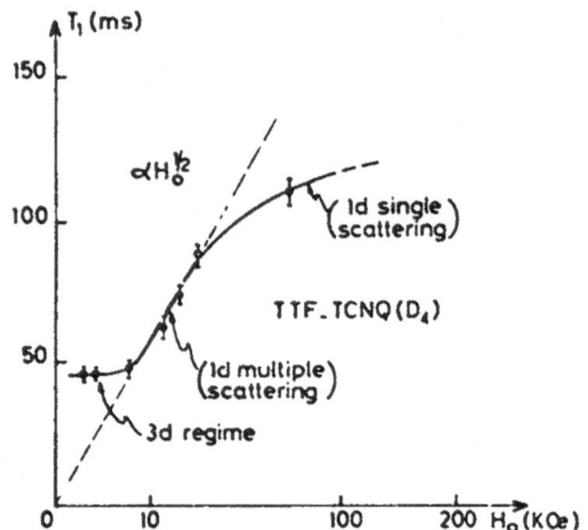

Fig. 2 Plot of T_1 vs $H_o^{1/2}$ in TTF-TCNQ /D_4/ demonstrating the existence of 3 relaxation regimes

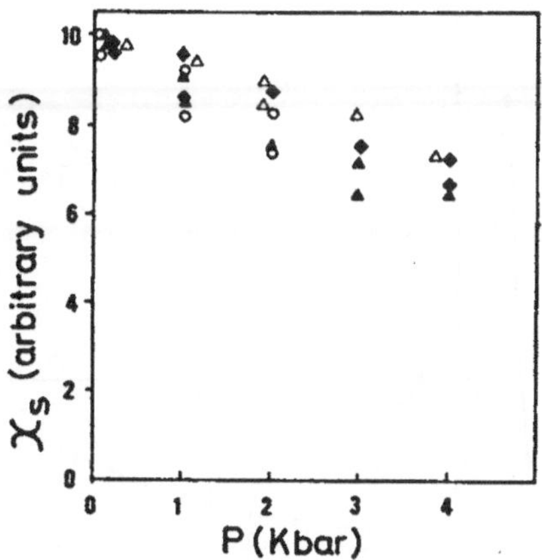

Fig. 3 Normalized spin susceptibility of TTF-TCNQ versus pressure at 34K △ , 295K ◆ , 220K ▲ , 170K O. The values of χ_s are normalized by the ambient pressure values at every temperature

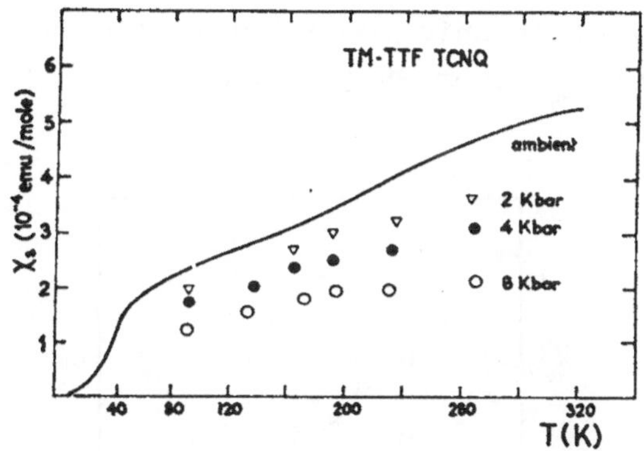

Fig. 4 The pressure dependence of the EPR spin susceptibility in TMTTF-TCNQ. The absolute value of the ambient pressure results have been taken from Ref. /48/

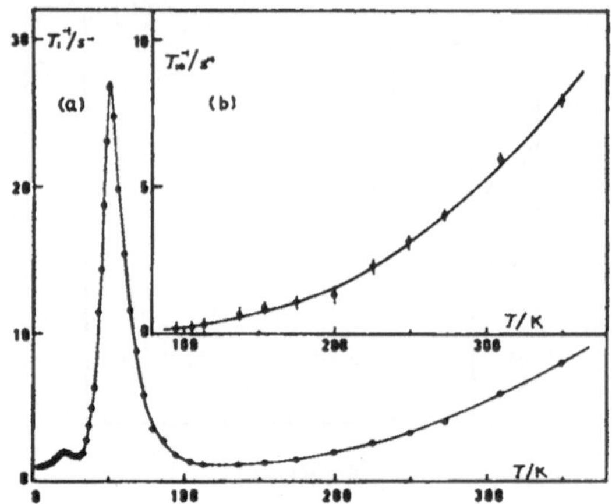

Fig. 5 /a/ Observed nuclear relaxation rate $1/T_1$ of protons in TMTTF-TCNQ /D_4/ at 18 MHz. The 52K peak is attributed to the classical reorientation of the CH_3 groups, while the 22K peak is due to the modulation of the tunneling splitting through the transitions of CH_3 group between the ground and first excited torsional states;

/b/ Electronic contribution to the relaxation rate T_{1e}^{-1} from 95 to 350K

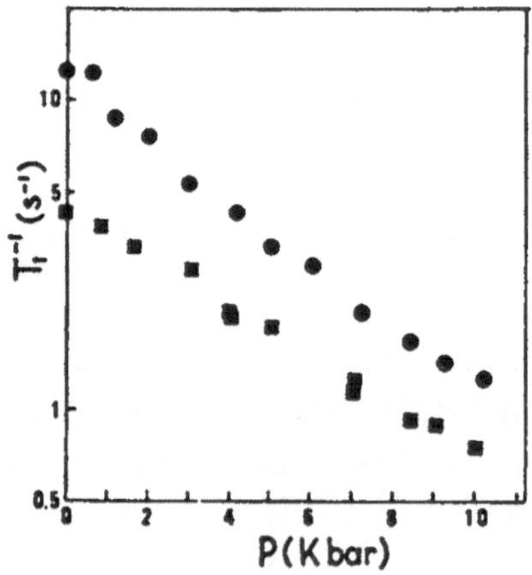

Fig. 6 Pressure dependence proton relaxation rates at 300K /24 MHz/ in TTF-TCNQ ●, TMTTF-TCNQ ■

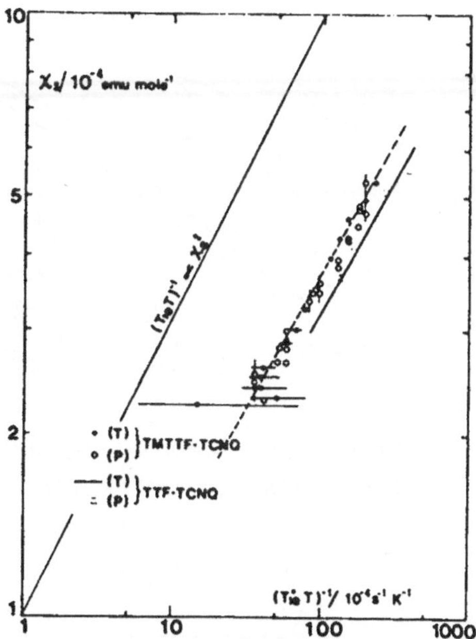

Fig. 7 A plot of the spin susceptibility versus $(T_{1e}T)^{-1}$ showing the law $\chi_s^{2.2} \sim (T_1T)^{-1}$ followed in TTF-TCNQ and TMTTF-TCNQ as a function of temperature or pressure. The ambient pressure data of TTF-TCNQ have been taken from Ref. /13/

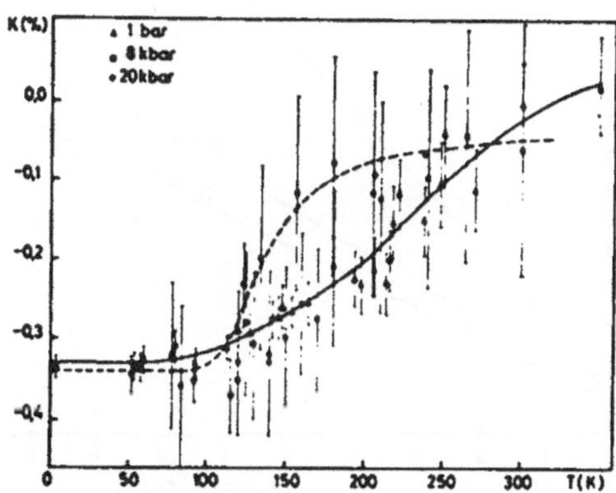

Fig. 8 Temperature dependence of the Pt^{195} Knight-shift at three pressures. Reference signal $H_2Pt\ Cl_6$. Zero pressure data taken from Ref. /35/

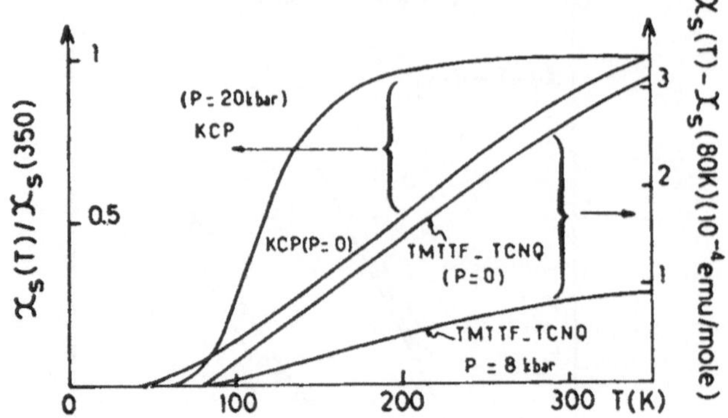

Fig. 9 Dependence of the observed χ_s under pressure; in KCP /8/ lhs scale; in TMTTF-TCNQ rhs scale

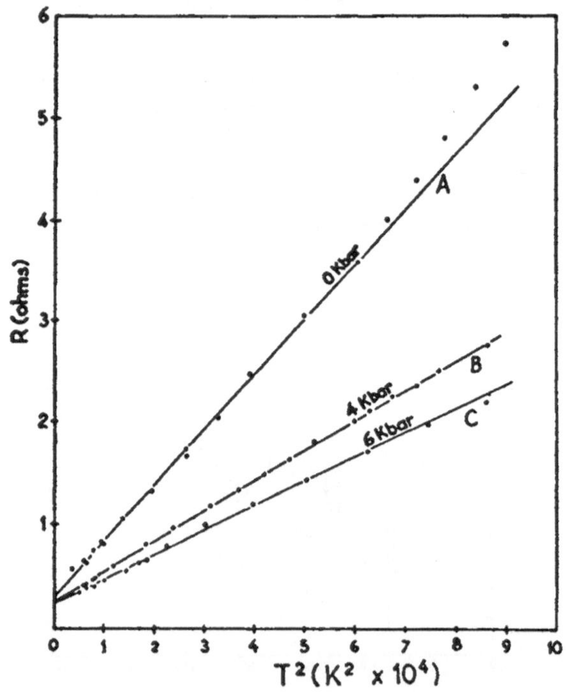

Fig. lo Examples of T^2 fits in the region 70-300K for TTF-TCNQ under pressure

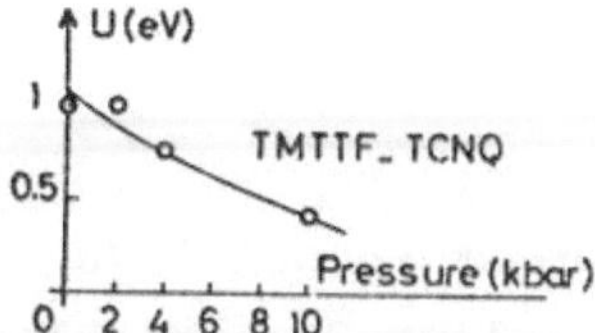

Fig. 11 Pressure dependence of the effective Coulomb repulsion in TMTTF-TCNQ or /TTF-TCNQ/

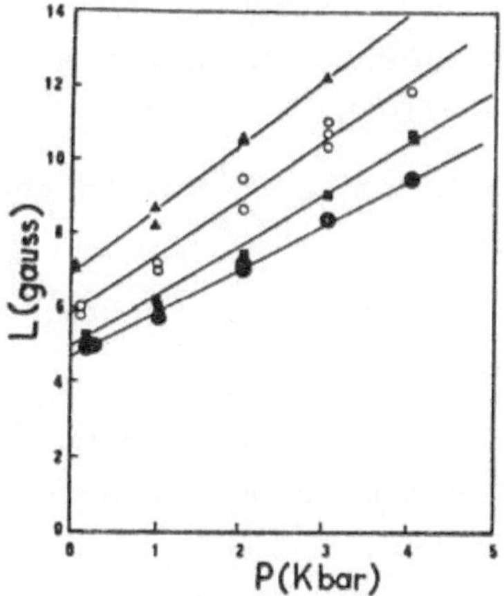

Fig. 12 EPR linewidth of TTF-TCNQ versus pressure at 343 K●, 295 K■, 220 K O, and 170 K▲

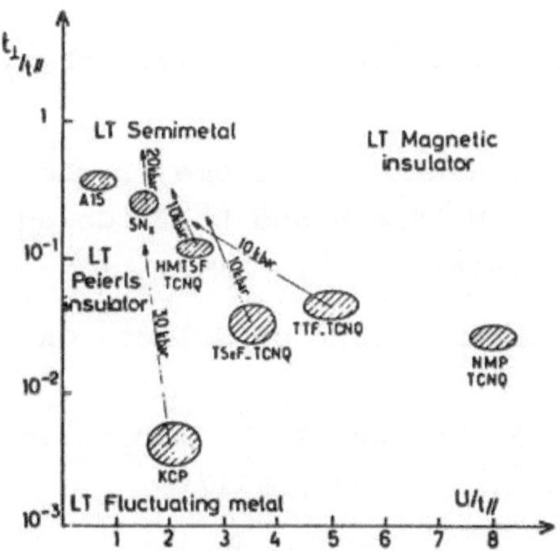

Fig. 13 A Summary of pressure effec ts on interchain coupling and Coulomb interaction of some quasi 1d conductors; see also Ref. /1/ for more details

REFERENCES

/1/ D. Jérome and M. Weger, see review article in the Proceedings of the Summer School 1976 on One dimensional conductors, H. E. Keller Editor, Plenum Press, New York 1977.

/2/ D. Jérome, W. Müller and M. Weger, J. Physique Lett. 35, L-77 /1974/.

/3/ J. R. Cooper, D. Jérome, M. Weger and S. Etemad, J.Physique Lett. 36, L-219 /1975/.

/4/ J. R. Cooper, D. Jérome, E. Etemad and E. M. Engler, /to be published/.

/5/ M. Thielemans, R. Deltour, D. Jérome and J. R. Cooper, Solid State Comm. 19, 21 /1976/.

/6/ G. Soda, D. Jérome, M. Weger, Communication at the Conference on One dimensional conductors, Siófok, 1976; also to be published.

/7/ W. H. G. Müller and D. Jérome, J- Physique Lett. 35, L-103 /1976/.

/8/ W. H. Müller, H. Niedoba and D. Jérome, Solid State Comm. 16, 655 /1975/.

/9/ B. Horovitz, H. Gutfreund, M. Weger, Phys. Rev. B12, 3174 /1975/.

/10/ J. R. Cooper, M. Weger, D. Jérome, D. Lefur, K. Bechgaard, A. N. Bloch and D. O. Cowan, Solid State Comm. 19, 749 /1976/.

/11/ R. Peierls, Quantum Theory of Solids, Oxford University Press, 1955.

/12/ A. J. Epstein, S. Etemad, A. F. Garito and A.J. Heeger, Solid State Comm. 9, 1803 /1971/.

/13/ E. F. Rybaczewski, A. F. Garito, A. J. Heeger and E. Ehrenfreund, Phys. Rev. Lett. 34, 524 /1975/.

/14/ E. F. Rybaczewski- L. S. Smith, A. F. Garito, A. J. Heeger and B. G. Silbernagel, Phys. Rev. /B/ /to be published/.

/15/ A. N. Bloch, D. O. Cowan, K. Bechgaard, R. E. Pyle, R. H. Banks and T. O. Poehler, Phys. Rev. Lett. 34, 1561 /1975/.

/16/ J. P. Ferraris, T. O. Poehler, A. N. Bloch and D. O. Cowan, Tetrahedron Lett. 27, 2553 /1973/.

/17/ D. W. Hone and P. M. Richards, Ann. Rev. Materials Sci. 4, 337 /1974/.

/18/ D. Hone, C. Scherer and F. Borsa, Phys. Rev. B9, 965 /1975/.

/19/ J. Villain, J. PhysiqueLett. 36, L-173 /1975/.

/2o/ F. Devreux, Phys. Rev. B13, 4651 /1976/.

/21/ F. Devreux and M. Nechtschein, Solid State Comm. 16, 275 /1975/.

/22/ A. Abragam, Nuclear Magnetism, Clarendon Press, Oxford 1961, p. 264.

/23/ G. Soda, D. Jérome, M. Weger, J. M. Fabre and L. Giral, Solid State Comm. 18, 1417 /1976/.

/24/ J. Gallice, J. P. Blanc, H. Robert, J. Alizon, private communication, Laboratoire d'Electronique, Université de Clermont-Ferrand; also report at Conference on One dimensional conductors, Siófok, 1976.

/25/ G. Soda, D. Jérome, M. Weger, K. Bechgaard and E. Pedersen, Solid State Comm. /in press/.

/26/ J. Winter, Nuclear Magnetic Resonance in Metals, Clarendon Press 1971.

/27/ The preparation of TTF-TCNQ and TMTTF-TCNQ is the matter of a report P. Calas, J.M. Fabre, M. Khalife-El-Saleh, A. Mas, E. Torreilles, L. Giral, C. R. Acad. Sci. 1975, 281O, 1O37 and 9O1

/28/ A. J. Berlinsky, J. F. Carolan, L. Weiler, Solid State Comm. 15, 795 /1974/.

/29/ A. A. Bright, A. F. Garito and A. J. Heeger, Phys. Rev. B1o, 1328 /1974/.

/3o/ A. J. Heeger and A. F. Garito, in Low Dimensional Cooperative Phenomena, ed. H. J. Keller, Plenum Press, New York, 1975.

/31/ F. Denoyer, R. Comés, A. F. Garito and A. J. Heeger, Phys. Rev. Lett. 35, 445 /1975/.

/32/ J. Gallice, J. Alizon, J. P. Blanc, H. Robert and G. Berthet, Université de Clermont-Ferrand, Private communication; also to be published.

/33/ S. K. Khanna, E. Ehrenfreund, A. F. Garito and A. J. Heeger, Phys. Rev. B1o, 22O5 /1974/.

/34/ H. Niedoba, H. Launois, D. Brinkmann and H. U. Keller, J. Physique Lett. 35, L-157 /1974/.

/35/ H. Launois and H. Niedoba, in Low Dimensional Cooperative Phenomena, ed. H. J. Keller, Plenum Press, New York, 1975.

/36/ P. A. Lee, T. M. Rice and P. W. Anderson, Phys. Rev. Lett. 31, 462 /1973/.

/37/ A. Bjelis and S. Barišić, J. Physique 36, L-169 /1975/.

/38/ C. Berthier, D. Jérome, G. Soda, C. Weyl, L. Zuppiroli, J. M. Fabre and L. Giral, Mol. Cryst. Liq. Cryst. 32, 261 /1976/.

/39/ C. Berthier, J. R. Cooper, D. Jérome, D. Soda, C. Weyl, J. M. Fabre and L. Giral, Mol. Cryst. Liq. Cryst. 32, 267 /1976/.

/4o/ D. Jérome and J. R. Cooper, /to be published/.

/41/ K. Holczer, /personal communication/.

/42/ R. Comés, S. M. Shapiro, G. Shirane, A. G. Farito and A. J. Heeger, Phys. Rev. Lett. 35, 1518 /1975/; H. A. Mook and C. R. Watson, Phys. Rev. Lett. 801 /1976/.

/43/ B. Welber, E. M. Engler, P. M. Grant and P. E. Seiden, Bull. Am. Phys. Soc. 35,311 /1976/.

/44/ J. Ferraris, D. O. Cowan, V. Walatka, J. H. Perlstein, J. Am. Chem. Soc. 95, 948 /1973/

/45/ R. P. Groff, A. Suna, R. E. Merrifield, Phys. Rev. Lett. 33, 418 /1974/.

/46/ S. Etemad, T. Penney, E. M. Engler, B. A. Scott, P. E. Seiden, Phys. Rev. Lett. 34, 741 /1975/.

/47/ S. Etemad, /personal communication/.

/48/ J. C. Scott, A. F. Garito and A. J. Heeger, Phys. Rev. Blo, 3131, /1974/.

/49/ C. Weyl, E. M. Engler, K. Bechgaard, G. Jehanno and S. Etemad, Solid State Comm. 19, 925 /1976/.

/5o/ P. E. Seiden, G. Cabib, Phys. Rev. B13, 1846 /1976/.

/51/ M. Weger, Solid State Comm. 19, 1143 /1976/.

/52/ J. R. Cooper, M. Weger, G. Delplanque, D. Jérome and K. Bechgaard, J. Physique Lett. /to be published/

/53/ J. C. Bonner and M. E. Fisher, Phys. Rev. 135, A64o, /1964/.

/54/ J. P. Pouget, H. Launois, T. M. Rice, P. Dernier, A. Gossard, G. Villeneuve and P. Hagenmuller, Phys. Rev. Blo, 1801 /1974/.

/55/ H. Shiba and P. A. Pincus, Phys. Rev. B5, 1966 /1972/.

/56/ H, Shiba, Phys. Rev. B6, 930 /1972/.

/57/ Y. Tomkiewicz, B. A. Scott, L. J. Tao and R. S. Title, Phys. Rev. Lett. 32, 1363 /1974/.

/58/ Y. Tomkiewicz, A. R. Taranko and J. B. Torrance, Phys. Rev. Lett. 36, 751 /1976/.

/59/ A. J. Epstein, S. Etemad, A. F. Garito and A. J. Heeger, Phys. Rev. B5, 952 /1972/.

/6o/ T. E. Phillips, T. J. Kistenmacher, A. N. Bloch and D. O. Cowan, JCS Chem. Comm. 334/1976

/61/ O. H. LeBlanc, J. Chem. Phys. 42, 4307 /1965/.

/62/ P. M. Chaikin, A. F. Garito and A. J. Heeger, J. Chem. Phys. 58, 2336 /1973/.

/63/ A. F. Garito and A. J. Heeger, Acc. Chem. Research. 7, 232 /1974/.

/64/ D. E. Schafer, G. A. Thomas and F. Wudl, Phys. Rev. B12, 5532 /1975/.

/65/ P. G. de Gennes and J. Friedel, J. Phys. Chem. Solids 4, 71 /1958/.

/66/ D. L. Mills and P. Lederer, J. Phys. Chem. Solids 27, 1805 /1966/.

/67/ R. Jullien, M. T. Béal-Monod and B. Coqblin, Phys. Rev. B9, 1441 /1974/.

/68/ Y. Tomkiewicz, E. M. Engler and T. D. Schultz, Phys.Rev. Lett. 35, 456 /1975/.

/69/ J. B. Torrance, B. A. Scott and F. B. Kaufman, Solid State Comm. 17, 1369 /1975/.

/70/ P. M. Richards in Low Dimensional Cooperative Phenomena, ed. H. J. Keller, Plenum Press, New York 1975.

/71/ P. A. Pincus in Low Dimensional Cooperative Phenomena, ed. H. J. Keller, Plenum Press, New York 1975.

/72/ I. F. Shchegolev, Phys. Stat. Sol. /a/ 12, 9 /1972/.

Conference on Organic Conductors and Semiconductors, Siófok, Hungary 1976

SOME REMARKS ON THE MICROWAVE PERMITTIVITY MEASUREMENTS OF ORGANIC SEMICONDUCTORS

M. JAWORSKI, Z. ROMASZEWSKI
Institute of Physics, Polish Academy of Sciences, 02-668 Warsaw, Poland

The application of microwave contactless techniques to the complex permittivity measurements of organic semiconductors is briefly discussed. Special attention is paid to the cavity perturbation technique of Buravov and Shchegolev and the dielectric resonance technique of Jaklevic and Saillant. A new uniform approach to the perturbation method valid for arbitrary complex permittivity is also suggested.

Microwave contactless techniques as applied to organic semiconductors have been widely developed during recent years /1-5/. In particular, a cavity perturbation method usually employed in dielectric measurements appears to be very useful also for the investigation of highly conducting anisotropic materials. In this paper applicability limits of the cavity perturbation method of Buravov and Shchegolev /1/ are shortly discussed and attention is drawn to some doubts and misunderstandings in interpretation of experimental results. A new uniform perturbation method valid for arbitrary complex permittivity regardless of the relation between the skin depth and sample dimensions is also suggested.

Let us consider a microwave cavity of volume V_c containing a small sample of volume V_s placed in locally homogeneous electric field. For $V_s \ll V_c$ the fundamental perturbation formula takes the following form /6/

$$\frac{\hat{\omega}_0 - \hat{\omega}}{\hat{\omega}} = (\hat{\varepsilon}_r - 1) \frac{\int_{V_s} \underline{E}\,\underline{E}_0^* dv}{2\int_{V_c} |\underline{E}_0|^2 dv} , \qquad /1/$$

where $\hat{\varepsilon}_r = \varepsilon_1 - i\varepsilon_2$, $\hat{\omega}$, $\hat{\omega}_0$ denote complex resonant frequency, $\underline{E}$, $\underline{E}_0$ - the electric field within the cavity with and without the sample, respectively.

The electric field inside the sample is usually evaluated from the quasi-static approximation

$$\underline{E} = \underline{E}_0 / [1 + n(\hat{\varepsilon}_r - 1)] , \qquad /2/$$

where n is the depolarization factor.

It should be noted, that Eq. /2/ is exactly correct only for an ellipsoid immersed in static homogeneous field. For real sample shapes, however, field homogeneity is violated by edge and corner effects and Eq. /2/ does not apply any more. Fortunately, according to Eq. /1/ we are interested in a total interaction between the sample and the field inside the cavity. Therefore, information about the field distribution at each point of the sample is not necessary and we can introduce an averaged $\tilde{n}$ value defined by

$$\int_{V_s} \underline{E}\,\underline{E}_0^* dv = \frac{1}{1 + \tilde{n}(\hat{\varepsilon}_r - 1)} \int_{V_s} |\underline{E}_0|^2 dv \qquad /3/$$

where $\tilde{n}$ is the generalized depolarization factor. Note that substitution of Eq. /2/ or Eq. /3/ into Eq. /1/ yields the same relation

$$\frac{\hat{\omega}_0 - \hat{\omega}}{\hat{\omega}} = \frac{\hat{\varepsilon}_r - 1}{1 + \tilde{n}(\hat{\varepsilon}_r - 1)} \cdot \frac{\int_{V_s} |\underline{E}_0|^2 dv}{2\int_{V_c} |\underline{E}_0|^2 dv} \qquad /4/$$

Thus, owing to the generalized $\tilde{n}$ definition, we can justify an application of Buravov's and Shchegolev's formalism also for nonellipsoidal sample shapes.

Intorducing dimensionless quantities $\delta = \frac{\omega_0 - \omega'}{\omega'}$, $\Delta/2 = \frac{1}{2}\left(\frac{1}{Q} - \frac{1}{Q_0}\right)$ $\alpha = \frac{1}{2}\int_{V_s} |E_0|^2 dv / \int_{V_c} |E_0|^2 dv$, , and solving Eq. /4/ for $\mathcal{E}_1$ and $\mathcal{E}_2$ one can obtain /1, 7/:

$$\mathcal{E}_1 = 1 + \frac{1}{\tilde{n}} \frac{\delta(\alpha/\tilde{n} - \delta) - (\Delta/2)^2}{(\alpha/\tilde{n} - \delta)^2 + (\Delta/2)^2} \qquad /5a/$$

$$\mathcal{E}_2 = \frac{\alpha}{\tilde{n}^2} \frac{\Delta/2}{(\alpha/\tilde{n} - \delta)^2 + (\Delta/2)^2} \qquad /5b/$$

The main problem in the practical application of Eqs. /5a b/ is the proper choice of n for real sample shapes. Unfortunately Eq. /5a/ is extramely sensitive to the assumed n value, particularly when $\mathcal{E}_2$ is comparable or greater than $\mathcal{E}_1$. To illustrate the numerical instability of Eq. /5a/ the dependence of the dielectric constant $\mathcal{E}_1$ on assumed depolarization factor $\tilde{n}$ has been computed. Results for the data corresponding to some extent to those of Khanna et al. /3/ /TTF-TCNQ, sample no. 1 at 40 K/ are shown in Fig. 1.

It can easily be seen that $\pm$ 2 % deviation from the assumed value of n leads to qualitative changes in the computed dielectric constant $\mathcal{E}_1$. Taking into account that the theoretical n value is expected to be accurate within $\pm$ 5o % /3/, one can conclude that $\mathcal{E}_1(T)$ dependence reported by Bloch et al. /2/ over the whole temperature range cannot be trusted. The experimental value $n_{exp} = \alpha / \delta_{RT}$ seems to provide a better approximation, nevertheless n_{exp} may be also incorrect since δ increases slightly with temperature /3, 8/. Therefore, the results of Khanna /Fig. 6 of Ref./3// which are reliable below 25 K /where $\mathcal{E}_2 \ll \mathcal{E}_1$ / become doubtful for higher temperatures when $\mathcal{E}_2$ increases and becomes comarable with $\mathcal{E}_1$

It is interesting to note that assuming $\tilde{n}$ below its real value /as in fact Bloch /2/ and Khanna /3/ did) always causes an apparent rise of the computed $\mathcal{E}_1$ value in the vicinity of metal-insulator transition. As a result, one can suppose that the temperature dependence of ε_1 reported for TTF-TCNQ /2, 3/ and related systems /9/ follows from the incorrect evaluation of effective depolarization factor rather than from physical reasons.

The unusually large dielectric constant ε_1 observed in cavity perturbation experiments may cause some doubts, therefore an attempt has been made to find that value independently by dielectric resonance technique /5, 1o/. Such a procedure turned out to be useful for $K_2Pt/CN/_4Br_{0.3}\cdot 3H_2O$, /5/, when the resonance of the Q-factor ranging from 3000 to 50 was observed below 20 K. However, similar results reported for TTF-TCNQ /10/ are controversial since the low-temperature conductivity is approximately two orders of magnitude higher than that of $K_2Pt/CN/_4Br_{0.3}\cdot 3H_2O$. Taking typical values for TTF-TCNQ at 4 K: $\varepsilon_2 \cong 100$, $\varepsilon_1 \cong 3000$, one can evaluate the Q-factor of the dielectric resonator to be of the order of 10. It can be shown /11/, that such a resonance is practically undetectable because of large bandwidth and weak coupling to the external field. Note, however, that the length of measured crystals /10/ was usually very close to the half-wavelength inside the guide. Therefore a supposition arises that the absorption line observed by Khanna et al. /10/ could be explained as dimension-type resonance of the half-wavelength waveguide section containing highly conducting sample.

In contrast to ε_1 , Eq. /5b/ determining ε_2 is numerically stable and can be used in quasi-static regime, even for $\varepsilon_2 > \varepsilon_1$. On the other hand, for the very high conductivity, when the skin depth is much smaller than the sample thickness, the surface impedance formalism can be employed /4/. Nevertheless, there exists an intermediate region of conductivity covering a few orders of magnitude, where both approximations fail.

Below we present a new uniform approach to the solution of the problem by perturbation method. The suggested perturbation formula, valid over the whole conductivity range. follows from the electrodynamic problem of a prolate spheroid immersed in a time-harmonic locally uniform field. The detailed derivation of the general perturbation formula goes beyond the scope of this paper and will be published elsewhere. Here we present the final relation only, written in the form analogous to Eq. /4/:

$$\frac{\hat{\omega}-\hat{\omega}}{\hat{\omega}} = \frac{\hat{\varepsilon}_r - 1 \cdot \varphi}{1+n(\hat{\varepsilon}_r-1)+(1-n)\,\varphi} \cdot \frac{\int_{V_s}|E_o|^2 dv}{2\int_{V_c}|E_o|^2 dv}, \qquad /6/$$

where

$$\varphi = N_{11}\frac{b}{2a}\left\{\sum_{n=1}^{\infty}\frac{1}{N_{1n}}\left[\int_{-1}^{+1}S_{1n}\left(\sqrt{\varepsilon_r}\,k\frac{d}{2}\eta\right)S_{11}\left(k\frac{d}{2}\eta\right)d\eta\right]^2 R_{1n}\left(\sqrt{\varepsilon_r}\,k\frac{d}{2},\frac{2a}{d}\right)\Big/ \Big/R'_{1n}\left(\sqrt{\varepsilon_r}\,k\frac{d}{2},\frac{2a}{d}\right)\right\}^{-1} - 1 \,;$$

R_{1n}, S_{1n} - spheroidal radial and angular wave function respectively /12/,

N_{1n} - normalization factor,

a, b - major and minor semiaxis of prolate spheroid respectively,

d - distance between focuses,

k - free space propagation constant

Assuming $\hat{\omega} \cong \hat{\omega}_o$ and introducing dimensionless quantities δ, $\Delta/2$, α , we obtain

$$\delta - i\Delta/2 = \frac{\hat{\varepsilon}_r - 1 - \varphi}{1+n(\hat{\varepsilon}_r-1)+(1-n)\,\varphi}\,\alpha \qquad /7/$$

Figure 2 shows $\Delta/2$ versus ε_2 dependence computed from Eq. /7/ for $\varepsilon_1 = 10^3$, $n = 10^{-3}$, $kb = 10^{-2}$.

It can easily be shown that for $|\sqrt{\varepsilon_r}\,kb| \ll 1$, φ is of the order of $|\sqrt{\varepsilon_r}\,kb|^2$. In other words φ is negligible in quasi-static regime and Eq. /7/ coincides with the perturbation formula of Buravov and Shchegolev. On the other hand, in the skin effect regime one can use asymptotic expansions of spheroidal wave functions /13/. Assuming $|\sqrt{\varepsilon_r}\,kb| \gg 1$ and $\varepsilon_2 \gg \varepsilon_1$ one obtains

$$\varphi \rightarrow \frac{1+i}{2\sqrt{2}}\sqrt{\varepsilon_2}\,kb$$

thus

$$\delta - i\Delta/2 \cong \frac{\alpha}{n} - i\,\frac{\alpha}{n^2}\,\frac{kb}{2\sqrt{2}\,\sqrt{\varepsilon_2}} \qquad /8/$$

The $\varepsilon_2^{-1/2}$ dependence is typical for the skin effect regime and can be also obtained by means of other methods /4/. The main conclusions of the paper are the following.

Measurements of ε_1 should be strictly limited to the dielectric region $/\varepsilon_2 \ll \varepsilon_1/$. For $\varepsilon_2 \gg \varepsilon_1$ reliable results cannot be obtained since Eq. /5a/ becomes extremely sensitive to experimental errors as well as to uncertainty of the n value. Therefore it seems that the temperature dependence of ε_1 reported in many papers /2, 3, 9/ is caused by numerical instability of Eq. /5a/ and one should be very careful in drawing any far-reaching physical conclusions on this basis.

Interpretation of observed dielectric resonance can be trusted in the case of $K_2Pt/CN/_4Br_{0.3}.3H_2O$[5], where $\varepsilon_2 \ll \varepsilon_1$ at helium temperature. Unfortunately, for TTF-TCNQ at 4 K is only one order of magnitude smaller than ε_1 and interpretation of observed effect as dielectric resonance is controversial. It should be stressed that the above doubts do not concern the unusually large value of dielectric constant. However, observed effects cannot be regarded as independent confirmation of the ε_1 value found by perturbation method and require a different interpretation.

As far as ε_2 measurements are concerned the perturbation method of Buravov and Shchegolev provides reliable results not only in the quasi-static region but practically at least one order of magnitude above the $|\sqrt{\varepsilon_r}\, kb| = 1$ limit. Moreover, for the spheroid of arbitrary complex permittivity we have found the universal perturbation relation valid for arbitrary wavelength inside the sample. This generalized approach not only covers various approximations used until now, but also strictly determines the limits of their applicability.

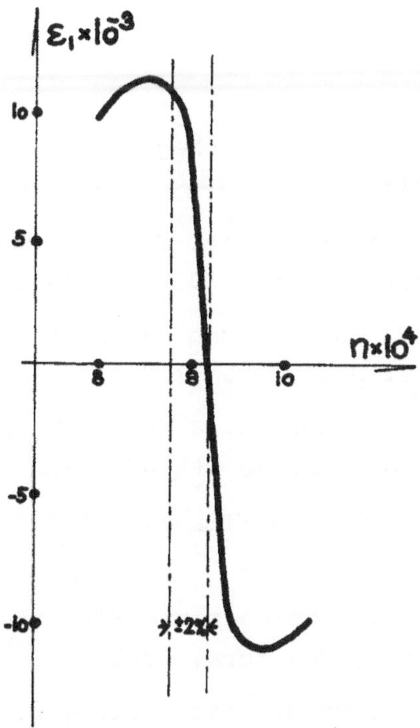

Fig. 1 Dielecrric constant ε_1 obtained from Eq. /5a/ versus assumed depolarization factor $\tilde{n}$

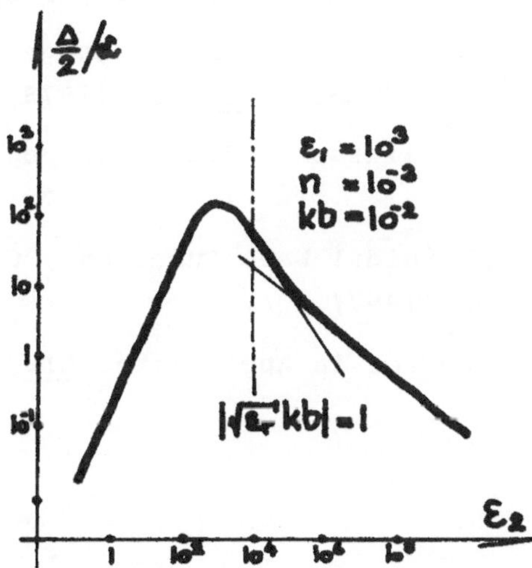

Fig. 2 Relative half-width of the cavity transmission band versus imaginary part of dielectric permittivity

REFERENCES

/1/ L. I Buravov and I. F. Shchegolev, Prib. Tekh.Eksp.2, 171 /1971/.

/2/ A. N. Bloch, J. P. Ferraris, D. C. Cowan, T. O. Poehler, Solid State Commun. 13, 753 /1973/.

/3/ S. K. Khanna, E. Ehrenfreund, A. F. Garito, A. J. Heeger, Phys. Rev. B10, 2205 /1974/.

/4/ M. Cohen, S. K. Khanna, W. J. Gunning, A. F. Garito, A. J. Heeger, Solid State Commun. 17, 367 /1975/.

/5/ R. C. Jaklevic, R. B. Saillant, Solid State Commun. 15, 307 /1974/.

/6/ R. F. Harrington, Time-harmonic Electromagnetic Fields, McGraw-Hill, New York, 1961.

/7/ I. F. Shchegolev, Phys. Stat. Sol. A12, 9 /1972/.

/8/ J. P. Ferraris, Solid State Commun. 18, 1169 /1976/.

/9/ S. K. Khanna, C. K. Chiang, A. F. Garito, A. J. Heeger, Solid State Commun. 18, 1405 /1976/.

/10/ S. K. Khanna, A. F. Garito, A. J. Heeger, R. C. Jaklevic, Solid State Commun. 16, 667 /1975/.

/11/ P. Affolter, B. Eliasson, IEEE Trans. on MTT, 21, 573 /1973/.

/12/ C. Flammer, Spheroidal Wave Functions, Stanford Univ. Press, USA/1957/.

/13/ J. W. Miles, Studies in Appl. Math. LIV, No. 4, 315 /1975/.

Conference on Organic Conductors and Semiconductors, Siófok, Hungary 1976

ORGANIC ALLOYS

YAFFA TOMKIEWICZ

IBM Thomas J. Watson Research Center Yorktown Heights, NY 10598, USA

Abstract: Comparison of the phase transitions of the two isostructural organic metals TTF-TCNQ and TSeF-TCNQ shows very significant differences. While TTF-TCNQ has been found to have two transitions at $38^{\circ}K$ and $53^{\circ}K$, TSeF-TCNQ has only one transition at $29^{\circ}K$. These differences clearly demonstrate the sensitivy of quasi-one-dimensional systems to small changes of their relevant parameters in contrast to the insensitivity of many three-dimensional systems to small changes of their relevant parameters, and raises questions about the predictability of their behavior. A systematic study of the corresponding alloys enables one to follow different physical properties such as conductivity, structure, magnetism and thermo-power and their dependence on continuous changes of parameters and thus to understand their unusually important effects on one-dimensional systems.

An interesting feature that is common to all the quasi-one-dimensional organic conductors exhibiting metallic conductivity down to low temperatures is the existence of two partially filled bands of donor and acceptor stacks. The metal-to-insulator transition is usually associated with the opening of Peierls gap in at least one of these bands. Therefore it is of utmost interest to study alloys created by selective doping of different stacks in order to evaluate the effects of the two stacks on various physical properties and on the metal-insulator phase transition. Conclusions with regard to stabilization of the metallic phase to low temperatures will be presented.

In the recent years an extensive research effort concentrated on organic metals. Compounds which belong to this family exhibit typically at room temperature conductivities of the order of 200-1000 $\Omega^{-1}cm^{-1}$. As the temperature is lowered different compounds have conductivities with different temperature dependences. Some of them go insulating at about 200°K like[1] NMP-TCNQ while others maintain their high conductivities to very low temperatures like[2] HMTSeF-TCNQ. The fact that these compounds are organic gives rise to a very large number of possible modifications due to the available skillful chemical engineering. In particular, the modifications of interest are those which will ultimately lead to a higher conductivity and stabilization of the metallic state to a very low temperature. The complication which arises upon making these modifications is the tremendous sensitivity of the metal-insulator transition to very slight changes. For example two isostructural[3] compounds TTF-TCNQ and TSeF-TCNQ, the difference between which is the replacement of the sulfur atoms on the TTF molecule by selenium ones, have very different metal-insulator transition temperatures-53°K in comparison[3] to 29°K. Additional major difference between these compounds is the fact that while in TTF-TCNQ there are three phase transitions at 38; 49 and 53°K in TSeF-TCNQ there is only 1 phase transition at 29°K. The ability to utilize the options of the chemical engineering depends upon understanding of the consequences of these modifications.

The subject of this talk is a system in which the modifications are such that at least 1 parameter-the-structure- is invariable. This system is $(TSeF)_x(TTF)_{1-x}(TCNQ)$ $0\leq x\leq 1$. In this talk I will try to relate the alloying to the observed effects on the metal-insulator transition temperature, shown in Fig. 1. In particular I will deal with the following subjects:

A. <u>TTF-TCNQ</u> (x=o). I will show that two of the phase transitions of TTF-TCNQ are related to two different 3-dimensional ordering temperatures of the donor and acceptor stacks undergoing Peierls distortions.

B. <u>TSeF-TCNQ</u> (x=1). I will relate the lowering of the metal-insulator transition temperature in TSeF-TCNQ in comparison to TTF-TCNQ to their different band structures. In particular the difference will be shown to relate to the magnitude of the overlap between the donor and acceptor wave functions, which will be referred to in this talk as the $\vec{a}$ axis hybridization.

C. $(TSeF)_x(TTF)_{1-x}(TCNQ)$ $0<x<1$. In this part of my talk I will compare and contrast effects of donor stack doping in TTF-TCNQ and in TSeF-TCNQ. While in TTF-TCNQ effects of donor stack doping (by TSeF) cause lowering of the metal-insulator transition temperature, in TSeF-TCNQ effects of donor stack doping (by TTF) cause an increase of the metal-insulator transition temperature. Moreover the relative lowering caused by doping in TTF-TCNQ is much smaller than the relative increase caused by doping in TSeF-TCNQ. I will relate these effects to the observation that TTF doping into TSeF-TCNQ interferes with the hybridization between the donor and acceptor stacks.

I. <u>TTF-TCNQ</u>

In order to study experimentally the role of each kind of stack in the 2 structural[4] phase transitions of TTF-TCNQ one can use either one of the following two techniques.

1. Measuring two combined properties (i.e. properties of both stacks)

of pure material, so chosen that they allow a decomposition of each into the respective contributions of the different stacks. Such a decomposition can be achieved with regard to the spin susceptibility.

2. Controlled modification of either the donor or the acceptor stacl accomplished by selective doping. In this section I will present both techniques and compare their conclusions.

A. Decomposition of the Magnetic Susceptibility

The Lorentzian shape of the single EPR absorption measured for TTF-TCNQ strongly suggests[5] that the magnetic excitations on both stacks are coherently mixed and therefore can be treated within the strong-coupling approximation. The g value measured at a given angle θ is then related to the g values of the TTF $[g_F(\theta)]$ and TCNQ $[g_Q(\theta)]$ stacks by

$$g(T,\theta)=\alpha(T)g_Q(\theta)+[1-\alpha(T)]g_F(\theta), \quad (1)$$

where $\alpha(T)$ is the fraction of the susceptibility on the TCNQ stack,

$$\alpha(T)=\chi_Q(T)/[\chi_Q(T)+\chi_F(T)]. \quad (2)$$

The values of $g_F(\theta)$ are known from g anisotropy measurements in the 25-35°K range, where the magnetic excitations are solely on the TTF chain. For $g_Q(\theta)$, we have chosen experimental values obtained from measurements on NMP-TCNQ for similar orientations of the TCNQ molecule. Evaluation of $\alpha(T)$, using the experimentally measured g values, and combining it with the measure‹ spin susceptibility data yield $\chi_F(T)$ and $\chi_Q(T)$ as shown in Fig. 2. For $T<53°K$, two distinct temperature regimes emerge:

<u>$38<T<53°K$:</u> χ_Q decreases strongly with decreasing temperature, while

χ_F is almost constant.

<u>$T<38^{o}K$</u>: χ_Q is negligibly small, while χ_F decreases strongly with decreasing temperature.

B. <u>Selective Doping of Donor and Acceptor Stacks</u>

An understanding of the phase transitions of TTF-TCNQ can be also obtained if one studies the effects of donor and acceptor doping on the position and sharpness of the transitions, and in particular, if one can compare the effects of the two kinds of doping. However, in order to make this comparison meaningful, one should have a measure for the relative strength of the perturbation of each impurity on its own kind of stack. As a rough experimental indication of the strength of the perturbation, we use the relative change in the value of the magnetic activation energy of the appropriate stack that results from the introduction of a fixed concentration of impurities on that stack. The physical reasoning behind this criterion is the following: the impurities modify the correlation length of the charge density waves, hence smear out the states near and in the gap and thus modify the measured effective activation energy.

Donor stack doping was accomplished by TSeF while acceptor stack doping was performed by introducing MTCNQ into the TCNQ stacks. We found[6] experimentally that the relative change in the magnetic activation energies as a consequence of doping is roughly the same. Therefore one can conclude that the perturbation produced by MTCNQ on the TCNQ stacks is similar in strength to that produced by TSeF on the TTF stacks. This conclusion makes meaningful the direct comparison of the effects of 3% donor- and acceptor-dopings on the two phase transitions of TTF-TCNQ.

The effects of doping on the position and sharpness of the phase transitions are shown in Fig. 3 presenting the derivative of tbe normalized conductivity of the pure, donor-doped and acceptor-doped material. The data indicate the existence of three phase transitions for TTF-TCNQ at 53°K, 46°K and 38°K. Acceptor- and donor-doping have similar effects on both the 46°K and 38K transitions, viz. they complete smear out the transitions. However, the metal-insulator transition at 53°K is affected differently by selective doping. While donor-doping smears it without shifting the temperature of the transition, acceptor-doping smears it markedly and in addition shifts the transition to about 47°K.

Let us now consider the picture that emerges from the decomposition and doping experiments about the roles of donor and acceptor stacks in the two phase transitions and in the intermediate and low temperature phases of TTF-TCNQ.

<u>The 53°K transition is on the TCNQ stacks, not on the TTF stacks</u>. This idea rests primarily on the observation of a large magnetic activation energy on the TCNQ stacks and a small or negligible activation energy on the TTF stacks below 53°K. (Result of the decomposition technique). It is also based on the fact that acceptor stack doping has a much greater effect on the width of this transition as well as on its position than donor stack doping.

What may seem surprising is that a TSeF impurity, while presumably inducing only a short-range charge density wave on its TTF stack is able to have as large an effect as it does on the TCNQ stacks. That such an effect should occur is consistent with the notion, which will be pursued later, that the wave-functions of the larger TSeF molecules hybridize much more effectively with the corresponding wave-functions of the TCNQ stacks.

The long-range order that develops in the 38°K transition is on the TTF stacks, and hence is driven by this development. This follows first the inference above that it is the TTF stacks that fail to have long-range order above 38°K. It was indicated by the TTF susceptibility that showed an appreciable magnetic activation energy developing on the donor stacks during this transition. The fact that the long-range order that develops is at least on the TTF stacks is confirmed by the decrease in the TTF magnetic activation energy by doping of the donor stack.

Perturbations in the TCNQ stacks have big effects on the TTF stacks, comparable with the effects of perturbations directly in the TTF stacks. This is in marked contrast to the effect of TTF stack perturbations on the TCNQ stacks. This finding can be explainable in terms of the following model: It is assumed that both kinds of stacks have a tendency to a Peierls distortion, that of the TCNQ stacks being the larger. When the TCNQ quasi-1D Peierls transition takes place at 53°K, the coupling to the TTF Peierls fluctuations is reduced essentially to zero by the symmetry of the TCNQ superlattice. As the temperature is lowered, this superlattice dimension begins to increase slowly, so that a single phase for the TTF Peierls fluctuations is preferred throughout, and some average order begins to develop throughout. Ultimately a quasi-1D Peierls transition on the TTF sublattice might be expected to take place, broadened by the presence of the ordering field of the TCNQ superlattice. The 38°-K transition can therefore be viewed as being driven by the TTF fluctuations, but according to this model it involves the TCNQ stacks in an average way.

Within this model, an impurity on one stack can affect the other stack both directly and also indirectly by inducing order (especially by pinning the phase of the order parameter) on its own stack over some distance, which then acts on the neighboring stacks of the other kind. For donor stack impurities near $53^{o}K$, this effect is small, because donor correlation lengths are presumably small, and because the amplitude of the order parameter on the donor stacks is never as large as on the acceptor stacks. For acceptor stack impurities, where correlation lengths and charge density waves are both much larger, this indirect perturbation of the donor stacks can be much larger.

II. TSeF-TCNQ

Since the TCNQ stack was shown in the previous discussion to drive the metal-insulator transition of TTF-TCNQ it is quite puzzling that the corresponding temperature in TSeF-TCNQ, having the same kind of stack, is lowered. Moreover the relationship between the existence of several phase transitions and the presence of 2 kinds of stacks in TTF-TCNQ makes the existence of a single phase transition in TSeF-TCNQ quite surprising. In the following I will show that the difference between the 2 compounds is caused by a different hybridization between the donor and acceptor electronic wave functions.

For comparison of the hybridization in TTF-TCNQ and TSeF-TCNQ one should use an experimentally measured quantity sensitive to this parameter like[7] the electron spin relaxation rates. In order to understand how the EPR linewidth has any bearing on the band structure, let us understand what relaxation processes contribute to the measured linewidth. The dominant process in three-dimensional metals is the spin-lattice relaxation produced by the scattering[8] of conduction electrons by acoustic phonons. However, the analysis of the linewidth and resistivity for an isotropic 3D metal must be modified[7] for a quasi-one-dimensional metal. In 3D metals, the dominant electron-phonon scattering processes contributing to both linewidth and resistivity scatter electrons within a shell of width 2kT straddling the Fermi surface, with or without spin-flip. The relative magnitudes of these processes depend only on the spin-orbit interaction strength through $(\Delta g)^2$, where Δg is the appropriate g-shift.

In electronically 1D systems (with flat energy surfaces) the scatterings from k to k ' are of two kinds, both of which are sharply reduced from 3D, much more in their contribution to the linewidth than to the resistivity:

(1) $\vec{k}\,' \approx \vec{k}$. The reduction in resistivity is less than in linewidth, since forward scattering contributes little to the resistivity in 3D, too.

(2) $\vec{k}\,' \approx -\vec{k}$. The spin-flip scattering matrix element is approximately zero by time-reversal invariance, further reducing the linewidth but not the resistivity

Deviations from this very small linewidth can arise from a relatively thick Fermi shell (high temperature or narrow bands) and/or from appreciable departures from one-dimensionality (interstack hybridization) associated with wave-function overlapping in directions perpendicular to the stacking axis.

Therefore whenever the spin-phonon mechanism is the dominant relaxation process, the magnitude of the linewidth, with proper normalization, can be used as a qualitative measure of the electronic dimensionality of the system. However, because the spin-phonon interaction might be negligible, one should check to see if indeed this is the spin-relaxation mechanism in the quasi-1D systems before drawing conclusions about the dimensionality of the system from the value of the linewidth. This can be done for TTF-TCNQ and TSeF-TCNQ in the following way: Since the free cations of these salts have large, yet significantly different, spin-orbit couplings, their solid solutions $(TSeF)_x(TTF)_{1-x}(TCNQ)$, $0 \leq x \leq 1$, provide a system in which the effective spin-orbit coupling of the donor stacks can be varied continuously while the structure is maintained. If the spin-phonon relaxation mechanism is the dominant one, the measured linewidth normalized to the square of the spin-orbit coupling should be independent of the composition x. Fig. 4 shows the linewidth of the single EPR absorption line as a function of the relative fraction x of TSeF on the donor stack. At 300°K the linewidth (in gauss) varies from 5 to 500 over the range from x=0 to x=1. Since the spin-orbit coupling on the acceptor stack is negligible in comparison to the spin-orbit coupling on the donor stack, the measured linewidth is expected to be proportional to the square of the spin-orbit coupling on the donor stack (if indeed the spin-phonon mechanism is dominant), which is measured by the deviation of the donor g-value from the free-electron value.

$$T_2^{-1} = \gamma_F (\Delta g_F)^2 \qquad (3)$$

The Δg_F-values were taken from low temperature (T<20K) g-values of the corresponding compounds, which correspond[9] to the donor stack g-values.

Fig. 5 shows γ_F as a function of the fraction of TSeF on the donor stack. In the regime $0 \leq x \leq 0.68$, the normalization to $(\Delta g_F)^2$ has helped in reducing by at least a factor of six the x-dependence of the linewidth for this isostructural set of compounds. This fact indicates that the spin-phonon relaxation phenomenon is important. The increase of γ_F with x at room temperature by a factor of sixteen in going from TTF-TCNQ to TSeF-TCNQ can therefore be interpreted in terms of spin-phonon relaxation as an increase of effective dimensionality or a thicker (normalized to the bandwidth) Fermi shell. The latter possibility does not apply to this particular case since it can be shown[9] that the bandwidth of TSeF in TSeF-TCNQ is larger than the bandwidth of TTF in TTF-TCNQ.

As to the direction along which the increased dimensionality occurs: Examination of the lattice parameters[10] of TSeF-TCNQ indicates that they are bigger than in TTF-TCNQ. However, the growth of the a-axis lattice parameter is over-compensated by the respective increase of the van der Waals radius of the selenium in comparison to the sulfur. Therefore the overlap in this direction between the wave functions of the TCNQ and TSeF molecules is larger than the overlap between TCNQ and TTF.

In conclusion we have shown that the hybridization gap is indeed larger in TSeF-TCNQ than in TTF-TCNQ. However, in order to affect the tendency of the system to undergo a Peierls transition, the hybridization gap should be at least of the order of kT in magnitude in the relevant temperature range. Whether this is indeed the case or not can be checked by comparing the densities of states at the respective Fermi energies of the TCNQ stack. Combining the total susceptibilities values for TTF-TCNQ[5] and TSeF-TCNQ[11] and the measured g-values yield the following conclusion: The TCNQ stack susceptibility at 300°K decreases from 3.2×10^{-4} emu/mole in TTF-TCNQ to 2.9×10^{-4} emu/mole in TSeF-TCNQ implying a decrease in the density of states. But at the same time, the TCNQ bandwidth is also decreasing because of an increased spacing of TCNQ molecules in TSeF-TCNQ, and this should increase the density of states. Thus we have experimental evidence that there must be a mechanism that more than offsets the narrowing of the band to produce a net decrease in the acceptor density of states. This mechanism is interstack hybridization.

As to the question why interstack hybridization is affecting a tendency of a system to undergo a Peierls distortion: In both investigated compounds the donor and acceptor bands are inverted[12]. Since the compounds are stoichiometric the bands cross at the Fermi energy. Most of the degeneracy is removed by the $\vec{a}$ axis hybridization. The part of the Fermi surface that can participate in the Peierls transition is therefore greatly reduced, thus decreasing the Peierls stabilization energy gained under a Peierls distortio of a given set of magnitudes and relative phases. The direct consequence is the reduction of the transition temperature.

Another interesting effect occurs in the temperature regime above the temperature of the 3D ordering. Since the interactions are quasi 1 dimensional a development of a long range order along the stack is a necessary precursor of a Peierls transition. If one kind of stack is more susceptible to a Peierls transition than the other kind a pseudo gap develops in its density of states, diminishing the hybridization, which involves both kinds of stacks. Thus the inhibiting effect that the hybridization has on the tendency of the second kind of stacks to undergo a Peierls transition is reduced. This effect is moving the transition temperatures of the 2 kinds of stacks together.

III. $(TSeF)_x(TTF)_{1-x}(TCNQ)$ $0<x<1$

Our understanding of the difference between TTF-TCNQ and TSeF-TCNQ will enable us to explain the effects of donor stack doping in these two compounds. Donor stack doping by TTF in TSeF-TCNQ interferes with the hybridization between the donor and acceptor stacks. Therefore the 3 dimensional ordering temperature of the undoped acceptor stack should increase as the hybridization is diminished. The donor stack ordering temperature would be determined on the other hand by competition between the doping effect which will tend to lower it and the diminished hybridization which will tend to increase it. Conductivity profile measurements in $(TSeF)_{0.99}(TTF)_{0.01}(TCNQ)$ indeed reveal[13] existence of 2 phase transitions. They were identified by g-shift measurements to correspond to donor and acceptor stacks ordering temperatures - the acceptor transition occurring at the higher temperature corresponding to the metal-insulator transition. The dopant concentration dependence of the hybridization gives rise therefore to the dependence of the metal-insulator transition temperature on doping. In TTF-TCNQ on the other hand the effects of hybridization are negligible and therefore donor stack doping affects the acceptor stack ordering temperature in the expected way-lowering it.

In conclusion we have shown that the series of alloys $(TSeF)_x(TTF)_{1-x}$ (TCNQ) can serve as a model system demonstrating the crucial role of the interstack hybridization between donor and acceptor stacks in determining the temperature of the metal-insulator transition. However not necessarily every replacement of the sulfur atoms on the donor stack by the bigger selenium ones guarantees reduction of the metal-insulator transition temperature. The similar[14] conductivity profiles of TMTTF-TCNQ and TMTSeF-TCNQ can serve as an example for the previous statement. But if the major change in the structure, achieved by the replacement, is in the inter donor-acceptor stack overlap a rather dramatic effect can be expected, as was shown in the presented work for TTF-TCNQ and TSeF-TCNQ.

ACKNOWLEDGMENTS

I am grateful to my colleagues at IBM who participated in the work summarized here and which is being published and will be published in more detailed articles: T. D. Schultz, R. A. Craven, E. M. Engler, J. B. Torrance and A. R. Taranko. I acknowledge S. Etemad for using his results prior to publication.

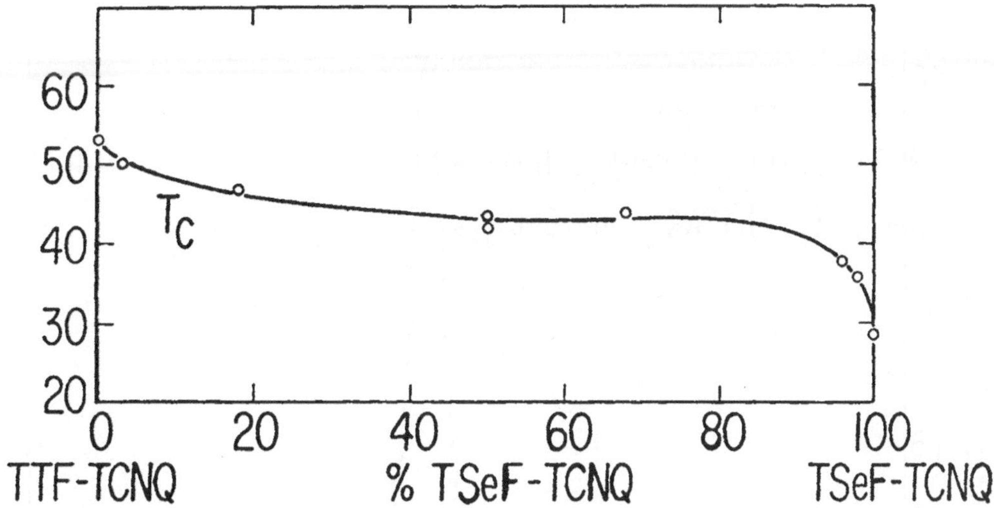

Figure 1 The metal-insulator transition temperature, T_c, as a function of the fraction of TSeF in the donor stack. This figure is taken, with the permission of the authors, from Ref. 10.

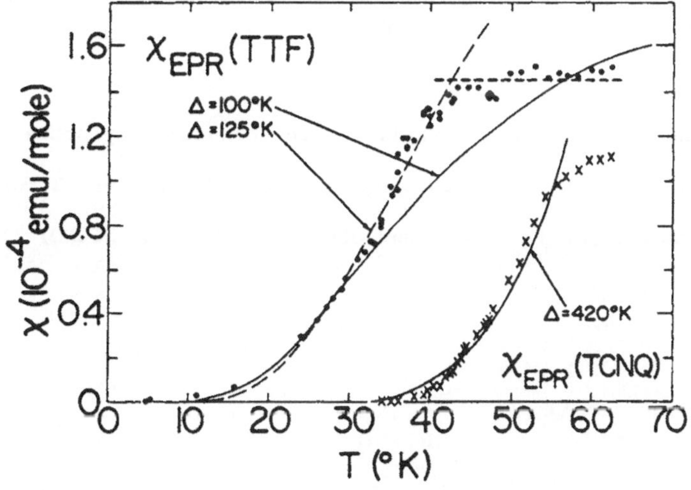

Figure 2 The respective susceptibilities χ_F and χ_Q of the (TTF) and (TCNQ) stacks. Several fits of χ_F and χ_Q to the expression $\chi = \frac{C}{T} e^{-\Delta/T}$, where Δ is the activation energy, are presented.

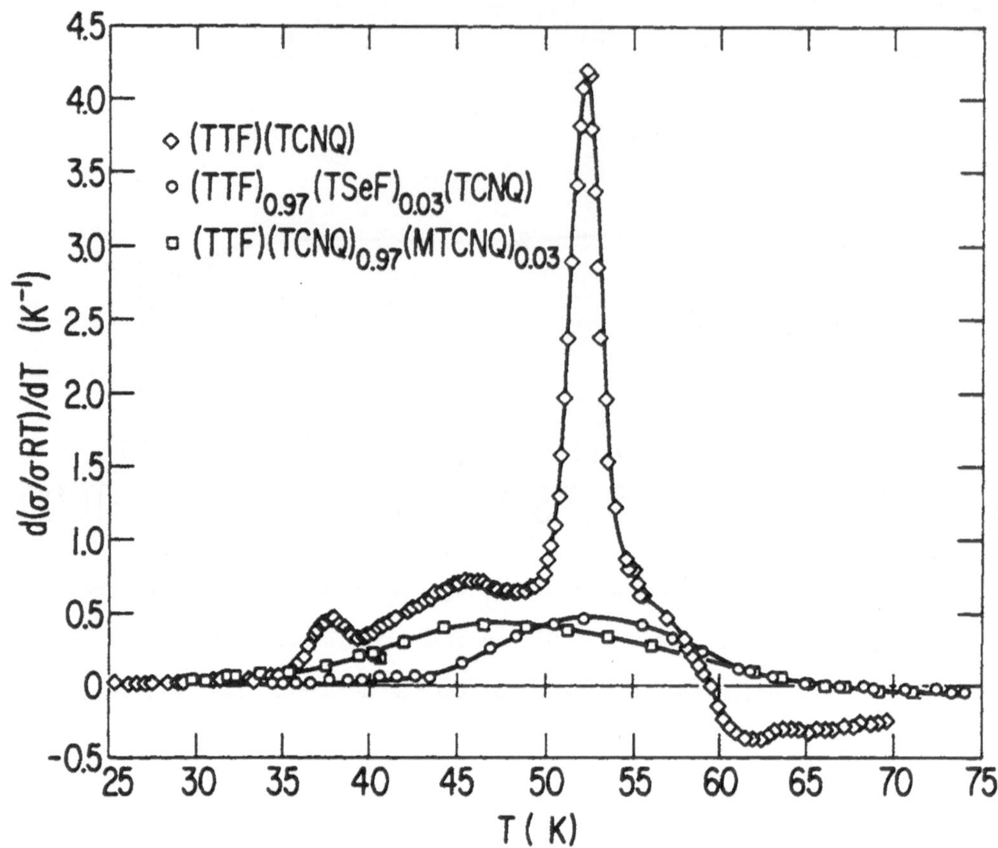

Figure 3 The temperature derivative of the conductivity normalized to the room temperature value of $d(\sigma/\sigma_{RT})/dT$ vs. T for TTF-TCNQ, donor doped (3% TSeF) and acceptor doped (3% MTCNQ) TTF-TCNQ.

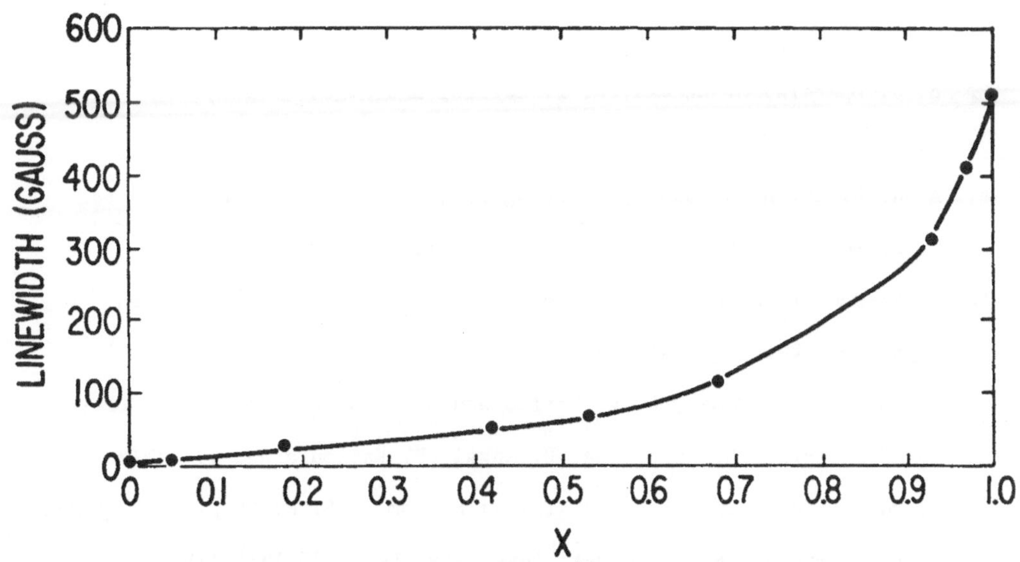

Figure 4 Linewidth dependence, for $H_{dc}||c^*$ direction, on x - the fraction of TSeF in the donor chain. The measurements were taken at 300°K.

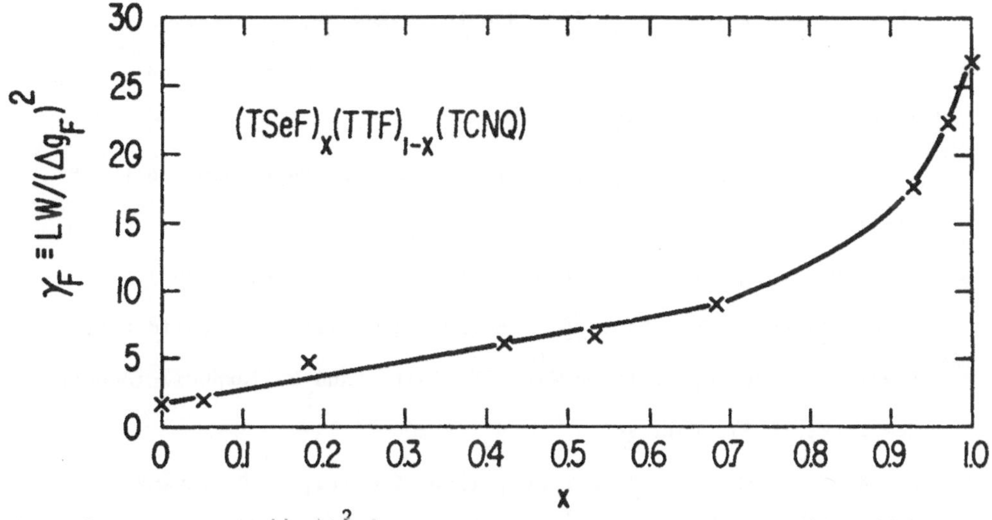

Figure 5 $\gamma_F = LW/(\Delta g_F)^2$ [where LW is the room temperature linewidth, $(\Delta g_F)^2 = (g_F - g_e)^2$, g_F is the donor-stack g value, and g_e is the free-electron g value] as a function of x- the fraction of TSeF in the donor stack. g_F values were taken from low-temperature (T <20°K) measurements.

REFERENCES

1. A. J. Epstein, S. Etemad, A. F. Garito, and A. J. Heeger, Phys. Rev. B5, 952, (1972).
2. A. N. Bloch, D. O. Cowan, K. Bechgaard, R. E. Pyle, R. H. Banks and T. O. Poehler Phys. Rev. Lett. 34 1561 (1975).
3. S. Etemad, T. Penney, E. M. Engler, B. A. Scott and P. E. Seiden, Phys. Rev. Lett. 34 741 (1975).
4. F. Denoyer, R. Comes, A. F. Garito and A. J. Heeger Phys. Rev. Lett. 35, 445 (1975); S. Kagoshima, H. Anzai, K. Kajimura and T. Ishiguro, J. Phys. Soc., Japan 39 1143 (1975); R. Comes, S. M. Shapiro, G. Shirane, A. F. Garito and A. J. Heeger, Phys. Rev. Lett. 35 1518 (1975).
5. Y. Tomkiewicz, B. A. Scott, L. J. Tao and R. S. Title, Phys. Rev. Lett. 32 1363 (1974); Y. Tomkiewicz, A. R. Taranko and J. B. Torrance, Phys. Rev. Lett. 36 751 (1976) and Phys. Rev. B, in press.
6. Y. Tomkiewicz, R. A. Craven, T. D. Schultz, E. M. Engler and A. R. Taranko to be published.
7. Y. Tomkiewicz, E. M. Engler and T. D. Schultz, Phys. Rev. Lett. 35 456 (1975).
8. See for example, A. W. Overhauser, Phys. Rev. 89, 689 (1953); R. J. Elliott, Phys. Rev. 96, 266 (1954); Y. Yafet, in Solid State Physics, edited by H. Ehrenreich, F. Seitz, and D. Turnbull (Academic New York, 1963), Vol. 14.
9. For discussion see Y. Tomkiewicz, T. D. Schultz, A. R. Taranko and E. M. Engler to be published.

10. S. Etemad, E. M. Engler, T. Penney, B. A. Scott and T. D. Schultz to be published.

11. J. C. Scott, S. Etemad and E. M. Engler to be published.

12. M. H. Cohen, J. A. Hertz, P. M. Horn and V. K. S. Shante, Bull. Am. Phys. Soc. 19 297 (1974); A. N. Bloch in Energy and Charge Transfer in Organic Semiconductors edited by K. Masuda and M. Silver (Plenum, New York, 1974), p. 159; U. Bernstein, P. M. Chaikin and P. Pincus, Phys. Rev. Lett. 34 271 (1975), A. J. Berlinsky, J. F. Carolan and L. Weiler, Sol. St. Com. 15, 795 (1974).

13. R. A. Craven, Y. Tomkiewicz and E. M. Engler to be published.

14. T. Carruthers, A. N. Bloch and D. O. Cowan, Bull. Am. Phys. Sol. 21, 313 (1976).

Conference on Organic Conductors and Semiconductors, Siófok, Hungary 1976

ORGANIC CONDUCTORS WITH LOW-SYMMETRIC ACCEPTORS

JAN R. ANDERSEN

Chemistry Department, Risø, DK-4000, Denmark

KLAUS BECHGAARD and CARSTEN BERG

H.C. Ørsted Institute, DK-2100, Denmark

CLAUS S. JACOBSEN

Physics Laboratory III, DTH, DK-2800, Denmark

Results are described for three new highly conducting organic solids, all based on substituted TCNQ-derivatives: TMTSF-2,5-dimethyl-TCNQ /TMTSF-DMTCNQ/ 1:1 shows "normal" conductivity vs.temperature behaviour, i.e. metallic conductivity at higher temperatures followed by a metal-semiconductor transition as the temperature is decreased below 5o K.

TTF-2,5-diethyl-TCNQ /TTF-DETCNQ/ 1:1 and TTF-methyl-TCNQ /TTF-MTCNQ/ 1:1 show hardly any temperature dependence in their conductivity in the high temperature range.

Results are discussed in the light of analogous systems and with emphasis on crystal structure.

Introduction

During recent years much work has been done on molecular design and synthesis of organic conductors in order to try to understand the behaviour of such compounds.

The most recent work has dealt with modifications of the donor in the prototype TTF-TCNQ /1/, including the notable successes TMTTF-TCNQ /2/, TSF-TCNQ /3/, TMTSF-TCNQ /4/ and HMTSF-TCNQ /5/.

The work presented here is the result of alterations in the acceptor molecule. The acceptors used are all of TCNQ-type, the difference from the unsubstituted compound being the introduction of one or two alkyl-groups.

2,5-dimethyl-TCNQ /DMTCNQ/, 2,5-diethyl-TCNQ /DETCNQ/ and methyl-TCNQ /MTCNQ/ were prepared according to the literature /6, 7/. The Table shows their molecular symmetries and solution reduction potentials as determined by cyclic voltametry together with the pertinent data for TCNQ.

TABLE 1

Compound	TCNQ	MTCNQ	DMTCNQ	DETCNQ
Molecular symmetry	D_{2h}	C_s	C_{2h}	C_{2h}
$E^1_{1/2}$	0.190	0.170	0.110	0.120
$E^2_{1/2}$	-0.350	-0.340	-0.350	-0.365

Electro-chemical data: volts vs SCE at Pt button electrode in

$CH_3CN/\underline{n}\text{-}Bu_4NBF_4$ /0.1M/

Discussion

The organic solid TMTSF-DMTCNQ /6/ /Fig. 1/ has a room-temperature doncudtivity of about 5oo $(\Omega\,cm)^{-1}$, and the temperature dependence of the conductivity shows "normal" behaviour i.e. the conductivity increases /by a factor of approximately lo/ tc a sharp maximum - in this case at 50 K. Below this temperature the material undergoes the characteristic metal-semiconductor transition.

TMTSF-DMTCNQ deserves comparison with the conducting form of the paren t system TMTSF-TCNQ /4/. The temperature dependences of the conductivities of both systems are very alike, but in TMTSF-TCNQ the maximum is situated at 71 K.

The crystal structures of the two systems are also very alike. Both are triclinic with comparable unit-cell parameters and with the "herring-bone" structure known from TTF-TCNQ /8/. The most pronounced crystallographic difference between the two systems is a rather elongated /6.1 %/ b-axis in TMTSF-DMTCNQ. This is to be expected because the b-direction is across the width of the molecules where the additional methyl-groups tend to push the chains apart thereb.y probably reducing the interchain coupling.

One of the features of reduced interchain coupling is the suppression of three-dimensional ordering and this may be responsible for the lower transition temperature seen in this material.

TTF-MTCNQ /9/ /Fig.2/ and TTF-DETCNQ /10//Fig. 3/ both have room-temperature conductivities of the same order of magnitude as TMTSF-DMTCNQ, but the increase in conductivity with decreasing temperature is merely 20 and 50 %, respectively, in the broad maxima at 210 and 180 K. In TTF-DETCNQ a sharp decline in conductivity occurs at 110 K, while TTF-MTCNQ gradually becomes insulating. As crystal data are only available at present for TTF-DETCNQ, the discussion will be restricted to this material.

TTF-DETCNQ crystallizes in a triclinic lattice with each acceptor molecule surrounded by four donor molecules and vice versa /lo, 11/. In this respect, TTF-DETCNQ resembles TTF-TNAP /12/ and HMTSF-TCNQ /13/, which also have the "four-nearest-neighbours" structure and not the usual "herring-bone" with only two nearest neighbours.

In addition, the normalized conductivity vs temperature behaviour of the three compounds seems comparable. No "peaked" curve is seen, at most there are very broad maxima.

Conclusion

It may be concluded that there is a strong correlation between conductivity vs temperature behaviour and structure for conducting solids of this kind.

The results strongly indicate that a "peaked" conductivity vs temperature curve is related to the "herring-bone" structure whereas the "four-nearest-neighbours" structure is characteristic for a solid with a less pronounced variation of the conductivity with temperature.

Whether TTF-MTCNQ follows these "rules-of-thumb" or not is at present very unclear because the material carries one additional feature, the permanent dipole introduced by MTCNQ. This seems likely to cause large disturbances in the electronic structures of materials of the kind mentioned above.

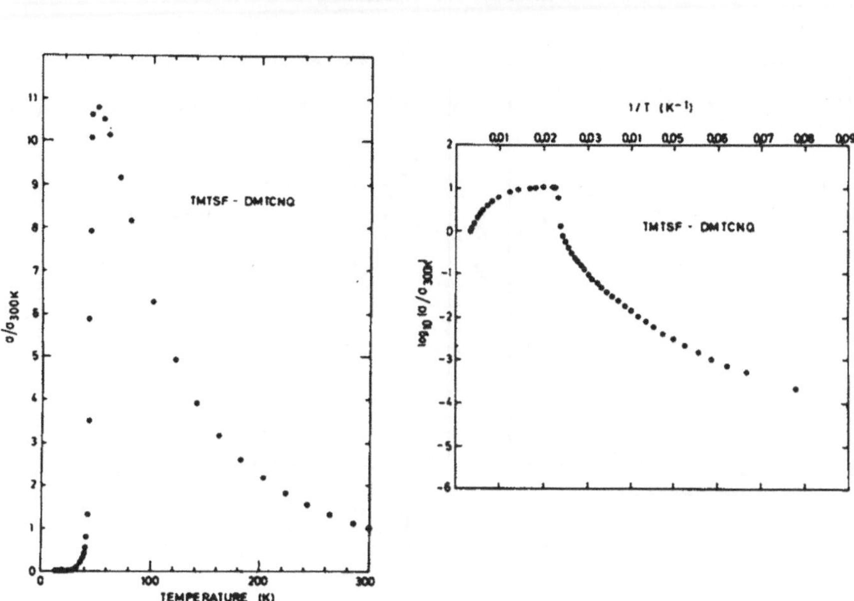

Fig. 1 Molecular design, temperature dependence of normalized conductivity and activation plot of the conductivity of TMTSF-DMTCNQ

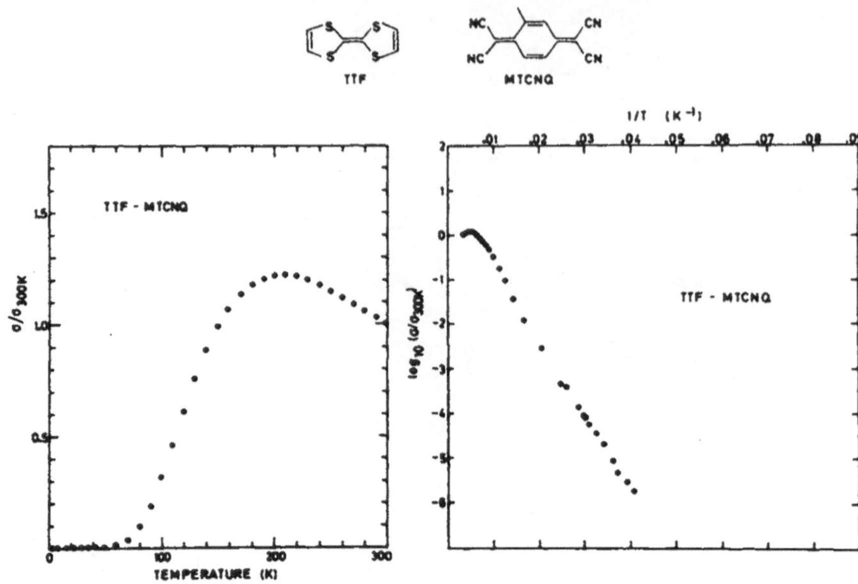

Fig. 2 Molecular design, temperature dependence of no lized conductivity and activation plot of the conductivity of TTF-MTCNQ

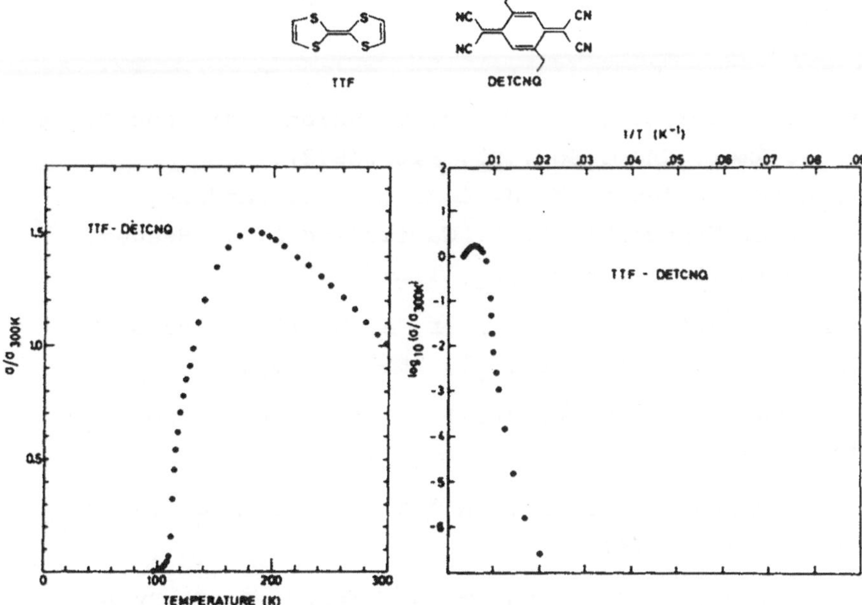

Fig. 3 Molecular design, temperature dependence of normalized conductivity and activation plot of the conductivity of TTF-DETCNQ

REFERENCES

/1/ /a/ J. Ferraris, D. O. Cowan, V. Walatka jr. and J.Pearlstein, J. Amer. Chem. Soc. 95, 948 /1973/;
/b/ L. B. Coleman, M. J. Cohen, D. J. Sandman, F. G. Yamagishi, A. F. Garito and A. J. Heeger, Solid State Commun. 12, 1125 /1973/.

/2/ J. P. Ferraris, T. O. Poehler, A. N. Bloch and D. O. Cowan, Tetrahedron Lett. 27, 2553 /1973/.

/3/ E. M. Engler and V. V. Patel, J. Amer. Chem. Soc. 96, 7376 /1974/.

/4/ K. Bechgaard, D. O. Cowan and A. N. Bloch, J.C.S. Chem. Commun. 937 /1974/.

/5/ A. N. Bloch, D. O. Cowan, K. Bechgaard, R. E. Pyle, R. H. Banks and T. O. Poehler, Phys. Rev. Letters 34, 1561 /1975/.

/6/ J. R. Andersen, C. S. Jacobsen, G. Rindorf, H. Soling and K. Bechgaard, J. C. S. Chem. Commun. 883 /1975/.

/7/ J. Diekmann, W. R. Hertler and R. E. Benson, J. Org. Chem. 28, 2719 /1963/.

/8/ T. J. Kistenmacher, T. E. Phillips and D. O. Cowan, Acta Cryst. B3o, 763 /1974/.

/9/ C. S. Jacobsen, J. R. Andersen, K. Bechgaard and C. Berg, Solid State Commun., /in press/.

/1o/ J. R. Andersen, C. S. Jacobsen, H. Soling and K. Bechgaard /unpublished results/.

/11/ A. J. Schultz, G. D. Stuky, R. Craven, M. J. Schaffman and M. B. Salamon, J. Amer. Chem. Soc. /in press/.

/12/ P. A. Berger, D. J. Dahm, G. R. Johnson, M. G. Miles and J. D. Wilson, Phys. Rev. B 12, 4085 /1975/.

/13/ T. E. Phillips, T. J. Kistenmacher, A. N. Bloch and D. O. Cowan, J. C. S. Chem. Commun. 334 /1976/.

Conference on Organic Conductors and Semiconductors, Siófok, Hungary 1976

NEUTRON AND X-RAY DIFFUSE SCATTERING STUDY OF TETRATHIOFULVALENE TETRACYANOQUINODIMETHANE (TTF · TCNQ)

R. COMÈS[a] and G. SHIRANE[b]

(a) - Laboratoire de Physique des Solides* Université Paris-Sud, 91405 ORSAY (France)

(b) - Brookhaven National Laboratory UPTON, New York 11973 (USA)**

Series of extensive X-Ray and Neutron scattering experiments[1-11] have been carried out in the last year on TTF-TCNQ and substantial agreement has now emerged. We can summarize the main results as follows :

A/ - Below the Peierls transition taking place at 54 K, two additionnal structural phase transitions are observed at 49 K and 38 K (figure 1). These 3 transitions correspond to a sequence of 3 different low temperature modulated phases which are thought to arise from the successive ordering of the charge density waves on the two different molecular species of TTF-TCNQ[12-17] leading to the successive modulations

$$2\,a \times 3{,}40\,b \times c \qquad (49 < T < 54\ K)$$
$$x(T) \times 3{,}40\,b \times c \qquad (38 < T < 49\ K)$$
$$4\,a \times 3{,}40\,b \times c \qquad (T < 38\ K)$$

B/ - Above the Peierls transition two different precursor 1-d fluctuations have been observed, as shown in the X-Ray pattern of figure 2.

*Laboratoire associé au C.N.R.S.

**Work performed under the auspices of the U.S. Energy Research and Development Administration.

1 - At the wave vector 0.295 b^* (attributed to $2k_F$) strongly temperature dependent X-Ray scattering was observed below 150 K. From the distribution of the intensity in the reciprocal space ($2k_F$ planes), the polarization of this modulation is found to have two components : a longitudinal component in chain direction and a transverse component in c^* direction which corresponds to the direction of tilt of the molecules. Inelastic neutron measurements reveal, in the same temperature range, an anomaly in the transverse mode with main polarization along c^* (figure 3). It is this fluctuation which diverges at the Peierls transition at 54 K.

The occurence of both polarizations (b^* and c^*) can be understood in terms of charge density waves as both of these components modify the intermolecular spacing in chain direction because of the tilt angle of the molecules in the stacks.

2 - Additional 1-d X-Ray scattering is observed at the wave vector 0.59 b^* (or 0.41 b^* in the reduced zone) which corresponds to twice the former 0.295 b^* wave vector or 4 k_F. This scattering has a very different temperature dependence : it persists to much higher tem-erature as shown in figure 4 and probably condenses below 49 K.

Several theoretical models have already been proposed in order to explain the simultaneous occurence of both $2k_F$ and $4k_F$ charge density wave anomalies and their different temperature dependence[18-21]. These models all assume important coulomb repulsion between electrons on at least one type of molecular stack.

C/ - The present study, combined with the independent X-Ray investigation of Kagoshima et al[9], rules out the presence of a giant Kohn anomaly at room temperature earlier reported by Mook and Watson[5]. As first suggested by Torrance, $2k_F$ spin density waves are now considered to be a possible origin of these extra cross sections at room temperature[11].

D/ - Future X-Ray and Neutron scattering studies should include :

1. the determination of the atomic motions involved in the charge density waves,
2. phonon studies at $4k_F$ wave vector,
3. further characterization of spin density waves.

The work reported here was done jointly with F. DENOYER, J.P. POUGET, S. KHANNA (Orsay, France), W.D. ELLENSON, V.J. EMERY, S.M. SHAPIRO (Brookhaven National Laboratory, USA) and A.F. GARITO, A.J. HEEGER (University of Pennsylvania, USA).

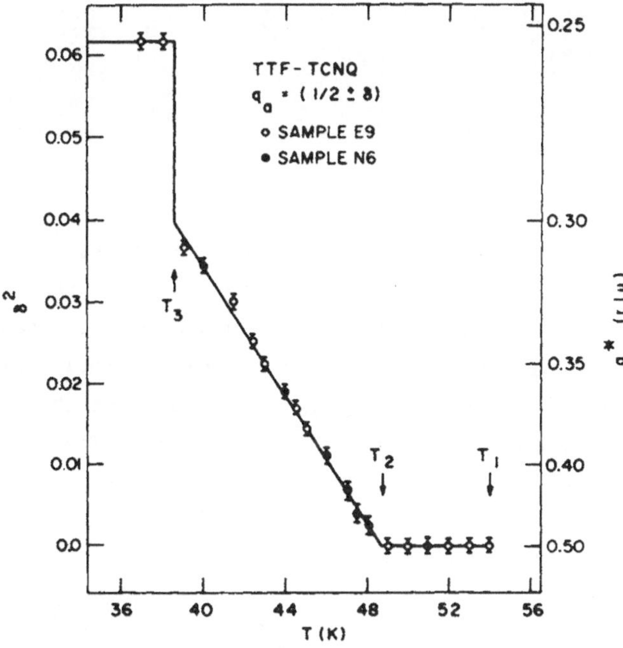

FIGURE 1 (from ELLENSON et al [7])

The occurence of 3 phase transitions in TTF-TCNQ is best visualized from the temperature dependence of the satellite Bragg peak position in reciprocal space when plotted as a function of $\delta^2 = (\frac{1}{2} - q_a)^2$, where q_a is the satellite wave vector component along a^*. This was first suggested by BAK and EMERY[12] who discovered the 49 K phase transition.

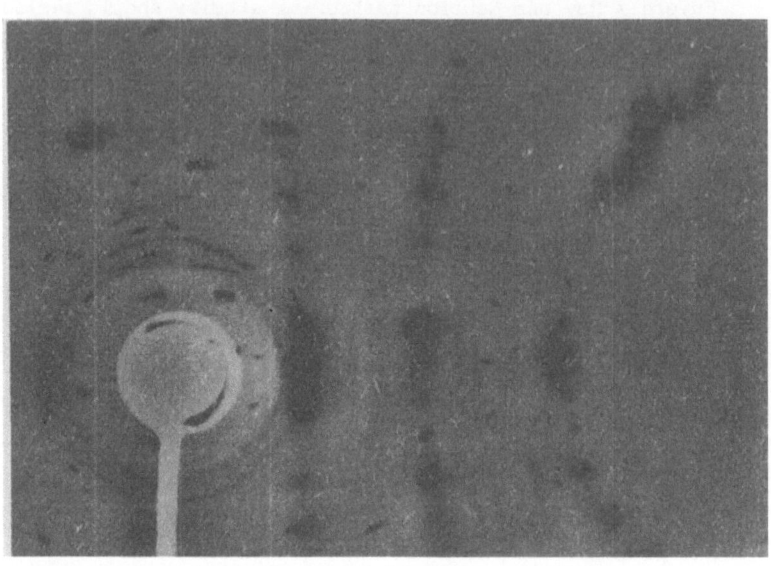

FIGURE 2 (from POUGET et al[8])

Diffuse X-Ray pattern of TTF-TCNQ at 60°K. Satellite reciprocal planes (1-d scattering) are clearly observed at the wave vectors 0.295 b* ($2k_F$) and 0.59 b* ($4k_F$). The comparable intensity of these two types of 1-d precursor at 60 K and their different temperature dependence (see figure 4) rules out that the $4k_F$ scattering might arise from a second order diffraction from $2k_F$.

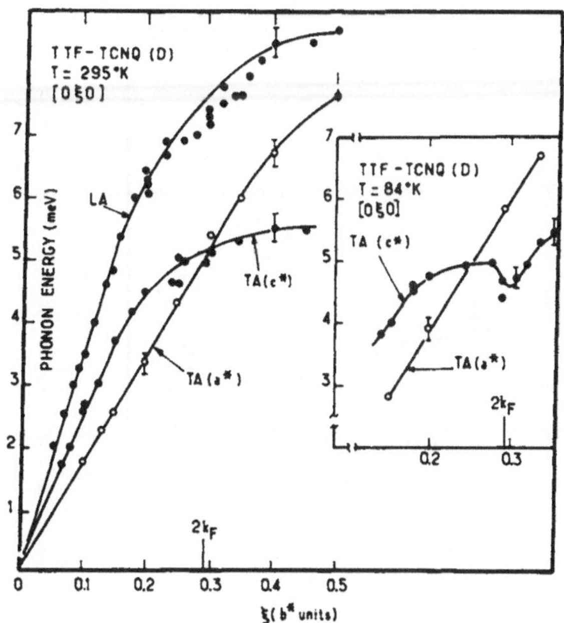

FIGURE 3 (from SHAPIRO et al)

Dispersion curves from TTF-TCNQ for modes propagating along [010], the chain direction, at 295°K.

The insert shows the transverse branches around $2k_F$ at 84°K. Note that the Kohn anomaly occurs in the branch main polarized along c^*.

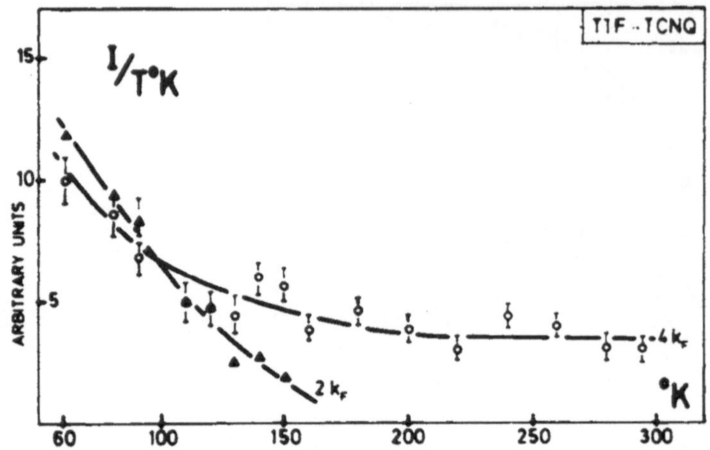

FIGURE 4 (from KHANNA et al)

Temperature dependence of the X-Ray diffuse intensity for the $2k_F$ and $4k_F$ scattering as estimated from microdensitometer readings of photographic patterns. This data is only semi-quantitative, but gives the general trend,. it shows in particular that the $2k_F$ scattering only developps below 150°K as observed by SHIRANE et al[6]. The plot corresponds to I/T in order to eliminate the temperature dependence of the phonon population factor which is proportional to kT for such low frequency phonons in this temperature range.

B I B L I O G R A P H Y

1. F. DENOYER, R. COMES, A.F. GARITO and A.J. HEEGER - Phys. Rev. Lett., 35, 445 (1975).

2. S. KAGOSHIMA, H. ANZAI, K. KAJIMURA and R. ISHIGORO - J. Phys. Soc. Jap., 39, 1143 (1975).

3. R. COMES, S.M. SHAPIRO, G. SHIRANE, A.F. GARITO and A.J. HEEGER - Phys. Rev. Lett., 35, 1518 (1975).

4. R. COMES, S.M. SHAPIRO, G. SHIRANE, A.F. GARITO and A.J. HEEGER - to be published : Phys. Rev. (1976).

5. H.A. MOOK and C.R. WATSON - Phys. Rev. Lett., 36, 801 (1976).

6. G. SHIRANE, S.M. SHAPIRO, R. COMES, A.F. GARITO and A.J. HEEGER - Phys. Rev. (1976). A preliminary report of these results was reviewed by G. SHIRANE at the Conference on Low Lying Vibrational Modes and their relationship to Superconductivity and Ferroelectricity, SAN JUAN, December 1-4, 1975 - to be published in Ferroelectrics.

7. W.D. ELLENSON, R. COMES, S.M. SHAPIRO, G. SHIRANE, A.F. GARITO and A.J. HEEGER - to be published in Sol. St. Comm., (1976).

8. J.P. POUGET, S.K. KHANNA, F. DENOYER, R. COMES, A.F. GARITO and A.J. HEEGER - in Phys. Rev. Lett., (1976). 37, 437

9. S. KAGOSHIMA, T. ISHIGURO and H. ANZAI - to be published.

10. S.K. KHANNA, J.P. POUGET, R. COMES, A.F. GARITO and A.J. HEEGER - to be published.

11. J.B. TORRANCE, H.A. MOOK and C.R. WATSON - to be published.

12. P. BAK and V.J. EMERY - Phys. Rev. Lett., 36, 978 (1976).

13. K. SAUB, S. BARISIC and J. FRIEDEL - Phys. Lett., 56 A, 302 (1976).

14. S. ETEMAD and T.D. SCHULTZ - Bull. Am. Phys. Soc., 21, 286 (1976).

15. P.M. CHAIKIN, R.L. GREENE, S. ETEMAD and E. ENGLER - Phys. Rev. B, 13, 1627 (1975).

16. Y. TOMKIEWICZ, A.R. TARANKO and J.B. TORRANCE - Phys. Rev. Lett., 36, 751 (1976).

17. E.F. RYBACZEWSKI, A.F. GARITO and A.J. HEEGER - Bull. Am. Phys. Soc., 21, 287 (1976) and to be published.

18. A.A. OVCHINNIKOV - Sov. Phys. JETP, 37, 176 (1973).

19. J. BERNASCONI, M.J. RICE, W.R. SCHNEIDER and S. STRÄSSLER - Phys. Rev., B 12, 1090 (1975).

20. V.J. EMERY - Phys. Rev. Lett., 37, 107 (1976).

21. H. SUMI - to be published.

22. P.A. LEE, T.M. RICE and R.A. KLEMM - to be published.

Conference on Organic Conductors and Semiconductors, Siófok, Hungary 1976

OPTICAL EVIDENCE FOR COULOMB INTERACTIONS IN TTF-TCNQ

JERRY B. TORRANCE

IBM Research Laboratory, Yorktown Heights, NY 10598, USA and IBM Research Laboratory, San Jose, California 95193 USA

A review is given of the optical properties of charge transfer salts of TCNQ, starting from the important examples of $(TCNQ^-)_2$ dimers in solution, K-TCNQ, and Cs_2TCNQ_3. In these cases, there is clearly a charge transfer band near 1.3 eV which gives a measure of the Coulomb interactions in these materials. The extension to more conducting TCNQ salts is not as clear, although according to one interpretation there is a similar charge transfer band in all TCNQ salts at the same energy. This model as well as another recently proposed will be described, along with evidence for and against each of them. The conclusion of both models is that the Coulomb interactions are strong in all TCNQ salts and that there is no evidence for excitonic screening.

It is becoming increasingly apparent[1] that Coulomb interactions play an important, if not dominant, role in the solid state properties of TTF-TCNQ. Recently evidence for these interactions has come from the enhancement[2] of the magnetic susceptibility, from the observation of spin waves[3] by inelastic neutron scattering, from the observation[4,5] and interpretation[1,3,6] of diffuse X-ray scattering at "$4k_F$" (or twice the expected wavevector), and finally from NMR relaxation rate measurements.[7] The first evidence and discussion of Coulor interactions in TTF-TCNQ was in the interpretation[8,9] of the near-infrared and visible absorption spectrum. In this paper, we present a review of the eviden for and against this interpretation, including more recent experiments and a second interpretation recently proposed.[10]

Optical spectroscopy is generally one of the most powerful probes of the basic electronic structure of solids. In the case of TTF-TCNQ, the complete evidence and arguments leading to the understanding of the optical properties between 0.2 and 4.0 eV are both detailed and lengthy.[9,10,11] Such an understanding must involve many comparisons with the spectra[8-23] of other TCNQ salts and non-TCNQ salts, as well as the spectra[24] of TCNQ anions in solution. In this paper there is insufficient space to discuss all of these comparisons and arguments or the basic theory of molecular excitons.[25] Thus, we shall giv only a rough summary of the optical properties, concentrating not on the detai of the far infrared,[26] but on the general nature of the major spectral feature of the entire 0.2-4.0 eV range. A completely analogous analysis has been give for the optical properties of the TTF^+ cation, with similar conclusions.

I. SPECTRUM OF K-TCNQ

The best place to start is with the comparison in Fig. 1 of the spectrum[24] of $(TCNQ^-)_2$ dimers in solution (solid line) with that[12,15,16,9] of a solid containing stacks of such dimers, as in K-TCNQ (dashed line). The remarkable similarity in these two spectra indicates that the three absorption lines of the solution dimer correspond to the three observed in the solid. (This similarly also strongly suggests that we should view the solid K-TCNQ (as well as other TCNQ salts) as composed of relatively weakly interacting TCNQ molecules.) The assignment of these peaks is readily obtained from polarized measurements on single crystals of K-TCNQ. These measurements[14,18] confirm the original assignment of Iida,[12] namely that the peaks labelled C and D in Fig. 1 are polarized within the molecular plane, indicating that these are intra-molecular excitations, i.e., molecular excitons. On the other hand, peak B is cleanly polarized along the stack, indicating that it is an inter-molecular excitation, i.e., a charge transfer band, associated with the transition:

$$(TCNQ^-)(TCNQ^-) \rightarrow (TCNQ^o)(TCNQ^{2-}) .$$

The energy of this excitation (peak B) is then $\sim(U-V_1)$, where U is the Coulomb repulsion energy between two electrons on the same molecule, i.e., $TCNQ^{2-}$ (right hand side of the above transition), and V_1 is the repulsion between electrons on neighboring $TCNQ^-$ molecules (left hand side). In a stack of such molecules, as in K-TCNQ, the electron excited in this transition is bound to the hole by an energy V_1 and thus the excitation should be viewed as a strongly bound exciton, extending out to nearest neighbors--an extended Frenkel exciton. It is important to note that the energy, ~1 eV, of this excitation indicates that the magnitude of the Coulomb interactions in K-TCNQ is significantly larger than estimates of ~0.5 eV for the bandwidth associated with delocalization down the stack. In this case, the Coulomb interactions give rise to an energy gap[28] in the spectrum, causing[8,9] K-TCNQ to be an insulator (a Mott insulator). (An important assumption made here is that the number, ρ, of electrons per TCNQ molecule is equal to 1. The situation for $\rho<1$ is considerably different, as we shall see.)

Thus, the interpretation[12,14,15,16,18,9,10] of the spectrum of K-TCNQ is quite clear and straightforward. Even the lower energy of peak B in K-TCNQ (compared to that of the solution dimer) is readily understood[11] in terms of molecular exciton theory.[25] At 300K, K-TCNQ is a solid with stacks of dimerized TCNQ anions with an almost eclipsed overlap. The spectrum of similar alkali-TCNQ salts with uniform stacks and those in which the molecular overlap is strongly slipped (as in TTF-TCNQ) all show[15] very similar spectra. This strong similarity indicates[11] that the spectrum of dimerized K-TCNQ in Fig. 1 is generally characteristic of any segregated stack of $TCNQ^-$ anions with $\rho=1$, i.e., one electron per TCNQ.

II. SPECTRUM OF Cs_2TCNQ_3

The major and unique difficulty of the optical properties of TCNQ salts is to understand the spectrum of a stack of mixed valence TCNQ molecules, i.e., one containing both $TCNQ^-$ anions and neutral $TCNQ^o$ molecules[29] (the case if $\rho<1$). One very special example of such a material is Cs_2TCNQ_3, for which the optical spectrum is not so difficult to interpret.[13,14] This material is specia because the X-ray structure[30] shows that the electrons ($\rho=2/3$ electron per molecule) in the stacks are localized on inequivalent TCNQ molecules, as schematically shown at the top of Fig. 2. Since these electrons are localized, we can view this stack as containing $(TCNQ^-)_2$ dimers separated by neutral $TCNQ^o$ molecules. Since the $TCNQ^o$ molecules absorb only above ~3 eV, they do not contribute absorption to peaks B and C. Thus, the spectrum of the stack should look like that of a dimer (without the decrease in energy of peak B observed in K-TCNQ). In fact, the peaks in the spectrum[19,20,9,10] of Cs_2TCNQ_3, shown in Fig. 2, have the same energies as peaks B, C, and D of the solution dimer (Fig. 1) and are given[13,14] the same interpretation: peaks C and D are intramolecular (molecular excitons), while peak B is a charge transfer band.

In addition, there is a new peak, labelled A in Fig. 2, which appears at lower energy (~0.5 eV). This transition is interpreted as the charge transfer excitation of an electron from $TCNQ^-$ to a neutral $TCNQ^o$ molecule, as shown schematically at the top of Fig. 2. We call this transition the mixed valence charge transfer band, since $TCNQ^o$ molecules will be present (presumably) only if the stack is mixed valence (i.e., only if $\rho<1$). This mixed valence charge transfer band (peak A) is clearly at a lower energy than the usual charge transfer band, in which an electron is excited from $TCNQ^-$ to another $TCNQ^-$, since the latter involves strong Coulomb interactions. (The finite energy of peak A is partly due to the inequivalence of the molecules in the initial and final states.) It should be remarked that the solution spectrum[24] of monomeric $TCNQ^-$ has an intramolecular transition at nearly the same energy, and with the same doublet structure, as peak B. This must be regarded as a coincidence, since peak B in Cs_2TCNQ_3 (Fig. 2) is definitely[13,14,20,9,10] a charge transfer band. The above assignment and interpretation of the four absorption peaks in Cs_2TCNQ_3 are confirmed by polarization measurements on single crystals.[14,19,9,10] Thus, there is general agreement on the spectra of K-TCNQ and Cs_2TCNQ_3, which indicate that $(U-V_1)$ ~1 eV in both compounds.

III. TSK MODEL

In other, more conducting, mixed valence TCNQ salts,[31] such as TEA-$TCNQ_2$ and Ad-$TCNQ_2$, the molecules are all equivalent and the electrons are not localized as they are in Cs_2TCNQ_3. Thus, the simple interpretation used for the latter salt was not extended to other TCNQ salts. More recently, Torrance, Scott, and Kaufman[9] (TSK) pointed out and emphasized the strong similarity of the spectra for a wide variety of conducting TCNQ salts to each other and to Cs_2TCNQ_3. As shown in Fig 2, this similarity also encompasses the metallic 1:1 salts TTF-, TSeF-, and NMP-TCNQ, as well as a large number of other complex salts[13,20,11] not shown in Fig. 2. It was suggested by TSK that the common electronic feature responsible for this striking similarity was a mixed valence stack of TCNQ molecules present in each material. This mixed valence character in TTF-, TSeF-, and NMP-TCNQ was attributed[9] to incomplete transfer of charge[32] from donor to TCNQ, while in the complex salts, a mixed valence TCNQ-stack is ensured by the non-1:1 stoichiometry. Recent diffuse X-ray scattering measurements[4,5,33,34] have confirmed that there is incomplete charge transfer in TTF-, TSeF-, and HMTSeF-TCNQ, with ρ=0.59, 0.63, and 0.74, respectively.

It was then proposed by TSK[9] that the interpretation and assignment of Cs_2TCNQ_3, could be extended to the other mixed valence TCNQ compounds, even if their electrons are not localized. In these materials, the localized picture of the excitations at the top of Fig. 2 can only be used as a crude guide; the peaks A and B should be described in terms of bands, as discussed in Refs. 1 and 9. Polarization measurements on single crystals of TEA-$TCNQ_2$ were presented[9,17] as evidence of this assignment. Subsequently, further polarization measurements[19,20,21,10] on a number of other complex TCNQ salts have confirmed the TSK assignment.

Such measurements in the metallic 1:1 TCNQ salts, however, are not so unambiguous. In NMP-TCNQ, for example, peak B is almost unpolarized,[9,10] while in TTF-TCNQ, some rather difficult measurements by Grant, Greene, and Castro,[22] suggest that peak B is predominantly polarized in the molecular plane, as expected for a molecular exciton. Similar results are more clearly seen in the very recent spectrum of HMTSeF-TCNQ by Jacobsen, Bechgaard, and Andersen.[23] The polarization alone can be misleading, however, since in these three materials the molecular plane is not orthogonal to the stacking axis. In this case, the charge transfer band (intermolecular transition, peak B) can borrow oscillator strength from the exciton (intramolecular transition, peak C) and hence its polarization might appear as if it were an exciton.[11] The importance of such interconfigurational interactions has been emphasized recently by Tanaka, et al[10] in their extensive experimental and theoretical study of the spectra of a variety of TCNQ salts. The argument[11] supporting the TSK interpretation is then that the energy of peak B is less sensitive than the polarization to these interconfigurational effects, which can complicate the picture. Therefore, the most significant feature of the data is taken to be the fact that all mixed valence TCNQ salts have a peak B at almost exactly the same energy (1.35±0.1 eV) with almost always the same doublet structure (Fig. 2). This striking similarity strongly suggests that peak B has a common origin and interpretation, which is best made for the simplest case Cs_2TCNQ_3 (or TEA- or $M\phi_3P$-$TCNQ_2$), where there is general agreement. Additional support for the TSK interpretation comes from the fact that it has been shown[27] to be valid in the unambiguous case of the mixed valence ($\rho<1$) and monovalence ($\rho=1$) halide salts of TTF.

IV. TANAKA MODEL

Other than the TSK model, the only complete interpretation for the optical properties of all the TCNQ salts is that presented very recently by Tanaka and coworkers.[10] (In addition, their paper contains the best and most complete set of single crystal optical measurements.) Although a number of other workers[22,23] have also suggested that peak B in TTF-TCNQ is a molecular exciton (intramolecular excitation), Tanaka, et al[10] describe a model for why this might be possible. Briefly, their model consists of approximating TTF- and NMP-TCNQ as well as Qn-$TCNQ_2$ by a stack with $\rho=0.5$ in which the electrons repel each other enough to form an ordered lattice with electrons on every other site, i.e., a Wigner crystal.[35] There are two important consequences of this assumption; first, since two electrons are never neighboring, there is a very low probability of exciting them onto the same molecule and hence no intensity of the charge transfer band corresponding to peak B in Cs_2TCNQ_3 (top of Fig. 2); secondly, for such $TCNQ^-$ molecules with $TCNQ^0$ molecules as neighbors, the energy of the molecular exciton is expected to be very near 1.3 eV. (This is the coincidence mentioned above). Thus, the charge transfer band would not be observed and the peak near 1.3 eV would be the exciton.

This model is logical and well reasoned and gives a self-consistent interpretation for the optical properties.[10] There are, however, a number of difficulties with this model. While for Qn-$TCNQ_2$ the assumption $\rho=0.5$ would be a good one, it is not for the metallic 1:1 TCNQ salts. For $\rho=0.6=3/5$ (recall $\rho=0.59$ for TTF-TCNQ), a simple picture shows that in an ordered electron lattice, half of the electrons have electrons on neighboring sites. Perhaps a better approximation of $\rho=0.59$ (for TTF-TCNQ) and $\rho=0.63$ (for TSeF-TCNQ) would be $\rho=0.67=2/3$, in which case every electron has another electron on a neighboring site. Then the spectra would be similar to Cs_2TCNQ_3, as observed (Fig. 2). The approximation of $\rho=0.5$ is even worse for HMTSeF- and NMP-TCNQ, where $\rho=0.74$[34] and ~0.95,[32,36] respectively, and yet the spectra of all these compounds are strikingly simiar[9,11] (Fig. 2). The Tanaka assignment is also inconsistent with the large increase in the energy of peak B under pressure.[37] (This increase is consistent with the TSK assignment.[11]) In addition, the interpretation of Tanaka, et al[10] has a number of problems with peak C as well as with the neglect of the contribution[25] of the exciton bandwidth to the exciton energy.

Summarizing, the TSK interpretation[9] involves a single assignment for all mixed valence TCNQ salts, although the borrowing of oscillator strength alters the polarization in a few cases (including TTF-TCNQ). According to the model of Tanaka, et al,[10] the TSK assignment works for most TCNO salts, but in the highly conducting cases the assignment is different due to a different internal electronic structure. The similarity of peak B in the different compounds in Fig. 2 is viewed[10] as a coincidence.

V. CONCLUSIONS

As we have described, there are some important differences between the two interpretations, which will need to be settled by future experiments. Nevertheless, it should be emphasized that the major physical conclusions of the extensive work of Tanaka, et al[10] are in substantial agreement with those of TSK:[9] (1) the optical properties of both TTF- and NMP-TCNQ are indicative of incomplete charge transfer; and (2) $(U-V_1)$ is large (1 - 1½ eV) in all TCNQ salts.

This conclusion indicates that the value of $(U-V_1)$ is not screened by the cation stack. Specifically, there is no evidence[9,10] of screening by the excitonic polarizability[38] of the cation,[39,40] which would have been dramatically different for TSeF and Cs, for example. This demonstrates that the Little mechanism of excitonic screening[38] is ineffective in these materials. We can also conclude that TTF- and NMP-TCNQ are much better conductors than K-TCNQ because the former are mixed valence ($\rho<1$) salts,[9] and not because the Coulomb interactions are screened, as had been widely supposed.[39,40]

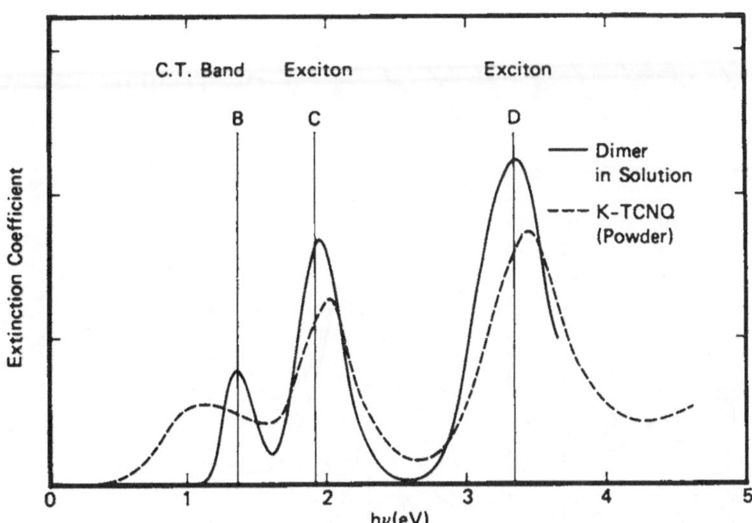

Fig. 1. A comparison of the extinction coefficient of the $(TCNQ^-)_2$ dimer in solution with that of the salt K-TCNQ (which contains stacks of dimers), showing the three characteristic peaks B, C, and D, present in both. (The extinction in the powdered K-TCNQ is x2).

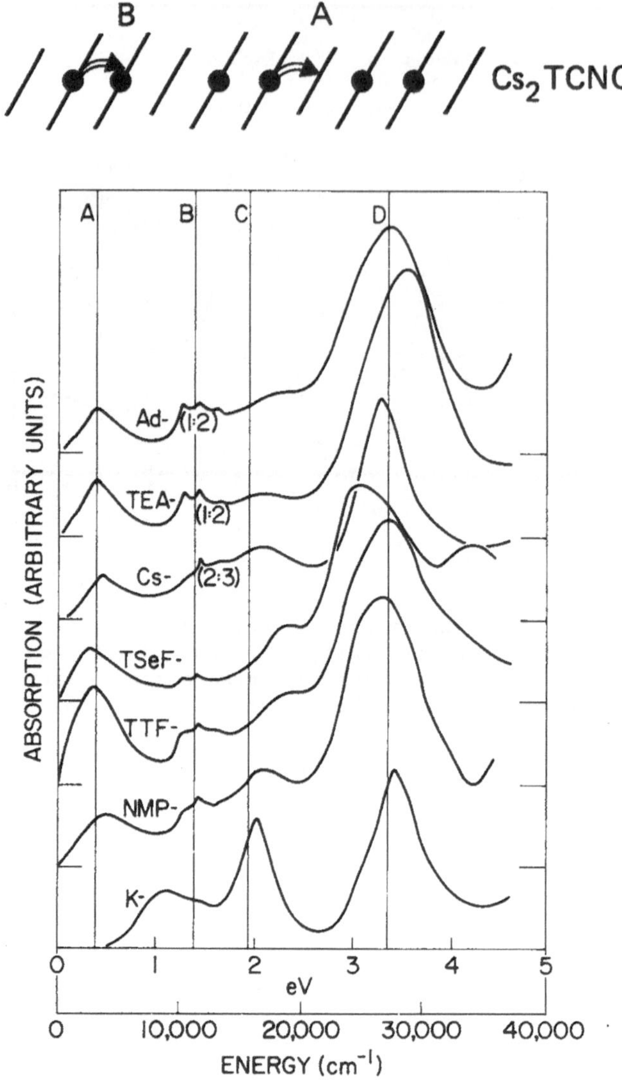

Fig. 2. At the top is a schematic view of the electrons (solid points) localized along the stacks in Cs_2TCNQ_3 and the two types of charge transfer excitations, A and B. Below are the absorption spectra for a wide variety of powdered TCNQ salts, dispersed into KBr disks. (After TSK, Ref. 9).

REFERENCES

† Present address.

1. J. B. Torrance, Proc. NATO Adv. Study Institute on "Chemistry and Physics of One-Dimensional Metals," Bolzano, Italy, Aug. 1976, to be published, H. J. Keller, ed. (Plenum Press, 1977).
2. J. B. Torrance, Y. Tomkiewicz, and B. D. Silverman, submitted to Phys. Rev. B.
3. J. B. Torrance, H. A. Mook, and C. R. Watson, submitted.
4. J. P. Pouget, S. K. Khanna, R. Comes, A. F. Garito and A. J. Heeger, Phys. Rev. Letters 35, 445 (1975); S. K. Khanna, J. P. Pouget, R. Comes, A. F. Garito and A. J. Heeger, to be submitted.
5. S. Kagoshima, T. Ishiguro, and H. Anzai, submitted to J. Phys. Soc. Japan.
6. V. J. Emery, Phys. Rev. Letters 37, 107 (1976); V. J. Emery, Proc. NATO Adv. Study Institute on "Chemistry and Physics of One-Dimensional Metals," Bolzano, Italy, Aug. 1976, to be published, H. J. Keller, ed. (Plenum Press, 1977).
7. D. Jerome and L. Giral, Proc. Conf. on Organic conductors and Semiconductors, Siofok, Hungary, Sept. 1976 (this issue); D. Jerome and M. Weger, Proc. NATO Adv. Study Institute on "Chemistry and Physics of One-Dimensional Metals," Bolzano, Italy, Aug. 1976, to be published, H. J. Keller, ed. (Plenum Press, 1977).
8. J. B. Torrance, B. A. Scott, D. C. Green, P. Chaudhari, and D. F. Nicoli, Solid State Comm. 14, 100 (1974).
9. J. B. Torrance, B. A. Scott, and F. B. Kaufman, Solid State Commun. 17, 1369 (1975).

10. J. Tanaka, M. Tanaka, T. Kawai, T. Takabe, and O. Maki, Bull. Chem. Soc. Japan 49, 2358 (1976).

11. J. B. Torrance (unpublished results).

12. Y. Iida, Bull. Chem. Soc. Japan 42, 71 (1969).

13. Y. Iida, Bull. Chem. Soc. Japan 42, 637 (1969).

14. S. Hiroma, H. Kuroda and H. Akamatu, Bull. Chem. Soc. Japan 44, 9 (1971).

15. N. Sakai, I. Shirotani, and S. Minomura, Bull. Chem. Soc. Japan, 45 3321 (1972).

16. Y. Oohashi and T. Sakata, Bull. Chem. Soc. Japan 46, 3330 (1973).

17. A. Brau, P. Breusch, J. P. Farges, W. Hinz and D. Kuse, Phys. Stat. Sol. (b) 62, 615 (1974).

18. S. K. Khanna, A. A. Bright, A. F. Garito, and A. J. Heeger, Phys. Rev. B10, 2139 (1974).

19. G. C. Wrighton, P. M. Grant, R. L. Greene, and G. Castro, Bull. Am. Phys. Soc. 18, 1578 (1973); G. C. Wrighton, Ph.D. Thesis, Stanford University (1974, unpublished).

20. Y. Oohashi and T. Sakata, Bull. Chem. Soc. Japan 48, 1725 (1975).

21. H. W. Helberg, Phys. Stat. Sol. (a) 33, 453 (1976).

22. P. M. Grant, R. L. Greene, and G. Castro, Bull. Am. Phys. Soc. 20, 496 (1975).

23. C. S. Jacobsen, K. Bechgaard, and J. R. Andersen, preprint (1976).

24. R. H. Boyd and W. D. Phillips, J. Chem. Phys. 43, 2927 (1965).

25. A. S. Davydov, Theory of Molecular Excitons (Plenum Press, N.Y., 1971).

26. D. B. Tanner, C. S. Jacobsen, A. F. Garito, and A. J. Heeger, Phys. Rev. B13, 3381 (1976).

27. J. B. Torrance, B. A. Scott, and B. Welber, to be submitted.

28. A. A. Ovchinnikov, Sov. Phys. JETP 30, 1160 (1970).

29. At frequencies faster than the frequency (~0.5 eV) of electrons hopping down the stack, the electrons may be viewed as localized and hence the stack contains either neutral or anion species.

30. C. J. Fritchie, Jr. and P. Arthur, Jr. Acta Cryst. 21, 139 (1966).

31. The chemical abreviations used in the text and their corresponding chemica names are: TCNQ (tetracyanoquinodimethane); TTF (tetrathiafulvalene); TSeF (the selenium analogue of TTF); NMP (N-methylphenazinium); Ad (acridinium) Qn (quinolinium); TEA (triethylammonium); HMTSeF (hexamethylene TSeF); $M\phi_3$ (methyl-triphenyl-phosphonium).

32. Z. G. Soos, Ann. Rev. Phys. Chem. 25, 121 (1974).

33. R. Comes, Proc. NATO Adv. Study Institute on "Chemistry and Physics of One-Dimensional Metals," Bolzano, Italy, Aug. 1976, to be published, H. J. Keller, ed. (Plenum Press, 1977).

34. C. Weyl, E. M. Engler, K. Bechgaard, G. Jehanno, and S. Etemaod, Solid State Commun. 19, 925 (1976), assuming that the observed scattering is at $2k_F$.

35. For a discussion of the Wigner crystal in TCNQ salts, see J. B. Torrance and B. D. Silverman, Phys. Rev. B (in press).

36. M. A. Butler, F. Wudl, and Z. G. Soos, Phys. Rev. B. 12, 4708 (1975).

37. B. Welber, P. E. Seiden, and P. M. Grant to be submitted.

38. W. A. Little, Phys. Rev. A134, 1416 (1964).

39. O. H. Le Blanc, J. Chem. Phys. 42, 4307 (1965).

40. A. F. Garito and A. J. Heeger, Accts. of Chem. Res. 7, 232 (1974).

Conference on Organic Conductors and Semiconductors, Siófok, Hungary 1976

ANISOTROPY IN THE CRITICAL BEHAVIOR OF TTF-TCNQ AND TSeF-TCNQ*

D. GUIDOTTI and P.M. HORN

The Department of Physics and The James Franck Institute
The University of Chicago, Chicago, Illinois 60637, USA

and
E.M. ENGLER+

IBM Thomas J. Watson Research Center Yorktown Heights, New York 10598, USA

ABSTRACT

We present the results of a detailed study of the critical behavior in the a and b axis resistivity of TTF-TCNQ and TSeF-TCNQ. We find that in TTF-TCNQ, $d\rho^a/dT$ and $d\rho^b/dT$ diverge as $T - T_c \rightarrow 0^+$ with the same critical exponent while in TSeF-TCNQ, $d\rho^a/dT$ and $d\rho^b/dT$ diverge with different critical exponents. These results are compared with various models for the origin of the critical behavior in the resistivity.

*Work supported by NSF under contract #NSF-DMR75-14360 and by the Louis B. Block Fund of the University of Chicago. We have also benefited from support of the Materials Research Laboratory by the NSF.

+Alfred P. Sloan Research Fellow.

It is now reasonably well established that the organic linear chain complexes TTF-TCNQ and TSeF-TCNQ undergo a structural transition which can be characterized as a three dimensional Peierl's distortion.[1, 2] Perhaps the most dramatic manifestation of this distortion is seen in the electrical properties which display a transition from metallic behavior for $T > T_c$ to semiconducting behavior for $T < T_c$.[3-5] In this paper we examine the contribution which structural fluctuations make to the electrical resistivity. While the experimental data and discussion pertain primarily to the region in the immediate vicinity of the three dimensional ordering temperature ($T \cong T_c$), many of the conclusions presented relate to the larger question of fluctuation behavior in the high temperature phase.

In Figures 1 and 2 we display the behavior of the a and b axis resistivity of TSeF-TCNQ in the vicinity of the three dimensional ordering temperature $T_c \cong 28^\circ K$. The ratio of the transverse to the longitudinal resistivity of TSeF-TCNQ is about $\rho_a/\rho_b \cong 70$ at room temperature. The sharp negative divergence in the temperature derivative of both the a and b axis resistivity is similar in form to resistive anomalies observed in metallic antiferromagnets near the Neel temperature[6] and can be expected to coincide with the three dimensional ordering temperature.[5] A critical exponent analysis of the TSeF-TCNQ data for $T > T_c$ is shown in Figure 3, where we plot $-\frac{dR}{dT}$ vs. $t = \frac{T-T_c}{T_c}$. The upper curve (closed circles) corresponds to the b-axis resistivity and the lower curve (open circles) to the a-axis data. The critical exponents of dR/dT are tabulated along with the TTF-TCNQ exponents in Table 1.

To illustrate the possible origins of these exponents, consider the simple one-electron form for the electrical resistivity[7]

$$\rho = \frac{m}{n e^2 \tau} \tag{1}$$

then

$$\frac{d\rho}{dT} = \frac{1}{e^2\tau}\frac{d}{dT}\left(\frac{m}{n}\right) + \frac{m}{n e^2}\frac{d}{dT}\left(\frac{1}{\tau}\right) \tag{2}$$

Near the three dimensional ordering, anomalies in the resistivity can arise from critical behavior in either the scattering rate $1/\tau$, or the density of states m/n. Sufficiently close to T_c, the resistivity will be dominated by the term in Eq. (2) which has the largest critical exponent.

The critical behavior of the scattering rate has been estimated by Horn and Guidotti[8] in a model which is similar to the critical scattering model developed by Suezaki and Mori[9] to describe the resistivity of metallic antiferromagnets. In this model, the critical scattering arises from electrons scattering from a soft phonon mode at wavevector $q = 2k_F$. The results of such a calculation for a highly anisotropic metal are shown in Table 1 . The agreement between experiment and theory for b-axis TSeF-TCNQ is good, while the behavior of TTF-TCNQ remains puzzling. We should point out that the predicted scattering exponents for the transverse (a and c axis) conductivity should not be directly compared with experiment since the Horn-Guidotti model treats the conduction in a band like picture whereas the transverse conductivity in these organic materials should undoubtedly be described as diffusive.

We now discuss the density of states term in Eq. (2). Neglecting again collective effects, the density of states contribution to ρ can be written as[8]

$$
m/n \propto \begin{cases} 1 + \dfrac{\langle |\Delta(x)|^2 \rangle}{(k_B T)^2} & (\langle |\Delta(x)| \rangle \lesssim k_B T) \quad \text{(3a)} \\[2ex] \left\langle e^{\frac{|\Delta(x)|}{k_B T}} \right\rangle & (\langle |\Delta(x)| \rangle \gtrsim k_B T) \quad \text{(3b)} \end{cases}
$$

where $\Delta(x)$ is the complex order parameter, k_B is the Boltzman constant and $\langle\ \rangle$ represents the usual statistical average over order parameter confirgurations. Equation (3) implies that m/n is sensitive to the local value of Δ (x) and hence will, in a highly anisotropic system, monitor the growth of the longitudinal correlation length, $\xi_\parallel(T)$, rather than the onset of long range order $\langle\Delta(x)\rangle$. Thus, if the fluctuation growth is highly anisotropic, ρ will be activated in the high temperature ($T > T_c$) phase and there will be no density of states contribution to the critical temperature dependence of ρ. Alternatively, if the fluctuation growth is relatively isotropic, ρ will appear metallic for $T > T_c$ and the critical behavior of m/n near T_c is obtained from Eq. (3a) and is given in Table 1.[8] Note that the density of states exponents are independent of direction and considerably weaker than the b-axis critical scattering exponents and hence they should not contribute to the experimentally observed b-axis exponents. Alternatively, they should be observable in a precision magnetic susceptibility experiment.

For completeness, we have listed in Table 1, other possible origins of the critical exponents in these organic materials. If, as has been suggested,[10] the conductivity for $T > T_c$ is dominated by collective Peierls-Frohlich fluctuations, then the critical behavior in the resistivity would arise from three dimensional pinning of the collective mode. While no detailed predictions exist for the temperature dependence in such a model, only the b-axis conductivity will display critical behavior since the collective mode propagates only along this axis of continuous symmetry breaking. We schematically illustrate this prediction in Table 1.

We should also mention the domain model suggested by J. C. Phillips[11] which describes the metal-nonmetal transition in TTF-TCNQ and related compounds as a percolation transition in a system composed of metallic and insulating domains. In the present context, the essential feature of such a model is that the critical behavior would be isotropic as illustrated in Table 1.

Neither the domain model, nor the collective mode pinning model appears to be consistent with the experimentally observed critical behavior. Furthermore, the domain model is inconsistent with the existence of critical behavior in the magnetic susceptibility of TTF-TCNQ.[12]

In conclusion, our experimental results suggest that the conductivity above T_c in TSeF-TCNQ is dominated by an anomaly in the electron-phonon scattering associated with the onset of a three dimensional Peierls distortion. While the data in TTF-TCNQ is more complicated, we expect that near T_c the same scattering mechanism is dominating the conduction in this material. These results along with the below T_c critical behavior,[8] suggests that the fluctuation growth is relatively isotropic in these materials and that for T above T_c there does not exist a large one dimensional correlation length. Recent diffuse x-ray studies[13] confirm this picture. Finally, it is difficult to reconcile these results with arguments which suggest the existence of a well defined (i.e., width $<<k_BT$) collective mode at room temperature since the sharp critical behavior in the resistivity suggests that relatively long range one dimensional order is not established even at much lower temperature.

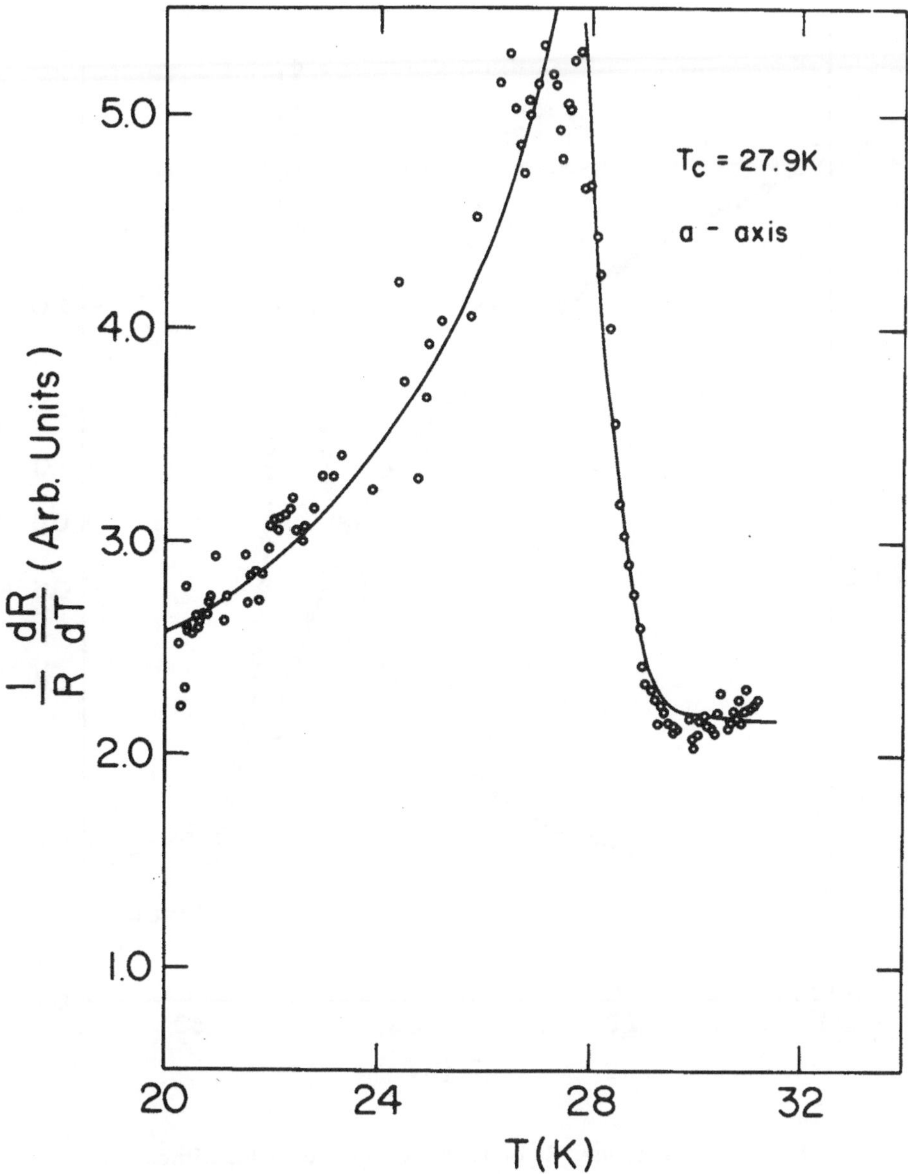

Fig. 1 Behavior of the logarithmic derivative of the a-axis resistance in TSeF-TCNQ near the critical temperature.

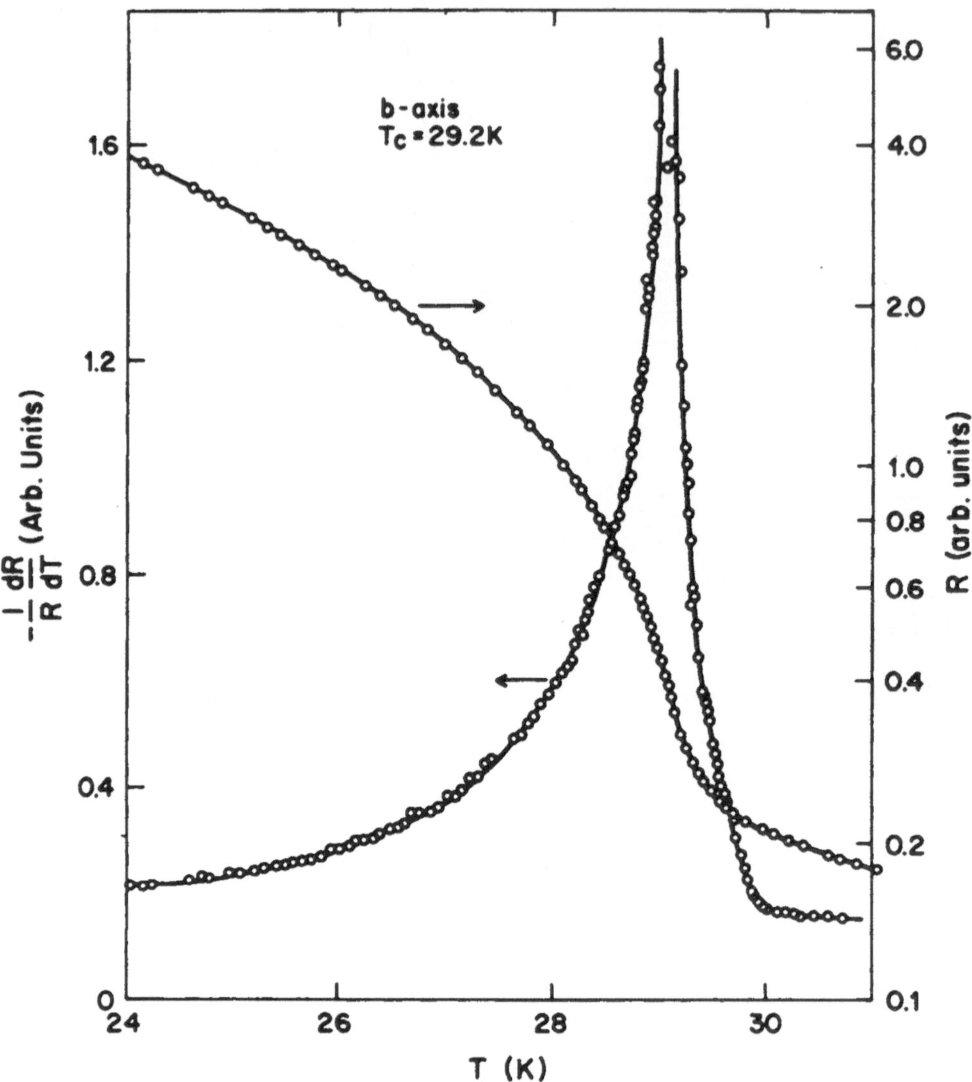

Fig. 2 The b-axis resistance, R, of TSeF-TCNQ and its logarithmic derivative in the vicinity of T_c.

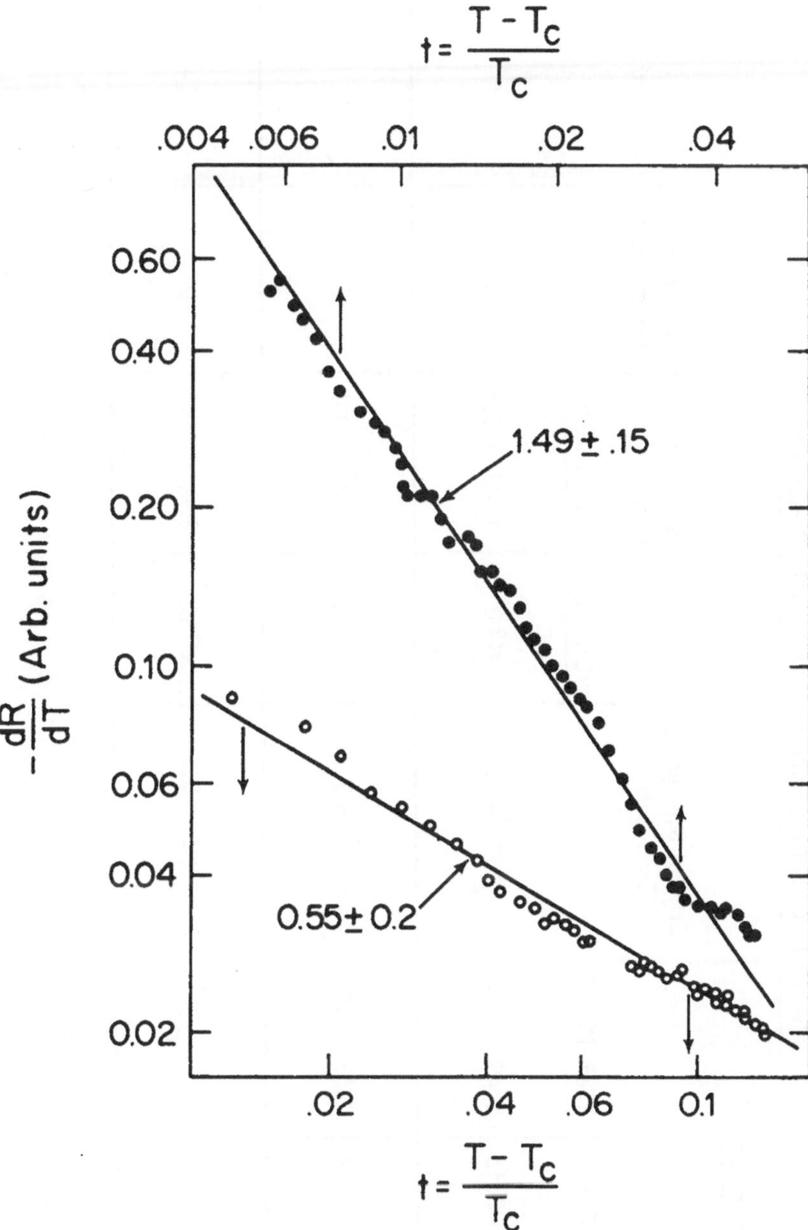

Fig. 3 The critical exponents of dR/dT above T_c for conduction in TSeF-TCNQ along the a-axis (open circles) and along the b-axis (closed circles).

TABLE I: CRITICAL EXPONENTS OF dR/dT FOR $T > T_c$

	Experimental		Theory				
			Critical Scattering Exponents	Density of States Exponents		Collective Mode Pinning	Percolation
	TTF-TCNQ	TSeF-TCNQ		Critical	Classical		
b-axis	1.09 ± 0.15	1.49 ± 0.13	$1+\eta+\nu = 1.5$	$\alpha \cong 0$	$1-\nu = 0.5$	X	Y
a-axis	1.04 ± 0.17	0.55 ± 0.2	$1-\eta-\nu = 0.5$	$\alpha \cong 0$	$1-\nu = 0.5$	0	Y
c-axis	—	—	$1-\eta-\nu = 0.5$	$\alpha \cong 0$	$1-\nu = 0.5$	0	Y

REFERENCES

1. F. Denoyer, R. Comes, A. F. Garito and A. J. Heeger, Phys. Rev. Letters 35, 445 (1975).

2. R. Comes, S. M. Shapiro, G. Shirane, A. F. Garito and A. J. Heeger, Phys. Rev. Letters 35, 1518 (1975).

3. L. B. Coleman, M. J. Cohen, D. J. Sandman, F. G. Yamagishi, A. F. Garito and A. J. Heeger, Solid State Commun. 12, 1125 (1973); A. N. Block, J. P. Ferraris, D. O. Cowan and T. O. Poehler, Solid State Commun. 13, 753 (1973).

4. M. J. Cohen, L. B. Coleman, A. F. Garito and A. J. Heeger, Phys. Rev. B 13, 5111 (1976); S. Etemad, Phys. Rev. B 13, 2254 (1976).

5. P. M. Horn and D. Rimai, Phys. Rev. Letters 36, 8-9 (1976).

6. See for example, R. A. Craven and R. D. Parks, Phys. Rev. Lett. 31, 383 (1973) and G. T. Meaden, N. H. Sze and J. R. Johnston in Dynamical Aspects of Critical Phenomena, edited by J. I. Budnick and M. P. Kawatra (Gordon and Breach, New York, 1972), p. 315.

7. It is well known that the simple one electron form for the resistivity is not valid in one dimensional systems. See e.g., A. Madhukar and Morrel H. Cohen, preprint.

8. P. M. Horn and D. Guidotti, Phys. Rev. B, in press.

9. Y. Suezaki and H. Mori, Prog. Theor. Phys. 41, 1177 (1969); also see S. Takada, Prog. Theor. Phys. 46, 15 (1971) and T. Kasuya and A. Kondo, Solid State Commun. 14, 249 (1974).

10. M. J. Cohen, L. B. Coleman, A. F. Garito and A. J. Heeger, Phys. Rev. B 10, 1298 (1974); D. B. Tanner, C. S. Jacobsen, A. F. Garito and A. J. Heeger, Phys. Rev. Lett. 33, 1559 (1974).

11. J. C. Phillips, preprint.

12. R. M. Herman, M. B. Solamon, G. De Pasquali and G. Stucky. Presented at the Conference on Organic Conductors and Semiconductors, Siófok, Hungary, 30 August - 3 September 1976.

13. J. P. Pouget, S. Kanna and R. Comes; and C. Weyl, E. M. Engler, K. Beckgaard, G. Jehanno, S. Etemad. Presented at the Conference on Organic Conductors and Semiconductors, Siófok, Hungary, 30 August - 3 September 1976.

Conference on Organic Conductors and Semiconductors, Siófok, Hungary 1976

MAGNETIC SUSCEPTIBILITY OF TTF-TCNQ SINGLE CRYSTALS IN THE PHASE TRANSITION REGION*

R.M. HERMAN, M.B. SALAMON, and G. de PASQUALI

Department of Physics and Materials Research Laboratory

G. STUCKY

Department of Chemistry and Materials Research Laboratory
University of Illinois, Urbana, Illinois 61801, USA

Abstract

The magnetic susceptibility of single crystals of TTF-TCNQ was measured along the principal directions over the temperature range from 4 K to 70 K. Definite anomalies were observed at 38 K and 54 K, the former showing hysteresis. No temperature dependent anistotropy was found. Fits to a two-gap model give a low temperature gap of 125 K and a high temperature gap of 370 K. A deviation from exponentially activated behavior was noted in the region near 48 K.

*Supported in part by the National Science Foundation Grant DMR 76-01058.

Changes in the magnetic susceptibility accompanying the 38 K and 54 K phase transitions in TTF-TCNQ have been observed in powder specimens by standard techniques[1/] and by means of ESR.[2/] The resolution in these studies has been rather low, and they have not sought to measure the anisotropy of the susceptibility in the ordered phase which would signal a possible antiferromagnetic or spin-density wave component in the phase transition. To address these questions, we have performed a series of high resolution susceptibility measurements using a small number (~10) aligned single crystals using the Faraday method.

In Fig. 1, we show the low-temperature portion of our data for magnetic field along the principal crystallographic axes. The data are assumed to contain the following components: a Curie paramagnetic term corresponding to approximately 0.04% spin-$\frac{1}{2}$ impurities; a temperature-independent background for each orientation,

$$\chi_a = -1.75 \times 10^{-4} \text{ emu/mole}$$
$$\chi_b = -2.35 \times 10^{-4} \text{ emu/mole}$$
$$\chi_c = -2.02 \times 10^{-4} \text{ emu/mole}$$

and finally, a temperature-dependent spin susceptibility.

A detailed examination of the region around the 38 K phase transition is shown in Fig. 2. The samples were mounted without clamping with their b-axes in the field direction. Clear evidence for the first order nature of this transition is seen in the 0.8 K hysteresis in the susceptibility data.

It has been suggested[2,3/] that the two phase transitions in TTF-TCNQ reflect different ordering temperatures for the two types of conducting chain. Despite the shortcomings of this model, we have attempted to fit our data to a two-gap model in which the susceptibility of the lower gap Δ_L saturates at 38 K. The result of such a fit is shown in Fig. 3, and leads to the following parameters:

$$\Delta_L \cong 125 \text{ K}, \quad \Delta_H \cong 370 \text{ K and} \quad \chi_{sat} = 0.96 \times 10^{-4} \text{ emu/mole.}$$

We note that a Pauli paramagnetism χ_{sat} corresponds to a density of states of 3.0 states/eV which is of the correct size for bands in this material.

A detailed examination of the region between 38 K and 54 K showed no anomaly in the susceptibility. However, the derivative of our data shows a pronounced change in the temperature dependence at about 48 K which would appear as a step in the second derivative. It has been suggested by Bak and Emery[3/] that one chain of this system begins to order at 54 K and the second at 47 K with a locking in of the entire structure at 38 K. Insofar as the spin susceptibility reflects the density of states at the Fermi level, our data strongly indicate that the 38 K transition results in the reduction of the electronic density of states toward zero, and that the 47 K transition has very little effect on the susceptibility.

A careful study of the anisotropy of these crystals has shown no additional anisotropy beyond that associated with the molecular diamagnetism. We conlude that there is no antiferromagnetic component to the phase transition, or, if so, that the flop field is lower than the lowest fields used in these measurements.

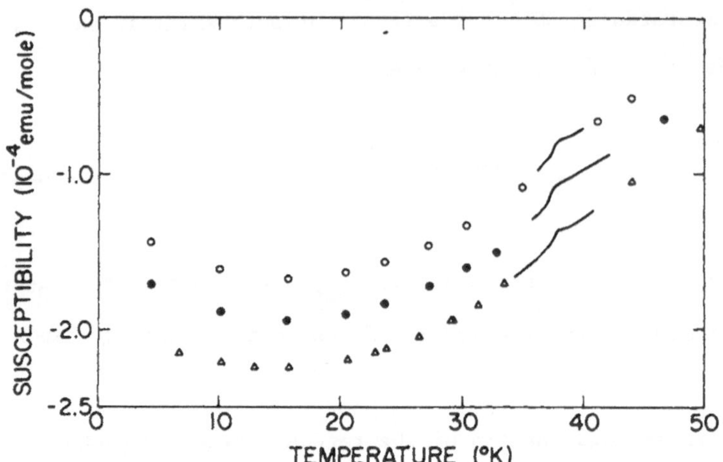

1. Temperature dependence of the total magnetic susceptibility. Magnetic field along the a-axis, open circles; along the b-axis, triangles; along the c-axis, closed circles.

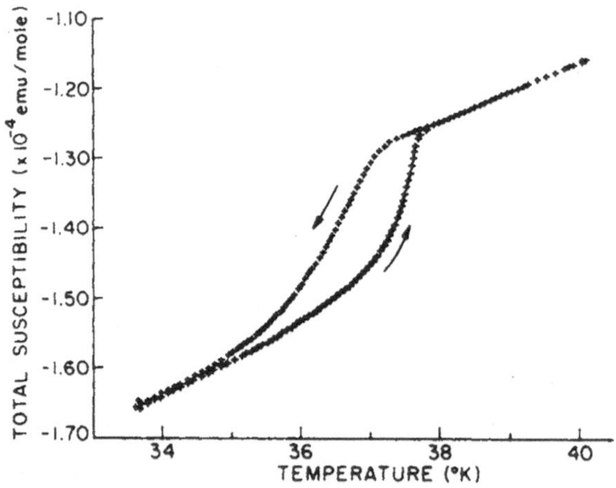

2. Hysteresis in the susceptibility near 38 K. Arrows indicate values for heating and cooling respectively.

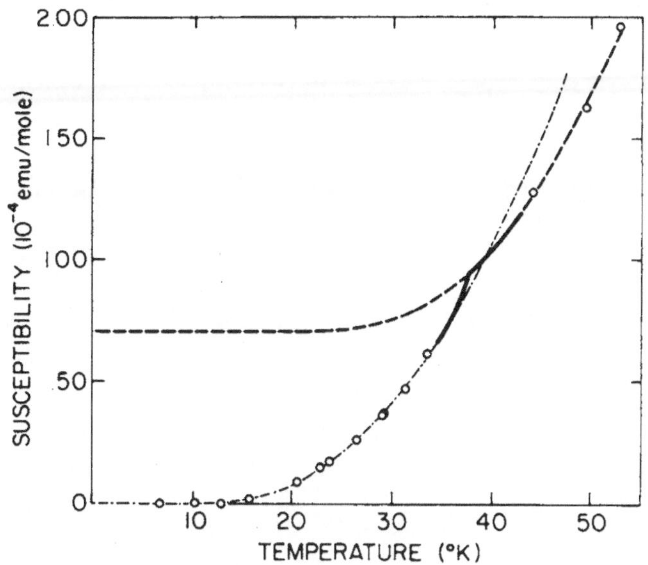

3. Results of a fit to the data using a two-gap model. Not all data points are shown.

References

1. J. C. Scott, A. F. Garito, and A. J. Heeger, Phys. Rev. B10, 3131 (1974).

2. Y. Tomkiewicz, A. R. Taranko, and J. B. Torrance, Phys. Rev Lett. 36, 751 (1976).

3. P. Bak and V. J. Emery, Phys. Rev. Lett. 36, 978 (1976).

VI

CHARGE TRANSFER SALTS OTHER THAN TTF-TCNQ AND ITS DERIVATIVES

Conference on Organic Conductors and Semiconductors, Siófok, Hungary 1976

NEW, HIGHLY CONDUCTIVE, ORGANIC SOLIDS

F. WUDL, G.A. THOMAS, W.M. WALSH, Jr. and
M.L. KAPLAN,

Bell Laboratories, Murray Hill New Jersey, USA

Several physical properties of four quasi-one-dimensional organic conductors are compared. Assuming that the intermolecular spacing along the conducting stacks and the magnitude of the susceptibility serve as indicators of the degree of electronic overlap, the data indicate that an increase in this overlap enhances the stability of the insulating ground state of these systems. As a result, the metal-insulator transition temperature is found to scale roughly linearly with an effective Fermi energy.

Studies of new materials based on substituted TTF and derivations of TTN will be presented.

Conference on Organic Conductors and Semiconductors, Siófok, Hungary 1976

THE NEW CLASS OF THE QUASI-ONE-DIMENSIONAL ORGANIC METALS - THE CATION-RADICAL SALTS OF THE TTT

I.F. SCHEGOLEV, M.L. KHIDEKKEL, E.B. JAGUBSKII and R. SCHIBAEVA

Institute of Chemical Physics, Chernogolovka, USSR

/ Title only /

Conference on Organic Conductors and Semiconductors, Siófok, Hungary 1976

PHYSICAL PROPERTIES OF A CHARGE TRANSFER COMPLEX WITH TCNQ AND IODINE: EVIDENCE OF A LATTICE DISTORTION

P. DELHAES, A. COUGRAND, S. FLANDROIS

Centre de Recherches Paul Pascal, C.N.R.S., Domaine Universitaire, 33405 TALENCE

D. CHASSEAU, J. GAULTIER, C. HAUW

Laboratoire de Cristallographie, Université de Bordeaux I, Domaine Universitaire, 33405 TALENCE

and

P. DUPUIS

Laboratoire de Chimie Physique Macromoléculaire 54000 NANCY (FRANCE)

The physical properties of the new complex and mixed salt $(TMA^+)_3(I_3^-)(TCNQ^-)_2(TCNQ^\circ)$ characterized recently are presented. Crystallographic study at room temperature has shown a diffuse scattering which indicates the existence of a 1-D superlattice. From dynamic (EPR) and static magnetism investigations and the existence of a specific heat anomaly an electronic PEIERLS distorsion is postulated. However the semi conductive behaviour of this salt is evidenced by the d.c. conductivity and the thermoelectric power results. A comparison with similar situations in TTF-TCNQ and mainly in KCP is presented.

I - INTRODUCTION

We have synthetised a new mixed radical anion of tetracyanoquinodimethane (TCNQ) and Iodine (I) with the developped formula (1) :

$(TMA^+)_3(I_3^-)\ (TCNQ)_3^{--}$ (TMA : Trimethylammonium)

This salt of stoichiometry 2-3 is a complex valency compound with an appreciable d.c. electrical conductivity at room temperature ($\sigma \sim 1(\Omega cm)^{-1}$). By assuming a full charge transfer this compound must present a 1/3 filled conduction band.

II - STRUCTURAL PROPERTIES

The crystallographic structure at room temperature has been determined (1). The unit cell is monoclinic with a symmetric group $C_{2/m}$ (a = 20,306 Å, b = 6,444 Å, c = 13,943 Å, β = 116°38). The black needle-like single crystals are growing along the direction of the stack or b-axis which presents the largest conductivity.

The principal features of this structure are :

- Regular stacks of TCNQ with a mean distance of 3,22 Å and a typical overlapping as in NMP-TCNQ or TTF-TCNQ.
- Iodine chains which alternate in the a direction with TCNQ stacks.
- Alternative hydrogen bonds (N...H-N) between quaternary ammoniums and TCNQ.

A one dimensional superstructure appears along the b-axis with a periodicity three times larger than the previous one (b' = 6,44 x 3 Å). This result indicates the presence of (I_3^-) as in other 1 D iodine chains (3). The theoretical analysis of structure factors shows that this diffuse scattering is caused by two-dimensionally disordered linear triiodide chains correlated through the TCNQ stacks (4). A dynamic disorder must be present. However, experiments at lower temperatures down to 150 K do not exhibit any variation of these diffuse lines and no occuring of a 3 D ordering (4).

A further investigation has allowed us to observe a diffuse scattering on a Bragg diagram as in KCP compound (2) (Figure 1).

III - PHYSICAL PROPERTIES

a - Transport properties :

The electrical d.c. resistivity has been measured along the needle axis. In spite of difference in the absolute values a similar behaviour has been observed (Figure 2) on several single crystals. This material behaves as a semi-conductor with a discontinuity around 150 K. The Seebeck coefficient measured on a polycrystalline sample confirms this result (1).

b - Paramagnetism :

The static susceptibility on a polycrystalline sample and the EPR intensity line on a single crystal show a similar thermal variation (Figure 3). Three temperature ranges can be distinguished :

- 350 > T(K) > 150 : a quasi constant paramagnetism which extrapolated to absolute zero gives $\chi_{p_n} \simeq 5.30\ 10^{-4}$ emu C.G.S./mole.
- 150 > T(K) > 20 : an exponential type decrease characteristic of an activated process
- 20 > T(K) > 2 : a Curie tail which depends on the chemical preparation (1).

Besides, by X band EPR spectroscopy a single Lorentzian line is observed whatever the single crystal position is in the resonance cavity. With the needle axis parallel to the static magnetic field the g-factor and linewidth thermal variations are presented (Figure 3) ; maxima values appear below 150 K; the origin of these is not understood.

c - Specific heat :

We have already shown that at low temperatures the lattice specific heat is only observed ($C_p \sim T^3$)(1). By adiabatic method on a polycrystalline sample we have looked for some anomaly around 150 K (Figure 4) : a reversible anomaly as in KCP (5) and in TTF-TCNQ (6), is found which indicates the presence of a phase transition.

IV - ANALYSIS OF EXPERIMENTAL RESULTS

From these experimental results a phase transition exists at 150 K. A comparison with similar quasi 1-D compounds which present an electronic Peierls transition is given in the following table.

Physical properties / 1 D Mixed valency compounds	Specific heat anomaly	Paramagnetism	d.c.electrical conductivity	Thermoelectric power	Crystallographic Structure (at room temperature)
KCP	$\frac{\Delta C_p}{R} = 3$ at T = 123 K	weak χ_p decreasing below 130 K	maximum value at T = 230 K	divergent behaviour at T < 100 K	regular stacks of $Pt(CN)_4$ disordered positions of K and Br
TTF-TCNQ *	$\frac{\Delta C_p}{R} = 0,3$ at T = 54 K	quasi constant χ_p and exponential decrease	maximum value at T = 58 K and semi conducting behaviour	divergent behaviour at T < 58 K	regular stacks of TTF and TCNQ
$(TMA\text{-}TCNQ\text{-}I)_3$	$\frac{\Delta C_p}{R} \simeq 6$ at T = 150 K	quasi constant χ_p and exponential decrease	no maximum but a jump at 150 K	semi conducting and divergent behaviour at T < 150 K	regular stacks of TCNQ disordered positions of (I_3^-), hydrogen bonds

* *Only the first transition (at the highest temperature) is considered here.*

We observe that just for the properties at thermal equilibrium, the results are similar. Because of the semi-conducting behaviour the possibility of a spin Peierls transition as observed on a 1-D magnetic system (7) cannot be precluded. From theoretical investigation (8) it is not possible to know which is the dominant mechanism because for a 1/3 filled band the superstructure must be the same, namely at 3 b (it is not the case for 1/4 filled band however where two different superstructures are proposed in agreement with experiments).

a - Transport properties :

The absence of an electrical conductivity maximum is troublesome. We assume that the intrinsic disorder effects are predominant :

- the alternative hydrogen bonding and the disordered triodide chains induced a disorder in the depth of the potential wells (Anderson's Model)
- the dynamic disorder must exist on TCNQ stacks which form triads: pseudo localized electronic states exist in the TCNQ energy band and a semi-conductive behaviour is expected.

b - Static properties :

A Peierls distorsion at zero Kelvin on purely 1 D system is a second order phase transition. The observed transition (Figure 4) might be of weakly first order transition ; it is interesting to note that the specific heat anomaly increases with the transition temperature in the three considered compounds. We suppose that this critical temperature increases in relation with the 3-D and the disorder effects.

This picture is confirmed by the paramagnetism behaviour. In the simplest case, the tight binding model, we can calculate from χ_{P_0} the integral transfer value(1) $t \simeq 0.020$ eV which is of the correct order of magnitude.

V - CONCLUSION

We have proposed an electronic Peierls distorsion ; by assuming a full charge transfer the superstructure will be commensurate : three times the lattice spacing in the stack direction. The main problem is to compare structural investigations and physical properties because the first ones furnish informations about iodine disorder mainly and the second ones have to deal with TCNQ only.

It will be necessary to prove the existence of dynamic or static triads of TCNQ and the appearance of the 3-D superstructure below 150 K to confirm our interpretation.

BIBLIOGRAPHY

(1) A. COUGRAND, S. FLANDROIS, P. DELHAES, P. DUPUIS, D. CHASSEAU, J. GAULTIER and J.L. MIANE - *Mol. Cryst. Liq. Cryst.* 32, p. 165 (1976).

A. COUGRAND - *"Thèse de 3ème cycle" - Bordeaux University (1976).*

(2) R. COMES, M. LAMBERT, H. LAUNOIS, H.R. ZELLER - *Phys. Rev.* B 8, p. 571 (1973).

(3) H. ENDRES, H.J. KELLER, M. MEGNAMISI-BELOMBE, H. PRITZKOW, J. WEISS and R. COMES - *To be published.*

(4) C. HAUW, J. GAULTIER and D. CHASSEAU - *To be published.*

(5) K. FRANULOVIC, D. DJUREK - *Physics Letters* 51 A, p. 91 (1975).

(6) R.A. CRAVEN, M.B. SALAMON, G. DE PASQUALI, R.M. HERMAN, G. STUCKY, A. SCHULTZ - *Phys. Rev. Letters* 32, p. 769 (1974).

(7) J.W. BRAY, H.R. HART, L.V. INTERRANTE, L.S. JACOBS, J.S. KASPER, G.D. WATKINS, S.H. WHEE, J.C. BONNER - *Phys. Rev. Letters* 35, p. 744 (1975).

(8) J. BERNASCONI, M.J. RICE, W.R. SCHNEIDER, S. STRASSLER - *Phys. Rev.* B 12, p. 1090 (1975).

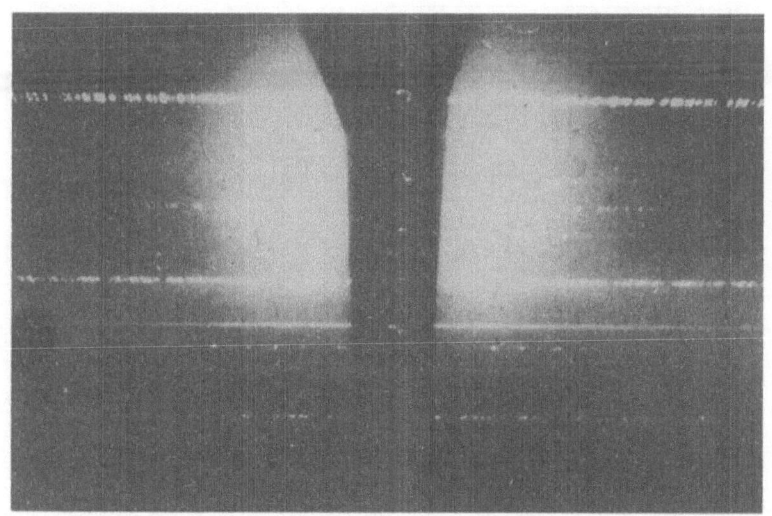

Figure 1 : *Scattering of X-rays from a* $(TMA-I-TCNQ)_3$ *rotating single crystal at room temperature. The thin continuous lines are due to the 1 D superstructure of* (I_3^-).

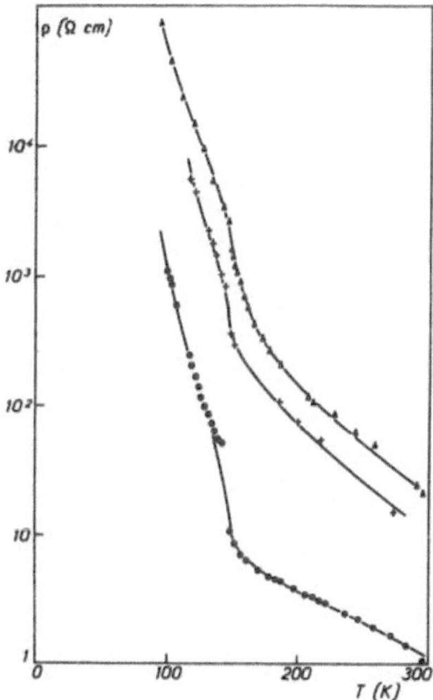

Figure 2 : *Thermal variation of D.C. electrical resistivity for different single crystals.*

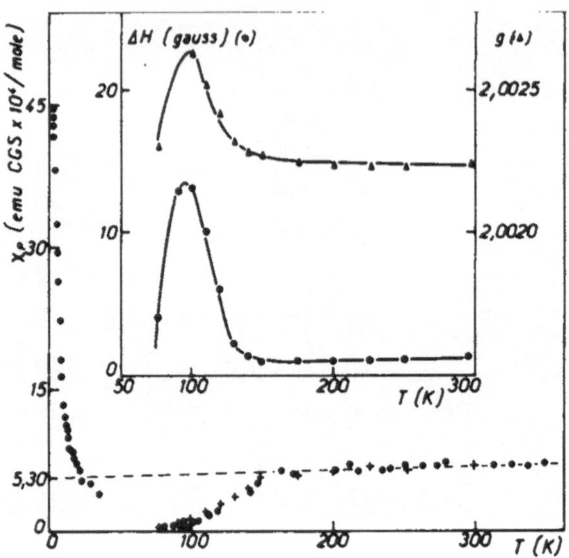

Figure 3 : *Thermal variations of paramagnetism, EPR linewidth and g-factor. The static measurements (X) are given after subtracting the diamagnetic component (χ_d = 5.91.10^{-4} uem C.G.S./mole). The dynamic measurements (.) are relative values adjusted to the static paramagnetism at 295 K.*

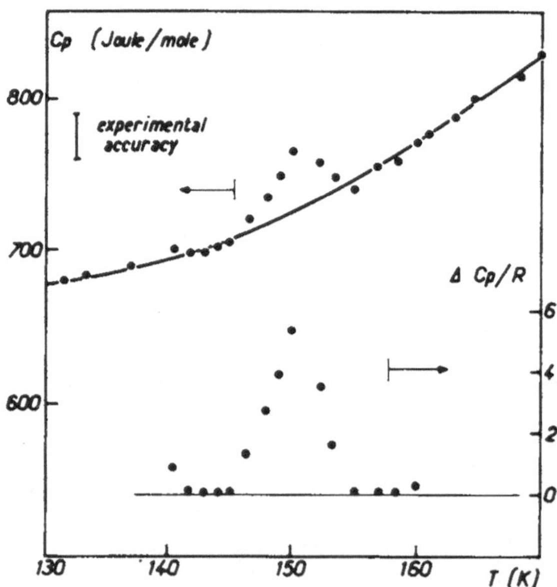

Figure 4 : *The specific heat anomaly around 150 K : The upper figure presents the reduced specific heat (R : perfect gas constant) after subtracting the monotonous thermal variation (full line in the lower figure).*

Conference on Organic Conductors and Semiconductors, Siófok, Hungary 1976

STRUCTURE, PHYSICAL PROPERTIES AND PHASE TRANSITION OF A TCNQ SALT OF 1 : 2 STOICHIOMETRY : DECA $(TCNQ)_2$

S. FLANDROIS, M.L. CHOUKROUN, P. DELHAES

Centre de Recherches Paul Pascal, Domaine Universitaire 33405 - TALENCE

D. CHASSEAU, J. GAULTIER, C. HAUW, M.T. HIRABOURE

Laboratoire de Cristallographie, Domaine Universitaire, 33405 - TALENCE

and
P. DUPUIS

Laboratoire de Chimie Physique Macromoléculaire, 54042 NANCY - FRANCE

In this paper, we discuss an irreversible phase transition occuring in the TCNQ salt which diethylcyclohexylammonium cation : DECA $(TCNQ)_2$, when heated above 350 K. The transition, which is of first order as evidenced by heat capacity measurements, is accompanied by very large discontinuities in properties such as magnetic susceptibility and d. c. electrical conductivity. The phase diagram $[P = f(T)]$ has been determined, as well as the crystal structures of both phases. The observed structural transformation allows us to explain the temperature behaviour of the physical properties.

I - INTRODUCTION

A few TCNQ salts are known to undergo phase transitions : the triphenylmethylphosphonium $^{+}(TCNQ)^{-}_2$ (1,2) the morpholinium $^{+}$ $TCNQ^{-}$ (3) and some simple alkali salts (4). We report here experimental determinations on a phase transition of the diethylcyclohexylammonium salt : $[(C_2H_5)_2(C_6H_{11})\ NH]^{+}$ $(TCNQ)_2^{-}$ which is distinguishable from the previous ones by large discontinuities of the observed physical properties (5).

The salt was prepared following the methods of ACKER and MELBY (6) from para bi(dicyanomethylbenzene). It was purified by recrystallisation in acetonitrile. The stoichiometry was controlled by elementary chemical analysis and spectroscopy.

II - PHYSICAL PROPERTIES

1) *Specific heat*

The thermal variation of specific heat measured by adiabatic calorimetry is shown in fig. 1. At about 347 K a jump of C_p occurs which corresponds to a first order transition. This is confirmed by D. S. C. measurement which gives an enthalpy of transition $\Delta H = 1160 \pm 100$ cal/mole. This transition which presents an endothermic signal is however irreversible, at least down to 4 K. To obtain the initial phase it is necessary to crystallize again the transformed product in acetonitrile. However the irreversibility is not due to solvent inclusion as evidenced by mass spectroscopy analysis.

2) *Magnetic susceptibility*

The thermal variation of paramagnetism has been determined for different samples (fig. 2). Curve ① corresponds to the initial phase before the transition (lower curve) and after (upper curve). Curves ② and ③ are related to mixtures of the two phases which can be obtained by other crystallization conditions (7).

Before the transition, the paramagnetic susceptibility can be interpreted by a singlet-triplet law, with an energy distance between the singlet and excited triplet states $J \simeq 0.08$ eV. After the transition the paramagnetism obeys within the experimental accuracy a pure Curie-law with one unpaired electron by molecule

3) *Electrical d. c. conductivity*

The thermal variation of the electrical resistivity exhibits at the transition temperature a large discontinuity (fig.3). A change of about four orders of magnitude is observed, the salt going from a state of good conductivity to an insulating one. This jump is exceptionnal with regard to the other known phase transitions occuring in anion radical salts and can be compared to the metal-insulator transitions found in inorganic compounds.

This discontinuity of resistivity was used to determine the phase diagram P = f (T) (fig. 4). The transition temperature change is very large with increasing pressure. The slope of the curve, $\frac{dP}{dT} \simeq 16$ Kg cm^{-2} deg^{-1}, allows us by using the Clapeyron relation to evaluate the volume change $\Delta V \simeq 15$ $\mathring{A}^3$ for one molecule. This value is in agreement with the crystallographic data : $\Delta V = 16.5$ $\mathring{A}^3$ per molecule. Morever this result confirms the first order phase transition.

III - CRISTALLOGRAPHIC STRUCTURES

Both phases have been studied :

1) *Conductive phase (low T)*

The unit cell is triclinic. The TCNQ stacks are organized in tetrads with two kinds of overlapping (fig. 5) : the same for each TCNQ pair belonging to a tetrad and the second one between two successive tetrads.

Furthermore from the measured bond lengths we can deduce that inside each tetrad the two central TCNQ molecules (A and $\bar{A}$ on the fig. 5) bear less charge than the two others. This distribution gives an explanation for the paramagnetic and semiconductor behaviour.

2) *Insulating phase (high T)*

The unit cell is also triclinic, but with an angle of almost 90° : the stack axis is very near from the c-axis. The TCNQ entities are organized in diads with respective distance 3.30 and 3.34 Å (fig. 6).

The bonds lengths examination shows that one TCNQ molecule (A on the figure) is neutral whereas the other one (B) bears one electron. Therefore the distance between two electrons is equal to the C distance : 6.64 Å and the electronic and magnetic interactions are negligible. So we get a Curie-law for the paramagnetism and an insulating behaviour.

IV - CONCLUSION

We have shown on this TCNQ salt that a kind of conductor-insulating phase transition exists. Several models have been proposed in order to explain the phase transitions occuring in ion-radical systems, for example by CHESNUT (8) and Mc CONNEL (9). However, these models do not seem valuable to explain the mechanism of the transition In particular, the observed entropy change (3.3. cal deg^{-1} $mole^{-1}$ is much larger than that one predicted by the models. An interesting fact must however be pointed out : by heating a change from a dimeric structure to a monomeric one is observed with a decrease of a factor two of the stacks axis (13.16 Å and 6.64 Å for the low and high temperature forms resp.). The system goes from a dimerized chain at low temperature with an activated paramagnetism to a regular chain of spins (Heisenberg antiferromagnet with a very weak interaction between the spins : $J \simeq 0$).

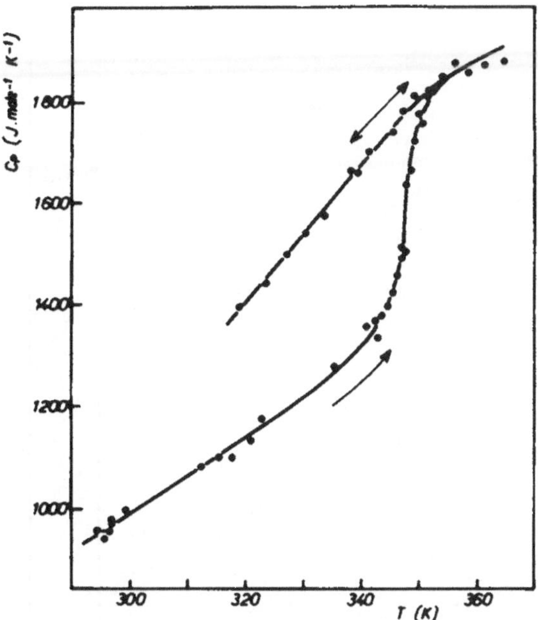

Fig. 1 : Thermal variation of the specific heat

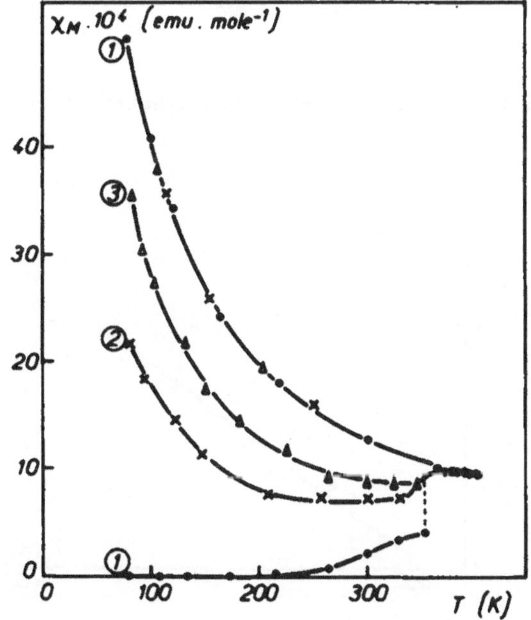

Fig. 2 : Thermal variation of the paramagnetism

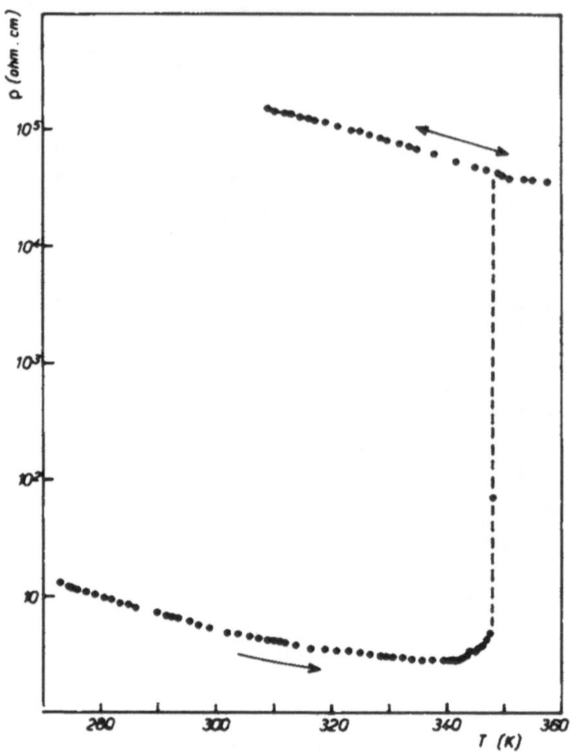

Fig. 3 : thermal variation of the electrical resistivity

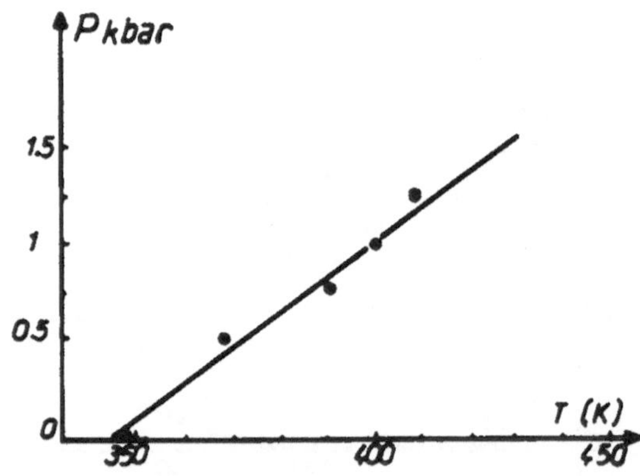

Fig. 4 : Phase diagram P = f (T)

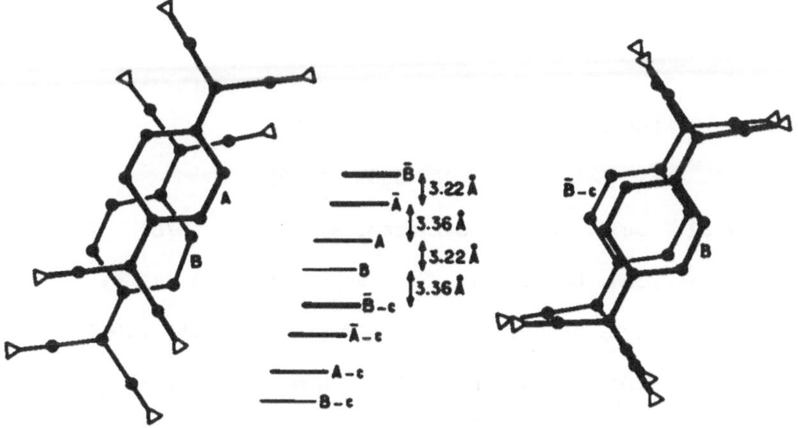

Fig. 5 : Crystal structure of the low T phase : tetrads with distances and overlappings

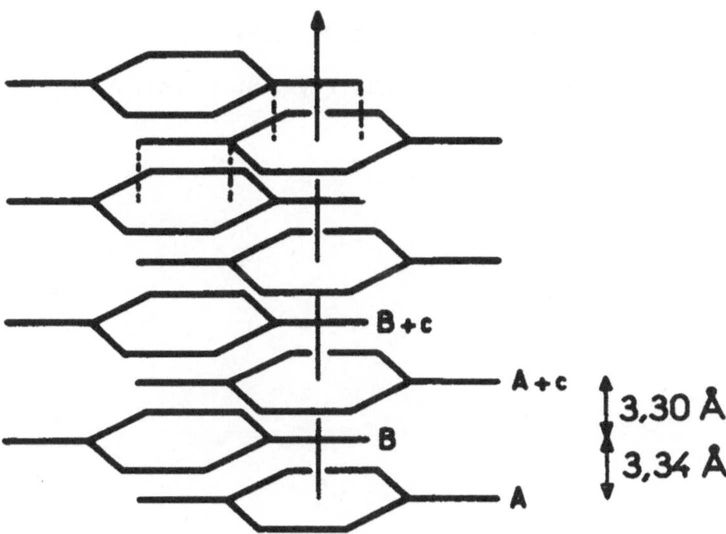

Fig. 6 : Crystal structure of the high T phase : diads with distance and overlapping

REFERENCES

1 - R. G. KEPLER, J. Chim. Phys., 39, 3528, (1963)

2 - Y. IIDA, Bull. Chem. Soc. Japan, 43, 3685 (1970)

A. KOSAKI et al., Bull. Chem. Soc. Japan, 43, 2280 (1970)

Y. SUSUKI and Y. IIDA, Bull. Chem. Soc. Japan, 46, 683 (1973)

3 - J. C. BAILEY and D. B. CHESNUT, J. Chim. Phys., 51, 5118 (1969)

4 - J. G. VEGTER, T. HIBMA and J. KOMMANDEUR, Chem. Phys. Lett. 3, 427 (1969)

J. G. VEGTER and J. KOMMANDEUR, Mol. Liq. Cryst., 30, 11 (1975)

5 - S. FLANDROIS, P. DELHAES, J. C. GIUNTINI and P. DUPUIS, Phys. Letters, 45 A, 339, (1973)

6 - D. S. ACKER et al., J. A. C. S., 82, 6408 (1960)

L. R. MELBY et al., J. A. C. S., 84, 3374 (1962)

7 - S. FLANDROIS, P. LIBERT and P. DUPUIS, Phys. Stat. Sol. A, 28, 411 (1975)

PHASE TRANSITION IN $P_y(TCNQ)_2$

K. HOLCZER, G. MIHÁLY, K. PINTÉR, A. JÁNOSSY,
G. GRÜNER* and W.G. CLARK**

*Central Research Institute for Physics, Budapest, Hungary
**Laboratoire de Spectrometrie Physique, Grenoble Cedex, France
Permanent address: Physics Department, University of California, Los Angeles, USA

A phase transition at 80 ^{o}K was investigated by magnetic susceptibility, low field ESR, dc and microwave conductivity and dielectric constant measurements. The magnetic properties are discussed in terms of a singlet-triplet model above, in terms of 1D Heisenberg model below the phase transition. A possible explanation in terms of the Zawadowski-Cohen model is presented.

Pyridine(TCNQ)$_2$ belongs to the intermediate conductivity TCNQ salts according to the classification of Kepler (1) and Shchegolev (2), its room temperature conductivity is $\sigma = 0.5\,\Omega^{-1}\text{cm}^{-1}$ the activation energy is $\Delta E = 0.1$ eV. We report a phase transition in this material between states having different one dimensional magnetic properties. The magnetic susceptibility, shown in Fig. 1 indicates two phase transitions, one at 150 ^{o}K and an other around 80 ^{o}K. The low frequency ESR signal undergoes a sharp change at the low temperature phase transition, but there is no change in the temperature dependence of ΔH around 150 ^{o}K. The dc and the microwave conductivity measured along the long axis reflect both phase transitions, a small hyteresis around 150 ^{o}K shows this to be of weakly first order. The microwave dielectric constant is higher below 80 ^{o}K and undergoes a sharp reduction above this temperature.

The weakly first order phase transition at 150 ^{o}K does not lead to sharp changes in the various physical properties; we believe at this temperature a small jump in the lattice parameter often observed in these types of salts (3) occurs leaving the overall behaviour unchanged.

The presence of a narrow and unshifted ESR signal at all temperatures excludes 3D antiferromagnetic ordering being an explanation of the phase transition at 80 ^{o}K. After substraction of the low temperature Curie tail the susceptibility is finite as $T \to 0$. We have fitted the low temperature part of

χ to the 1D Heisenberg model. At high temperatures the behaviour is indicative of a singlet-triplet behaviour

$$\chi(T) = \frac{2}{3}\frac{N\mu^2 g^2}{kT}\frac{1}{(1+\frac{1}{3}e^{-J/kT})} \quad .$$

Both fits describe well the susceptibility in the whole temperature range in question. In the low temperature range, where the regular Heisenberg antiferromagnet gives an adequate fit we obtain $J = 64\ ^{o}K$, in the high temperature singlet-triplet region $J = 153\ ^{o}K$.

The dc conductivity shows a reduction of the single particle gap in the low temperature region. Somewhat below room temperature the gap obtained from the fit $\sigma = \sigma_o \exp(-\Delta E/kT)$, $\Delta E = 0.1$ eV, below 80 ^{o}K $\Delta E = 0.05$ eV. The large dispersion of the conductivity suggests that hopping between localized states also contributes to the conductivity at low temperatures, however the reduction of the gap, as determined from the dc conductivity is well demonstrated (Fig. 2).

The dielectric constant (Fig. 3) $\epsilon' = 20$ at room temperature and is weakly temperature dependent down to 100 ^{o}K. The change of a factor of 4 of ϵ' at the transition and the more rapid temperature dependence below the transition indicates a rearrangement of the electronic states.

In the absence of available crystal structure of $Py(TCNQ)_2$ we can only speculate on the reasons for this peculiar phase transition which leads to a reduction of the collective and single particle gaps (we note that this behaviour is the opposite of what is observed in spin - Peierls systems (3)).

We believe that due to the large donor dipoles the donors are alternating (with the N-H group up or down). As there are 1N TCNQ and N donor chains in these types of materials, the model proposed by Zawadowski and Cohen (4) can be applied. In this model (Fig. 4) there are boxes containing two TCNQ sites and one donor (as required by the chemical formula $D(TCNQ)_2$). The electrons transfered by the donors to the TCNQ-s can not move out from their boxes. Hopping is possible only between the two TCNQ sites in the same box with probability t. The other parameters of the model are the nearest neighbour exchange and Coulomb interactions J and U between the electrons of neighbouring boxes. Depending on the relative magnitude of these parameters the model has different ground states.

In the case of $t \ll 3|J|-U$ the ground state is that of weakly coupled singlet pairs and the susceptibility shows a singlet-triplet temperature dependence. The opposite limit $3|J|-U \gg t$ leads to a 1D Heisenberg model with an effective exchange interaction J_{eff}. The lower limit for J_{eff} is $J/4$. Al-

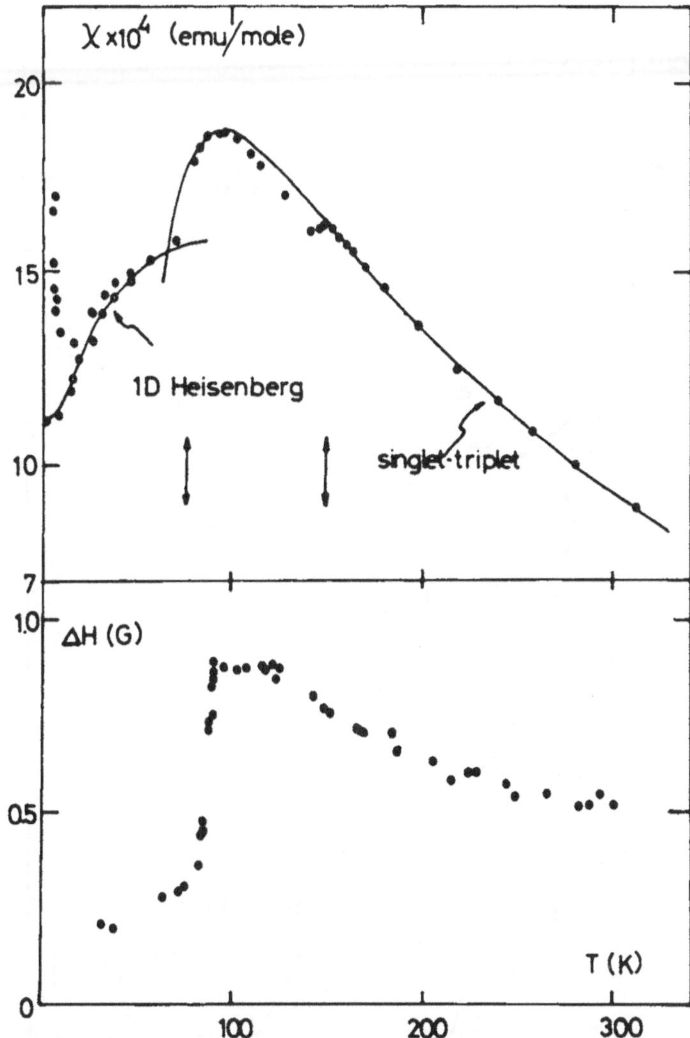

Fig. 1 a/ Susceptibility of $Py(TCNQ)_2$. The small Curie term has been subtrated at low temperatures ($T < 20\ ^oK$). The full lines are fits to 1D Heisenberg and singlet-triplet model.

b/ Low frequency ESR linewidth (peak to peak value of the derivative ESR signal).

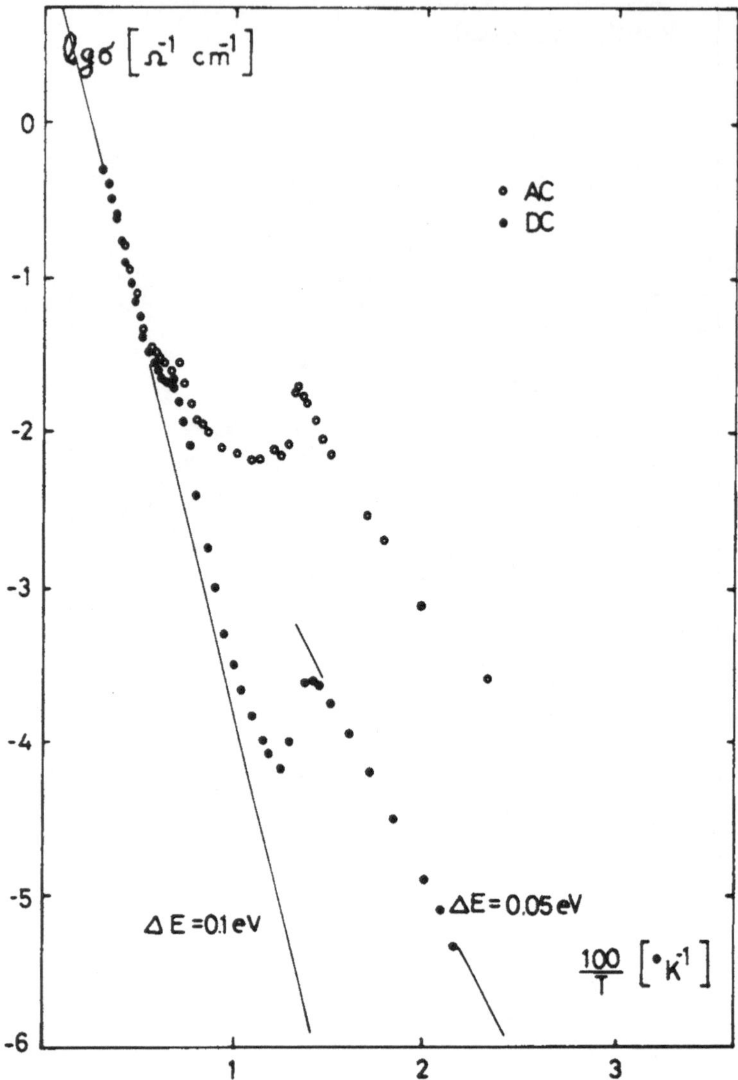

Fig. 2 dc and microwave conductivity versus T^{-1}

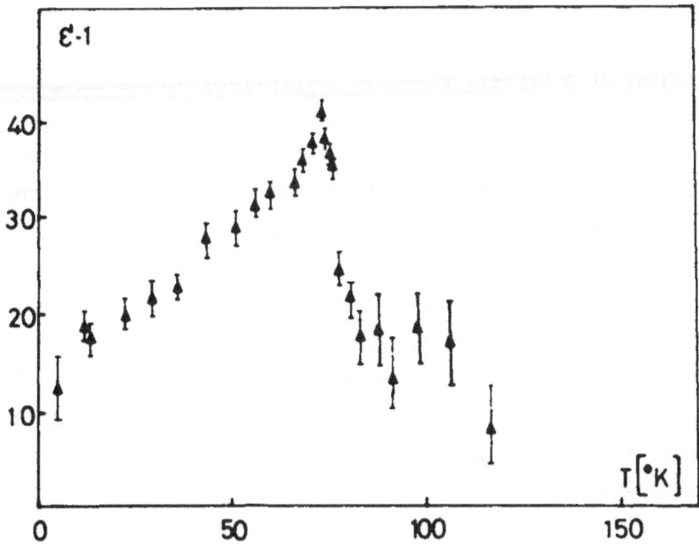

Fig. 3 Microwave dielectric constant versus temperature

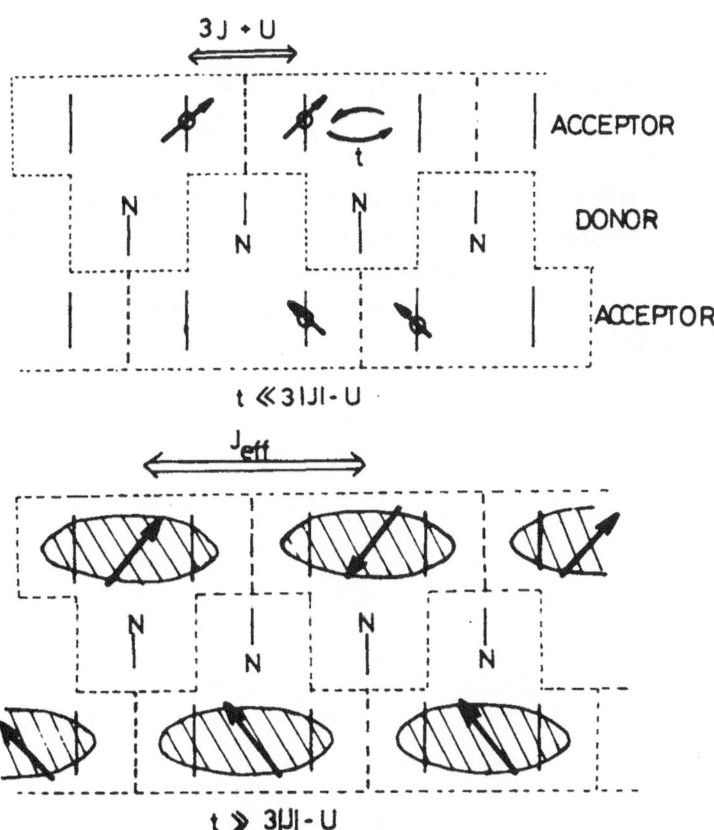

Fig. 4 Ground state configurations of the Zawadowski-Cohen model. The donor arrangement is also shown.

though the intermediate case $t \sim 3|J|-U$ has not been solved it is natural to assume that there is a critical value of $^{t}/(3|J|-U)$ where the ground state changes from one limit to the other. Thus one can interpret the phase transition as a crossover from one limit to the other with $^{t}/(3/J/-U)$ growing as the temperature decreases as a consequence of lattice contraction or interactions with phonons (5).

This explanation is consistent with the relatively small change in J between the two phases, and explains qualitatively the peculiar behaviour of the dielectric constant as well as the conductivity.

Susceptibility experiments were performed in cooperation with Dr. M. Miljak and Dr. J. Cooper. Two of us (W. G. C. and G. G.) are grateful to Dr. M. Minier and Dr. P. Ségransan for the hospitality in Grenoble, where the ESR experiments were performed.

REFERENCES

(1) R. G. Kepler, J. Chem. Phys. 39, 352 (1963)

(2) I. F. Shchegolev, Phys. Stat. Sol. 12, 9 (1972)

(3) I. W. Bray, H. R. Hart, Jr., L. V. Interrante, T. S. Jacobs, J. S. Kasper, G. D. Watkins, S. H. Wee, J. C. Bonner, Phys. Rev. Lett. 35, 744 (1975)

(4) A. Zawadowski, M. H. Cohen, to be published

(5) M. H. Cohen, private communication

(6) P. I. Kuindersma, G. A. Sawatzky and J. Kommandeur, J. Phys. C. 8, 3005 (1975)

Conference on Organic Conductors and Semiconductors, Siófok, Hungary 1976

DYNAMIC NUCLEAR POLARIZATION IN TTF-TCNQ AND K. TCNQ

J. ALIZON, G. BERTHET, J.P. BLANC, J. GALLICE and H. ROBERT

Laboratoire d'Electronique et Résonance Magnétique,[†]
Université de Clermont, B.P. 45 - 63170 - AUBIERE - France

The temperature dependence of the dynamic nuclear polarization enhancement (A_L) of the protons in tetrathiafulvalinium tetracyanoquinodimethane (TTF-TCNQ) and potassium TCNQ (K.TCNQ) is reported. Theoretical expression of A_L is given ; experimental results are discussed and allow us to conclude about the behaviour of $\tilde{R}(\omega)$, the Fourier transforms of the dynamic autocorrelation functions $< S_i^z \, S_i^z \, (t) >$.

The scheme of the apparatus is shown in the figure 1.

The sample, inside of the nuclear coil, is put in the microwave cavity working in the T.E. 112 mode at 9.6 GHz. The nuclear signal is observed with a pulse spectrometer at about 14.6 MHz . By saturating the electron resonance, the nuclear signal is enhanced and we measure the temperature dependence of this enhancement.

In practice, the electron resonance is only partially saturated and the nuclear signal enhancement at an infinite microwave power, A_L, is obtained by extrapolating the curves $[I(P) - I(o)]^{-1}$ versus P^{-1} . I_o and I(P) are respectively the nuclear signals without and with polarization ; P is the microwave power.

The results are plotted on the figures 2 and 3 for K.TCNQ and TTF-TCNQ. In the case of the potassium salt, the accuracy is about 10% but in the case of TTF-TCNQ the saturation of the electron resonance is very small and then the accuracy very poor. However, we can conclude that :

[†]Equipe de Recherche Associée au C.N.R.S. N° 90

- for K.TCNQ, A_L is a temperature dependent
- for TTF-TCNQ, A_L is temperature independent.

We shall now discuss the experimental results.

Theoretically, A_L and the proton spin-lattice relaxation time may be written[1] :

$$A_L = \frac{|\omega_S|}{|\omega_I|} \frac{(|A|^2 - |B|^2)\ \tilde{R}(\omega_S)}{(|A|^2 + |B|^2)\ \tilde{R}(\omega_S) + \frac{|C|^2}{2}\ \tilde{R}(\omega_I)}$$

$$\frac{1}{T_1} = \alpha\ 8\ \pi\ [(|A|^2 + |B|^2)\ \tilde{R}(\omega_S) + \frac{|C|^2}{2}\ \tilde{R}(\omega_I)]$$

- A, B, C are the hyperfine constants ; the scalar part is included in A only, B and C represent the anisotropic part (dipolar part). In the case of TTF-TCNQ, because of the presence of protons and paramagnetic electrons in both TTF and TCNQ chains, $|A|^2 = \frac{1}{2}[|A_Q|^2 + |A_F|^2 + 2\ |A_{QF}|^2]$ where A_Q, A_F A_{QF} describe respectively the interaction on TCNQ, TTF, and TCNQ-TTF. We have assumed the strong coupling limit of the two chains for electronic [2,3] and nuclear spins [4].

- $\alpha = 1$ for K.TCNQ ; $\alpha \simeq 0,7$ for TTF-TCNQ [5]

These expressions have been calculated in the case where the system obeys the following conditions :

- the Zeeman relaxation arises mainly from electron-proton interaction,
- the mechanisms of nuclear diffusion are very fast and may be neglected,
- nuclear and electronic Zeeman reservoirs are defined respectively by only one spin temperature .

We shall investigate the various cases which may occur :

1°) $\tilde{R}(\omega_S)$ and $\tilde{R}(\omega_I)$ have the same behaviour

a - $\tilde{R}(\omega_S) = \tilde{R}(\omega_I) = $ ct.

then $\frac{1}{T_1} \alpha \left(|A|^2 + |B|^2 + \frac{|C|^2}{2} \right) \tilde{R}(\omega_I)$

and

$$A_L = \left(\frac{\omega_S}{\omega_I}\right) \frac{|A|^2 - |B|^2}{|A|^2 + |B|^2 + \frac{|C|^2}{2}}$$

T_1 is frequency independent

A_L is temperature independent

b - $\tilde{R}(\omega_S)$ and $\tilde{R}(\omega_I)$ are proportional to $\nu^{-1/2}$

then $\frac{1}{T_1} \alpha \left(|A^2| + |B|^2 + \frac{1}{\sqrt{660}} \frac{|C|^2}{2} \right) \tilde{R}(\omega_S)$

and

$$A_L = \left(\frac{\omega_S}{\omega_I}\right) \frac{|A^2| - |B|^2}{|A|^2 + |B|^2 + 26 \frac{|C|^2}{2}}$$

T_1 is frequency dependent

A_L is temperature independent

2°) $\tilde{R}(\omega_S)$ and $\tilde{R}(\omega_I)$ have not the same behaviour

a - If $\frac{|C|^2}{2} \tilde{R}(\omega_I)$ is not negligible against

$\left(|A|^2 + |B|^2 \right) \tilde{R}(\omega_S)$,

T_1 is frequency dependent

A_L is temperature dependent

b - If $\frac{|C|^2}{2}\tilde{R}(\omega_I)$ is negligible, then

- T_1 has the same variation as $\tilde{R}(\omega_S)$

- A_L is temperature independent

c - If $\frac{|C|^2}{2}\tilde{R}(\omega_I)$ is very important, then

- T_1 has the same variation as $\tilde{R}(\omega_I)$

- A_L is temperature dependent.

All these assumptions and the conclusions to which they lead are summarized in table I.

Table I

Assumptions		Frequency depend. of T_1	Temperature depend. of A_L
$\tilde{R}(\omega_e)$ and $\tilde{R}(\omega_n)$ have the same behaviour	$\tilde{R}(\omega_e) = \tilde{R}(\omega_n)$	NO	NO
	$\tilde{R}(\omega_e) \propto \omega^{-1/2}$, $\tilde{R}(\omega_n) \propto \omega^{-1/2}$	YES	NO
$\tilde{R}(\omega_e)$ and $\tilde{R}(\omega_n)$ have not the same behaviour	$\frac{|C|^2}{2}\tilde{R}(\omega_n)$ is not negligible	YES	YES
	$\frac{|C|^2}{2}\tilde{R}(\omega_n)$ is negligible	like $\tilde{R}(\omega_e)$	NO
	$\frac{|C|^2}{2}\tilde{R}(\omega_n)$ is very large	like $\tilde{R}(\omega_n)$	YES

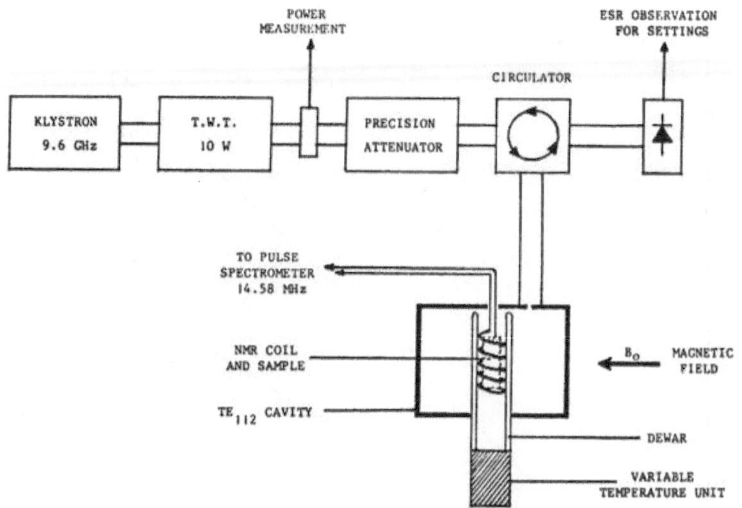

Fig 1 : Scheme of the apparatus

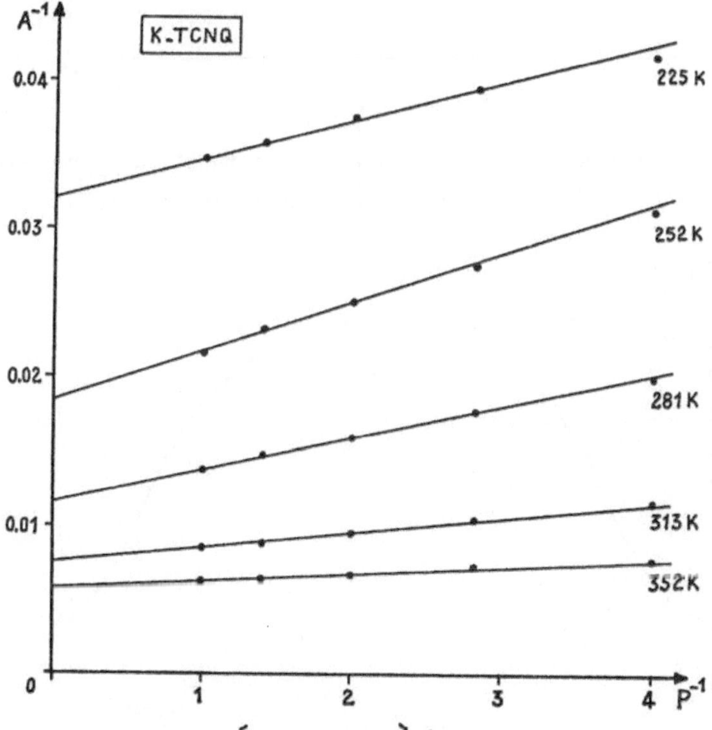

Fig 2 : Enhancement $A = \left[\frac{I(P) - I_o}{I_o} \right]^{-1}$ as a function of reciprocal power for K. TCNK.

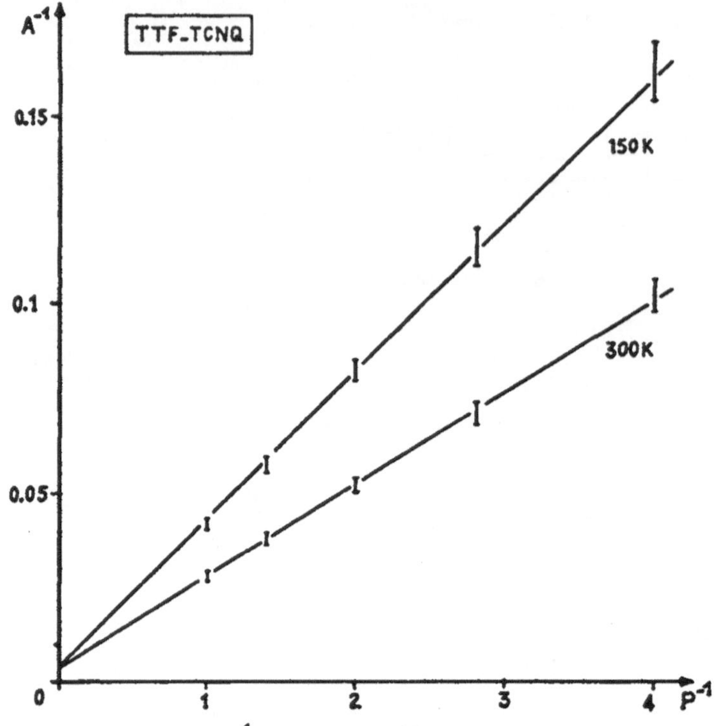

Fig 3 : Enhancement $A = \left[\frac{I(P) - I_o}{I_o}\right]^{-1}$ as a function of reciprocal power for TTF-TCNQ.

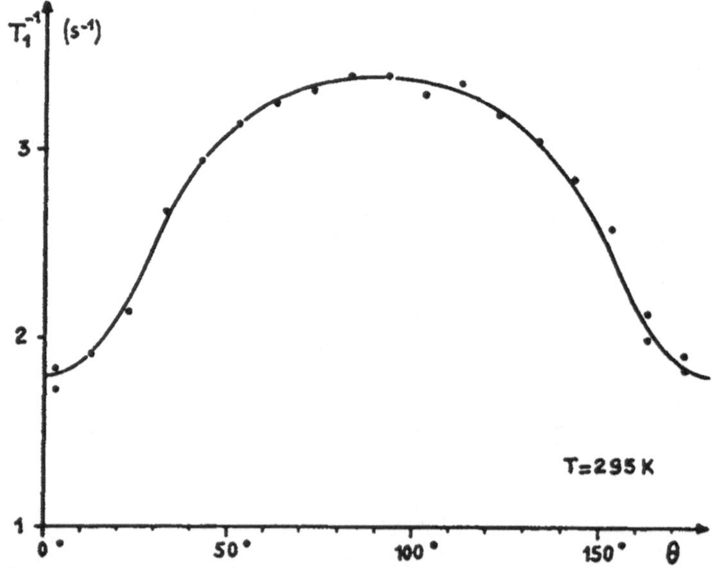

Fig 4 : T_1 anisotropy in K.TCNQ.

In the case of TTF-TCNQ, T_1 measurements [6] show a frequency independence in the frequency range about 14 MHz ; moreover, we have seen that A_L is temperature independent. Only the first assumption leads to conclusions in agreement with experimental results. So, we can conclude that in TTF-TCNQ, $\tilde{R}(\omega_S) = \tilde{R}(\omega_I)$. This result confirms that the interchain coupling is larger than 10 GHz [7].

In the case of K-TCNQ, the third assumption is consistent with experimental results. $\tilde{R}(\omega_S)$ and $\tilde{R}(\omega_I)$ have not the same behaviour and $\frac{|C|^2}{2} \tilde{R}(\omega_I)$ is not negligible. C being anisotropic, one may expect T_1 to be anisotropic ; because of the large enhancement observed by dynamic nuclear polarization, it is possible to measure T_1 on about twenty single crystals aligned along their long axis. The results plotted in the figure 4 confirm our conclusion.

REFERENCES

[1] J. ALIZON, et al. to be published (1976)

[2] Y. TOMKIEWICZ, A.R. TARANKO and J.B. TORRANCE, Phys. Rev. Letters 36, 13, 751 (1976).

[3] E.F. RYBACZEWSKI, L.S. SMITH, A.F. GARITO, A.J. HEEGER and B.G. SILBERNAGEL, Preprint (1976).

[4] This fact is proved in NH_4^+ $TCNQ^-$
J. ALIZON, G. BERTHET, J.P. BLANC, J. GALLICE and H. ROBERT, Phys. Stat. Sol.(b) 65, 577 (1974).

[5] F. DENOYER, F. COMES, A.F. GARITO and A.J. HEEGER, Phys. Rev. Lett. 35, 445 (1975).

[6] G. SODA, D.JEROME, M. WEGER, J.M. FABRE and L. GIRAL, Solid State Comm. 18, 1417 (1976).

[7] D. JEROME and L. GIRAL,
G. SODA et al.
To be published in the proceeding of the "Conference on Organic Conductors and Semiconductors" Siofok (1976).

Conference on Organic Conductors and Semiconductors, Siófok, Hungary 1976

ELECTRICAL AND MAGNETIC PROPERTIES OF COMPLEXES OF 2,2',6,6'-TETRAMETHYL-4,4'-BIPYRYLENE (TMBP) AND 2,2',6,6'-TETRAMETHYL-4,4'-BITHIOPYRYLENE (TMBTP) WITH HEXACYANOBENZENE (HCNB)

A. CHYLA

Institute of Physical and Organic Chemistry, Wroclaw Technical University, Poland

Z. ROMASZEWSKI

Institute of Physics, Polish Academy of Sciences, Warszawa, Poland

Electrical and magnetic properties of complexes of TMBP and TMBTP with HCNB are presented. If seems that the observed paramagnetism of TMBP HCNB complex can be separated into an extrinsic Curie term and the intrinsic component, the latter being only slightly temperature dependent.

A comparison of experimental data with results of quantum-chemical calculations in PPP approximation is also attempted.

Introduction

During the last years much attention has been given to the investigation of ion-radical TCNQ complexes possessing many interesting electrical and magnetic properties. These organic molecular crystals are generally built of donor and acceptor stacks. Such a structure facilitates current transport and therefore these substances exhibit large electrical conductivities. In the search for new well conducting organic compounds we have synthesized complexes of 2,2',6,6'-tetramethyl - 4,4'-bipyrylene /TMBP/ and 2,2',6,6'-tetramethyl-4,4'-bithiopyrylene /TMBTP/ with hexacyanobenzene /HCNB/ as am acceptor /electron affinity - 2.5 eV/. The components were purified and amorphous complexes crystallized from CH_3CN solution and then by recrystallization from DMF single crystals /0.5^x1^x4mm/ were obtained. The measurements of various physical properties including d.c. and microwave conductivity, static magnetic susceptibility. EPR, IR and UV spectra were carried out on the obtained materials.

Experiment

Measurements on d.c. conductivity were carried out by four-probe technique in the temperature range 77 to 350 K. The microwave investigations were performed in the same temperature range. The temperature dependence of both d.c. and microwave conductivities of TMBTP.HCNB complex exhibit an exponential character over the whole temperature range with the same activation energy of 0.45 eV, with the conductivity changing from $10^{-4}om^{-1}cm^{-1}$ at room temperature to $10^{-7}om^{-1}cm^{-1}$ at liquid nitrogen temperature /see Fig. 1/. The room temperature conductivity of TMBP.HCNB complex equals $2\cdot10^{-2}om^{-1}cm^{-1}$. The temperature dependence of the d.c. conductivity is exponential with activation energy of 0.19 eV; the microwave conductivity follows this dependence only to about 170 K, becoming temperature independent below this temperature /see Fig. 2/.

In all cases no difference between the measurements on poly- and single crystalline material was noticed.

The dielectric constant of TMBTP.HCNB determined from microwave measurement amounts to 3.9 and is temperature independent whereas that of the TMBP.HCNB complex increases from 12 at 77 K to about 15.5 at 220 K, an d is almost temperature independent above this temperature /see Fig. 3/.

A similar behaviour of the microwave conductivity and dielectric constant was reported for many well- and moderately conducting TCNQ complexes /1/. It is often attributed to the hopping of electrons between localized states in the band gap, in a similar way as observed in amorphous semiconductors.

The static magnetic susceptibility of the two complexes determined by the Faraday method in the temperature range 4.5 to 350 K was positive. The susceptibility of TMBTP.HCNB was nought /experimental error - $0.06 \cdot 10^{-6}$ emu/mole/ and temperature independent, whereas the susceptibility of TMBP.HCNB was about $0.7 \cdot 10^{-4}$ emu/mole and did not depend on temperature above 5o K. Below this temperature it increased according to the Curie law /c = 0.0058 emu.K/mole/ /see Fig.4/.

It seems that we can separate the observed paramagnetism as an extrinsic Curie term and into intrinsic component which is only slightly temperature dependent.

Temperature independent paramagnetism often observed /2 - 5/ in highly conducting complexes is usually explained as being due to Pauli paramagnetism of a degenerated electron gas. In our case, however, the low conductivity values make such an explanation rather dobious.

Unusual magnetic and electrical properties might also arise from quasi one-dimensional crystal structures of these compounds. The acceptor stacks /especially TCNQ/ may be either regular, i. e. with equally spaced molecules, or alternating when composed of diads, triads or tetrads. In the latter case some substances exhibit EPR spectrum characteristic of mobile, thermally activated triplet states /triplet excitons/. The spectrum may result from the excitation of two or more coupled TCNQ entities /6, 7/. The triplet character of the paramagnetic excitation is shown by the anisotropic two-lines EPR spectrum which results from a zero field splitting of the triplet levels being described by the spin Hamiltonian /8/:

$$H = DS_z^2 + E/S_x^2 + S_y^2/$$

where D and E are the zero field splitting parameters, and S_x, S_y, S_z are spin components of the triplet state. The EPR spectrum of the TMBP.HCNB complex obtained by us consists of two lines, these lines allow us to assume the triplet exciton approximation for the interpretation of paramagnetic susceptibility. The appropriate analysis of χ /T/ dependence and EPR spectrum yields the coupling integral /9/ J=0.045 eV and the splitting parameter D=34 G.

The above results suggest the possibility of the chain structure of the TMBP.HCNB complex. In the absence of X-ray data we have simulated the probable crystallographic structure of this complex using typical values of the bond lengths and we have performed quantum-chemical calculations in the PPP approximation. It was found that the ground state of this complex should be of triplet character, the charge transfer surpasses 1. The calculated value of the S - T energetic distance varied between 0.25 and 0.30 eV depending on the assumed molecular distances. It must be realized however, that the Mataga - Nishimoto /1o/ interpolation employed in our calculations, giving the best agreement with experiment for S - S translations, yields considerably underestimated energies of triplet states. The calculations suggest the presence of a bandrich spectrum of electronic transitions up to 3.5 eV, the band intensities being relatively weak. The calculated values of dipole moment indicate that the bands correspond to the charge transfer. Two of the experimentally found absorption bands /11/ positioned at 2.07 and 2.27 eV and having considerable intensities seem to correspond to the calculated CT bands at 2.00 and 2.16 eV.

It follows from the calculations that the TMBP.HCNB complex should be very strong. This is confirmed by the IR spectrum and by the low value of the D parameter together with large value of J as stated previously.

The authors wish to express their indebtedness to Dr. L.Syper for the synthesis of donors, and to Mr. J. Lipinski for the execution of numerical calculations.

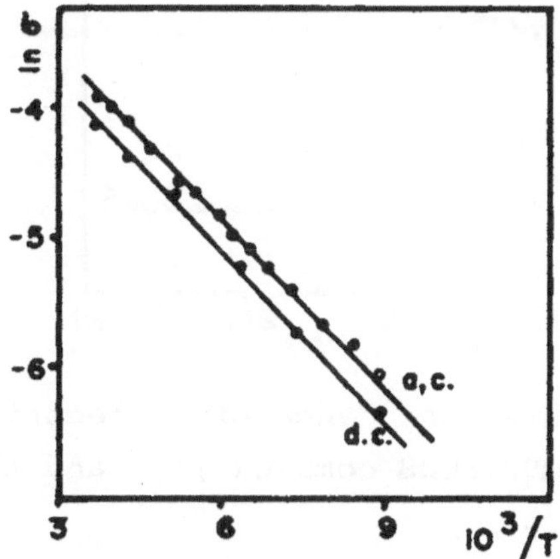

Fig. 1 Temperature dependence of the electrical conductivity of the TMBTP.HCNB complex. Empty circles - microwave conductivity, full circles - d.c. measurements

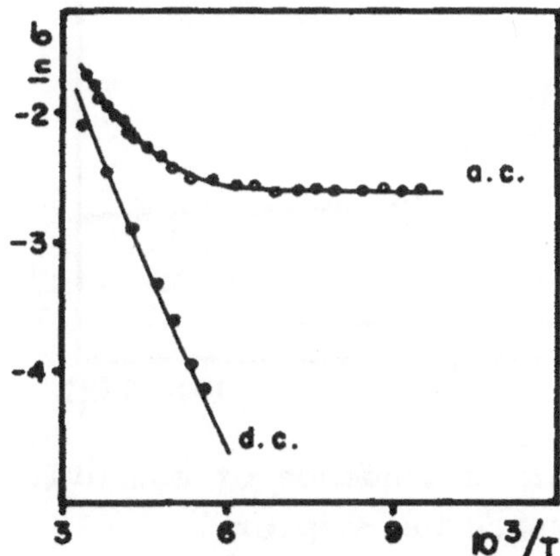

Fig. 2 Temperature dependence of the electrical conductivity of the TMBP.HCNB complex. Empty circles - microwave conductivity, full circles - d.c. measurements

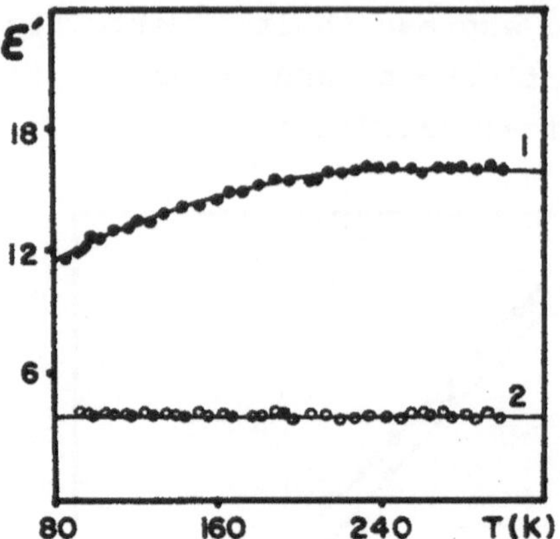

Fig. 3 Temperature dependence of dielectric permittivity of the TMBP.HCNB complex /1/, and the TMBTP-HCNB complex /2/

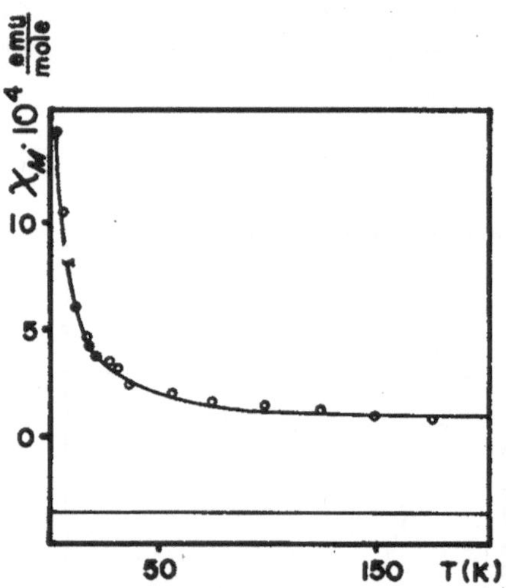

Fig. 4 Temperature dependence of magnetic susceptibility of the TMBP.HCNB complex

REFERENCES

/1/ I.F. Shchegolev, Phys. Stat. Sol. /a/ 12 , 9 /1972/.

/2/ P. Delhaes, S. Flandrois, G. Kreyer, Mat. Res.Bull. lo, 825 /1975/.

/3/ R. G. Kepler, J. Chem. Phys. 39, 3528 /1963/.

/4/ M. Miljak, A. Jánossy, G. Grüner, KFKI Report 75-38 /1975/.

/5/ K. Holczer, G. Mihály, A. Jánossy, G. Grüner, KFKI Report 75-41 /1975/.

/6/ D. B. Chesnut, J. Chem. Phys. 51, 5118 /1969/.

/7/ G. J. Ashwell, D. D. Eley, M. R. Willis, J. Woodward, J.C.S. Faraday II, 71, 1785 /1974/.

/8/ S. Flandrois, J. Amiell, F. Carmona, P. Delhaes, Solid State Commun. 17, 287 /1975/.

/9/ L. N. Bulayevskii, Phys. Tverd. Tela /USSR/ 11, 1132 /1969/.

/lo/ A. Golebiewski, in "Chemia Kwantowa Zwiazków Organicznych" ed. by /PWN, 1973/, p. 245.

/11/ L. Syper, A. Sucharda-Sobczyk, Bull. Pol. Acad. Sc. XXIII, 563 /1975/.

Conference on Organic Conductors and Semiconductors, Siófok, Hungary 1976

INTENSITIES OF THE INFRARED ABSORPTION IN TCNQ SALTS

K. KRÁL, L. DOBIÁŠOVÁ, N. RYŠAVÁ

Institute of Physics, Czechoslovak Academy of Sciences, Prague, Czeschoslovakia

Our work was stimulated by the question as to what is the magnitude of the electronic transfer $\langle n \rangle$ per one TCNQ molecule from the metallic atoms to the TCNQ molecules in the simple alkaline TCNQ salts.

The influence of the magnitude of $\langle n \rangle$ on the optical spectra has recently been discussed in two theoretical papers (1, 2). In Ref. (1) it was assumed that the concentration of the unpaired electrons $\langle n \rangle$ on TCNQ molecules in simple alkaline TCNQ salts is equal to one and the TCNQ chains were represented by a collection of independent dimers of $TCNQ^-$ ions. Then it was found that the strongest absorption in the range of molecular vibrations should correspond to the creation of pairs of intramolecular vibrational quanta in the course of absorption of a single photon. These transitions were shown to be, on average, much more intense than the vibrational absorption of neutral TCNQ ions. The large intensity was shown to be due to a strong admixture of the electronic charge transfer states in the dimers to the vibrational states of the TCNQ molecules. This absorption pattern should be roughly similar, as for the line positions, to the first-overtonic absorption of an isolated TCNQ molecule vibrational transitions, but the intensities should greatly differ from the intensities of the absorption of an isolated TCNQ molecule. The average molar integral extinction coefficient, ε_1 , of an average line in this spectrum was estimated (1) to be about 8.7×10^{17} cm^{-1} sec^{-1} mol^{-1} litre. We understand here that mol means one mol of TCNQ molecules either neutral or ionized.

In paper (2) an interpretation is developed of the absorption spectra of the TCNQ salts with $\langle n \rangle < 1$. Two distinct kinds of motion are supposed to be available for an unpaired electron, either a motion in which the electron moves coupled to the vibrational deformation of a single molecule, or the free motion of an electron, which moves, unbound to any molecular deformation, along the TCNQ stack. The energies of these two kinds of motion may or may not be separated by an energy gap, dependent on the quantities like $\langle n \rangle$, on the depth of the polaronic hole of a single TCNQ molecule, on the magnitude of the charge transfer matrix element along the stack of TCNQ molecules, and so on. In paper (2) the optical absorption was considered for the case of the presence of the gap, namely, for such salts in which the electrons are localized by the interaction with the intramolecular vibrations. Two kinds of light absorption processes were considered. The first one was the absorption due to the electronic transfer from a localized state into a state localized on a nearest neighbour neutral molecule. The other process was the optical transfer from a localized electronic state into the band of the free, or extended, motion of the unpaired electron. The average integral molar extinction coefficient, ε_{ℓ}, of an average line of the spectrum corresponding to the transitions into the localized states can be estimated with the help of the formula for ε_{ℓ} found in Ref. (2) to obtain $\varepsilon_{\ell} = 1.6 \times 10^{17}$ cm^{-1}sec^{-1} mol^{-1} litre, for $\langle n \rangle = 1/2$. In this case the integral molar extinction coefficient ε_{H} can be estimated (2), from the absorption into the high energy half of the free motion band, to obtain $\varepsilon_{H} \approx \varepsilon_{\ell}$. This is in accord with the infrared measurements in the complex salts. For example, in TEA $(TCNQ)_2$ (3) the neighbouring transitions, one of them being identified with the vibrational transition corresponding to the transition into the localized state and the other having the shape of a broad band, have apparently comparable integrals over the absorption curves.

We have measured the optical infrared absorption on the simple alkaline TCNQ salts, Na:TCNQ, K:TCNQ, purple-red RbTCNQ, and also the complex salt Cs_2TCNQ_3 dispersed in KBr pellets. The magnitudes of the integral molar extinction coefficient for the absorption peak attributed to the absorption due to the cyano-group vibrations $C \equiv N$ are: 1.5×10^{16}, 1.7×10^{16}, 1.6×10^{16} and 0.96×10^{16}, respectively, in the units of cm^{-1} sec^{-1} mol^{-1} litre. The strongest absorption peaks observed in the simple alkaline salts can apparently be attributed to the transitions corresponding to the excitation of the vibrational stretching modes according to the identification found in Ref. (4). According to Ref. (2) this means that we observe the optical transitions corresponding to $\langle n \rangle \neq 1$ rather than $\langle n \rangle = 1$. From the close resemblance of the overall shapes of the absorption patterns, similar to those found in Ref. (5), of the measured simple alkaline salts and of the complex Cs salt and from the near coincidence of their absolute intensities we conclude that the simple salts absorb in the infrared like the complex salts. For this reason we conclude that the electronic density $\langle n \rangle$ is perhaps seriously different from unity in the simple salts, in contrast to previous expectations (6).

If the dimers with $\langle n \rangle = 1$ prevailed in the simple salts, the electronically enhanced double phonon absorption (Ref. (1)) would be at least easily observable on the background of the purely vibrational absorption of the salts. This is not observed, however.

According to the above conclusion about the magnitude of $\langle n \rangle$ in the simple salts and according to the theory (2), the presence of broad absorption maximum in the range about 3000 to 5000 cm^{-1} would not then be unexpected in the simple alkaline salts. The presence of such maxima in these salts cannot be excluded on the basis of our measurements on KBr pellets.

REFERENCES

(1) K. Král, Czech. J. Phys. B., (to be published)

(2) K. Král, Czech. J. Phys. B., (to be published)

(3) A. Brau, P. Bruesh, J. P. Farges, W. Hinz and D. Kuse, phys. stat. sol. (b) 62, 615 (1974)

(4) B. Lunelli and C. Pecile, J. Chem. Phys., 52, 2375 (1970)

(5) Z. Iqbal, C. W. Christoe and D. K. Dawson, J. Chem. Phys., 63, 4485 (1975)

(6) I. I. Ukrainskii, V. E. Klymenko and A. A. Ovchinnikov, Preprint ITP-75-89E, Kiew 1975.

Conference on Organic Conductors and Semiconductors, Siófok, Hungary 1976

TRIPLET EXCITONS IN SIMPLE TCNQ SALTS

T. HIBMA and J. KOMMANDEUR

Laboratory for Physical Chemistry University of Groningen
The Netherlands

Electron-electron repulsion implies electron-hole attraction. If the crystal symmetry is such that a band gap occurs, one expects this attraction to lead to the formation of excitons.

When the bandwidth is large these will be Wannier, when it is small these will be Frenkel excitons. The excitons can be of singlet and of triplet character. The latter species can be observed in ESR.

In Rb^+TCNQ^- and TMB^+ /tri-methyl-benzimidazolium/$TCNQ^-$, the former having a dimerized chain-type $TCNQ^-$-structure, the latter having isolated dimers, such triplet excitons were studied. The observed spin-dipolar interaction gives information on the electron-hole separation, the behavior of the linewidths as a function of temperature and angle of the magnetic field yields accurate values of the anisotropy of the exciton motion and shows that the excitons are self-trapped by a lattice distortion.

Comparison of the thermal energy required to excite an exciton with the activation energy for electronic conductivity allows determination of U and t_1, the on-site electron-electron repulsion and the intra-dimer one-electron transfer integral. We find $U \approx 1{,}3$ and $1{,}1$ eV; $t_1 \approx 0{,}4$ and $0{,}35$ eV for the Rb^+ and TMB^+ salt, respectively.

Rb^+TCNQ^- shows a first order Peierls transition at 376°K from a dimeric to a monomeric chain structure. As expected, the triplet excitons dissociate, i.e. their ESR signals merge with those of the electrons and holes, but neither the magnetic susceptibility nor the electronic conductivity then start showing metallic behavior. The Peierls-induced one-electron gap is apparently replaced by a gap due to the two-electron term U. In this region the spin susceptibility varies roughly proportional to temperature in the range where it could be measured. This is in qualitative agreement with numerical results obtained by Shiba and Pincus for the one-dimensional half-filled Hubbard model.

Conference on Organic Conductors and Semiconductors, Siófok, Hungary 1976

INTERCHAIN COUPLING AND DISORDER IN COMPLEX TCNQ SALTS WITH AROMATIC DONORS

G. MIHÁLY, K. RITVAY-EMANDITY, G. GRÜNER

Central Research Institute for Physics, 1525 Budapest, P.O.B. 49., Hungary

The various properties of complex TCNQ salts with aromatic donors are discussed. The donors are asymmetric and the permanent dipole moments were changed systematically from high to low values. It is argued, that large donor dipoles are alternating and a charge ordered state develops. Smaller donor dipoles are randomly oriented. For random donor-acceptor interaction energies larger than the gap which develops due to interplay of intra and interchain Coulomb couplings the materials are disordered 1d metals. In the opposite case small bandgap semiconductors are obtained.

TCNQ forms charge transfer salts[1] with a broad variety of closed shell donors, with composition $D^+/TCNQ/_2^-$. The salts have widely different physical properties in spite of similar structural features, i.e. segregated donor and acceptor stacking. Various models were found to be succesful for accounting for the physical properties of restricted groups of these salts. The basic distinction between the different groups, in particular the relation between the properties of donors and the cooperative phenomena is however only poorly understood.

We attempt, by systematically varying the donor molecules, to correlate the electric and magnetic properties of these salts to the intramolecular donor properties.

The donors we have investigated /and most of the aromatic closed shell donors/ are asymmetric and have large permanent dipole moments in the charge transfer state. In the case of segregated donor and acceptor stacks the electrostatic interaction between the donor dipoles and electrons on the conducting TCNQ chains leads to large site energies which may be random or regular depending on the arrangement of the donor dipoles. We demonstrate below, by discussing experiments performed on complex TCNQ salts with various donors, that this difference is of primary importance in establishing a charge ordered state, a disordered one-dimensional material or a small bandgap semiconductor.

Entropy favours a random donor arrangement along the chain but dipole-dipole interactions may stabilize an ordered donor stacking, shown schematically in Fig.1, in the case of large donor dipoles. As a rough estimate we take two dipoles, with charges $e(1+x)/2$ and $e(1-x)/2$ and length $\ell \sim 2.5$ Å at distance $a \sim 4$ Å apart (ℓ is the size of the donors along the direction of the dipoles, a the interchain distance between donors). With $V = V_{\uparrow\downarrow} - V_{\uparrow\uparrow} \sim 0.4\ X^2$ eV , the critical temperature for disordered donor stacking $kT_c \sim V$. With $T_c \sim 400^\circ K$, the temperature where the materials are prepared we get $X_c \sim 0.4$ for the critical charge difference above which ordered dipole arrangement is expected. For smaller dipoles disorder is more likely. Steric factors may also influence this tendency towards alternating stacking, for example $-CH_3$ groups at one side of the molecules tend to avoid each other, thus also favouring alternating donor arrangement.

The donor dipoles lead to site energies at the TCNQ chain, which is alternating for ordered dipoles and random for disordered dipole stacking. The site energy is of the order of

$$E_{DA} \sim 0.6\ X\ \text{eV} \qquad /1/$$

for typical donor-acceptor distance $r_{DA} \sim 7$ Å.

The intra- and interchain Coulomb interactions between electrons on the TCNQ chains are the subject of extensive study. For a quarter filled band the on-site and nearest neighbour interactions $U_o \sim 1$ eV and $U_1 < 1$ eV are considered in the framework of the extended Hubbard model. The interchain Coulomb coupling is given by[2]

$$V = \frac{e^2}{r_\parallel} K\left(\pi \frac{r_\perp}{r_\parallel}\right) A \qquad /2/$$

where A is the amplitude of the charge density wave developed as a consequence of both on-chain and interchain interactions. For A = 1 the typical interchain Coulomb coupling is of about 0.3 eV. In the absence of donor-acceptor interactions the resulting gap is determined by the interplay of on-chain and interchain correlation, we call this Coulomb gap E_c.

For donors with large X (and with $-CH_3$ groups) the alternating stacking leads to alternating site energies. The large on-site Coulomb interaction excludes double occupancy of the TCNQ sites. The ground state for a quarter filled TCNQ chain (appropriate for the complex salts with composition 1:2) depends on the geometrical arrangement of the donor and acceptor chains, but for localized electrons ($t_\parallel = 0$) the ground state is either that in which every second TCNQ site is occupied by one electron or that of electron pairs separated by empty pairs of sites, as shown in Fig.1 .

With X decreasing a random donor stacking is expected to occur (evidences for this in the case of the NMePh salt are discussed by Theodoru[3]) and then the resulting solid state properties are determined by the balance of random site energy E_{DA} and Coulomb gap E_c. Broadly speaking for $E_{DA} > E_C$ the interchain coupling is washed out resulting in decoupled chains with disorder leading to Anderson localization and hopping conductivity, to random exchange between the electrons, etc. In the opposite case interchain forces are dominating and may lead to a ground state in which charge density waves on neighbouring chains are correlated. The disorder leads then to band tailing, and perhaps to mobility edges.

We have investigated the complex $D^+/TCNQ/_2^-$ salts with a series of donors shown in Fig.2, where the charge distribution (obtained from a tight binding calculation[4]) is also indicated. X is the difference of charge located on the lower and upper part of the molecule. The permanent dipole moment strongly decreases with increasing number of rings; the dipole moment of Py is high, and previous arguments suggest an ordered dipole chain. Qn and Ad have intermediate dipole moment, while for 2-3 BAd the dipole moment is fairly small. The methyl substituion leaves the charge distribution nearly unchanged, but acts as an important steric factor which enhances the probability of alternating donor arrangement. The salts were prepared by the method of Melby[5], the composition was determined both by chemical analysis and optical spectroscopy in solution. The external appearance of single crystals indicates three well distinguished class of materials. In the beginning of the series (salts with donors Pyridine, N-Methyl-Pyridine and N-Methyl--Quinoline) large, brittle crystals were obtained (typical dimensions 5x0.2x0.05 mm). Multiple phases and compounds with various compositons were also found in these salts. The Quinolinium, Acridinium and N-Methyl-Acridinium salts have a rope--like behaviour (typical dimensions 5x0.2x0.02 mm) indicative of loosely bound chains, this is also evidenced by diffuse streaks observed in electron diffraction patterns[6]. Always a 1:2 composition is obtained. The 2-3 Benzacrinidium salt is obtained in the form of small plates and needles (dimensions 1x1x0.1 and 1.5x0.05x0.05 mm, respectively) both forms having a composition 1:2.

Single crystal dc conductivity was measured by four contact method along the long axis of the crystals. The conductivity measured in other directions in a few cases was always lower by a few orders of magnitude confirming the high anisotropy of electron transport. The magnitude and temperature dependence of the conductivity, shown in Figs 3, 4 and 5 are widely different, but can be divided in three classes. The NMePy, NMeQ (and also Py which will be discussed elsewhere[7]) salts are semiconductors with low conductivity, and with a large and well defined gap. The existance of a large single particle gap is further supported by the absence of the dispersion of the conductivity, and also by the small dielectric constant measured at microwave frequencies; $\varepsilon = 3.5$ for the NMeQn and 20 for the Py salt.

The Qn, Ad and NMeAd salts (Fig 4) have large conductivities with a smeared out maximum below room temperature. The dc and microwave conductivities are different below the maximum[1,8] the dielectric constant is enormous and increases with increasing temperature[1]. The low temperature part of the dc conductivity can not be described by a simple activation process[22] and is also sensitive to crystal perfection[13]. At high temperatures, well above the maximum the resistivity was found to be proportional to the temperature over a broad range of temperatures in $Ad(TCNQ)_2$[14].

Both crystal modifications of the 2-3 BAd salt are semiconducting with a well defined activation energy at high temperatures, in the low temperature region (see insert of Fig 5) the data are also well represented by an activated conductivity with a lower activation energy. The thermoelectric power also displays a behaviour typical for a band semiconductor and is well described by $S = \mathrm{const}\left(\frac{\Delta E}{kT} + B\right)$.

We discuss these differences in the transport properties together with the magnetic properties of these compounds on the basis of interfierence of Coulomb coupling between the TCNQ chains and dipole effects described above.

Although we do not have direct crystallographic evidence at present, we suggest that due to large dipole moments at the beginning of the series, and also due to steric factors associated with the $-CH_3$ groups, the donor dipoles are alternating in $Py(TCNQ)_2$, $NMePy(TCNQ)_2$ and $NMeQn(TCNQ)_2$. For the NMe derivatives there is no observable lattice distortion[6], thus the single particle gap is not associated with phonon effects. The alternating site energy E_{DA} splits the band, and a large on-site Coulomb interaction together with a small bandwidth leads to a semiconducting behaviour in a quarter filled band, in the strong coupling limit U_o, $E_{DA} \gg t_{\parallel}$, where $t_{\parallel}$ the on-chain transfer integral, with a gap given by U_o or E_{DA} whichever is the smaller. $U_o \sim 1$ eV[9] in these salts and therefore the alternating site energy is expected to be associated with the observed gap. With $X \sim 0.7$, we obtain $E_{DA} \sim 0.4$ eV, which accounts well for $E_g = 2\Delta E \sim 0.6$ eV. In this "charge ordered" state each second site is occupied by electrons in the quarter filled band assuming full charge transfer . Evidence for this comes from the susceptibility which shows a Curie-Weiss behaviour[10] in NMeQn$(TCNQ)_2$ with weakly coupled magnetic moments in agreement with the ground state shown in Fig. 1a. In Py$(TCNQ)_2$ the singlet-triplet behaviour of the high temperature susceptibility[7] suggests a ground state depicted in Fig 1b. A different stacking geometry may be responsible for this and also for the smaller bandgap /$\Delta E = 0.1$ eV/ observed in this salt.

In this group of materials the electrostatic donor-acceptor interaction is of primary importance in localizing the electrons at particular lattice sites. Both the donor-acceptor and the on-site Coulomb interaction is much larger than the on-chain transfer integral $t_{\parallel} \sim 0.1$ eV, this leads to strong charge ordering and low conductivity. The magnetic properties of these salts can be accounted for by the model proposed by Zawadowski and Cohen[11] in which the effect of donors is explicitly considered. The model leads both to magnetic and non-magnetic ground states, and is also capable of accounting for phase transitions observed in this class of materials. The extension of this model should account also for the transport properties in this strong coupling regime.

In the middle of the series, the Qn, Ad and NMeAd salts have large conductivities and display a behaviour rather similar to that found in the salt NMP TCNQ, and discussed widely in the literature (for references see Refs. 3 and 15). X-ray experiments point to a random donor arrangement[14]; with $X \sim 0.3$ and 0.5 then we obtain a random site energy $E_{DA} \sim 0.2-0.3$ eV. This leads to decoupled chains and no phase transition occurs down to $T = 0$[15]. The basic mechanism of this depression is the smearing of the density of states near band edges which supresses the instabilities and reduces the energy gain due to gap formation. Therefore these materials can be regarded as one dimensional no correlation between the chains with properties determined solely by on-chain correlations and disorder[16]. Additional evidence for the lack of interchain coupling has been obtained from experiments[10] performed in NMeQn(TCNQ)$_2$. Solvent inclusion is observed in this material, the solvent molecules destroy the interchain correlations, and the material displays a behaviour similar to that shown in Fig 4. When the solvent is evaporated, the charge ordered state discussed before develops. Both the transport[12] and magnetic[3] properties can be described by the 1d random Hubbard model

$$H = \sum_{l\sigma} \varepsilon_l a^+_{l\sigma} a_{l\sigma} + U_0 \sum_l n_{l\uparrow} n_{l\downarrow} + \sum_{lm\sigma} t_{lm} a^+_{l\sigma} a_{m\sigma} \qquad /3/$$

where ε_l the random site energy. The model has been applied to NMP-TCNQ[15] (half filled band) but is probably suitable to complex salts which have quarter filled bands.

The random disorder results in an increased conductivity in these materials, due to the absence of electrostatic forces which would lead to a charge ordered state. The disorder however leads also to electron localization and hopping conduction[12]; the combination of the two sets a limit to $\sigma \sim 100\,\Omega^{-1}\mathrm{cm}^{-1}$ in this class of salts, with variable range hopping occuring at low temperatures.

In 2-3 BAd$(TCNQ)_2$ the donor asymmetry is small, and the random site energy is smaller than the interchain Coulomb coupling. The well defined semiconducting gap arises then from interchain coupling which leads to coupled charge density waves (CDW) on the neighbouring chains[2]. The observed gap is comparable to the intrachain bandwidth as $t_{\parallel} \sim 0.1$ eV, therefore the interchain coupling is not strong enough to localize charges completely. This intermediate coupling is reponsible for the high conductivity. The high temperature part of the conductivity can be analyzed in terms of highly anisotropic band semiconductor, where

$$n = N_0 \left(\frac{2m^*}{\hbar}\right)^{1/2} (kT)^{1/2} \exp\left\{-\frac{E_G}{2kT}\right\} \qquad /4/$$

and $\sigma = ne(\mu_e + \mu_h)$. Assuming $\mu_e = \mu_h$ we obtain mobility $\mu \sim 20\ \mathrm{cm^2 V^{-1} sec^{-1}}$ for both crystal forms, and a mean free path $\lambda \gg a$ for $m = m^*$, and even for a large effective mass m^* is larger than the lattice constant. The low temperature conductivity, which is also indicative to an activated process can be explained by polaron formation or impurity effects, but the most likely explanation is band tailing due to the weak inherent disorder and a sharp mobility gap separating extended and localized states. The magnetic properties have not been investigated yet, but a similar material 4-4'BIP$(TCNQ)_2$ strong correlation effects are evident from the behaviour of the susceptibility.

In conclusion we argue that in various complex TCNQ salts with asymmetric donors the interplay of interchain Coulomb forces and donor disorder leads to different physical properties depending on the donor dipoles and also on steric factors. For large dipoles the donor stacking is regular, the alternating site energy together with the on-site Coulomb interaction leads to a charge ordered (magnetic or non-magnetic) ground state and to a low conductivity. Smaller dipoles are randomly oriented, and when the random site energy is larger than the interchain interactions, the chains are decoupled and the electric and magnetic properties can be analyzed in terms of random 1d Hubbard model. If $E_{DA} < E_C$, the interchain coupling leads to a CDW state and to semiconducting behaviour which can be analyzed in terms of band theory. In the first case the materials are obviously poorly conducting; in the 1d disordered limit the conductivity is limited by Anderson localization. For small (or zero) disorder the small interchain hopping and relatively large interchain

Coulomb coupling leads to a semiconducting state[17,18]. It is highly unlikely therefore that materials with still much higher conductivity will be found in these types of charge transfer salts.

The picture we propose relies entirely on the Coulomb forces, with no substantial screening. It has been demonstrated recently that polarizabilities, orders of magnitude larger than that observed are needed to reduce the Coulomb effects substantially[19]. Electron-phonon coupling is also neglected. This point of view is supported by the estimation of relevant energies which indicate that the elastic energy associated with the Peierls distortion is too small to account for gaps of the order of a few tenths of eV[2], in other words lattice distortion, if present, is most likely to be also of Coulombic origin.

Useful discussions with A.Jánossy, K.Holczer and M.H.Cohen are acknowledged.

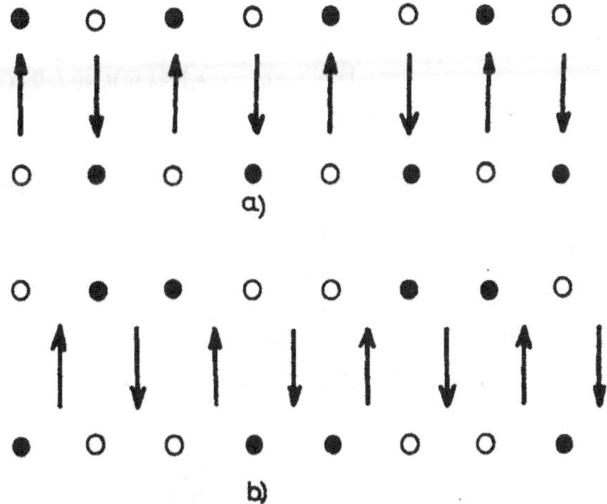

Fig 1 Ground state configurations for complex TCNQ salts with ordered donors

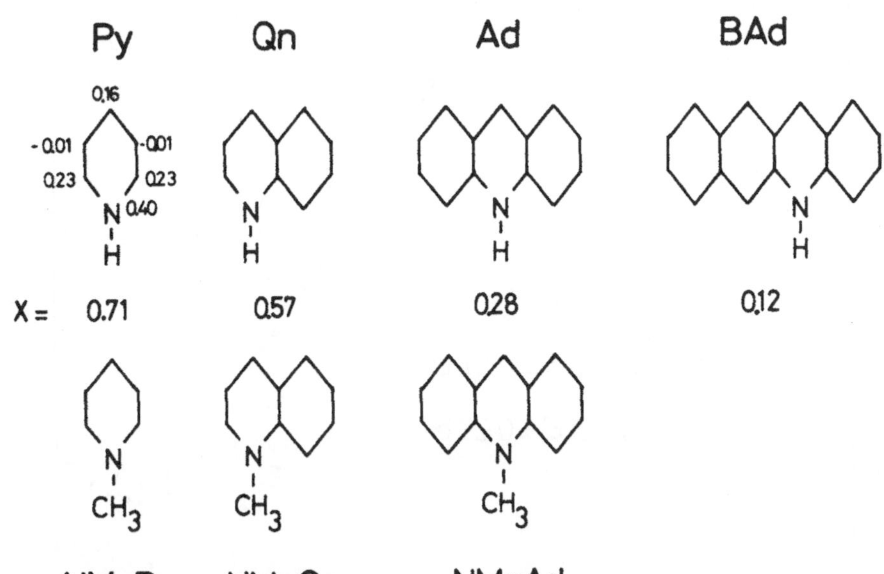

Fig 2 Molecular structures of the donor molecules. The numbers indicate the charge localized on each atom obtained from a tight binding calculation. X is the difference of charges localized on the lower and upper part of the molecules

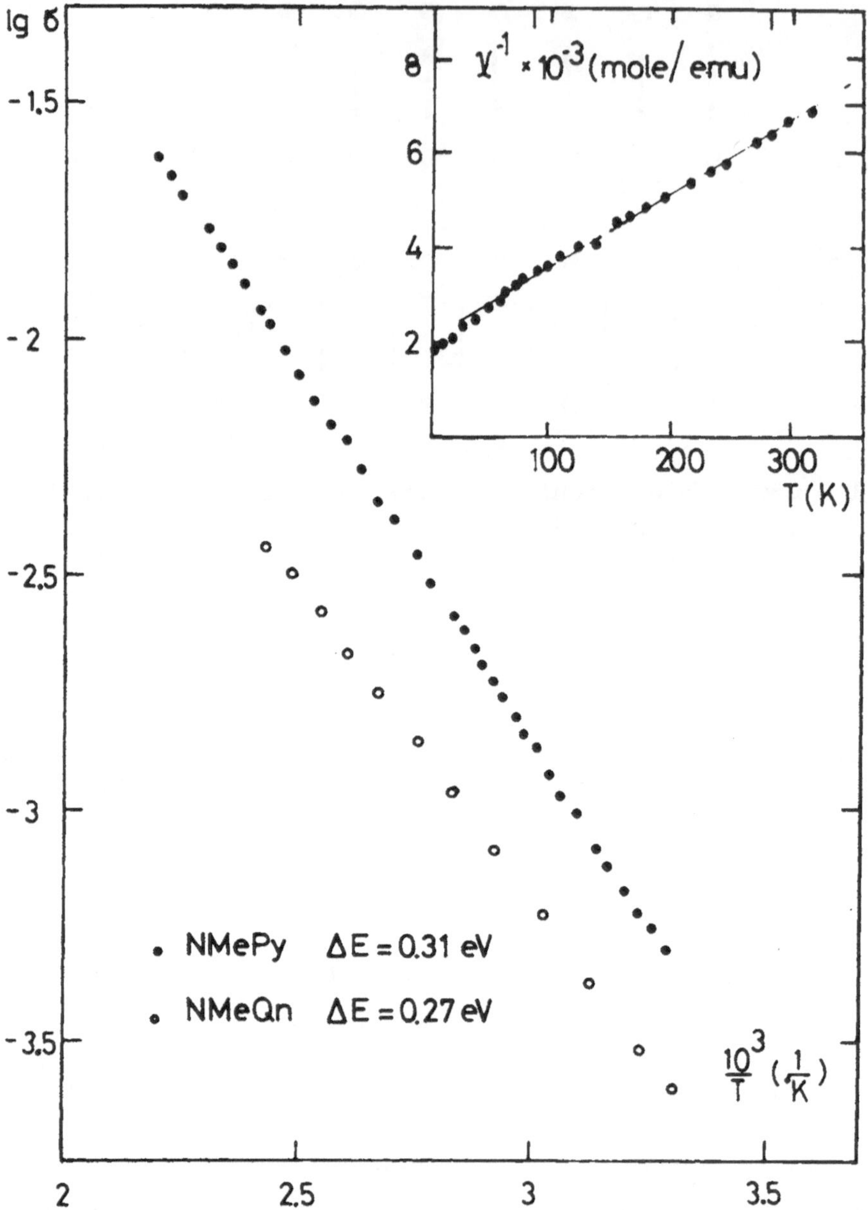

Fig 3 NMePy.(TCNQ)$_2$ and NMeQn(TCNQ)$_2$. The insert shows the susceptibility of NMeQn(TCNQ)$_2$

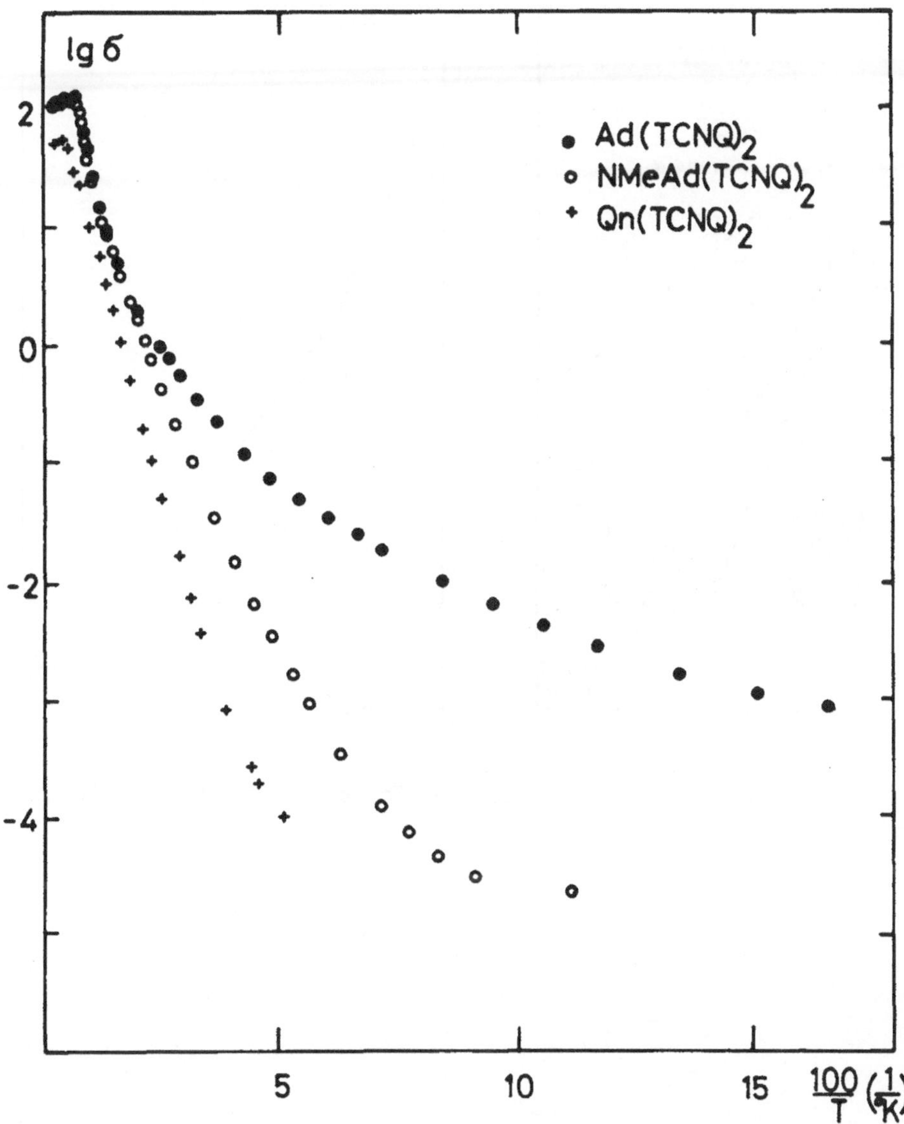

Fig 4 Single crystal conductivity of $Qn(TCNQ)_2$, $Ad(TCNQ)_2$ and $NMeAd(TCNQ)_2$. The temperature of the conductivity maximum T_m = 240, 150 and 180°K for the Qn, Ad and NMeAd salt respectively

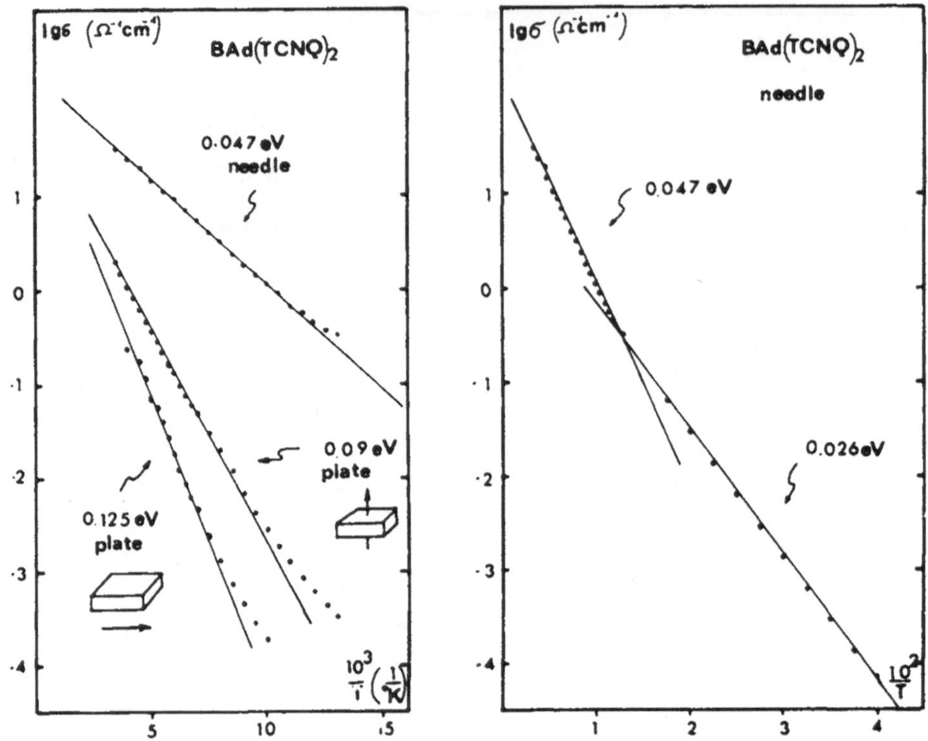

Fig 5 Conductivity of the two forms of 2-3 $BAd(TCNQ)_2$.

REFERENCES

1 I.F.Shchegolev: Phys.Status Solidi 12, 9 /1972/

2 F.Woynarowich, L.Mihály, G.Grüner: Solid State Comm. 19, 1189 /1976/

3 G.Theodoru /to be published/

4 M.Kertész /private communication/

5 L.R.Melby et al.: J.Am.Chem.Soc. 84, 3374 /1962/

6 D.Kunstelj, K.Pintér, G.Grüner: /to be published/

7 K.Holczer, G.Mihály, A.Jánossy, G.Grüner, G.Clark: /in this volume/

8 K.Holczer, G.Mihály, A.Jánossy, G.Grüner: Mol.Cryst. Liq.Cryst. 32, 199 /1976/

9 J.B.Torrance /in this volume/

10 G.Mihály, K.Holczer, G.Grüner, D.Kunstelj: Solid St. Comm. 19, 1091 /1976/

11 A.Zawadowski, M.H.Cohen: Phys.Rev.B. /to be published/

12 A.N.Bloch, R.B.Weisman, C.M.Varma: Phys.Rev.Lett. 28, 753 /1972/; - V.K.Shante: /to be published/

13 G.Mihály, K.Ritvay-Emandity, G.Grüner: /to be published/

14 G.Mihály, K.Ritvay-Emandity, G.Grüner: J.Phys.C. 8, 361 /1975/

15 M.H.Cohen /this volume/
E.I.Rashba /this volume/

16 This situation should clearly be distinguished from the weak random potential limit for example $-CH_3$ substitution in the TCNQ /Bechgaard et al to be published/, where only a smearing of an otherwise sharp phase transition is observed.

17 L.Mihály, J.Sólyom: J.Low Temp.Phys. /to be published/

18 G.Mihály, K.Holczer, A.Jánossy, G.Grüner: Solid State Comm. /to be published/

19 R.L.Bush: Phys.Rev. B12, 5698 /1975/

Conference on Organic Conductors and Semiconductors, Siófok, Hungary 1976

TCNQ SALTS WITH SYMMETRIC AND ASYMMETRIC DONORS

G. MIHÁLY, K. HOLCZER, A. JÁNOSSY,
G. GRÜNER* and M. MILJAK**

*Central Research Institute for Physics, Budapest, Hungary
**Institute of Physics of the University, Zagreb, Yugoslavia

ABSTRACT

Transport and magnetic properties of complex TCNQ salts with symmetric and asymmetric bipyridine donors are presented. The salt with symmetric donor is a small bandgap semiconductor with large anisotropy in the conductivity, the susceptibility indicates strong electron correlations. The asymmetric donor leads to a material in which both the electric and magnetic properties can be discussed by electron localization due to disorder.

Certain complex salts of TCNQ, of which $Qn(TCNQ)_2$ is the prime example (1), have a high conductivity and a rounded "metal-insulator" transition. Inherent disorder (2) is believed to play a fundamental role in preventing the development of a 3d semiconducting ground state, limiting the mean free path and in determining the low temperature magnetic properties (3).

To investigate the role played by disorder connected with the asymmetric donors, we have prepared $BIP(TCNQ)_2$ salt with symmetric and asymmetric bipyridine molecules shown in Fig. 1. An inherent feature of BIP molecules is that they can occur both in doubly ionized and in neutral form. In the salts we have investigated, the ratio of neutral and doubly ionized BIP is 1:1 this also corresponds to the $TCNQ^-/TCNQ^0$ ratio. Thus the TCNQ chain is quarter filled.

The dc and ac conductivity of 44' $BIP(TCNQ)_2$ is shown in Figs 2 and 3. The material is a highly anisotropic semiconductor with a fairly sharp gap $E_g = 2\Delta = 0.16$ eV.

At low temperatures hopping conductivity may take place and most probably this causes the curvature in the ac conductivity shown in Fig. 2. The

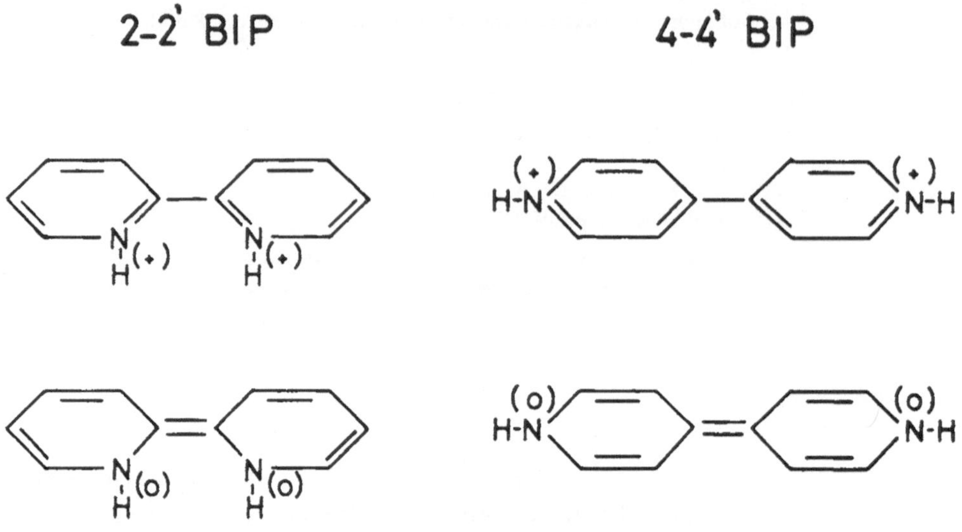

Fig. 1 Structure of the symmetric (2-2') and the asymmetric (4-4') bipyridine molecules. Upper : doubly ionized, lower: neutral form

high conductivity and the absence of dispersion indicates small gap semiconducting behaviour. For a highly anisotropic semiconductor the number of excited electrons

$$n_e = N_o \left(\frac{2m^*}{\hbar}\right)^{1/2} (kT)^{1/2} \exp\left\{-\frac{E_G}{2kT}\right\} \tag{1}$$

(N_o = number of chains per unit cross section) and the conductivity is given by

$$\sigma = en(\mu_e + \mu_h) \tag{2}$$

The TEP for a band semiconductor

$$S = \frac{k}{e}\left[\frac{\mu_e - \mu_h}{\mu_e + \mu_h} \; \frac{E_G}{2kT} + A\right] \tag{3}$$

where the constant A contains terms coming from the electron and hole scattering processes and also depends on the electron and hole effective mass ratio.

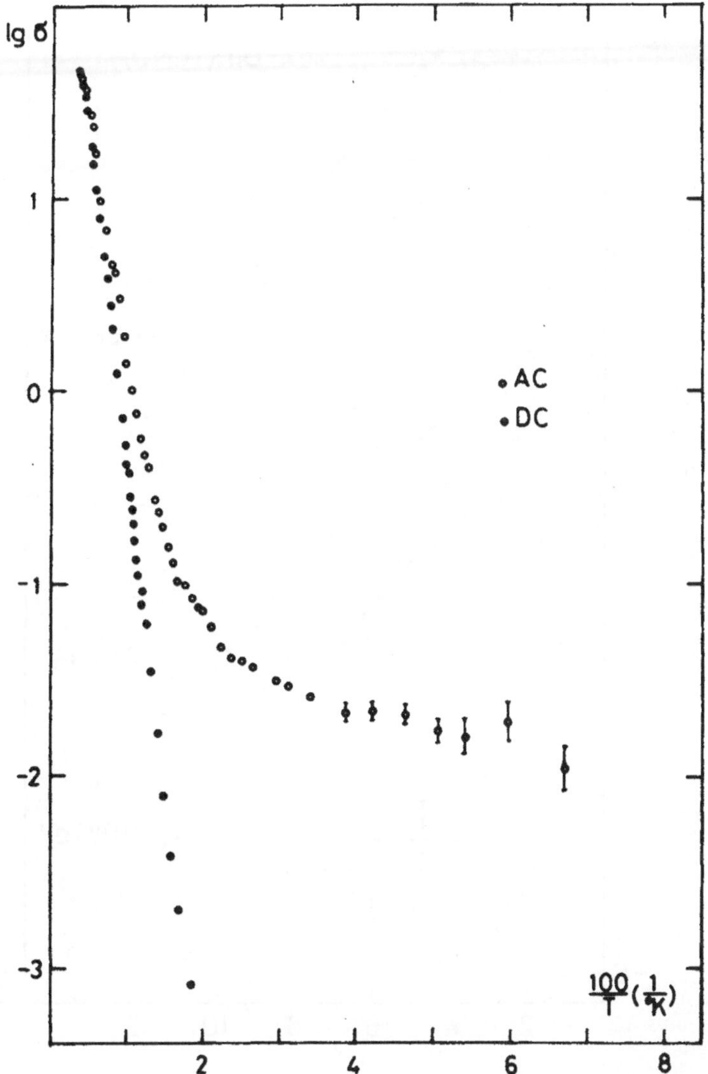

Fig. 2 Conductivity of 44' BIP(TCNQ)$_2$ dc, o microwave data

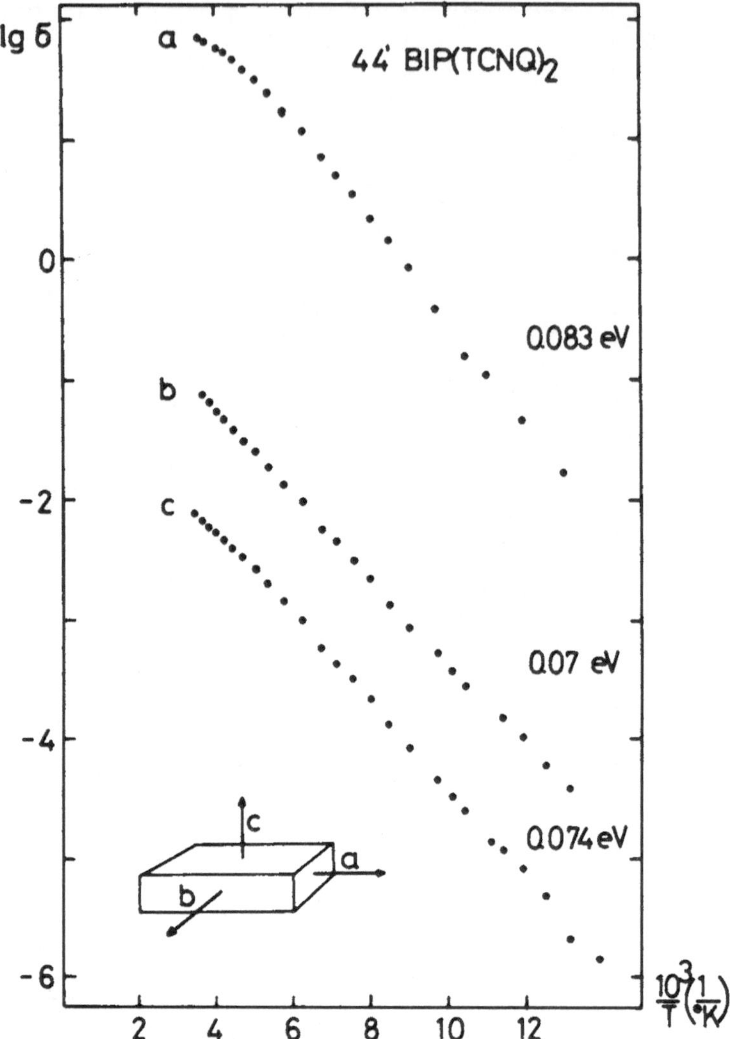

Fig. 3 Anisotropy of dc conductivity of 44' $BIP(TCNQ)_2$

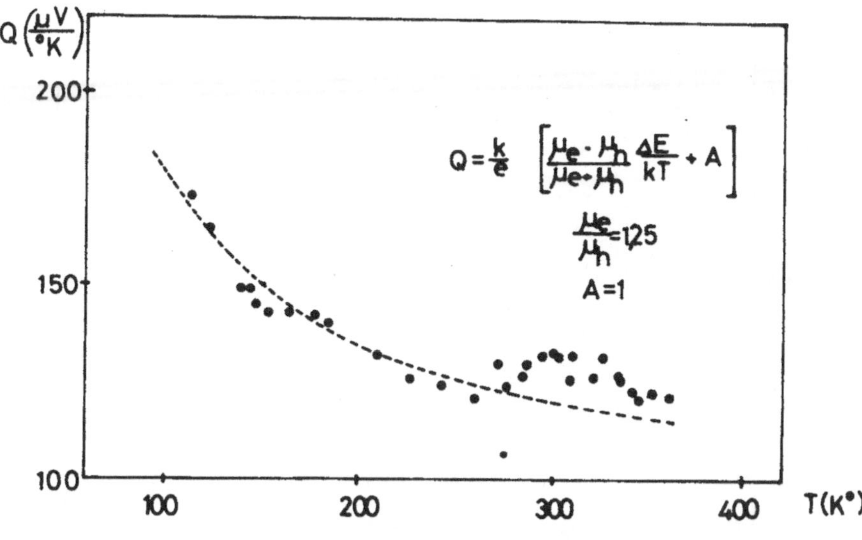

Fig. 4 Thermoelectric power of 44' BIP$(TCNQ)_2$

The observed thermopower of 44' BIP$(TCNQ)_2$ can be represented by Eq (2), and gives for the ratio of the mobilities $\frac{\mu_e}{\mu_h} \approx 1.25$ [Fig. 4] . Using Eq (2), with parameters shown in Fig 2 we obtain $\mu_e \approx \mu_h \approx 20\ cm^2 V^{-1} sec^{-1}$. The $\lambda = \frac{\pi\hbar}{2ea}\mu$ relation gives for the mean free path $\lambda \approx$ 100 Å. This is certainly an overestimate, as m^x may be much larger than the free electron mass, for example a value $m^x = 10\ m$ usually observed in these materials (4) leads to $\lambda \sim 30$ A, a value similar to that found in TCNQ salts where disorder is not present.

The material, however, is not an uncorrelated semiconductor. This is confirmed by the behaviour of the susceptibility which displays a behaviour characteristic to a singlet-triplet excitation with broad exciton band [Fig. 5]. A simple singlet-triplet model with a sharp excitation energy J

$$\chi_{ST}(T) = \frac{\mu_{eff}^2}{T} \quad \frac{1}{1+\frac{1}{3} e^{J/kT}} \tag{4}$$

does not fit the susceptibility, the measured $\chi(T)$ has a broader and smaller maximum than that one would obtain from Eq. (4). (This is not due to variations of J from crystal to crystal, as confirmed by ESR measurements on several single crystals.) This indicates a bandwidth of the same order as J itself. Ap-

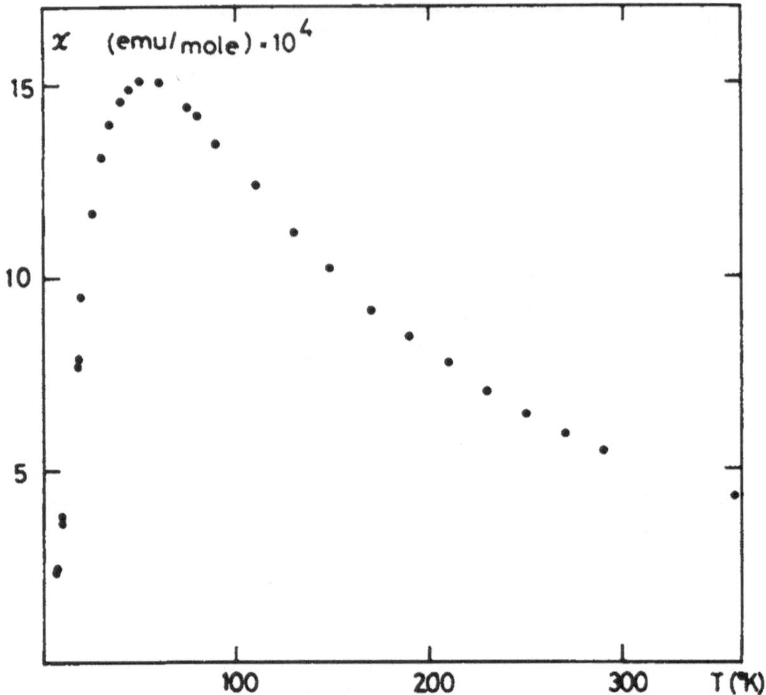

Fig. 5 Magnetic susceptibility of 44' BIP(TCNQ)$_2$

proximate expressions for a finite bandwidth D are available only in the $kT << J$ limit (5), due to the low J value however an analysis in terms of this model is not possible here. We have fitted $\chi(T)$ therefore with the expression

$$\chi(T) = \frac{1}{2D} \int_{J_0-D}^{J_0+D} \chi_{ST}(J)\, dJ \tag{5}$$

which corresponds to a distribution of temperature independent excitation energies in the region J_0 - D and J_0 + D with same probability.

A computer fit gives J = 170 $^{\circ}$K and D = 150 $^{\circ}$K, thus the bandwidth is comparable to the excitation energy.

The existence of a collective mode below the one-electron excitation band suggests that the properties of 44' BIP(TCNQ)$_2$ are determined by the

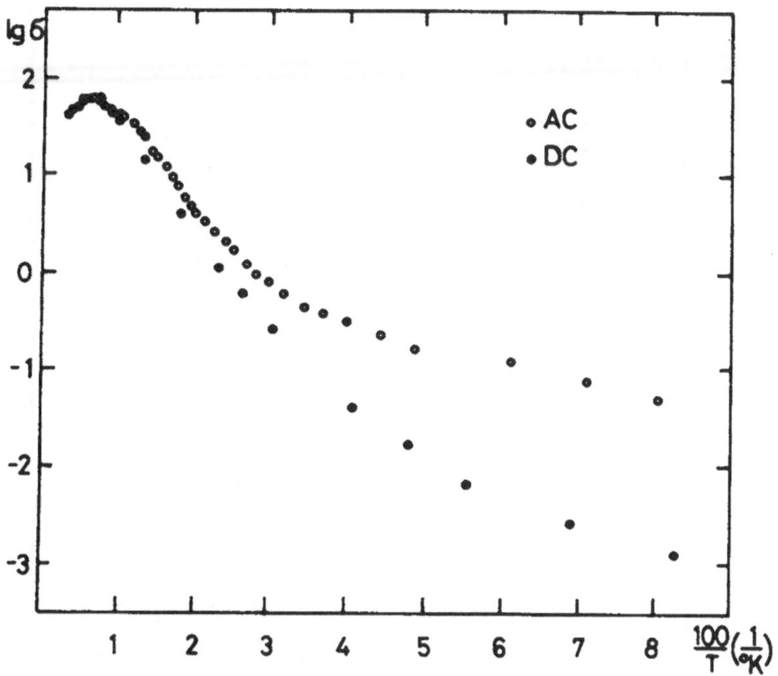

Fig. 6 Conductivity of 22' $BIP(TCNQ)_2$. • dc, o microwave data

strong e-e interaction. The fairly large interchain Coulomb interaction ($U_\perp \sim 0.2$ eV) together with the on-chain e-e interaction develops coupled CDW-s along the chains and leads to a semiconductor. The single particle gap is small, comparable to the bandwidth along the chains. No strong localization is expected to arise under such circumstances, and the material is highly conducting.

22' $BIP(TCNQ)_2$ displays a completely different behaviour. The rounded maximum in σ as well as the dispersion of the conductivity below the maximum is characteristic to that found in other well conducting "disordered" materials, also the susceptibility is similar (Figs 6 and 7), to that found in $Qn(TCNQ)_2$ and $Ad(TCNQ)_2$ for example.

These properties are interpreted on the basis of a 1d random Hubbard model. The random site energy arises from the random donor dipole arrangement. For the pyridine molecule $D \approx 1$ eÅ and the random site energy for characteristic donor acceptor distance $r \approx 7$ Å (is given by the electrostatic interaction between donor dipole and electron on the TCNQ chain)
$\Delta E_s \approx 0.5$ eV.

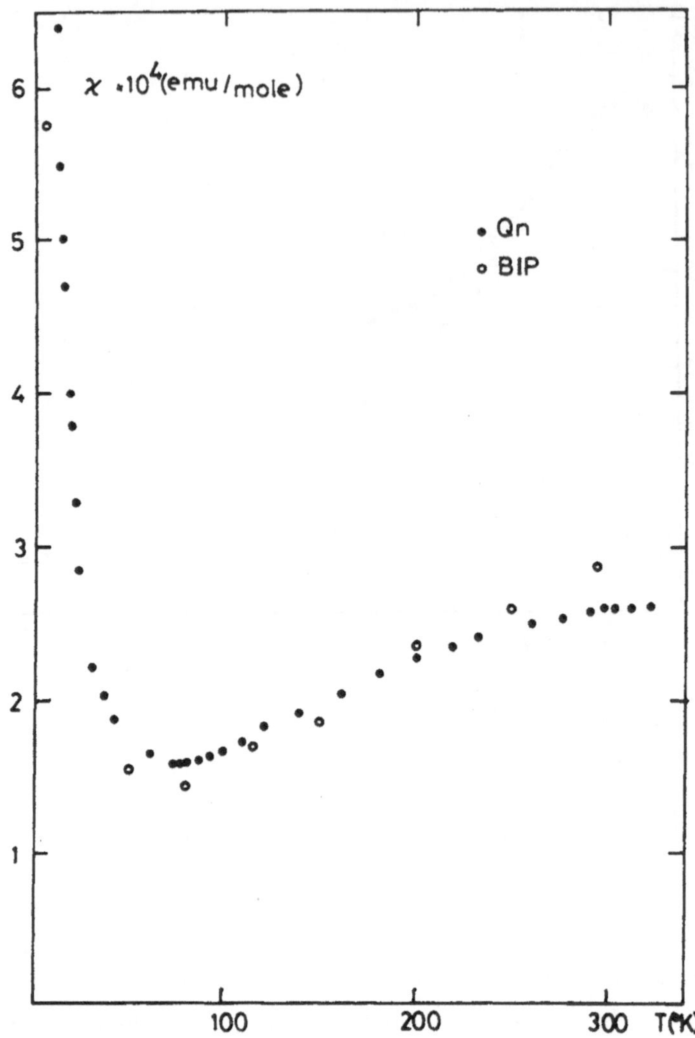

Fig. 7 Magnetic susceptibility of 22' BIP(TCNQ)$_2$ compared to Qn(TCNQ)$_2$ data

Thus ΔE_s is larger than the interchain Coulomb forces, and this depresses the coupling to $T \to 0$.

In conclusion, with no disorder present, the interchain coupling leads to a CDW state, where electron correlations on neighbouring chains are correlated and the material is a highly anisotropic small bandgap semiconductor with electron correlation effects playing a decisive role.

Disorder destroys this coupling and for large disorder the material can be described in terms of models where only on chain correlation effects and random disorder are involved.

REFERENCES

(1) I. F. Shchegolev: Phys. Stat. Sol. A 12, 9 (1972)

(2) A. N. Bloch, R. B. Weismann and C. M. Varma: Phys. Rev. Lett. 28, 753 (1971)

(3) G. Theodoru: Phys. Rev. (to be published)

(4) D. B. Tanner, C. S. Jacobsen, A. F. Garito and A. F. Heeger: Phys. Rev. B13, 3381 (1976)

(5) L. N. Bulayevskii Fiz. tverd. Tela 11, 1132 (1969)

(6) F. Woynarowich, L. Mihály and G. Grüner: Solid State Comm. (to be published)

Conference on Organic Conductors and Semiconductors, Siófok, Hungary 1976

INVESTIGATIONS OF HIGH CONDUCTIVITY OF DIPYRANYLIDENE AND DITHIADIPYRANYLIDENE - -TCNQ COMPLEXES

J. ALIZON, J. BLANC, J. GALLICE and H. ROBERT

LERM, Université de Clermont-Ferrand, 63170 Aubière, FRANCE

C. FABRE and H. STRZELECKA

CNRS, Gr 12 Reactivité et Mécanisme en Chimie Organique,
2, rue Dunant, 94320 Thiais, FRANCE

J. RIVORY

Laboratoire d'Optique des Solides, Université Paris VI,
4, Place Jussieu 75230 Paris, FRANCE

C. WEYL

Laboratoire de Physique des Solides, Université Paris-Sud
91 405 - Orsay, France

ABSTRACT

New complexes with DIP ϕ_4 or DIPS ϕ_4 as donors and TCNQ as an acceptor have been synthetized. Single crystals exhibit high conductivity along the needle axis at room temperature : 40 $(\Omega cm)^{-1}$ for 1 and 250 $(\Omega cm)^{-1}$ for 2. X-ray analysis, temperature dependences of conductivity, susceptibility and relaxation time T_1 for both complexes are reported.

Recently [1], chemical and physical properties of a new series of charge transfer salts (1:1) with dipyranylidene (DIP R_4) as donors and TCNQ as an acceptor have been studied. The best 1-D conductor is obtained with aromatic substituent as $R = C_6H_5 = \phi$. In this communication , we report a more detailed investigation on complexes synthetized from dipyranylidene (DIP ϕ_4) and dithia-dipyranylidene (DIPS ϕ_4) as donors* and TCNQ as an acceptor.

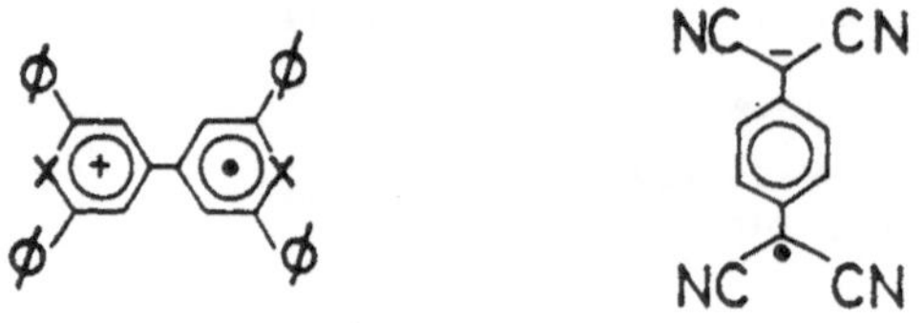

when X = O DIP $\phi_4^{+\cdot}$, TCNQ $^{\mp}$ 1

X = S DIPS $\phi_4^{+\cdot}$, TCNQ $^{\mp}$ 2

We have chosen these donor molecules because they are isoelectronic with TTF donor, both having 14 π electrons,

1 and 2 exhibit high symmetry (C_{2h} as point group) and large size. The distances between the heteroatoms are unusually large for such a donor molecule : slightly greater than 7 and 8 Å between O-O or S-S atoms.

Electrochemical analysis [2-4] shows weak half-wave potentials as those of TTF cations,

	ε_1	ε_2	$\Delta E_{1/2}$	Ref.
DIP ϕ_4	+ 0.15	+ 0.47	0.32 volt	[2]
DIPS ϕ_4	+ 0.22	+ 0.41	0.19	[2]
TTF	+ 0.47	+ 0.84	0.37	[2]
	+ 0.33	+ 0.70	0.37	[3]
	+ 0.32	+ 0.70	0.38	[4]
TTF ϕ_4	+ 0.40	+ 0.73	0.33	[4]

* Synthesis of Se, Te analogues are in progress.

These electrochemical data show that DIPS ϕ_4 is a very promising candidate for a donor in organic metals and this is confirmed by d.c. measurements on single crystals of 2 [5] . We obtained, at room temperature, for 1 25-40 $(\Omega cm)^{-1}$
2 250-300 $(\Omega cm)^{-1}$.

This last value being higher than those given for TTN-TCNQ [6] and TTT-TCNQ or TTT$(TCNQ)_2$ [7] .

The main feature of the temperature dependence of the conductivity of 1, which is shown on the fig. 1, is a reversible transition at 240 K, which is clearly seen on the lower conductivity sample. Fig. 2 shows a semiconductor-semiconductor transition with a sharp decrease of the conductivity of 10^2. The activation energy in the high temperature region is 0.04 eV. In the low temperature region an activation energy cannot be defined.

The behaviour of 2 is quite different. On the fig. 3, the resistance vs. T curves exhibit a small decrease typical of metallic behaviour followed by a sharp metal-semiconductor transition at 145 K. On figs. 3 and 4, we can see that the metal-semiconductor transition which is sharp at the first cooling cycle, broadens when repeating heating and cooling cycles.

The crytals are monoclinic. The stacking-axis corresponds to the binary axis with the following parameters :

$$\underline{1}\ |\vec{b}| = 3.87\ \text{Å}$$
$$\underline{2}\ |\vec{b}| = 3.82\ \text{Å}$$

The sulfur compound has a smaller intermolecular spacing in spite of the fact that the Van der Waals radius of sulfur is 0.44 Å larger than that of the oxygen [8] . This seems to favour conductivity when the open valence shell ions are equispaced closer than their Van der Waals radii.

X-ray analysis reveals a perfectly long range order along the stacking-axis as it can be seen from oscillating crystal photograph around the needle axis. But in both cases, at large Bragg angles there are only weak Bragg spots, so weak as to become nonobservable, indicating some disorder in a plane perpendicular to the stacking-axis. This was confirmed by the examination of (h0l) plane in the reciprocal space. This disorder seems to correspond to small angular rotation of the molecules around the stacking-axis between each adjacent stack. So we can say that this material has some fibrous character, which induces larger uncertainties in the determination of the parameters $|\vec{a}|$ and $|\vec{c}|$

and the angle β. However we can give accurate interplanar distances in the direct space :

$$d_{100} = (31.9 \pm 0.2)\ \text{Å}$$
$$d_{001} = (25.0 \pm 0.2)\ \text{Å}$$

The main crystallographic parameters of DIPS ϕ_4-TCNQ are :

$$a = (34.7 \pm 0.4)\ \text{Å}$$
$$b = (\ 3.82 \pm 0.01)\ \text{Å}$$
$$c = (27.2 \pm 0.4)\ \text{Å}$$
$$\beta = 113°3 \pm 0°5$$

The space group is C 2/c with Z = 4.

We have also studied the magnetic properties by high and low field ESR and the proton nuclear spin lattice relaxation time T_1. These results show quite different behaviour for the two materials :

- the susceptibility of 1 is temperature independent [1]while the sulfur compound of 2 shows a strong temperature dependence as can be seen on fig.5, very similar to the behaviour of TTF-TCNQ and its analogues and also $(TTT)(TCNQ)_2$ [7].
- the first compound behaves according to the KORRINGA law with a good agreement at high temperature [1], while for the second one, where a metal-semiconductor transition occurs, it is not the case. This fact can be attributed to the 1-D character of this system.

Conclusion

These new series of donors show that in the case of sulfur compounds reported here, high d.c. conductivity can exist together with large donor molecules. In other words, dipyranylidene and dithiadipyranylidene - TCNQ complexes present high conductivity and also very pronounced 1-D character.

Acknowledgements

We are grateful to Dr.M.T. LE BILHAN for aid in X-ray work and the technical assistance of S.DEMIANIV is appreciated.

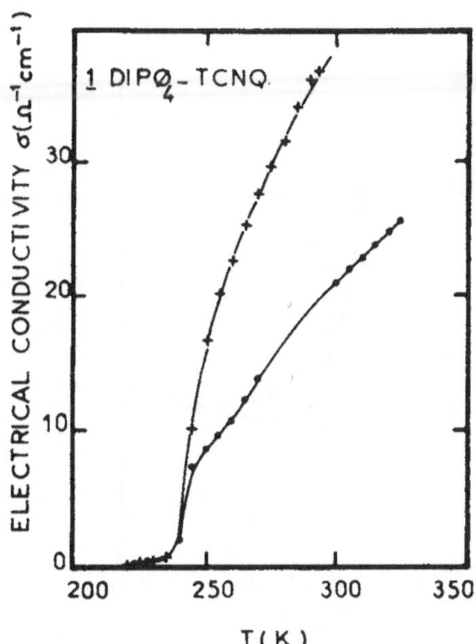

Fig. 1 : Temperature dependence of electrical conductivity σ for two $DIP\phi_4$-TCNQ single crystals.

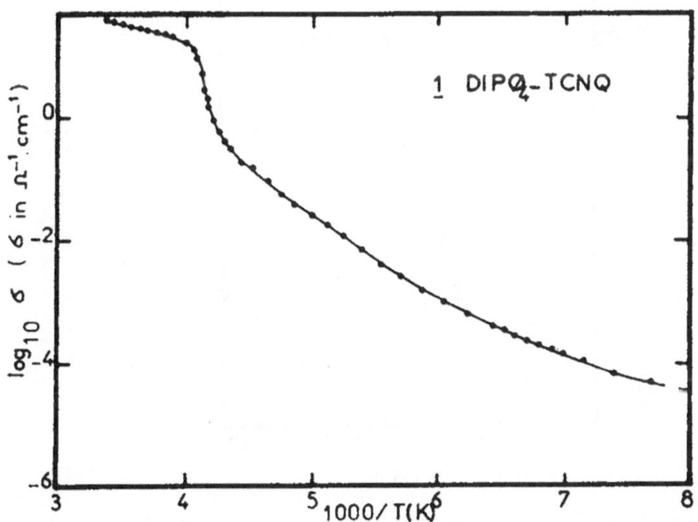

Fig. 2 : Log_{10} σ versus T^{-1}(K) for one DIP ϕ_4-TCNQ single crystal.

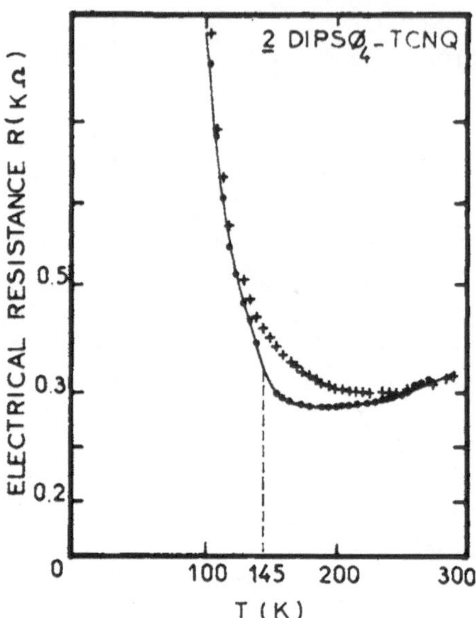

Fig. 3 : Temperature dependence of electrical resistance for one DIPS ϕ_4-TCNQ single crystal : (•) first cooling, (+) return to room temperature.

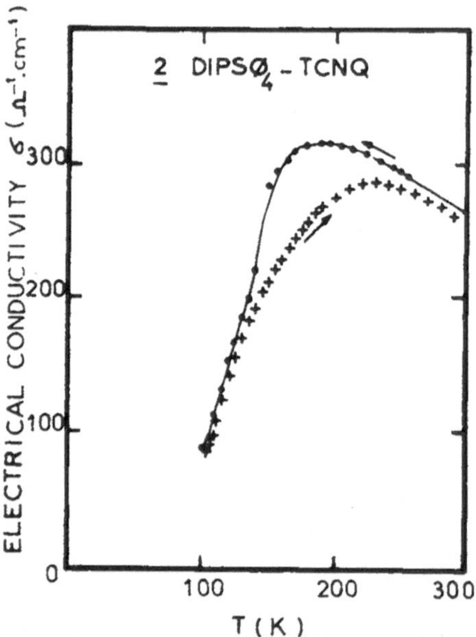

Fig. 4 : Temperature dependence of electrical conductivity for one DIPS ϕ_4-TCNQ single crystal showing degradation of the conductivity maximum : (•) first cooling, (+) return to room temperature.

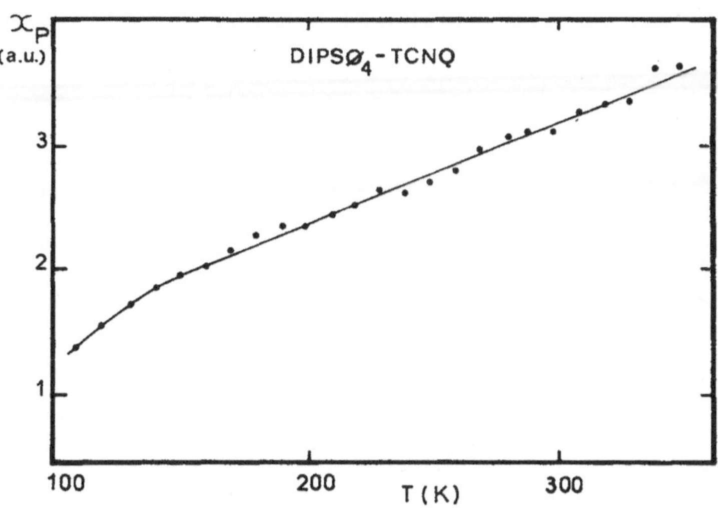

Fig. 5 : Temperature dependence of X_p (arbitrary units) of DIPS ϕ_4- TCNQ.

REFERENCES

[1] - ALIZON J.,GALLICE J., ROBERT H., and DELPLANQUE G.,WEYL C., and FABRE C., STRZELECKA H., Mol.Cryst. Liq.Cryst.,33, 91, [1976] .

[2] - HÜNIG S.,GARNER B.J., RUIDER G., and SCHENK W., Liebigs Ann.Chem.,1036, [1973].

[3] - HÜNIG S., KIESSLICH G., QUAST H. and SCHEUTZOW D., Liebigs Ann. Chem., 310, [1973].

[4] - CALAS P., FABRE J.M., TORREILLES E., and GIRAL L., C.R.Acad.Sci., 280, 901, [1975] and MAS A. Thesis Montpellier [1976].

[5] - Single crystal growth

a) DIP ϕ_4-TCNQ : in a three-chamber diffusion apparatus (M.L.KAPLAN Int. J.for Crystal growth [1976]) with CH_3CN + CH_2Cl_2 (1 : 1) as solvant during a weak at 40°C; reactifs concentration 10^{-4} M/l.

b) DIPS ϕ_4 - TCNQ : slow evaporation from CH_2Cl_2 + CH_3 CN (2 : 1).

[6] - PERLSTEIN J.H., FERRARIS J.P., WALATKA V.V., COWAN D.O., and CANDELA G.A., AIP Conference Proceedings No 10, Vol.3, DENVER, Colo., 1494, [1972] .

[7] - BURAVOV L. I., EREMENKO O.N., LYOBOVSKII R.B., ROSENBERG L.P., KHIDEKEL M.L., SHIBAEVA R.P., SHEGOLEV I.F. and YAGUBSKII E.B., JETP lett, 20, 208, [1974] .

[8] - PAULING L., The Nature of the Chemical Bond, p.514, Third Edition by Cornell University [1960].

VII

POLYSULFUR NITRIDE, $(SN)_x$

Conference on Organic Conductors and Semiconductors, Siófok, Hungary 1976

ELECTRONIC PROPERTIES OF ORGANIC AND POLYMERIC METALS

A.F. GARITO

Department of Physics, University of Pennsylvania, Philadelphia, Pennsylvania 19104, USA

Studies of the electronic and structural properties of the quasi-one dimensional organic metal TTF-TCNQ continue to experimentally characterize the Peierls soft mode instability associated with conducting one dimensional chains. The effects on the behavior of the soft mode instability due to specific changes in the molecular sites along the chains can be systematically studied with related TTF-TCNQ derivatives. The polymer, polymeric sulfurnitride, $/SN/_x$, is providing a case where the conducting linear chains are strongly coupled electronically leading to a suppression of the instability as evidenced by its electronic properties. Studies of these properties are reviewed and compared with results for TTF-TCNQ.

Conference on Organic Conductors and Semiconductors, Siófok, Hungary 1976

ELECTRONIC STRUCTURE AND OPTICAL PROPERTIES OF POLYSULFUR NITRIDE, $(SN)_x$

P.M. GRANT, W.E. RUDGE and I.B. ORTENBURGER

IBM Research Laboratory San Jose, California 95193, USA

ABSTRACT: We have made theoretical and experimental studies of the electronic structure and optical properties of (SN)x based on the use of OPW and pseudopotential calculational techniques and normal incidence polarized reflectance measurements. The principal finding of the calculation is the existence of a Fermi surface containing a closed direction which stabilizes (SN)x against static Peierls-Frohlich distortions thus permitting the superconducting state to occur. From the Fermi surface and band structure, we have computed the frequency-dependent dielectric tensor, $\epsilon_{\mu\nu}(\omega)$, and plasma tensor, $(\omega_p^2)_{\mu\nu}$. The calculated optical spectrum compares favorably with the results of the reflectance measurements.

Polymeric sulfur nitride, (SN)x, was first synthesized by Burt[1] in 1910, apparently as part of an effort to fix the the atomic weight of nitrogen. Over the intervening years it was recognized that (SN)x possessed an unusually high conductivity for a polymeric material. Interest quickened when it was found that for crystalline (SN)x, the conductivity had a metallic temperature dependence down to 1.2 °K and was highly anisotropic with regard to crystallographic direction.[2,3] In fact, the chain-like nature of the crystal structure[4,5] strongly suggested that (SN)x might display some of the quasi-one-dimensional characteristics of (TTF)(TCNQ) and KCP, notably the Peierls-Frohlich instability at low enough temperatures. However, it was subsequently found that (SN)x went superconducting at 0.3 °K[6] and the principal physical question thus became one of how (SN)x escaped the lattice instability usually associated with one-dimensional structures.

We decided to investigate this question by doing a thorough calculation of the one-electron properties of (SN)x. That these properties might best be represented in the Bloch representation of an OPW calculation was affirmed by the relatively large conduction bandwidths ($\sim$ 2-3 eV) inferred from specific heat[3] and optical[2] measurements. Furthermore, the experimental uncertainty introduced by electron-hole mean free path considerations into any k-space features to be uncovered in the calculation seemed tolerable. That is, using $\Delta k\, \ell \gtrsim 1$ as our uncertainty relation connecting direct and reciprocal space, we find $\Delta k \lesssim$ 10% of a reciprocal lattice vector given a mean free path estimate of $\ell > 10$ Å. We feel 10 Å represents a conservative realistic lower bound on the mean free path of a carrier in an arbitrary direction in a typical (SN)x crystal. From the results of the computation and concurrent comparison with experiment, we were able to select from the following three possibilities the most likely explanation for the low temperature metallic stability of (SN)x: (1) incommensurability of the Fermi vectors with the reciprocal lattice; (2) many-body interchain effects according to the model of Klemm and Gutfreund;[7] or (3) topology of the (SN)x Fermi surface.

Details of the calculational technique, the resulting band dispersions, and comparisons with other (SN)x band structures have been presented elsewhere.[8] For our purposes, that part of the OPW band structure given in Fig. 1 will be sufficient. The directions $\Gamma\Lambda Z$ and $E U A$ are co-linear with the polymer chain axis (the crystallographic b-axis), the direction of highest conductivity in (SN)x, while ZE and $A\Gamma$ represent reciprocal space duals to an interchain direction. Note that the intersection of the Fermi level with bands along $\Gamma\Lambda Z$ results in two possibly incommensurate Fermi vectors in accordance with explanation (1) above. Kamimura, et al.[9], have in fact argued that this is indeed the source of electronic stability in (SN)x. Berlinsky,[10] on the other hand, has since shown that a rigorously one-dimensional (SN)x band structure will always produce Fermi vectors commensurate with the reciprocal lattice. From a purely empirical point of view, nature seems to "rationalize" incommensurateness in one-dimensional systems, as witnessed by the examples of (TTF)(TCNQ) and KCP where no special reason for wave vector rationality is expected and yet the Peierls-Frohlich state occurs. We believe (1) is not the answer for (SN)x. More specifically, we draw attention to the intersection of the Fermi level with bands along ZE. Intersections of this kind at other points in the Brillouin zone produce the electron-hole Fermi surfaces shown in Fig. 2. It is immediately apparent that a lowering of ground state energy through the introduction of a gap at every point on these surfaces is inconsistent with any topologically possible lattice distortion. We thus find point (3) to be the most plausible explanation for the stability of (SN)x crystals to Peierls-Frohlich effects. Nonetheless, we cannot completely rule out the mechanisms of point (2) (Ref. 7). We do feel, however, that the simplicity of point (3) and the general support of the OPW band structure properties by experiment, to be discussed next, argue effectively for our choice.

Two measureable quantities which can be directly calculated from the OPW results are the plasma tensor and the frequency dependent dielectric tensor given by the following expressions:

$$(\omega_p^2)_{\mu\nu} = 4\pi e^2 \int_{S_F} dS \, \frac{v_\mu v_\nu}{|v|} \, , \quad v_\mu = \frac{1}{\hbar} \frac{\partial E}{\partial k_\mu} \, ; \qquad (1)$$

$$\epsilon_{2\mu\nu}(\omega) = \frac{(\omega_p^2)_{\mu\nu}\tau}{\omega(1+\omega^2\tau^2)} + \frac{e^2}{\pi m^2\omega^2}\sum_{nn'}\int d^3k$$
$$\times \langle nk|p_\mu|n'k\rangle\langle n'k|p_\nu|nk\rangle$$
$$\times\ \delta(E_{n'}(k) - E_n(k) - \hbar\omega)\,, \quad (2)$$

where the momentum matrix elements are taken with respect to the OPW Bloch functions and τ is a phenomenological Drude scattering time. The actual calculation of these quantities is effected via a pseudopotential interpolation of the OPW results followed by Gilat-Raubenheimer integration techniques. The theoretical plasma tensor, referred to its principal axes with respect to the (SN)x crystallographic coordinates, is given by Fig. 3. We see that these principal axes lie satisfyingly close to those interchain directions along which we anticipate the strongest interchain interaction to exist. The plasma tensor, in addition to determining the Drude optical properties of a metal, also determines the normal state dc conductivity and the intrinsic superconducting critical field anisotropy. It is one of the most fundamental physical properties of a metal which is directly computable from its band structure. To our knowledge, our calculation of $(\omega_p^2)_{\mu\nu}$ for (SN)x is the first for any metal. The average ratio of the b-axis component to the orthogonal components is roughly 9.5:1 which would represent the effective mass ratio for an idealized ellipsoidal Fermi surface. That these values are in reasonable agreement with experiment can be seen from Figs. 4

and 5 where we compare measured normal incidence polarized reflectivities with those calculated using Eqs. (1) and (2) along with the usual Fresnel reflectance equations. Note the close agreement between the calculated and experimental plasma reflectance minima. In computing the contribution of the Drude term in Eq. (2), we have used the experimental values of γ found by Cohen[11] and Grant, et al.[12] The Drude-like regions of the spectrum can be clearly identified; however, it is apparent that interband transitions contribute significantly to the extent of being visibly embedded in the Drude region of the $R \perp b$ spectrum. The origin of this structure can be seen in Fig. 6 where we have plotted only the interband transition terms in ϵ_{2yy} and ϵ_{2xx}. At approximately 0.6 eV, a strong transition appears in ϵ_{2xx} which is manifested as the peak superposed on the weak Drude background of Fig. 5. We assign this structure to an M_1-type van Hove singularity arising from the parallelism of the bands intersecting the Fermi level along ZE in Fig. 1.

In the region 2-4 eV, we also observe the effects of interband transitions in both Figs. 4 and 5. Transitions between parallel bands above and below the Fermi level along ΓA are a possible source of this structure. For energies above 4 eV, we compare with the (SN)x film data of Cohen[11] reproduced in Fig. 7. These data were taken on a non-oriented film using unpolarized incident light. Various methods of averaging $\epsilon_{\mu\nu}$ were tried in an attempt to emulate experimental conditions, but the most favorable comparison occured when we simply took half the calculated value of R‖b. This suggests that the film structure is random between polymer chains and that critical point structure from zone directions perpendicular to the chains is washed out. The observed structure most likely arises from parallel bands along DVB (not shown in Fig. 1,; see Fig. 2 of Ref. 8).

In summary, the existence of zone directions in which the Fermi surface of (SN)x closes indicates that structurally perfect (SN)x is not quasi-one-dimensional in the sense of (TTF)(TCNQ) and KCP and essentially explains why Cooper pairing, and not a Peierls-Frohlich mode, is the mechanism which actually destabilizes the metallic state in the crystalline form of this material. The Fermi surface closure derives directly from band dispersion perpendicular to the polymer chain axis. From a one-electron point of view, the implications for the production of superconductivity in linear organic polymer or stacked donor-acceptor conducting systems are clear: sufficient interchain interaction must be engineered into their crystallographic structures to produce enough Fermi surface curvature to inhibit the lattice instabilities which have been the hallmark of such systems up to now.[13] That this criterion may only be marginally met in real (SN)x samples is underscored by the recent discovery of a weak Kohn anomaly at room temperature by Pintschovius and coworkers.[14] As can be seen from Fig. 2, the Fermi surface contains some fairly planar regions over much of its surface area. The effect of interchain interaction is to "pinch off" these planar regions and form separate electron-hole volumes. Indeed, the apparent lack of long-range order in the interchain directions of $(SN)_x$ films may be connected with the failure to date to observe superconductivity in even highly oriented layers.[15] Returning to our earlier uncertainty principle argument, we note that if the mean free path in the interchain direction becomes sufficiently small (due to disorder or size effects arising from extreme fibrosity), then k-space will no longer be continuous and the carriers will not "see" the closure of the Fermi surface. Thus, gaps may begin to appear in the one-electron density of states originating either from disorder, size effects, or, if the individual chains actually become independent and isolated (hence one-dimensional), even from Peierls-Frohlich effects taking place on the microscopic scale.

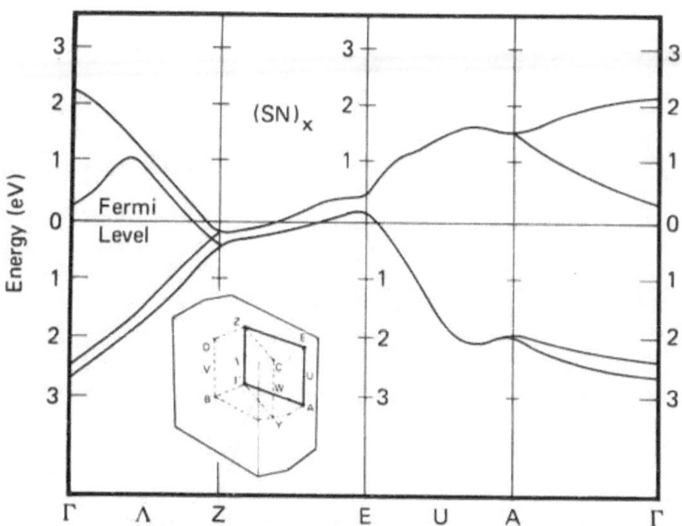

Fig. 1. Detail of the OPW band structure near E_F over the indicated path (heavy line) through the Brillouin zone.

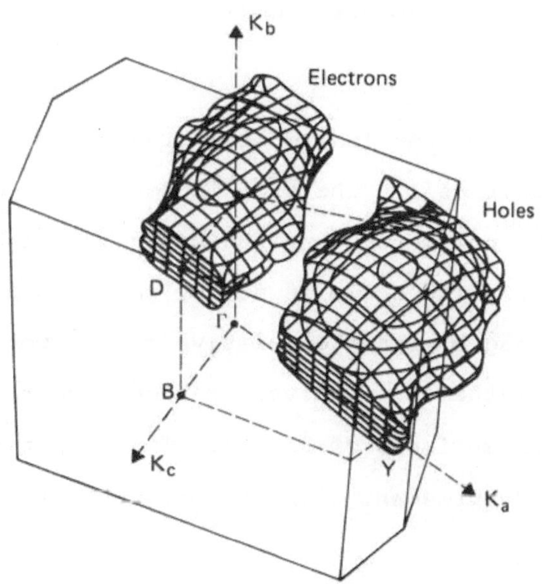

Fig. 2. The outermost electron and hole Fermi surfaces of (SN)x. Not shown is a second set of surfaces nested under those given above.

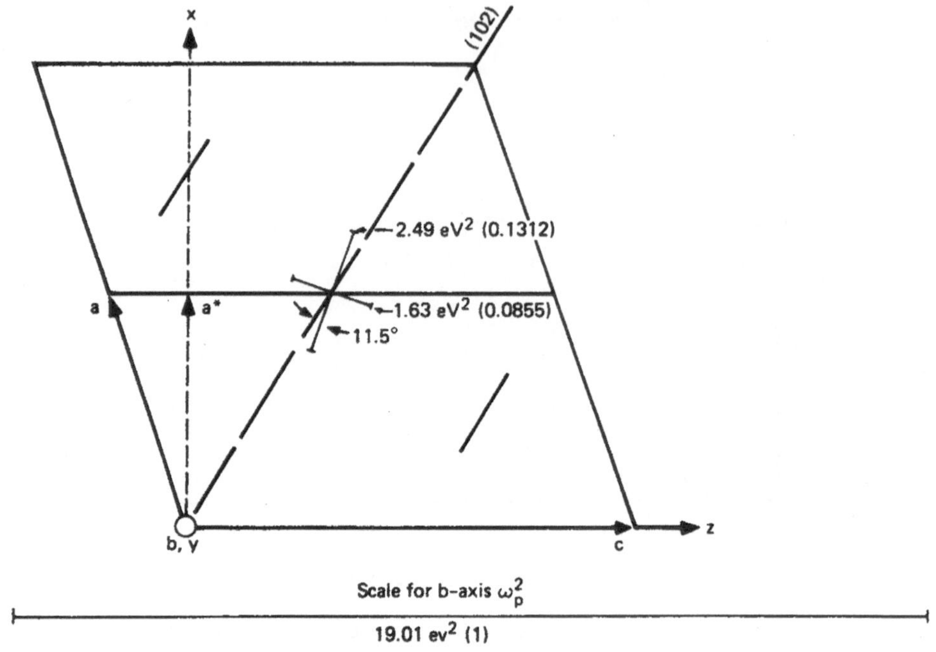

Plasma Tensor Principal Axes for Combined Holes and Electrons

Fig. 3. Principal axes of the (SN)x plasma tensor with respect to the monoclinic unit cell. The view is down the b-axis with the planes of the polymer chains indicated by the short lines in the a-b plane. The scale is set by the plasma tensor component along b with the relative ratio among components given in parentheses. The values shown are for the combined hole-electron surfaces of Fig. 1. The result $\sqrt{\langle\omega_p^2\rangle_{yy}} = 4.4$ eV is in good agreement with the measured value 4.6±1 eV from Ref. 12.

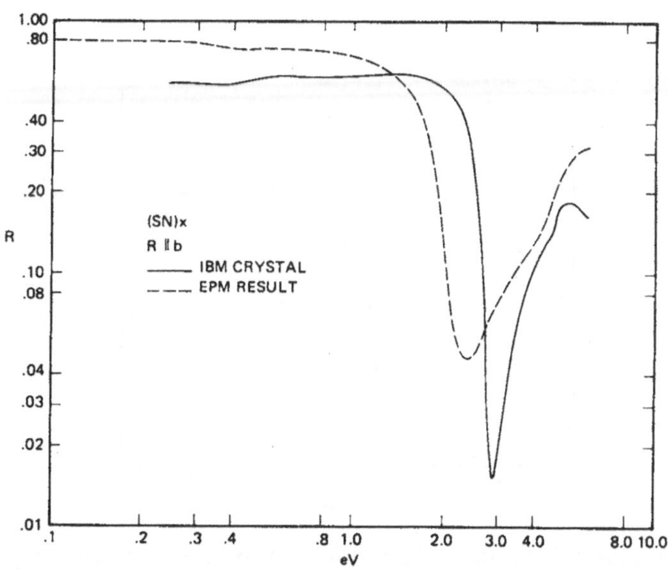

Fig. 4. Calculated and measured reflectivities for light polarized parallel to the b-axis. $\tau_{\parallel}/h$ = 1.8 eV^{-1} (Ref. 12).

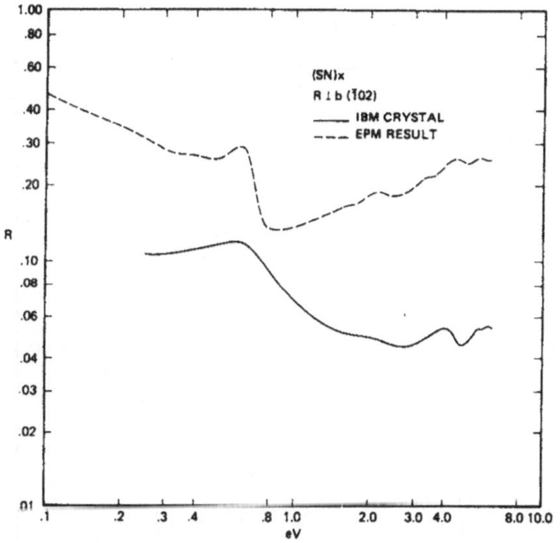

Fig. 5. Calculated and measured reflectivities for light polarized perpendicular to the b-axis and incident on the (102) plane. $\epsilon_{\mu\nu}$ was rotated to the proper frame of reference. $\tau_{\perp}/h$ = 0.6 eV^{-1} (Ref. 11) was assumed for all perpendicular directions.

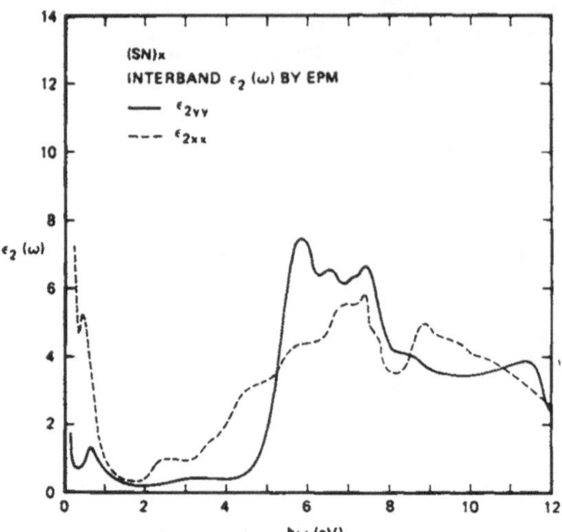

Fig. 6. Interband transition contributions to the ϵ_2 -tensor.

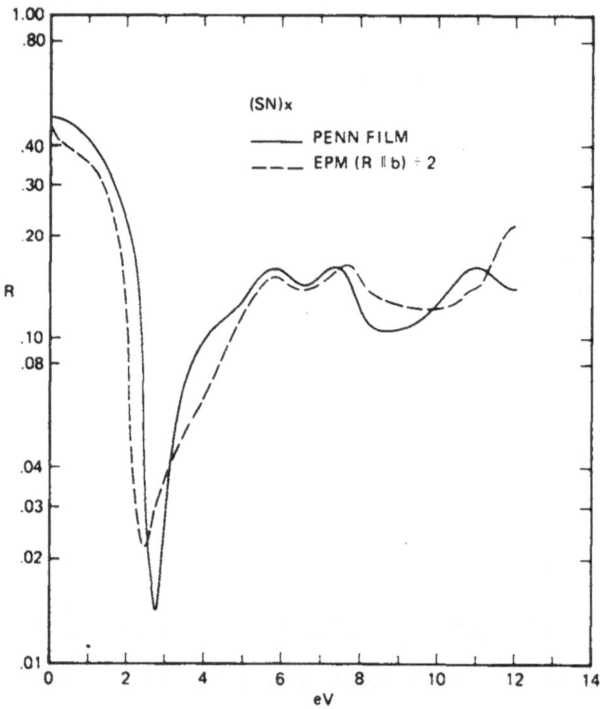

Fig. 7. Comparison between calculated and measured (SN)x unoriente film reflectivities. Film data from Cohen.[11]

REFERENCES

1. F. P. Burt, J. Chem. Soc. (London), 1171 (1910).

2. V. V. Walatka, M. M. Labes and J. H. Perlstein, Phys. Rev. Letters 31, 1139 (1973).

3. R. L. Greene, P. M. Grant and G. B. Street, Phys. Rev. Letters 34, 89 (1975).

4. M. Boudeulle, Cryst. Struct. Commun. 4, 9 (1975).

5. C. M. Mikulski, P. J. Russo, M. S. Saran, A. G. MacDiarmid, A. F. Garito and A. J. Heeger, J. Am. Chem. Soc. 97, 6358 (1975).

6. R. L. Greene, G. B. Street and L. J. Suter, Phys. Rev. Letters 34, 577 (1975).

7. R. Klemm and H. Gutfreund, to appear in Physical Review.

8. W. E. Rudge and P. M. Grant, Phys. Rev. Letters 35, 1799 (1975); W. E. Rudge, I. B. Ortenburger and P. M. Grant, (to be published in Physical Review).

9. H. Kamimura, A. J. Grant, F. Levy, A. D. Yoffe and G. D. Pitt, Solid State Commun. 17, 49 (1975).

10. A. J. Berlinsky, J. Phys. C: Solid State Phys. 9, L238 (1976).

11. M. J. Cohen, Ph.D. Thesis (University of Pennsylvania, 1975).

12. P. M. Grant, R. L. Greene and G. B. Street, Phys. Rev. Letters 35, 1743 (1975).

13. P. M. Grant, R. L. Greene, W. D. Gill, W. E. Rudge and G. B. Street, Mol. Cryst. Liq. Cryst. 32, 171 (1976).

14. L. Pintschovius, H. Wendel and H. Kahlert, Conference on Organic Conductors and Semiconductors (Hungarian Academy of Sciences, Siofok, 1976).

15. F. de la Cruz and H. J. Stolz, Solid State Commun. 20, 241 (1976); W. D. Gill, unpublished data and private communication.

Conference on Organic Conductors and Semiconductors, Siófok, Hungary 1976

ACOUSTIC PHONONS OF POLYMERIC SULFUR NITRIDE, $(SN)_x$

L. PINTSCHOVIUS[a], H. WENDEL[b]
and H. KAHLERT[c]

a) KFZ, Inst. f. Angew. Kernphysik, Karlsruhe, FRG
b) MP-Inst. Für Festkörperforschung, Stuttgart, FRG
c) Ludwig-Boltzmann-Inst. Für Festkörperforschúng, Wien, Austria

The acoustic phonons of $(SN)_x$ have been studied by inelastic neutron scattering. These investigations are extremely difficult because of the small size and poor quality of the crystals and therefore had to be restricted to the long wavelength region.

The dispersion relation of phonons traveling in chain direction shows a large difference between the slopes of the longitudinal and the transverse branches which points to a clear distinction in the strength of intra- and interchain forces. This is confirmed by the upward curvature of the lower transverse branch and by a comparison of the longitudinal velocities of sound for different crystallographic directions.

A valence force model has been used to analyze our data and the Raman and IR data of Stolz et al. [1]. It turns out that there is a threefold hierarchy in the strenght of the binding forces: 1. The by far strongest forces connect atoms within the same chain. 2. The strongest interchain binding forces connect atoms within the $(\bar{1}02)$-plane, which is the plane of the polymer chains. The corresponding force constants are by one order of magnitude smaller than the largest intrachain force constants. 3. In directions approximately perpendicular to the $(\bar{1}02)$-plane the binding forces are rather weak.

The corresponding force constants are by a factor of about 3 smaller than the largest interchain force constants.

Therefore, with respect to the lattice dynamics $(SN)_x$ can be regarded either as a quasi one-dimensional system or as a layer-type crystal.

[1] H. J. Stolz, H. Wendel, A. Otto, L. Pintschovius, and H. Kahlert, to be published.

Conference on Organic Conductors and Semiconductors, Siófok, Hungary 1976

X-RAY AND ULTRAVIOLET PHOTOEMISSION OF POLYMERIC SULFUR NITRIDE $(SN)_x$

P. MENGEL

Institut für angewandte Physik, Universität Karlsruhe, FRG

I.B. ORTENBURGER, W.E. RUDGE, and P.M. GRANT

IBM Research Laboratory, San Jose, California 96193, USA

ABSTRACT

We have determined, both experimentally and theoretically, the one-electron density of states surrounding the Fermi level in polymeric sulfur nitride, $(SN)_x$. The experimental measurements were performed using X-ray and ultraviolet photoemission spectroscopy (XPS and UPS), while the theoretical studies employed calculations based on OPW and pseudopotential band structures of $(SN)_x$. The XPS measurements included observations of N and S core level shifts which yielded values for the number of charged-transferred electrons of about 0.4. This is compared with our recent XPS data on a related compound, Se_4N_4, yielding a charge transfer of 0.3 e only. One of the major observations from both XPS and UPS was that of low photoemissive yields near the Fermi level. Our calculations of the photoemissive response tensor, $D_{\mu\nu}(E,h\nu)$, show that this low yield near E_F is primarily a result of dipole matrix element energy dependence.

In order to gain information on the elecronic structure of polymer sulfur nitride, $(SN)_x$, we investigated the photoemission properties of this interesting material. The measurements were obtained on nonoriented $(SN)_x$ films using x-ray and variable ultraviolet photoemission spectroscopy (XPS and UPS) (1), (2). Table 1 shows the nitrogen and sulfur core levels in $(SN)_x$ and in neutral elements. Also included in table 1 is the nitrogen core level energy of Se_4N_4 from our current XPS measurements on this material (3). Siegbahn (4) has found an experimental relationship between the binding energy and net charge as determined by the chemical shifts of the S(2p) and N(1s) lines in several nitrogen - and sulfur - containing compounds. Applying his criteria to our N(1s) core level energies, we obtain a net charge transfer figure in $(SN)_x$ and Se_4N_4 . The results are shown in table 2 together with results of charge transfer in $(SN)_x$ and S_4N_4 taken from XPS data of Salaneck et al (5).

Comparing the results of the Se_4N_4 and S_4N_4 structures, we note that only half as much charge is transferred in the SeN bond. Salahub and Messmer (6) have calculated a charge transfer of 0.5 e for $(SN)_x$ by the self consistent-field $X\alpha$ standing wave method, resulting in a large dipolemoment for the SN bond. Further it is shown, that this dipolemoment is the determining factor for the square planar geometry of S_2N_2, from which by thermal excitation the polymerization process is very likely to occur. In the important question of analogs one would expect that substitutions for S and N such that a larger difference in electronegativity results are more favorable. One realistic possibility here is believed to be a replacement of S by Se. However, the decrease of the charge transfer in SeN as determined from our XPS data indicates a much less stable hypothetical Se_2N_2 structure in a S_2N_2 geometry. This may explain why so far all attempts by the chemists to prepare Se_2N_2 and $(SeN)_x$ analogs have failed.

The XPS spectrum for the valence band region of $(SN)_x$ is shown in Fig.1. In the lower half of the figure the experimental results are directly compared with two theoretically derived single-particle density of states (DOS). A one-dimensional tight-binding extended Hückel (1DTB) computation has been published by Thomas and Parry (7) and a three dimensional OPW (3DOPW) calculation was done by Rudge and Grant (8). Both theoretical DOS show good overall qualitative agreement with the experimental data and in addition the 3DOPW DOS gives good quantitative agreement in peak energy positions and relative peak intensity ratios. The 3DOPW DOS presented here is based on the older Lyon or Boudeulle crystal structure. When using the newer crystal structure with less S-S coupling, referred to as Penn structure, the theoretical DOS remains remarkably unchanged. The only difference is a shift of peak 1 by 0.8 eV to lower energies. It is shown in reference (8) that this shift arises from a displacement of certain intrachain bands and none of the DOS structure appears directly related to interchain bands. Thus, one cannot expect the XPS data to be sensitive to interchain coupling, which explains the overall good agreement with the pure 1DTB DOS.

A low density of states at the Fermi level was found in the experiment. A simple area calibration to 44 electrons yields a value of 0.04 states/(eV x spin x molecule). This number should be considered as a crude estimate only, due to different photoionization cross section of sulfur and nitrogen orbitals and other experimental errors.

In the UPS measurements we were able to prepare the $(SN)_x$ films inside of the UHV chamber. X-ray analysis and electron reflection patterns show the $(\bar{1}02)$ plane of $(SN)_x$ to lie parallel to the substrate plane in a textured orientation. A family of electron distribution curves (EDCs) for photon energies of 7.6 eV $\leq h\nu \leq$ 11.1 eV is shown in Fig.2. The EDCs do not show much structure but one would not expect that from the XPS data and the calculated DOS. We see well resolved now the first peak at 1 eV below E_F, also a peak at 4.5 eV and a further structure at 2.5 eV. The most surprising result is the very low electron yield at the Fermi level, which in fact lies more or less within the sensitivity limitation of the spectrometer.

In order to relate the experimental sturcture to the density of occupied states and to clarify the low electron yield at the Fermi level, an extended calculation of the photoemissive response tensor $D_{\mu\nu}(E,h\nu)$ was undertaken based on pseudopotential band structure calculation (9). In Fig.3 we compare experimental and calculated EDCs for some vlaues of exciting photon energy. We like to point out that we also accounted for the observed textured film structure. This was achieved by applying an arbitrary rotation on the $D_{\mu\nu}(E,h\nu)$ tensor around an axis perpendicular to the sample surface and averaging over all angles.

For low photon energies we find a reasonable agreement between theory and experiment. The entire experimental structure is revealed in the theoretical data. We used here the Penn crystal structure - which we believe is proofed to be the more accurate one - but that leads to a disagreement in energy position of about 0.6 eV for the first peak below E_F. A better agreement for this peak was obtained when using the Lyon crystal structure. No agreement between experimental and calculated EDCs is found for photon energies greater 10 eV as shown in Fig.3 for $h\nu = 11.1$ eV. We think it is unreasonable to assume, that the pseudopotential method is sufficiently accurate to determine the conduction bands far above the Fermi level. Of overriding importance is the fact, that the calculated EDCs show the same low photoemissive yield at E_F. To investigate the reasons for this low electron contribution at E_F, we calculated the energy dependence of the average matrixelement (9). This is shown in the lower half of Fig. 4 for one particular photon energy of $h\nu = 7.9$ eV. In the upper half of the figure we computed the distribution curve based on a simple joint density model with constant matrixelement assumption for the same photon energy.

The explanation is obvious. In the joint density model we obtain an electron yield of 0.03 states/(eV x spin x molecule) at E_F comparable with the XPS data. However, the matrixelement dependence shows that the transition probability at E_F is only 1/5 of the average value ($\equiv 1$).

The product of both matrixelement energy function and joint density of states leads then directly to a neglecting small electron yield at the Fermi level in excellent agreement with the experimental results.

We note that no explanations based on relaxation or localization effects are used to explain this low photoemissive yield at E_F. Thus, as far as the analysis of photoemission data in general is concerned, a constant matrixelement assumption can easily lead to wrong conclusions about the importance of such effects.

Table 1: XPS determined core level energies of sulfur and nitrogen in $(SN)_x$, in Se_4N_4 and in neutral elements.

Element	Level	Binding Energy (eV) in $(SN)_x$	Binding Energy (eV) in Se_4N_4	Binding Energy (eV) for neutral Elements
N	$1S_{1/2}$	- 397.3	- 398	- 399
S	$2S_{1/2}$	- 228.5		- 229
	$2P_{3/2}$	- 163.5		- 162
	$2P_{1/2}$	- 164.5		- 163
	$2P^{a)}$	- 164.5		

a) After long exposure (> 4h) to air or the XPS X-ray beam.

Table 2: Charge transfer values δ for $S^{+\delta}N^{-\delta}$ and $Se^{+\delta}N^{-\delta}$ in $(SN)_x$, S_4N_4 and Se_4N_4 from XPS data.

Structure	Reference	Charge Transfer δ(e)
$(SN)_x$	this work	0.42
	Salaneck et al 5	0.5
S_4N_4	Salaneck et al 5	0.6
Se_4N_4	this work	0.29

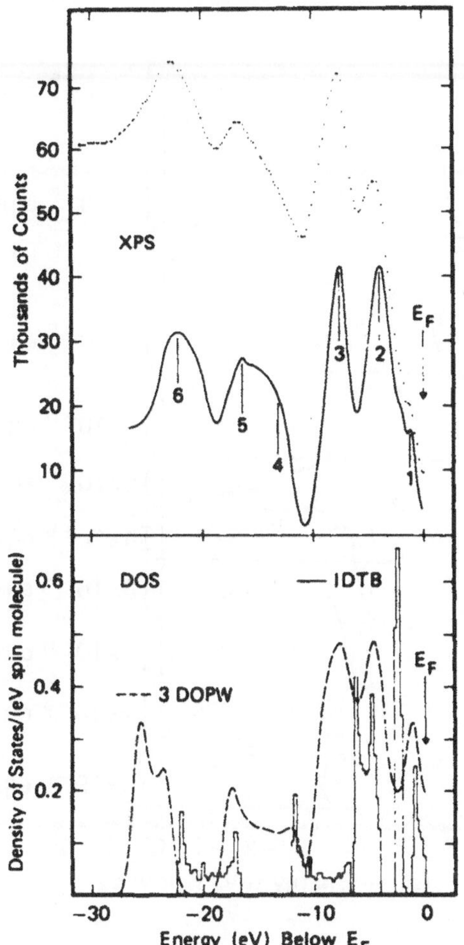

Figure 1: Top: X-ray photoemissions spectrum, raw data (dotted line) and background corrected data (solid line).
Bottom: One-electron valenceband density of states for a one-dimenstional planar chain (1DTB, solid line), and for a three-dimensional OPW calculation based on the Lyon crystal structure (3DOPW, broken line). The ordinate scale refers only to the OPW result which was obtained using an 0.25 eV Gaussian broadening function.

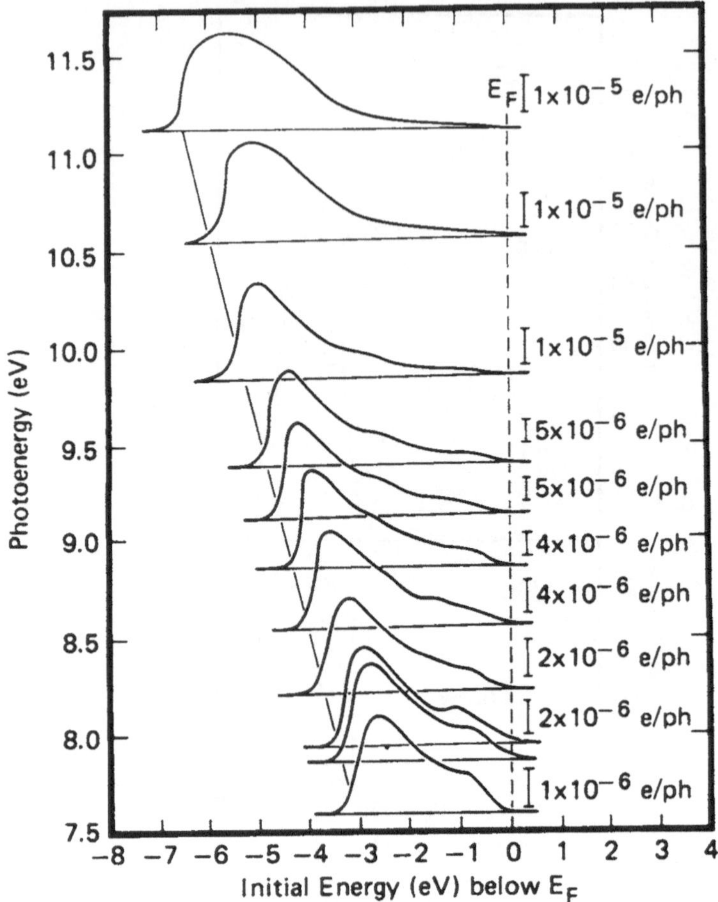

Figure 2: A family of EDCs for 7.6 eV $h\nu \leq 11.1 \leq$ eV from an about 1 μ thick $(SN)_x$ film. Each curve is normalized to its maximum. In absolut values the quantum yield for the highest photon energy $h\nu = 11.1$ eV is more than a factor of 10 greater than for the lowest photon energy. On a more sensitiver scale the distribution curve for $h\nu = 11.1$ eV reveals all the structure shown in EDCs for the lower photon energies. The diagonal line marks zero kinetic energy.

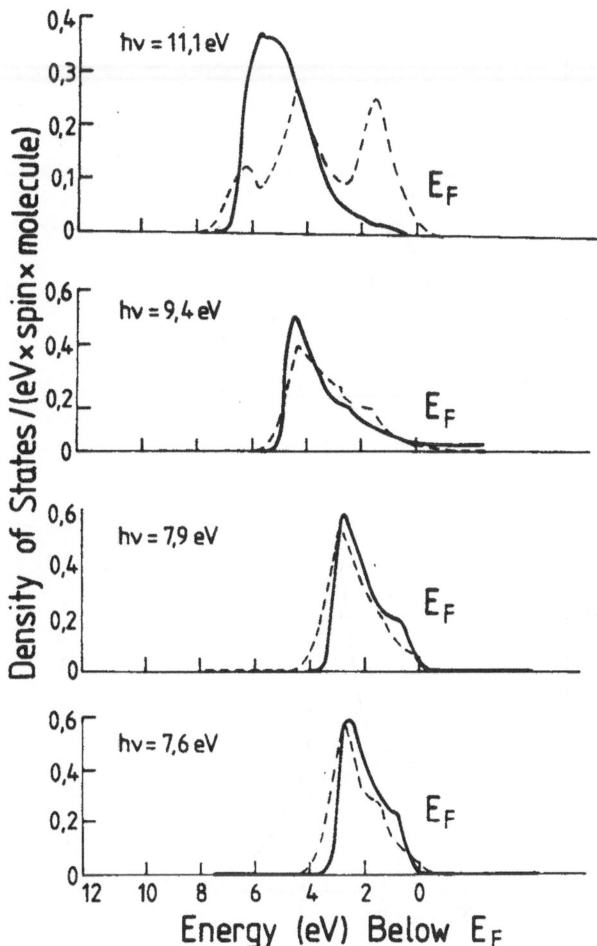

Figure 3: A comparison between calculated (broken line) and experimental (solid line) EDCs for hν = 7.6, 7.9, 9.4 and 11.1 eV. All experimental and theoretical curves are normalized to the total quantum yield.

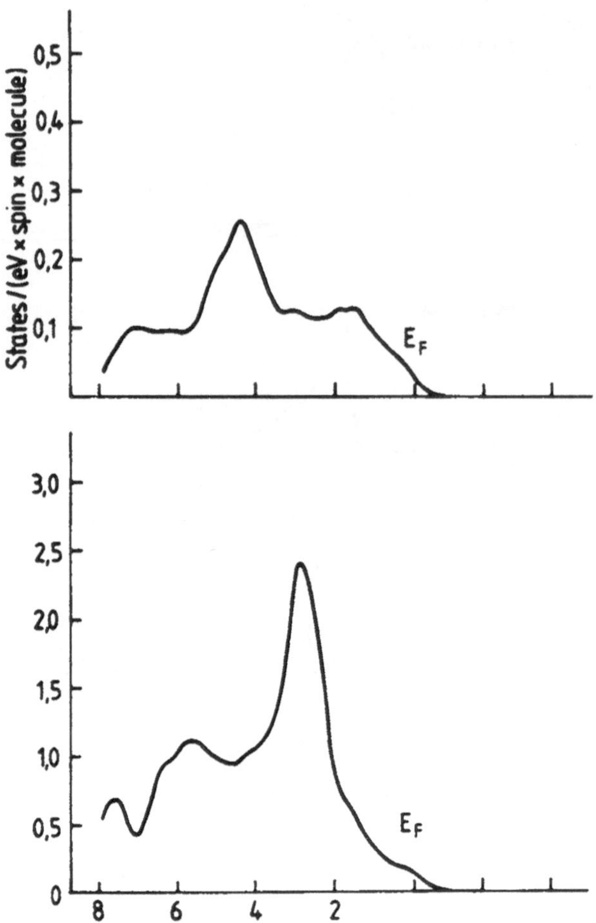

Figure 4: Top: EDC for hν = 7.9 eV based on simple joint density model.
Bottom: Average dipolematrixelement function M (E, hν) for the same photon energy.

REFERENCES

1) P.MENGEL, P.M. GRANT. W.E. RUDGE, B.H.SCHECHTMAN and D.W. RICE
Phys.Rev.Lett. 35, 1803 (1975)

2) P.MENGEL, I.B.ORTENBURGER and P.M.GRANT
Submitted to phys.rev.

3) P.MENGEL, C.BRULET and G.B.STREET
to be published

4) K.SIEGBAHN, C.NORDLING, A.FAHLMAN, R.NORDBERG,K.HAMRIN, J.HEDMAN, G.JOHANSSON, T.BERGMARK, S.E.KARLSSON, I.LINDGREN and B.LINDBERG,
in ESCA; Atomic, Molecular and Solid State Structure Studies by means of Electronic Spectroskopy (Almquist and Wiksells, Uppsala, Sweden, (1967)

5) W.R.Salaneck, J.W.p Lin, and A.J.EPSTEIN
Phys.Rev.B 13, 5574 (1976)

6) D.R.SALAHUB and R.P.MESSMER
submitted to J.ChemPhys.

7) D.E.PARRY and J.M.THOMAS
J.Phys.C: Solid State Phys. 8, L 45 (1975)

8) W.E.RUDGE and P.M.GRANT
Phys.Rev.Lett. 35, 1799 (1975)

9) For details see ref. 2 and P.M.Grant, W.E.RUDGE and I.B.ORTENBURGER this volume.

REFERENCES

Conference on Organic Conductors and Semiconductors, Siófok, Hungary 1976

SUPERCONDUCTING PROPERTIES OF $(SN)_x$

R.L. GREENE and W.D. GILL

IBM Research Laboratory San Jose, California, USA

L.J. AZEVEDO, W.G. CLARK and G. DEUTSCHER

Department of Physics UCLA Los Angeles, California, USA

ABSTRACT: The superconducting properties of polysulfur nitride, $(SN)_x$, are reviewed. It is shown that superconductivity is a bulk property of $(SN)_x$, but that a detailed understanding of many properties of the superconducting state must take account of the fibrous morphology and imperfect nature of $(SN)_x$ crystals.

Since the initial discovery[1] of superconductivity in the sulfur nitride polymer, $(SN)_x$, with transition temperature $T_c \approx 0.3°K$, a variety of experiments have been performed to learn more about the superconducting state in this unusual material. In this paper we will very briefly review this work and discuss our present understanding of the superconducting properties of $(SN)_x$. Several reviews of the normal state properties of $(SN)_x$ have recently appeared.[2] The superconducting experiments which have been reported up to the present time are: 1) The variation of T_c with crystal quality,[3,4] 2) the angular and temperature dependence of the critical magnetic field[5], 3) the hydrostatic pressure dependence of T_c[6], 4) the observation of superconducting fluctuations[7], and 5) the specific heat from 0.11--1.5°K[8].

In discussing these measurements on $(SN)_x$ one cannot ignore the effect of crystal imperfections[2] which seem to be an inevitable result of the solid state polymerization process. Scanning electron microscope studies show that every so-called single crystal is in reality an oriented bundle of fibers each with a diameter of a few hundred Å. Diffraction measurements show that all crystals are twinned with (100) as the twin plane and have a high degree of disorder in directions perpendicular to the polymer (fiber) axis. There are also some imperfections along the polymer chains. In addition, about five atomic percent of hydrogen is present which presumably is randomly bonded to nitrogen atoms.[2b] One definitive effect of these crystal imperfections is seen in the variation in T_c with crystal quality. Crystals with higher resistivity ratio $[\rho_{\parallel}(300°K)/\rho_{\parallel}(4.2°K)]$ have higher and sharper (ΔT_c) transition temperatures[2c, 3]. Even though the higher quality crystals have a fibrous morphology and perpendicular disorder, they appear to have fewer imperfections along the chains, since $\rho_{\parallel}(4.2°K)$ is small and no ESR signal is observed. Emission spectrographic analysis shows no measurable metal (including magnetic) impurities (<5ppm) even in crystals with a low (<5) resistivity ratio. The influence, if any, of the hydrogen on the superconductivity is not understood.

The temperature dependence of the critical magnetic field is shown in Fig. 1.[5] Several features are unusual and have been discussed in detail in Ref. 5. The extrapolated value of $H_{c2\parallel}$ at T = 0°K exceeds the paramagnetic limit of $H_p = 18.4\ T_c \approx 5.4$ kOe, the temperature dependence of $H_{c2\perp}$ is anomalous and different from $H_{c2\parallel}$, the critical field near T_c is only weakly dependent on $\rho_{\parallel}(4.2°K)$ and $H_{c2\parallel}/H_{c2\perp}$ is large and temperature dependent. The anisotropy in H_{c2} can be explained from a combination of the intrinsic anisotropy in the semi metallic band structure and the anisotropy in scattering time (caused by tunnelling between the fibers) leading to an anisotropy in the coherence length, $\xi(T)$. On the other hand, the weak dependence of H_{c2} on ρ_n, the angular dependence of H_{c2}, and the temperature dependence of $H_{c2\parallel}/H_{c2\perp}$ suggest that a model of weakly coupled[9] superconducting fibers of diameter d ≈ 200Å is more appropriate. The latter view is supported by the work of Civiak et al[7] who claim to observe superconducting fluctuations in the conductivity of $(SN)_x$ just above T_c. Fluctuation effects would not be expected in the case of large fibers ($d > \xi_{\perp}(T)$) or for coupled fibers of any size. It should also be noted that similar critical field effects have been observed in granular superconducting films[10, 11] where the coupling between grains plays an important role.

Under quasi-hydrostatic pressure T_c increases from $\gtrsim 0.3°K$ at atmospheric pressure to $\approx 0.5°K$ at 8 kbar[6]. T_c has not been measured at higher pressure but at 100 kbar[12] there is no superconductivity above 1.2°K so the pressure dependence of T_c must saturate or change direction. Since it is unusual for T_c to increase under pressure in an s-p superconductor, it was suggested[6] that electronic band structure changes with pressure must be significant. However, a calculation[13] of the electron density of states with different lattice parameter showed very little change and it was proposed[13] that some phonon modes must soften under pressure. We consider this latter explanation rather unlikely and suggest that another possible origin of the increase in T_c is a stronger coupling of the superconducting fibers under pressure. The work on granular metals[11] has shown that T_c can be decreased from its intrinsic value by weak coupling between grains. This explanation makes the pressure results consistent with the critical field and fluctuation studies.

Finally the specific heat measurements below 1°K[8] suggest that the superconductivity is a bulk effect. A broad hump in specific heat is observed starting below 0.4°K. The entropy associated with this hump is comparable to that of a simulated BCS anomaly with $T_c = 0.26°K$. The smearing of the superconducting transition may be caused by localized magnetic moments at the ends of broken $(SN)_x$ chains, by other sample inhomogeneities or by fluctuation effects. The specific heat anomaly is quite sensitive to strain and magnetic field and more work is needed to completely understand the details of such data in a complicated material like $(SN)_x$.

In summary, our present view of the superconductivity in $(SN)_x$ is the following. Polysulfur nitride is a semi-metal with bulk type II superconductivity caused by the usual BCS mechanism. The transition temperature depends strongly on the polymer chain perfection and other unknown types of non-magnetic disorder. The temperature and angular dependence of the critical field, the pressure dependence of T_c and the transition width as determined from specific heat and conductivity measurements results partly from intrinsic properties of $(SN)_x$ and partly from the fibrous morphology and other imperfections present in $(SN)_x$ crystals grown to date. The task of quantitatively separating the intrinsic properties from the extrinsic ones, so as to learn more about the superconducting and normal state properties of $(SN)_x$, remains an important problem for future study.

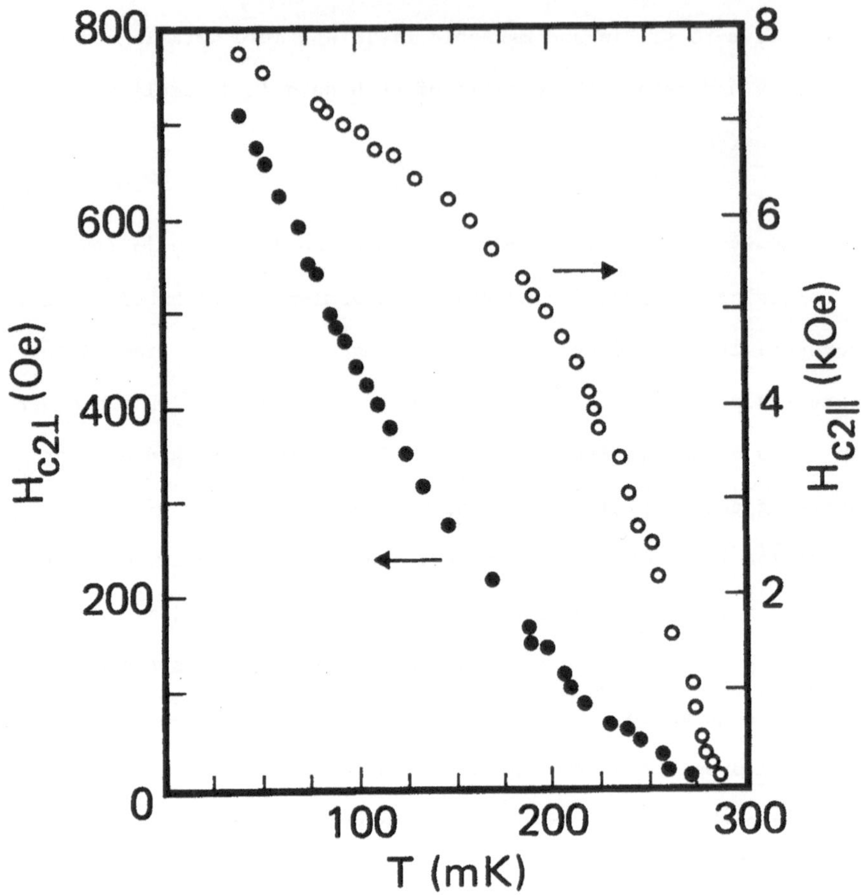

Fig. 1. Upper critical field H_{c2} parallel and perpendicular to the b (fiber) axis vs. temperature for crystalline $(SN)_x$.

REFERENCES:

1. R. L. Greene, G. B. Street and L. J. Suter, Phys. Rev. Lett. 34, 577 (1975).

2. a) H. P. Geserich and L. Pintschovius, Festkörperprobleme 16, 65 (1976).

   b) G. B. Street and R. L. Greene, IBM Journal of Res. and Dev., to be published.

   c) R. L. Greene and P. M. Grant, Proceedings of the NATO-ASI on "Chemistry and Physics of One-Dimensional Metals", Bolzano, Italy, August 1976 (edited by H. J. Keller, Plenum Press 1977).

3. G. B. Street, H. Arnal, W. D. Gill, P. M. Grant, and R. L. Greene, Mat. Res. Bull. 10, 877 (1975).

4. P. Civiak, W. Junker, C. Elbaum, H. I. Kao and M. M. Labes, S. S. Comm. 17, 1573 (1975).

5. L. J. Azevedo, W. G. Clark, G. Deutscher, R. L. Greene, G. B. Street and L. J. Suter, S. S. Comm. 19, 197 (1976).

6. W. D. Gill, R. L. Greene, G. B. Street and W. A. Little, Phys. Rev. Lett. 35, 1732 (1975).

7. R. L. Civiak, C. Elbaum, W. Junker, C. Gough, H. I. Kao, L. F. Nichols and M. M. Labes, S. S. Comm. 18, 1205 (1976) and to be published in Phys. Rev.

8. L. F. Lou and A. F. Garito, preprint.

9. By weak coupling we mean $\xi_\perp$(isolated fiber) $>$ d and $\xi_\perp$(crystal) $<$ d. For strong coupling $\xi_\perp$(crystal) is $>$ d.

10. G. Deutscher and S. A. Dodds. submitted to Physical Review.

11. H. C. Jones, App. Phys. Lett. 27, 471 (1975).

.2. C. W. Chu, private communication.

.3. M. Schluter, J. R. Chelikowsky and M. L. Cohen, Phys. Rev. Lett. 36, 452 (1976).

Conference on Organic Conductors and Semiconductors, Siófok, Hungary 1976

QUANTUM CHEMICAL STUDIES ON $(SN)_x$

M. KERTÉSZ*, A. AŽMAN**, Á.I. KISS*** and J. KOLLER**
*Central Research Institute for Chemistry of the Hungarian Academy of Sciences, Budapest, Hungary
**B. Kidrič Chemical Institute, University of Ljubljana, Yugoslavia
***Department of Physical Chemistry, Technical University, Budapest, Hungary

ABSTRACT

The self-consistent-field (SCF) ab initio Hartree-Fock crystal orbital method is applied with success to polysulfur nitride $(SN)_x$, chains using non-local exchange and evaluating all integrals over atomic orbitals within 5 atomic neighbours accurately.

Furthermore, on the basis of a systematic combination of chemical arguments and results of molecular orbital calculations the $(HP - CH)_2$ ring is suggested as possibly the most stable analogue of $(S - N)_2$, the precursor of $(SN)_x$.

INTRODUCTION

Quantum chemical calculations may contribute to the understanding of the electronic structure of solids in many ways, e.g. by band structure calculations, evaluation of other observables from the wave functions, stability analysis, predictions for hypothetical substances, etc. The specificity of the system is crucial, as one starts from a given chemical model, usually from the geometry of the system. If the unit cell consists of several atoms - a situation typical for substances covered by the name organic conductors and semiconductors - quantum chemical approaches are perhaps even more useful than for the "atomic" solids due to the large number of interactions of comparable order even if merely the electronic part of the Hamiltonian is concerned. In discussing the applicability of certain model Hamiltonians (like the extended Hubbard and many others) one needs reliable estimates of different electronic interactions and for such problems the quantum chemical approach should be used. Furthermore, in trying to design solids having desired properties one can make similar estimates as the calculations go along the same lines for hypothetical systems, provided reasonable guesses for the geometry can be made.

In present day quantum chemistry the linear-combination-of-atomic orbitals (LCAO) calculations play a predominant role including simple semi-empirical (SE) one electron methods up to sophisticated correlated ab initio ones.

BAND STRUCTURE CALCULATIONS

A few words for the justification in applying ab initio (a.i.) methods for the band structure calculations seem to be in order. We do not have any SE method at present which is able to give reasonable results for several physical properties. For example, the popular extended Hückel (EH) method (first applied to a 1D $(SN)_x$ model in Ref. (1) and to 3D models in Refs (2,3) is unable to predict charge distribution and stability correctly, especially for such highly polar systems as $(SN)_x$. Self-consistent methods as e.g. the complete neglect of differential overlap (CNDO) methods are better for charge distribution but bad at predicting level spacing for σ and π orbitals, and so on.

Unfortunately, a.i. methods are very expensive in computer time and cannot be applied to very large unit cells. (The limit of applicability is about 10^6 - 10^7 electron-electron repulsion integrals over atomic orbitals to be included at present.) Therefore, a.i. calculations will in our opinion merely supplement SE band methods for a long time. Nevertheless, their results will help in improving SE methods.

We have applied the a.i. Hartree-Fock crystal orbital method to three nuclear configurations: I was an equidistant chain with r_{SN} = 1.62 Å and all bond angles taken as 115°; II was a chain taken from Boudelle's structure (4); chain III corresponded to the "Penn" structure (5).

π-electrons seem to play a predominant role in determining the electronic structure of $(SN)_x$ and the crossing of σ and π type orbitals at the Fermi level (E_F), a conclusion emerging from a recent non-self-consistent calculation (6), is in contrast to general chemical trends. However, in our a.i. band structure calculations (7-9), pure half-filled π bands resulted. This result is confirmed by some XPS evidence (10). Furthermore, increasing $S2_p$ character towards E_F indicated by XPS (10) is also in agreement with our wave function. It may be noted that a σ - π band crossing at E_F was also noticed in a non-self-consistent study of polyene (11) which also turned to half-filled pure π -band using SCF procedures.

There are two further inevitable successes of our a.i. calculations: the charge transfer (CT) from S to N was predicted quite well (7) before the measurements were performed (12, 13). We obtained values between 0.30e and 0.36e depending on geometry and the number of neighbours included. The stability was also in accord with experiment viz. the "Penn" structure is the most stable among the three models considered (II is 0.011 a.u. higher than III). The molecular variant of the method was able to reproduce the geometry of S_2N_2 also very well (14).

In Table 1a and 1b some typical results from different band structure calculations for $(SN)_x$ are compared.

The a.i. method seems to be the best for the overall description and deficiences, in our opinion, are mainly connected with the omittance of the correlation (15): the conduction band width is too large and the effective mass as well as the density of states at E_F are therefore probably accidentally close to experiment. A 3D calculation using the a.i. SCF method is for numerical reasons out of the question today, but we think that our Fock matrix elements should be used in a 3D EH type work. Nevertheless, successes of a 1D model indicate that - at least as far as stability of the chain, charge distribution, the deeper valence bands and the type of the wave function at E_F are concerned - 3D couplings are unimportant.

We mention briefly two further variational calculations (9) using the a.i. method and a one-determinantal many-electron wave function. We have looked for broken-symmetry solutions: one corresponding to a charge-density-wave (CDW) and another using different orbitals for different spins (DODS) corresponding to a spin-density-wave (SDW). We have found both types of instabilities to occur, gaps of 0.120 a.u. and 0.535 a.u. caused 0.007 and 0.116 a.u. energy gains corresponding to CDW and SDW, respectively. The amplitude of the waves may be characterized by the following quantities (only the largest alternations are given here): $\pm$ 0.04e charge alternation was found on S for CDW and $\pm$0.45 was the alternation in the diagonal elements of the densities corresponding to spin 1 and spin 2 of the N2p orbitals. The very large gap found for the SDW indicates, that the DODS wave function is completely unsuited for $(SN)_x$, while the small energy gain and wave-amplitude of the CDW may be easily lifted even by small interchain couplings, depressing a Peierls-type distortion in $(SN)_x$.

Table 1a

Summary of selected results from different band structure calculations for $(SN)_x$

Method, model and Ref.	Conduction band width (eV)	m_{eff} (m_e)	$N(E_F)$ $\frac{\text{states}}{\text{eV spin molec.}}$	Charge transfer (e)	Stability
EH, 1D (1)	~ 1				
TB, 1D (6)	~ 3		0.7		
CNDO, 1D (16)	~5		0.06	0.18	
ab initio, 1D (7 - 9)	5.6	~2.2	0.10	0.36	Penn (5) is more stable than (4)
EH, 1D, 3D (2 - 3)	1.5	m = 1.5 m = 4.5			
OPW, 3D (17)	2-3		0.13		
Pseudopotential, 3D, (18)	~ 2.5		0.3	0.9	
Experiment	1.6 ± 0.3 (10) 2-3 (19)	~1 (10) ~2 (19)	0.12 (10) 0.18 (20)	0.5 (13) 0.30-0.42 (12)	Penn (5) is most probably correct

Table 1b

Method, model and Ref.	Optical transitions (eV)	Wave function at E_F	Type of conduction band	Comment
EH, 1D (1)			$\tilde{\pi}$	overall agreement with XPS spectra (10)
TB, 1D (6)	0.5, 2.45		σ and π	
CNDO, 1D (16)				no CDW instability found
ab initio, 1D (7 - 9)		increasing SZ_p π-type character towards E_F	$\sim\tilde{\pi}$	unstable against SDW and CDW; overall correspondance of deeper valence bands to XPS peaks
EH, 1D, 3D (2, 3)				semimetal, electron and hole pockets in the Fermi surface
OPW, 3D (17)				good agreement with XPS except conduction band
Pseudopotential, 3D (18)	1.5, 5.5	mainly SZ_p (σ, π)		deformation potential ~1 - 3 eV, conventional superconductivity claimed
Experiment	1.8, 4-6 (21)	increasing SZ_p character towards E_F (10)		

SELECTION OF POSSIBLE ANALOGUES

Both $(SN)_x$ and its precursor, the four membered planar ring molecule, S_2N_2 are the only compounds exhibiting certain special electronic properties of a planar delocalized π -electronic structure which leads in both systems to an extra stabilization and to the metallic properties in $(SN)_x$. It is perhaps worthwhile to speculate on possible analogues. $(SeN)_x$ has been suggested by Salahub and Messmer (22) and Györffy (23) since Se is the next number of Group VI. Our aim, in a very recent study (24), was to consider the possibilities of analogues of S_2N_2 in a more systematic way.

In considering the possible candidates we have taken the following characteristics as basic in the structure of S_2N_2 (14):

1. It consists of a planar ring

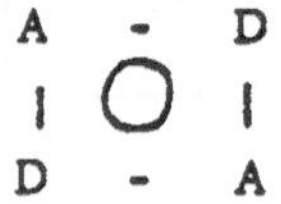

```
A - D
| O |
D - A
```

where D is a π -donor and A a π -acceptor; D contributing two and A one π -electron to the delocalized π-electron system.

2. The highest occupied and the lowest empty orbitals are π -type the corresponding transition lying in the UV or visible. Those transitions seem to be preferable which connect an orbital mainly localized on D with an orbital extending over D and A.

3. The ring is energetically more stable than the isolated D - A units.

The electronegativity and number of π-electrons plus the minumum number of σ -electrons for closing the ring rule out the majority of the elements from the Periodic Table. As for A the only realistic candidates are nitrogen and carbon, for D sulfur, selenium and phosphorus are reasonable to be considered. It is necessary to attach side groups to carbon as well as to P in order to attain the appropriate valence state. For side groups those are preferable which increase the stability of the ring and the CT while keeping a reasonably small size. Altogether 15 molecules have been selected.

EH molecular orbital calculations have been performed for these systems as well as for the isolated D - A units. Charge distributions, highest occupied and lowest empty levels, π -resonance energies have been considered.

On the basis of the results we concluded that besides Se - N, the P - C combination is the most promising one. The P - N combination seems to be less stable. Regarding the substituents, H atoms seem to be the most suitable, neither methyl nor mercapto groups tend to increase the stability.

Unfortunately, due to lack of experimental information, our speculations on the analogues should be considered with proper reservations. The larger part of the work remains to be done by preparative chemists.

REFERENCES

(1) D. E. Parry and J. M. Thomas, J. Phys. C., 8, 145 (1975)

(2) W. L. Friesen, A. J. Berlinsky, B. Bergersen, L. Weiler and T. M. Rice, J. Phys. C., 8, 3549 (1975)

(3) A. A. Bright and P. Soven, Solid State Commun., 10, (1975)

(4) M. Boudelle and P. Michelle, Acta Crystallogr. A28, S199 (1972)

(5) C. M. Mikulski, P. J. Russo, M. S. Saran, A. G. MacDiarmid, A. F. Garito and A. J. Heeger, J. Amer. Chem. Soc., 97, 6358 (1975)

(6) V. T. Rajan and L. M. Falicov, Phys. Rev. B12, 1240 (1975)

(7) M. Kertész, J. Koller, A. Aźman and S. Suhai, Phys. Lett. A55, 107 (1975)

(8) M. Kertész, J. Koller and A. Ažman (submitted to phys. stat. sol /b/)

(9) M. Kertész, J. Koller and A. Ažman (in preparation)

(10) L. Ley, Phys. Rev. Lett., 35, 1796 (1975)

(11) J. E. Falk and R. J. Fleming, J. Phys. C8, 627 (1975)

(12) P. Mengel, P. M. Grant, W. E. Rudge, B. H. Schlechtman and D. W. Rice, Phys. Rev. Lett. 35, 1803 (1975)

(13) W. R. Salaneck, J. W. Lin and A. J. Epstein, Phys. Rev. B13, 5574 (1976)

(14) M. Kertész, A. Ažman, D. Kocjan, S. Suhai and Á. I. Kiss (submitted to Chem. Phys. Lett.)

(15) See e. g. S. T. Pantelides, D. J. Mikish and A. B. Kunz, Phys. Rev. B10, 2602 (1974)

(16) A. Zunger, J. Chem. Phys., 63, 4854 (1975)

(17) W. E. Rudge and P. M. Grant, Phys. Rev. Lett., 35, 1799 (1975)

(18) M. Schlüter, J. R. Chelikowsky and M. L. Cohen, Phys. Rev. Lett., 35, 869 (1975)

(19) R. M. Grant, R. L. Greene and G. B. Street, Phys. Rev. Lett., 35, 1743 (1975)

(20) R. L. Greene, P. M. Grant and G. B. Street, Phys. Rev. Lett., 34, 89 (1975)

(21) D. Chapman, R. J. Warn, A. G. Fritzgerald and A. D. Yoffe, Trans. Faraday Faraday Soc., 60, 294 (1964)

(22) D. R. Salahub and R. P. Messmer, J. Chem. Phys., 64, 2039 (1976)

(23) B. Györffy (private communication, 1975)

(24) Á. I. Kiss and M. Kertész (in preparation)

Conference on Organic Conductors and Semiconductors, Siófok, Hungary 1976

GALVANOMAGNETIC EFFECTS IN POLYMERIC SULFUR NITRIDE, $(SN)_x$

W.D. GILL, W. BEYFR* and G.B. STREET

IBM Research Laboratory San Jose, California USA

ABSTRACT: Magnetoresistance of $(SN)_x$ crystals has been measured for conductivity both parallel and perpendicular to the crystalline b-axis. A measure of the intrinsic conductivity anisotropy is obtained from the dependence of $\Delta\rho_\perp/\rho_\perp$ on the orientation of the crystal in the magnetic field. The galvanomagnetic tensor has been calculated from the theoretical Fermi surface anisotropy and compared with experimental results. The electron and hole mobilities parallel to the $(SN)_x$ chains are 430 cm^2/V-sec and 600 cm^2/V-sec respectively with a scattering time of $\sim 10^{-13}$ sec.

* Permanent address: Fachbereich Physik der Universitat Marburg, Marburg, F.R.G.

Polymeric sulfur nitride is a highly anisotropic crystalline solid in which the measured conductivity anisotropy $\sigma_{\parallel}/\sigma_{\perp}$ is of order 10^2 at 300°K and increases to greater than 10^4 at 4°K. However optical properties and band structure calculations show much less intrinsic anisotropy suggesting that extrinsic effects due to the fibrous nature of the material dominate the measured perpendicular conductivity. In an attempt to determine the intrinsic conductivity anisotropy of $(SN)_x$ fibers we have measured the magnetoresistance of $(SN)_x$ crystals at 4°K for current both parallel and perpendicular to the b-axis. The experimental results have been compared with the behavior predicted from the galvanomagnetic tensor obtained from the band structure[1] and Fermi surface calculations[2] of Rudge et al.

Resistivity measurements were carried out at 23 Hz using 4-probe geometry with the samples immersed in liquid helium. The sample could be rotated with respect to the magnetic field. Sample resistance was continuously recorded with increasing field strength up to fields of 70 kG At the lower fields negative magnetoresistance was observed, however, a positive quadratic effect was dominant at higher fields. Only this positive magnetoresistance component is discussed in this paper.

The most useful experimental configuration for measuring the conductivity anisotropy was with the current flow perpendicular to the b-axis. The transverse magnetoresistance could then be measured for $H \parallel b$ and $H \perp b$ where the greatest anisotropy is expected. The angular dependence of the observed transverse magnetoresistance follows a

$\cos^2\theta$ distribution as shown in Fig. 1. The observed anisotropy ratio was 3.1 for the case of $H \perp b$ over $H \parallel b$. The magnitude of the magnetoresistance was $\Delta\rho/\rho B^2 = 4.1 \times 10^{-4}$ $(cm^2/V\text{-sec})^2$ for $H \perp b$. With current flow parallel to the fibers (along b-axis) a transverse magnetoresistance of the same magnitude was also observed. For current parallel to the b-axis a longitudinal magnetoresistance $\Delta\rho/\rho B^2 \simeq 10^{-4}$ $(cm^2/V\text{-sec})^2$ was measured.

Using the Fermi surface obtained from band structure calculations[1], Rudge et al.[2] have determined the plasma tensor principal axes for the electron and hole pockets. We have calculated the galvanomagnetic tensor of $(SN)_x$ using these plasma tensor anisotropies in a two-carrier conductivity model assuming a single isotropic scattering time. The predicted angular dependence of magnetoresistance is shown as the solid curve in Fig. 1. The calculated anisotropy is 5.7 in reasonable agreement with the measured anisotropy. From the normalization of the calculated curve to the experimental data at $\theta = 90°$ we can calculate the carrier mobilities. The electron and hole mobilities parallel to the $(SN)_x$ chains are 430 $cm^2/V\text{-sec}$ and 600 $cm^2/V\text{-sec}$ respectively and the scattering time is about 1.5×10^{-13} sec.

Magnetoresistance anisotropy is found to be in good agreement with predictions of the galvanomagnetic tensor obtained from the calculated Fermi surface. This result shows that the low perpendicular conductivity of $(SN)_x$ is an extrinsic effect due to interfiber barriers. A more detailed report of the galvanomagnetic effects and their implications on the properties of $(SN)_x$ will be presented elsewhere.

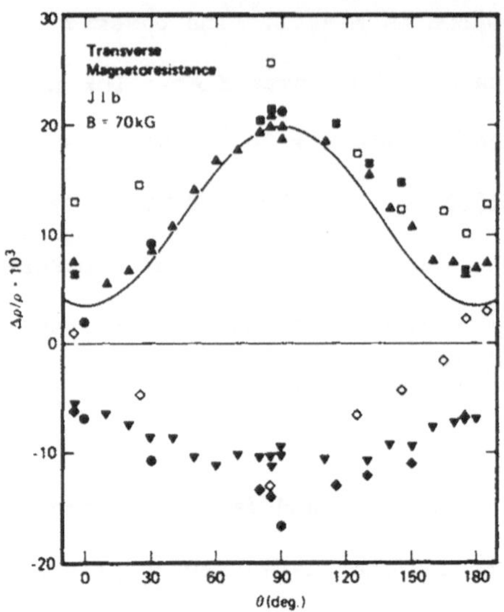

Fig. 1: Angular dependence of magnetoresistance in $(SN)_x$ crystals. The upper part refers to the positive quadratic component while the lower part shows the angular dependence of the negative magnetoresistance. Solid points are for several samples all at 4.2°K. The open points were for one sample at ∿1.4°K. The solid curve shows the predicted anisotropy for the calculated galvanomagnetic tensor.

REFERENCES

1. W. E. Rudge and P. M. Grant, Phys. Rev. Lett. 35, 1799 (1975).
2. W. E. Rudge, I. B. Ortenburger and P. M. Grant, to be published.

VIII

METAL COMPLEXES AND ORGANIC SEMICONDUCTORS

Conference on Organic Conductors and Semiconductors, Siófok, Hungary 1976

EXPERIMENTAL EVIDENCE FOR SOLITARY-WAVE EXCITATIONS IN GCP*

M.J. SCHAFFMAN, M.B. SALAMON and G. de PASQUALI

Department of Physics and Materials Research Laboratory, University of Illinois at Urbana-Champaign, Urbana, Illinois, USA

and

A.J. SCHULTZ and G.D. STUCKY

Department of Chemistry and Materials Research Laboratory, University of Illinois at Urbana-Champaign, Urbana, Illinois, USA

A new pseudo-one-dimensional compound, $[C(NH_2)_3]_2Pt(CN)_4Br_{0.23}\cdot H_2O$, which we denote GCP, has been prepared and its crystal structure and physical properties measured. This material is similar to KCP, with the potassium ion replaced with a large, asymmetric, organic cation. The structure of GCP was determined, and diffuse x-ray scattering revealed a Peierls distortion in the lattice at room temperature, corresponding to a superlattice of 7.9 ± 0.2 $\underline{c}$, where $\underline{c}$ is the Pt-Pt distance along the chain. The marked variation in intensity along the diffuse sheets even at room temperature indicates that the Peierls transition temperature of GCP is higher than that of KCP. Measurements of the electrical conductivity reveal both a temperature dependence and a non-ohmic behavior which suggest the presence of solitary-wave excitations. At low electric field strengths, the conductivity varies as

$$\sigma = \sigma_o T^{-y} \exp(-E_o/k_B T),$$

where $y=1.5$ and $E_o/k_B=370$ K, while at higher fields non-ohmic behavior is observed, becoming noticeable below about 160 K. The current was found to vary with electric field according to $I = I_o \sinh(eE\lambda/k_B T)$; $\lambda \doteq 1.2 \times 10^4$ Å at 77 K and increases with temperature. A number of tests were performed to establish that the non-ohmic behavior is intrinsic to the samples and not due to a heating effect. Thus, GCP appears to be the first known example of a soliton conductor.

*Work supported in part by National Science Foundation Grant DMR-72-03026.

Conference on Organic Conductors and Semiconductors, Siófok, Hungary 1976

SYNTHESIS OF LINEAR CHAIN TRANSITION METAL COMPLEXES

H. ENDRES, H.J. KELLER, I. LEICHERT,
M. MÉGNAMISI-BELOMBÉ, H. van de SAND and J. WEISS

Anorganisch-Chemisches Institut der Universität Heidelberg,
D-6900 Heidelberg, FRG

Nickel(II), palladium(II) and platinum(II) form planar complexes with 1,2-diondioximato ligands. The compounds crystallize in columnar stacks with different angles of inclination between the molecular planes and the stacking direction. The metal-metal distances depend strongly on the electronic and steric properties of the ligands. Upon oxidation with molecular iodine mixed valence compounds can be obtained. The stacking direction becomes perpendicular to the molecular planes, in these solids and the metal-metal distances decrease considerably. I_3^--anions are incorporated into the lattice to form linear arrays parallel to the metal chains (1).

---

(1) H. Endres, H. J. Keller, M. Mégnamisi-Belombé, W. Moroni, H. Pritzkow, J. Weiss and R. Comès, Acta Cryst., A 32, (1976) in press

Conference on Organic Conductors and Semiconductors, Siófok, Hungary 1976

PHONON-ASSISTED MOBILITIES IN ORGANIC MOLECULAR CRYSTALS (ANTHRACENE)

I. VILFAN

J. Stefan Institute, University of Ljubljana, Ljubljana, Yugoslavia

The phonon-assisted mobility due to intermolecular vibrations is investigated. It is shown that the phonon-assisted mobility due to translational vibrations is important whenever the charge-carrier bandwidth is smaller than the vibrational energies and the electron-phonon interaction is strong. The phonon-assisted mobility caused by rotational vibrations is important in anthracene, where it increases the mobility of electrons in the c′ direction as much as by one order of magnitude.

1. Introduction

The charge-carrier mobility in organic molecular crystals is governed by charge-carrier transfer integrals and by phonon scattering. The charge-carrier transfer integrals, beind of the order of $\lesssim$ 0.1 eV /1, 2/, clearly promote the charge-carrier transport, while the electron-phonon interaction generally reduces the mobility /3 – 5/. It has been shown by Gosar and Choi /3/ and by Gosar and Vilfan /6/ that phonons on the other hand also give rise to additional contributions to mobility and thus enhance it. The increase in mobility due to phonons is called the phonon-assisted mobility. In order to evaluate the phonon-assisted mobility we have to start from the definition of the current density operator, $j = \dot{P} = \frac{i}{\hbar}\left[H, P\right]$, where H is the hamiltonian and $P = e\, r$ is the dipole moment of the moving charge. Apart from the usual procedure the electron position is not approximated by the equilibrium molecular positions, but by instantaneous positions which include lattice vibrations in r /6/, $r = R + u$. An extra term j_1 in the expression for the current density j , $j = j_o + j_1$, is obtained, which is proportional to the displacements u. The mobility, being proportional to the current densiti correlation function $\langle j /t/ j/0/ \rangle$ is thus composed of the standard term $\langle j_o j_o \rangle$ and of phonon-assisted mobilities proportional to $\langle j_o j_1 \rangle$ and $\langle j_1 j_1 \rangle$. An estimate of the last term has been made by Gosar and Choi /3/.

In this paper the phonon-assisted mobilities are investigated in more detail. In the first part we shall discuss the phonon-assisted mobilities due to intermolecular translational vibrations and in the second part the phonon-assisted mobilities due to intermolecular rotational vibrations, which give a good explanation for the electron mobility in the c′ direction of anthracene.

2. Phonon-assisted mobility due to translational vibrations

The charge-carrier located at a molecule in the crystal lattice induces dipole moments at nearest neighbouring molecules. Between the charge-carrier and the induced dipole moments there exists an attractive force which causes a decrease in intermolecular distances around the charged molecule. The change in distance Δ is found by minimizing the potential V acting between two neighbouring molecules /7/ /see Fig. 1/

$$V = Au^2 - Bu \qquad /1/$$

The first term is the harmonic potential V^o acting on molecules in the neutral crystal. The constant A is related to mean square lattice displacements $\langle u^2 \rangle$ through the equipartition theorem

$$A = \frac{kT}{2\langle u^2 \rangle} \qquad /2/$$

The second term describes the attractive potential between the charge-carrier and the induced dipole moments. Since the displacements u are small compared to intermolecular distance, the potential can be expanded into series and only the linear term is retained. The constant B is then equal to

$$B = \frac{4}{r}\left[\frac{\alpha}{2}\left(\frac{e}{4\pi\varepsilon_o}\right)^2 \frac{1}{r^4}\right] \qquad /3/$$

where α is the mean molecular polarizability along the line joining two nearest molecules. Due to thermal fluctuations of the lattice the linear potential Bu causes some randomness in the local electron energies. The spread Γ in the local electron energies is related to the constant B through

$$\Gamma^2 = 2B^2 \langle u^2 \rangle \qquad /4/$$

In this way we get for the change of the equilibrium molecular separation around the charged molecule

$$\Delta = \frac{B}{2A} = \frac{\Gamma}{2\,kT}\langle u^2 \rangle^{\frac{1}{2}} \qquad /5/$$

In organic molecular crysrals the charge-carrier transfer integrals w are sensitive to the changes in intermolecular distance. The relation between w and the lattice displacements can be linearized and written in the form

$$w = w_o + \frac{\partial w}{\partial r} u \qquad /6/$$

The above outlined reduction in intermolecular distance causes an increase of the tranfer integrals and of the charge-carrier mobility, which is proportional to the square of the transfer integrals /5/,

$$\mu = \frac{e}{\hbar kT} \frac{\sqrt{\pi}}{2\Gamma} \sum_{\text{neighbours}} R^2 w^2 \qquad /7/$$

The standard mobility μ_0 is reproduced if w is replaced by w_o. Subtracting μ_0 from /7/ we finally obtain for the leading term in the phonon-assisted mobility, which is proportional to $\langle j_o j_1 \rangle$, the following expression

$$\mu_1 = \mu_0 \frac{2}{w_0} \left| \frac{\partial w}{\partial r} \right| \Delta = \mu_0 \frac{1}{w_0} \left| \frac{\partial w}{\partial r} \right| \cdot \frac{\Gamma}{kT} \langle u^2 \rangle^{1/2} \qquad /8/$$

The derivation above is a classical treatment, but a quantum mechanical calculation of μ_1 has also been done and the same result obtained. The phonon-assisted mobility μ_1 is thus proportional to the derivative of the transfer integral and to the change Δ in the intermolecular distance. The change Δ as well as $\partial w/\partial r$ are temperature independ ent. Therefore μ_1 has the same temperature dependence as μ_o and it cannot be separated from μ_o on the basis of the measured temperature dependence of the mobility.

Using for Γ , $\langle u^2 \rangle$, and $\partial w/\partial r$ the data of anthracene /5, 8/, which are the only available, we find for the change in the intermolecular distance $\Delta = 0.2$ Å and $\mu_1 = 0.9\ u_o$.

The expression for μ_1 , Eq. /8/, is valid if the transfer integral is smaller than the energy of intermolecular vibrations. On the other hand, if the transfer integrals are of the order of or greater than the phonon energies, the electron escapes before the deformation of the lattice reaches the equilibrium value Δ. Then the value of μ_1 is diminished.

In this case the phonon-assisted mobility is most suitably treated with the Green function method. The mobility u_1 is represented by the graphs as shown in Fig. 2. Topologically different graphs contribute to μ_1 a factor

$$\frac{n+1}{z-\mathcal{E}-\hbar\omega}+\frac{n}{z-\mathcal{E}+\hbar\omega}\,) \qquad /9/$$

where n is the thermal equilibrium number of phonons, z is the energy of the electron line, $\mathcal{E}$ is the electron energy, and ω is the phonon frequency. If the electron energy is real and z is replaced by $\mathcal{E}$, the expression /9/ is equal to $-1/\hbar\omega$ and the result /8/ can be obtained. If, however, the transfer integrals are not small as compared to $\hbar\omega$, they cause the decay of the electron state i before the lattice deformation Δ takes place. This is taken into account by complex electron energy, $\Sigma = \mathcal{E} - iW$ for electrons, and $\Sigma = \mathcal{E} + iW$ for holes, where W is the half-width of the charge-carrier energy band. Again by replacing z by $\mathcal{E}$, Eq. /9/ gives the factor

$$\frac{\hbar\omega}{\hbar^2\omega^2+W^2} \qquad /10/$$

We see that the final life time $\tau = \hbar/W$ of the electron state reduces the mobility μ_1, Eq. /8/, by the factor

$$\frac{(\hbar\omega)^2}{(\hbar\omega)^2+W^2} \qquad /11/$$

The same factor has been derived by Šunjić and Lucas /9/ for the reduction of phonon effects in X-ray photoemission linewidths.

3. The explanation of the electron mobility in the c direction in anthracene by rotational vibrations

All the reported calculated mobilities /3 - 5/ of electrons in the c′ direction of anthracene are two orders of magnitude smaller than the experimental values. One reason for this discrepancy can be the neglection of hydrogen atoms in the transfer integral calculations. On the other hand the transfer integral that determines $\mu_{c'}$ is extremely sensitive to rotations of the molecules along the short molecular axis

N /8/. Therefore we expect that the phonon-assisted mobility due to rotational vibrations could contribute essentially to the total calculated mobility.

The electron transfer integrals are expanded to the second derivative with respect to the angles of rotation along the N axis of the initial /i/ and final /f/ molecule, between which the electron transfer occurs.

$$w = w_0 \left[1 + k(\varphi_i + \varphi_f) + \tfrac{1}{2} \ell (\varphi_i^2 + \varphi_f^2) \right], \qquad /12/$$

where $k = \partial w/(w\partial\varphi)$ and $\ell = \partial^2 w/(w\partial\varphi^2)$. The electron mobility, given by Eq. /7/ is therefore composed of six contributions of which all the terms containing an odd power in φ_i or φ_f become zero, because the average values $\langle\varphi_i\rangle$ and $\langle\varphi_f\rangle$ are zero. The phonon-assisted mobility is then equal to

$$\mu_{1c'} = \mu_{oc'} \left[2k^2 \langle\varphi^2\rangle + 4\ell \langle\varphi^2\rangle + 8\ell^2 \langle\varphi^2\rangle^2 \right] \qquad /13/$$

The values for the derivatives k and ℓ are obtained from the work of Delacote and Tiberghien /8/,who calculated the transfer integrals at different molecular orientations. We get $k = 0.4$ deg^{-1} and $\ell = 0.05$ deg^{-2}, while $\langle\varphi^2\rangle = 10.5$ deg^2 /10/. Using these data, $\mu_{1c} = 7.7\,\mu_{oc'}$. We see that the phonon-assisted mobility increases the total mobility of electrons in the c' direction of anthracene by almost one order of magnitude. Nevertheless the calculated mobility is still one order of magnitude too low. As mentioned, the inclusion of hydrogens in the calculation of w may increase the transfer integrals and improve this result.

The temperature dependence of the electron mobility $\mu_{c'}$ which increases slightly with temperature, can be explained with the ponon-assisted mobility. The squared bracket in /13/ contributes to $\mu_{1c'}$ a factor $T^{1.3}$, which almost cancels the T dependence of $\mu_{oc'}$ /5/, yielding thus a temperature independent electron mobility in the c' direction of anthracene.

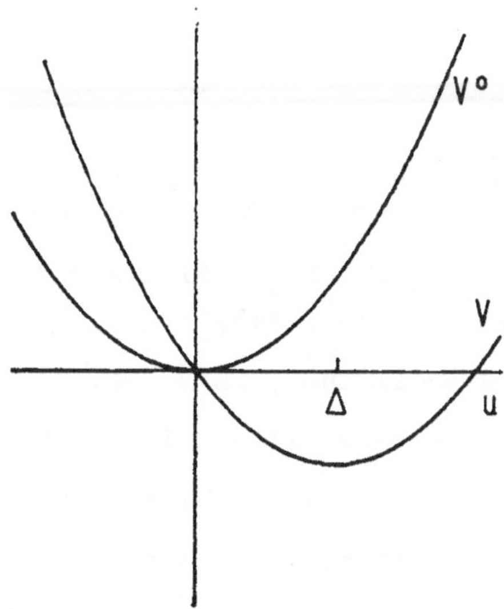

Fig. 1 Intermolecular potential of the neutral /V^o/ and charged /V/ molecule

Fig. 2 An example of the graph describing the phonon-assisted mobility μ_1. —— electron line, ----- phonon line

REFERENCES

/1/ R. M. Glaeser and R. S. Berry, J. Chem. Phys. 44, 3797 /1966/.

/2/ R. Silbey, J. Jortner, S. A. Rice and M. T. Vala Jr., J. Chem. Phys. 42, 733 /1965/.

/3/ P. Gosar and Sang-il Choi, Phys. Rev. 150, 529 /1966/.

/4/ L. Friedman, Phys. Rev. 14C, A 1649 /1965/.

/5/ I. Vilfan, Phys. Stat. Sol. /b/ 59, 351 /1973/.

/6/ P. Gosar and I. Vilfan, Mol. Phys. 18, 49 /1970/.

/7/ I. G. Austin and N. F. Mott, Adv. Phys. 18, 41 /1969/.

/8/ G. Delacote and A. Tiberghien, Proc. 4th Mol. Cryst. Symp., Enschede, Netherlands,1968. /unpublished/.

/9/ M. Šunjić and A. Lucas, "On the Phonon Contribution to the X-ray Photoemission Linewidths in Polar Crystal Films", /to be published/.

/lo/ G. S. Pawley, Phys. Stat. Sol. 20, 347 /1967/.

Conference on Organic Conductors and Semiconductors, Siófok, Hungary 1976

ON THE POSSIBLE MECHANISMS OF CONDUCTIVITY IN BIOPOLYMERS AND CHARGE TRANSFER MOLECULAR CRYSTALS

S. SUHAI and G. BICZÓ

Central Research Institute for Chemistry of the Hungarian Academy of Sciences, 1525 Budapest 114, POB. 17, Hungary

Theoretical results are presented for the electronic transport properties of polynucleotides, proteines and TCNQ molecular crystals. Experimental results and theoretical predictions are compared in the latter case. The role of asymmetry is emphasized in creating an attractive interaction between the pairs of delocalized electrons. The effect of optical phonons and excitonic polarons is also discussed.

The main object of this lecture is to emphasize the importance of asymmetry in creating attractive interactions between the delocalized electrons of biopolymers and charge transfer crystals and to prove the reality of our former results obtained for DNA models by presenting theoretical results obtained for the experimentally carefully investigated TCNQ complexes.

The complexity of organic charge transfer crystals and, in first place, biopolymers makes possible the formation of several different more or less separated electronic systems. Among these there may be very good "channels" for electronic conduction. Treating their conductivity properties in the first approximation separately, we may obtain unwaited results which seems to be contradict to the usual chemical (?) conceptions. For example, in the parallel-chain β conformation of polyglycine, values of 1.3-1.9 m_e (free electron mass) for effective masses were found in the direction of H-bonds and 0.2-0.3 m_e along the polypeptide backbone[1] using MINDO/2 CO (Crystal Orbital) method. Next, for the sugar-phosphate chain of a helical DNA model 10-25 m_e and for its different periodic single- and double-stranded poly-base models different values between 5 and 450 m_e were obtained from CNDO/2 CO calculations[2]. Thus, if the environment and the interactions of these more or less separated electronic systems cause further modifications in their state, then a variety of interesting phenomena may be produced. Going in the treatment to a second and third approximation, we recall former results obtained for simple polymers[3] and helical DNA models[4-6] obtained by a slightly generalized version of Little's original idea[7]. Namely, the Coulomb repulsion between the delocalized π-electrons may be strongly reduced (in contradiction to a conclusion of Davis, Gutfreund and Little[8] presented in their lecture that "screening by electrons in all filaments is not very effective". Their models, in reality, had not enough high asymmetry[9]) and, for some of the scattering processes, overcompensated in a single stranded poly C DNA model by the polarization of its side chain-type σ-electronic systems[4]. This effect may be enhanced by taking into account the polarization of its neighbouring delocalized π-electronic system in the poly G part of the poly (G-C) double helix[5-6] arranged according to the Watson-Crick model. On the other hand, some evidences has been found by Cope[10] and discussed by Ladik and Biczó[11] for the existence of superconductive tunnelling in biological systems. Furthermore, it was shown very recently by an interesting experiment[12] that the occurrence of superconductive-type collective states cannot be ruled out in biological molecules.

Regarding the complexity of the most interesting systems and the absence of experimental results for these, we turned to the organic charge transfer crystals to check our former results. First we treated independent linear columns of $TCNQ^-$ and NMP^+ molecules in the geometrical arrangement given by Fritchie[13]. Applying $\sigma - \pi$ separability, we calculated the wave function of the one-dimensional π-electronic system with the aid of the PPP CO method[14]. The bare Coulomb repulsion was determined in the same way as in[4]. In evaluating we applied Slater-type atomic orbitals, substituted the eigenvectors obtained in[15] and made use of the integral formulae given in[4]. We obtained in this way for the repulsion of two conduction electrons at the Fermi level (FL) of the $TCNQ^-$ chain C(FL) = 18.77 eV.

As the next step, using a generalized version[4] of Little's original proposal, we have to take into account that the atomic cores of the TCNQ molecules (containing besides the nuclei and the 1s electrons also the σ-type valence shell electrons in the PPP approximation) provide a localized polarizable "side chain" electron system which may substantially reduce through its screening the repulsion between the π-electrons. For the description of these electrons we applied LCAO molecular orbitals localized each on two centers the coefficients of which were determined by Del Re's method[16]. The indirect attractive term between two delocalized π-electrons arising from this polarization effect is obtained as a sum of contributions from different σ-bonds of the TCNQ molecule as it is given more detailed in[4]. Taking again the Bloch functions at the Fermi level of the $TCNQ^-$ column we obtain for this attractive term $A^{\sigma}(FL) = -16.54$ eV.

A second mechanism[5] promoting the attraction between conduction electrons of the $TCNQ^-$ chain may be the polarization of the π-electrons in neighbouring NMP^+ chains. Since in the band structure of the NMP^+ chain* the rather closely lying filled bands are separated by a large forbidden gap from the empty bands it seemed to be difficult to separate a dominant $t \rightarrow t'$ scattering process which has been possible for the previously mentioned poly(G-C) model. Instead we had to sum over all the possible virtual excitations within the NMP^+ chain. Using again the eigenvectors of the $TCNQ^-$ column at the Fermi level we obtain for this attraction $A^{\pi}(FL) = -1.90$ eV. So we have

---

*Since also the lower lying bands have been shown to play a non-negligible role in the electronic processes of the TCNQ crystals[15] investigations of the above type should be extended in the future for other scattering matrix elements involving states in these bands.

$$\Gamma = C + A^{\sigma} + A^{\pi} = 0.33 \text{ eV}$$

for the effective Coulomb repulsion between two conduction electrons at the Fermi level of TCNQ. Experiments confirm this small value of the interelectronic repulsion in TCNQ. Analysing the low-temperature magnetic susceptibility values of NMP-TCNQ Coleman et al.[17] obtained for the effective Coulomb repulsion 0.14 eV while the measurement of the polarographic half-wave potentials in the highly polar solvent CH_3CN resulted 0.16 eV for the energy change in the reaction 2 $TCNQ^{-} \rightarrow TCNQ^{0} + TCNQ^{--}$[18]. Similar results were obtained by Heeger and his coworkers[19,20]. In the same time, we gave here a quantitative verification of one of Heeger's conclusion told in his lecture[21] that phase transitions indicate importance of interchain interactions (see term A^{π} above).

It is interesting to mention hare an indirect experimental evidence for the reliability of our calculations.

Here we obtained for $C + A^{\sigma}$ the value 2.23 eV. In the same time, applying the PPP method for the treatment of π-electrons, the widely used Mataga-Nishimoto formula provides values between 2.5-3.0 eV for the screened Coulomb interaction integrals which are very near to the above value.

Since the above discussed reduction of the Coulomb term is of decisive importance for the conductive properties it is of great interest to look for further possibilities which could contribute to the attractive term. Inspection of the scattering matrix elements given in[3-5] we may see that they contain besides the symmetric ground state also the antisymmetric excited states of the polarized electronic systems. Therefore, in highly symmetrical systems the contributions to the attractive term may be strongly diminished by cancellation effects. Accordingly, the highest contributions to the above calulated attractive term were obtained in TCNQ from the C-N and C-H bonds and in the poly C model from the N-H bonds. Increasing of the molecular asymmetry could be thus very useful in producing well conducting organic materials.

It was almost wonderful for us to see during this conference the extremely many picturesquely symmetrical compounds investigated experimentally. Why would we stop atthis stage?! If we should like to find such strange effects as, e.g., organic superconductivity, we must give up at least partly this nice symmetry. For illustration, we propose the experimental investigation of compounds something like

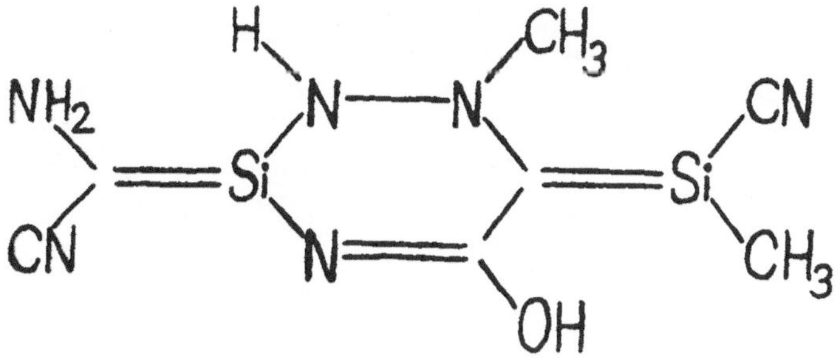

H
CH_3
NH_2
N
N
CN
Si
Si
CN
N
CH_3
OH

instead of the highly symmetrical TCNQ, or

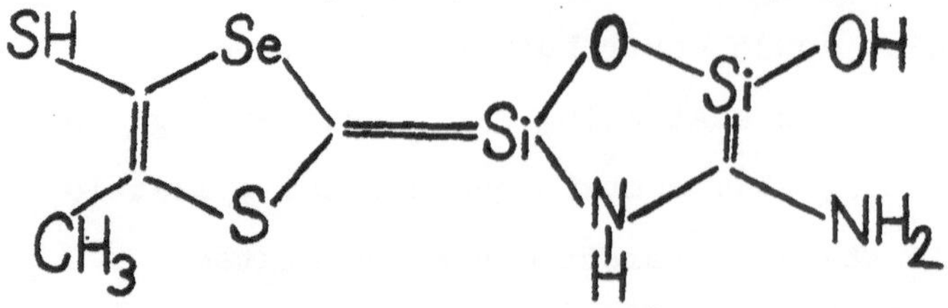

instead of TTF, etc., etc. Naturally, taking into account their preparability, crystallization properties, and so on: we give these examples only to clarify how the symmetry must be destroyed. The best asymmetrical examples, however, are given by the Nature one self and DNA must be mentioned in the first place. Its experimental investigation is highly proposed.

References

1. S. Suhai, Theoret. Chim. Acta (Berl.) 34 (1974) 157.
2. S. Suhai, Biopolymers 13 (1974) 1739.
3. J. Ladik, G. Biczó and A. Zawadowski, Phys. Letters 18 (1965) 257.
4. J. Ladik, G. Biczó and J. Rédly, Phys. Rev. 188 (1969) 710.
5. G. Biczó and S. Suhai, Phys. Lett. 51A (1975) 223.
6. G. Biczó and S. Suhai, Studia Biophys. (Berlin) 55 (1976) S85.
7. W.A. Little, Phys. Rev. 134A (1964) 1416.
8. D. Davis, H. Gutfreund and W.A. Little, this issue.
9. D. Davis, H. Gutfreund and W.A. Little, Phys. Rev. B13 (1976) 4766.
10. F.W. Cope, Physiol. Chem. Phys. 3 (1971) 403.
11. J. Ladik and G. Biczó, ibid 4 (1972) 495.
12. N.A.G. Ahmed et al., Phys. Lett. 53A (1975) 129.
13. C.J. Fritchie, Acta Crystallogr. 20 (1966) 892.
14. J. Ladik and G. Biczó, Acta Chim. Hung. 67 (1971) 297.
15. S. Suhai, J. Phys. C, to be published.
16. G. Del Re, J. Chem. Soc. (1958) 4031.
17. L.B. Coleman, J.A. Cohen, A.F. Garito and A.J. Heeger, Phys. Rev. B7 (1973) 2122.
18. L.R. Melby, R.J. Harder, W.R. Hartler, W. Mahler, R.E. Benson and W.E. Mockel, J.Am. Chem. Soc. 84 (1962) 3374.

19. A.J. Epstein, S. Etemad, A.F. Garito and A.J. Heeger, Phys. Rev. B5 (1972) 952.
20. S.K. Khanna, A.A. Bright, A.F. Garito and A.J. Heeger, Phys. Rev. B10 (1974) 2139.
21. A.J. Heeger, this issue.

Conference on Organic Conductors and Semiconductors, Siófok, Hungary 1976

ON THE CONDUCTIVITY OF SOME ORGANIC DYES

Ch. IVANOVA and V. PENEVA

Department of Physics and Biophysics, Medical Academy, Sofia, Bulgaria

Organic dyes are subject to different investigations since they render a possibility for practical application. On the other hand organic dyes are most simple as organic semiconductors. Therefore the results from such experiments can be analogically ascribed to some more complex organic semiconductors.

A significant part of these investigations refers to the electric conductivity of organic dyes, the remaining part concerns electromotive voltage.

Usually the organic dyes are amorphous. Their electric properties depend on the presence of gases such as hydrogen, oxygen, etc.

Results from our investigation on numerous (1, 2, 3) dyes, are given in our previous papers: ruby S, Bismark brown, water blue, basic fuchsin, eosin bluish, eosin water-soluble, methyl red, crystal violet, alisarin yellow GG, diamant fuchsin, malachite green, etc.

The resistance of the considered organic dyes are of the order of $10^9 \cdot 10^{11} \Omega$. No electromotive force is observed both in dark and when illuminated in all our specimens when two equal electrodes are used.

The electromotive force in the dark is approximately 450-1200 mV when different electrodes are used. It is our opinion that this electromotive force in darkness can be explained assuming that the specimen (organic dye layer with two different electrodes) is considered as galvanic element with rigid electrolyte. Electromotive force 10-300 mV is observed from illumination of a specimen with two different electrodes (Au-Al). Sometimes it shows the same polarity as the electromotive force in the dark though for some organic dyes it is vice versa (1).

To determine the character of the contacts and the type of electric conductivity in organic dyes we use I/V characteristics in the present paper.

A thin-layer semiconductor with high-resistivity placed between metals with a different work function may be expected to possess rectifying properties according to (12). The rectifying action of a given semiconductor suggests the presence of injecting and blocking contacts.

A number of authors (4-8) have studied the phenomena connected with the injection of current carriers in organic semiconductors. It was established that when equal electrodes are used the current does not depend on the polarity of the applied voltage. When the electrodes are different the I/V characteristics become non-linear and the degree of non-linearity depends on the sign of the potential of one of the electrodes.

The type of contacts in organic dyes in which part of the conductivity is conditioned by ions, is a problem involving additional complications wherefor it has not been sufficiently elucidated.

The following nine dyes were investigated: ruby S, Bismark brown, water blue, basic fuchsin, eosin bluish, eosin water-soluble, methyl red, alisarin yellow GG, crystal violet.

The specimens were prepared as follows: four electrodes, two gold and two aluminium ones, at a 1 mm distance were evaporated in vacuo on a glass substrate. The organic dye layer was deposited on the electrodes by means of evaporation from an aqueous solution. The method of displacing the electrodes employed by us made it possible to take the I-V characteristics on one and the same specimen with equal as well as with different electrodes.

The I-V characteristics were taken in darkness by transmitting constant voltage to the specimen and the current was measured electrometrically by the voltage drop on standard resistances.

RESULTS AND DISCUSSION

According to the results obtained, the investigated dyes can be conventionally divided into two types: type A and type B. In type A, which includes violet, alisarin yellow GG, rubyn S and methyl red, the I-V characteristics obtained with equal electrodes are symmetric and pass through the beginning of the coordinate system (Fig. 1a), while those obtained with different electrodes also pass through the beginning of the coordinate system, but the change of the current in an inverse direction does not follow Ohm's law - the aluminium contact proves blocking in this case.

In the B type of dyes, which includes water blue, basic fuchsin, eosin bluish, eosin water-soluble and Bismark brown, the I-V characteristics with equal electrodes are also symmetric, but do not pass through the beginning of the coordinate system (Fig. 1b). Here the current begins at a certain voltage different from zero, which in the different dyes varies from 50 to 300 mV.

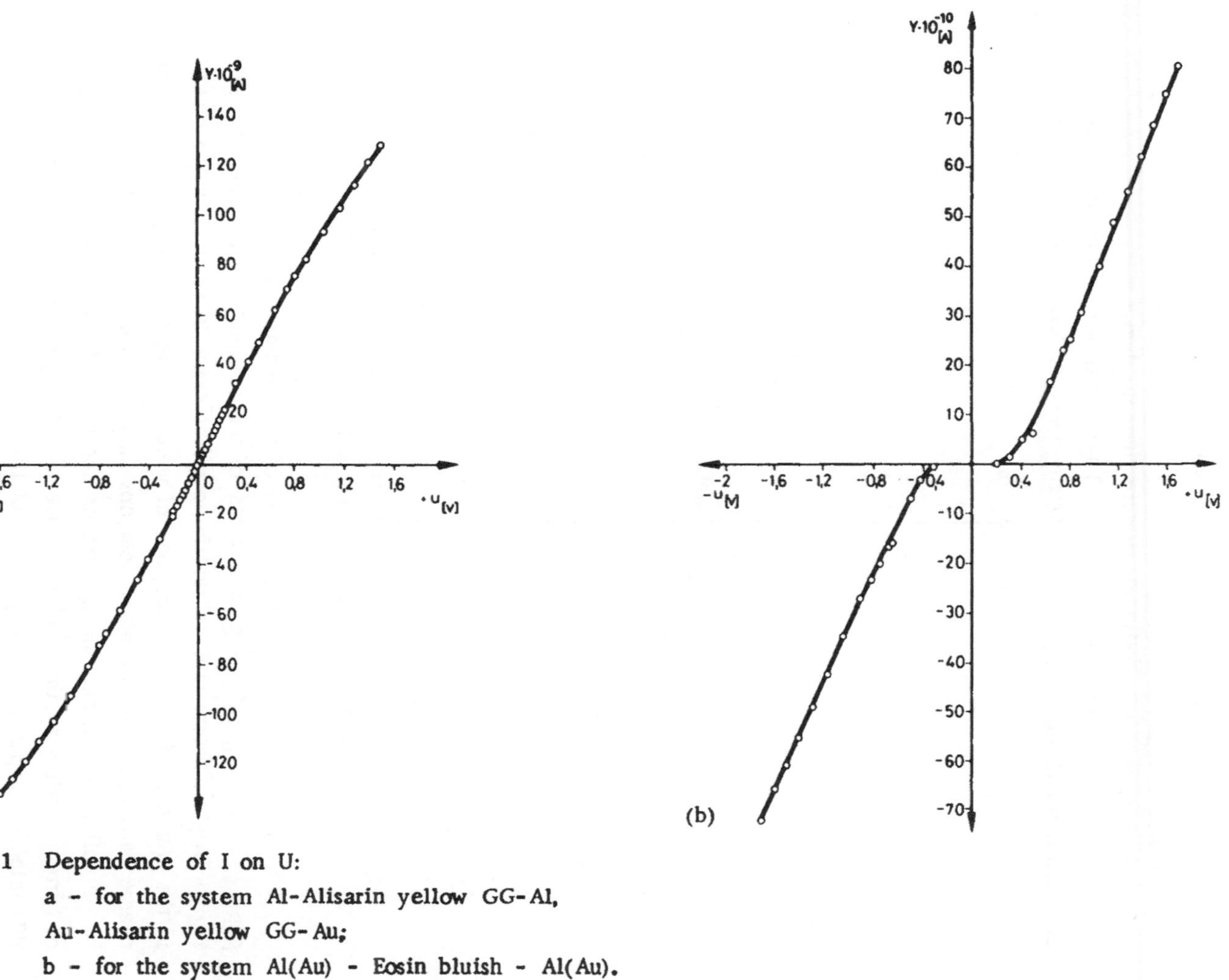

Fig. 1 Dependence of I on U:
a - for the system Al-Alisarin yellow GG-Al, Au-Alisarin yellow GG-Au;
b - for the system Al(Au) - Eosin bluish - Al(Au).

These results show that B type dyes with Al and Au electrodes have an ion conductivity. With the investigated metal contacts these dyes form a solid galvanic element with a solid electrolyte (organic dye).

The I-V characteristics obtained in B type dyes with different electrodes are asymmetric, the change of the current does not follow Ohm's law (Fig. 3).

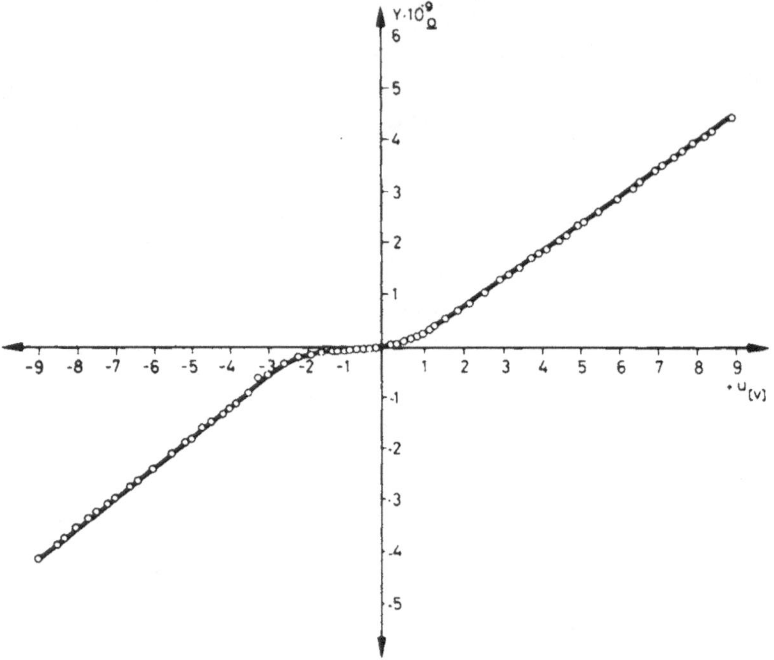

Fig. 3 Dependence of I on V for specimen Eosin bluish with different electrodes (Al and Au) after aging

With different electrodes the degree of non-linearity of I-V characteristics depends on the sign of the gold and not of the aluminium electrode. According to Wright's theory (10), a rectifying effect exists in semiconductor layers in the presence of two contacts, one of which is capable of injecting electric charge carriers, while the other does not have this property or only slightly so.

It follows therefrom that our results can be connected with the injection of current carriers by the gold electrode at the organic dye. The degree of this injection depends on the polarity of the potential applied to the gold electrode.

Besides non-linearity, the I-V characteristics taken in B type dyes with different electrodes manifest a typical feature: they do not pass through the beginning of the coordinate system (Fig. 2a, b). At zero outer voltage fed to

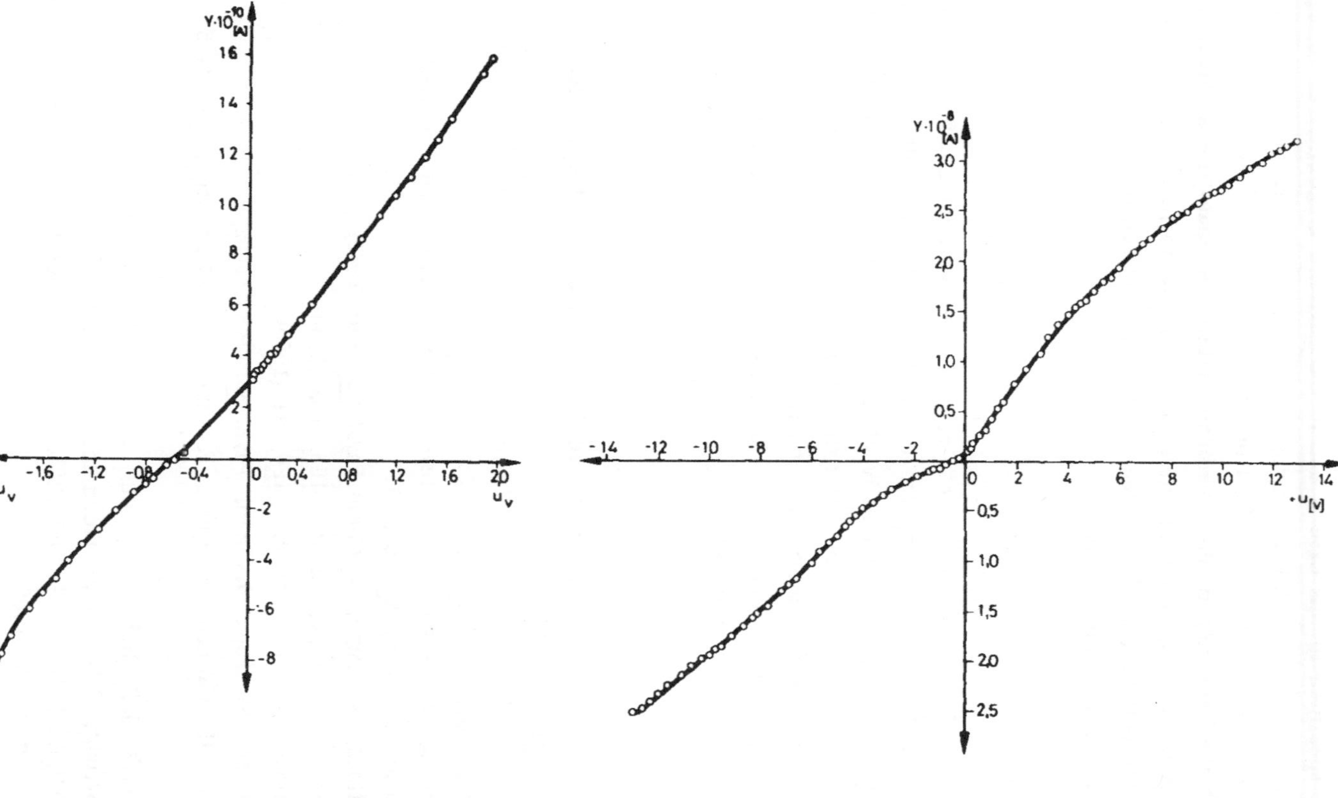

Fig. 2 Dependence of I on V:

a - Eosin bluish-different electrodes, U from 0-2 V;

b - Eosin bluish-different electrodes, U from 0-14 V

the specimen, the current differs from zero. The value of the current passing through the specimen in this case is due to the galvanic electromotive voltage in dark of the specimen. With the aging of the specimen, this voltage decreases to zero. The I-V characteristics of an old specimen, taken after six months, pass through the beginning of the coordinate system. At that, the aluminium contact becomes blocking when negative voltage is fed to it. This can be explained by the obtainment of an oxide layer on the surface of the aluminium electrode with a rectifying action.

CONCLUSIONS

The type of I-V characteristics can serve for the determination of the type of contacts as well as for detecting an ion component in the conductivity. The dyes investigated by us, conventionally classified as type A on the basis of the above deductions, have a purely electronic conductivity, while the B type dyes have an ion component in their conductivity.

Acknowledgement are due to Prof. R. Andreychin for his interest in this work and for his critical remarks.

REFERENCES

(1) H. Ivanova, V. Peneva , Comtes rendus de l'Academie bulgare des Sciences, 23, 10 (1970)

(2) H. Ivanova, V. Peneva, R. Andreychin, Comtes rendus de l'Academie bulgare des Sciences, 26, 9 (1973)

(3) H. Ivanova, V. Peneva, R. Andreychin, Comtes rendus de l'Academie bulgare des Sciences, 28, 5 (1975)

(4) H. Kalman, M. Pope, J. Chem. Phys. 32, 300 (1960)

(5) H. Kalman, M. Pope, Nature, 185, 753 (1960)

(6) H. Kalman, M. Pope, Nature, 186, 31 (1960)

(7) H. Kalman, M. Pope J. Chem. Phys. 36, 2482 (1962)

(8) M. Pope, H. Kalman, A. Chen, P. Gordon, J. Chem. Phys. 38, 2486 (1963)

(9) F. A. Haak, J. P. Nolta, J. Chem. Phys. 38 2648 (1963)

(10) G. T. Wright, Solid State Electronics, 2, 165 (1961)

(11) I. A. Eligülashvili, T. A. Nakashidze, L. D. Rozenstein, A. A. Hatiashvili, Elektrochimiya, 2, 1966, 107

(12) Z. A. Rottenberg, S. D. Levina - Elektrochimiya, 1 1965, 993

AUTHOR INDEX

Selected Issues from

Lecture Notes in Mathematics

Vol. 431: Séminaire Bourbaki – vol. 1973/74. Exposés 436–452. IV, 347 pages. 1975.

Vol. 433: W. G. Faris, Self-Adjoint Operators. VII, 115 pages. 1975.

Vol. 434: P. Brenner, V. Thomée, and L. B. Wahlbin, Besov Spaces and Applications to Difference Methods for Initial Value Problems. II, 154 pages. 1975.

Vol. 440: R. K. Getoor, Markov Processes: Ray Processes and Right Processes. V, 118 pages. 1975.

Vol. 442: C. H. Wilcox, Scattering Theory for the d'Alembert Equation in Exterior Domains. III, 184 pages. 1975.

Vol. 446: Partial Differential Equations and Related Topics. Proceedings 1974. Edited by J. A. Goldstein. IV, 389 pages. 1975.

Vol. 448: Spectral Theory and Differential Equations. Proceedings 1974. Edited by W. N. Everitt. XII, 321 pages. 1975.

Vol. 449: Hyperfunctions and Theoretical Physics. Proceedings 1973. Edited by F. Pham. IV, 218 pages. 1975.

Vol. 458: P. Walters, Ergodic Theory – Introductory Lectures. VI, 198 pages. 1975.

Vol. 459: Fourier Integral Operators and Partial Differential Equations. Proceedings 1974. Edited by J. Chazarain. VI, 372 pages. 1975.

Vol. 461: Computational Mechanics. Proceedings 1974. Edited by J. T. Oden. VII, 328 pages. 1975.

Vol. 463: H.-H. Kuo, Gaussian Measures in Banach Spaces. VI, 224 pages. 1975.

Vol. 464: C. Rockland, Hypoellipticity and Eigenvalue Asymptotics. III, 171 pages. 1975.

Vol. 468: Dynamical Systems – Warwick 1974. Proceedings 1973/74. Edited by A. Manning. X, 405 pages. 1975.

Vol. 470: R. Bowen, Equilibrium States and the Ergodic Theory of Anosov Diffeomorphisms. III, 108 pages. 1975.

Vol. 474: Séminaire Pierre Lelong (Analyse) Année 1973/74. Edité par P. Lelong. VI, 182 pages. 1975.

Vol. 484: Differential Topology and Geometry. Proceedings 1974. Edited by G. P. Joubert, R. P. Moussu, and R. H. Roussarie. IX, 287 pages. 1975.

Vol. 487: H. M. Reimann und T. Rychener, Funktionen beschränkter mittlerer Oszillation. VI, 141 Seiten. 1975.

Vol. 489: J. Bair and R. Fourneau, Etude Géométrique des Espaces Vectoriels. Une Introduction. VII, 185 pages. 1975.

Vol. 490: The Geometry of Metric and Linear Spaces. Proceedings 1974. Edited by L. M. Kelly. X, 244 pages. 1975.

Vol. 503: Applications of Methods of Functional Analysis to Problems in Mechanics. Proceedings 1975. Edited by P. Germain and B. Nayroles. XIX, 531 pages. 1976.

Vol. 507: M. C. Reed, Abstract Non-Linear Wave Equations. VI, 128 pages. 1976.

Vol. 509: D. E. Blair, Contact Manifolds in Riemannian Geometry. VI, 146 pages. 1976.

Vol. 515: Bäcklund Transformations. Nashville, Tennessee 1974. Proceedings. Edited by R. M. Miura. VIII, 295 pages. 1976.

Vol. 516: M. L. Silverstein, Boundary Theory for Symmetric Markov Processes. XVI, 314 pages. 1976.

Vol. 518: Séminaire de Théorie du Potentiel, Proceedings Paris 1972–1974. Edité par F. Hirsch et G. Mokobodzki. VI, 275 pages. 1976.

Vol. 522: C. O. Bloom and N. D. Kazarinoff, Short Wave Radiation Problems in Inhomogeneous Media: Asymptotic Solutions. V. 104 pages. 1976.

Vol. 523: S. A. Albeverio and R. J. Høegh-Krohn, Mathematical Theory of Feynman Path Integrals. IV, 139 pages. 1976.

Vol. 524: Séminaire Pierre Lelong (Analyse) Année 1974/75. Edité par P. Lelong. V, 222 pages. 1976.

Vol. 525: Structural Stability, the Theory of Catastrophes, and Applications in the Sciences. Proceedings 1975. Edited by P. Hilton. VI, 408 pages. 1976.

Vol. 526: Probability in Banach Spaces. Proceedings 1975. Edited by A. Beck. VI, 290 pages. 1976.

Vol. 527: M. Denker, Ch. Grillenberger, and K. Sigmund, Ergodic Theory on Compact Spaces. IV, 360 pages. 1976.

Vol. 532: Théorie Ergodique. Proceedings 1973/1974. Edité par J.-P. Conze and M. S. Keane. VIII, 227 pages. 1976.

Vol. 538: G. Fischer, Complex Analytic Geometry. VII, 201 pages. 1976.

Vol. 543: Nonlinear Operators and the Calculus of Variations, Bruxelles 1975. Edited by J. P. Gossez, E. J. Lami Dozo, J. Mawhin, and L. Waelbroeck, VII, 237 pages. 1976.

Vol. 552: C. G. Gibson, K. Wirthmüller, A. A. du Plessis and E. J. N. Looijenga. Topological Stability of Smooth Mappings. V, 155 pages. 1976.

Vol. 556: Approximation Theory. Bonn 1976. Proceedings. Edited by R. Schaback and K. Scherer. VII, 466 pages. 1976.

Vol. 559: J.-P. Caubet, Le Mouvement Brownien Relativiste. IX, 212 pages. 1976.

Vol. 561: Function Theoretic Methods for Partial Differential Equations. Darmstadt 1976. Proceedings. Edited by V. E. Meister, N. Weck and W. L. Wendland. XVIII, 520 pages. 1976.

Vol. 564: Ordinary and Partial Differential Equations, Dundee 1976. Proceedings. Edited by W. N. Everitt and B. D. Sleeman. XVIII, 551 pages. 1976.

Vol. 565: Turbulence and Navier Stokes Equations. Proceedings 1975. Edited by R. Temam. IX, 194 pages. 1976.

Vol. 566: Empirical Distributions and Processes. Oberwolfach 1976. Proceedings. Edited by P. Gaenssler and P. Révész. VII, 146 pages. 1976.

Vol. 570: Differential Geometrical Methods in Mathematical Physics, Bonn 1975. Proceedings. Edited by K. Bleuler and A. Reetz. VIII, 576 pages. 1977.

Vol. 572: Sparse Matrix Techniques, Copenhagen 1976. Edited by V. A. Barker. V, 184 pages. 1977.

Lecture Notes in Physics

Vol. 44: R. A. Breuer, Gravitational Perturbation Theory and Synchrotron Radiation. VI, 196 pages. 1975.

Vol. 45: Dynamical Concepts on Scaling Violation and the New Resonances in e^+e^- Annihilation. Edited by B. Humpert. VII, 248 pages. 1976.

Vol. 46: E. J. Flaherty, Hermitian and Kählerian Geometry in Relativity. VIII, 365 pages. 1976.

Vol. 47: Padé Approximants Method and Its Applications to Mechanics. Edited by H. Cabannes. XV, 267 pages. 1976.

Vol. 48: Interplanetary Dust and Zodiacal Light. Proceedings 1975. Edited by H. Elsässer and H. Fechtig. XII, 496 pages. 1976.

Vol. 49: W. G. Harter and C. W. Patterson, A Unitary Calculus for Electronic Orbitals. XII, 144 pages. 1976.

Vol. 50: Group Theoretical Methods in Physics. 4th International Colloquium. Nijmegen 1975. Edited by A. Janner, T. Janssen, and M. Boon. XIII, 629 pages. 1976.

Vol. 51: W. Nörenberg und H. A. Weidenmüller, Introduction to the Theory of Heavy-Ion Collisions. IX, 273 pages. 1976.

Vol. 52: M. Mladjenović, Development of Magnetic β-Ray Spectroscopy. X, 282 pages. 1976.

Vol. 53: D. J. Simms and N. M. J. Woodhouse, Lectures on Geometric Quantization. V, 166 pages. 1976.

Vol. 54: Critical Phenomena. Sitges International School on Statistical Mechanics, June 1976. Edited by J. Brey and R. B. Jones. XI, 383 pages. 1976.

Vol. 55: Nuclear Optical Model Potential. Proceedings 1976. Edited by S. Boffi and G. Passatore. VI, 221 pages. 1976.

Vol. 56: Current Induced Reactions. International Summer Institute, Hamburg 1975. Edited by J. G. Körner, G. Kramer, and D. Schildknecht. V, 553 pages. 1976.

Vol. 57: Physics of Highly Excited States in Solids. Proceedings 1975. Edited by M. Ueta and Y. Nishina. IX, 391 pages. 1976.

Vol. 58: Computing Methods in Applied Sciences. Proceedings 1975. Edited by R. Glowinski and J. L. Lions. VIII, 593 pages. 1976.

Vol. 59: Proceedings of the Fifth International Conference on Numerical Methods in Fluid Dynamics. 1976. Edited by A. I. van de Vooren and P. J. Zandbergen. VII, 459 pages. 1976.

Vol. 60: C. Gruber, A. Hintermann, and D. Merlini, Group Analysis of Classical Lattice Systems. XIV, 326 pages. 1977.

Vol. 61: International School on Electro and Photonuclear Reactions I. Edited by C. Schaerf. VIII, 650 pages. 1977.

Vol. 62: International School on Electro and Photonuclear Reactions II. Edited by C. Schaerf. VIII, 301 pages. 1977.

Vol. 63: V. K. Dobrev et al., Harmonic Analysis on the n-Dimensional Lorentz Group and Its Application to Conformal Quantum Field Theory. X, 280 pages. 1977.

Vol. 64: Waves on Water of Vabriable Depth. Edited by D. G. Provis and R. Radok. 231 pages. 1977.

Vol. 65: Organic Conductors and Semiconductors. Proceedings 1976. Edited by L. Pál, G. Grüner, A. Jánossy and J. Sólyom. 654 pages. 1977.